Lecture Notes in Computer S

Commenced Publication in 1973
Founding and Former Series Editors:
Gerhard Goos, Juris Hartmanis, and Jan van Lee

Ming-Yang Kao   Xiang-Yang Li (Eds.)

Algorithmic Aspects in Information and Management

Third International Conference, AAIM 2007
Portland, OR, USA, June 6-8, 2007
Proceedings

 Springer

Volume Editors

Ming-Yang Kao
Northwestern University
Department of Electrical Engineering and Computer Science
McCormick School of Engineering and Applied Sciences
2145 Sheridan Road, Room M324, Evanston, IL 60208, USA
E-mail: kao@cs.northwestern.edu

Xiang-Yang Li
Illinois Institute of Technology, Chicago, USA
and Microsoft Research Asia, Beijing China
and Nanjing University, Nanjing China
E-mail: xli@cs.iit.edu

Library of Congress Control Number: 2007927664

CR Subject Classification (1998): F.2.1-2, E.1, G.1-3, J.1

LNCS Sublibrary: SL 3 – Information Systems and Application, incl. Internet/Web and HCI

ISSN   0302-9743
ISBN-10   3-540-72868-6 Springer Berlin Heidelberg New York
ISBN-13   978-3-540-72868-9 Springer Berlin Heidelberg New York

Springer is a part of Springer Science+Business Media

springer.com

© Springer-Verlag Berlin Heidelberg 2007
Printed in Germany

Typesetting: Camera-ready by author, data conversion by Scientific Publishing Services, Chennai, India
Printed on acid-free paper   SPIN: 12071777   06/3180   5 4 3 2 1 0

Preface

The papers in this volume were presented at the 3rd International Conference on Algorithmic Aspects in Information and Management (AAIM 2007), held June 6–8, 2007 in Portland, Oregon, USA. This conference is intended for original algorithmic research on immediate applications and/or fundamental problems pertinent to information management and management science, broadly construed.

Submissions to the conference this year were conducted electronically. A total of 120 papers were submitted, of which 40 were initially accepted, and 1 paper was withdrawed later. The papers were evaluated by an international Program Committee consisting of Hee-Kap Ahn, Takao Asano, Amotz Bar-Noy, Hans Bodlaender, Peter Brucker, Leizhen Cai, Gruia Calinescu, Jianer Chen, Siu-Wing Cheng, Marek Chrobak, Yang Dai, Rudolf Fleischer, Jie Gao, Joachim Gudmundsson, Bhaskar DasGupta, Gregory Gutin, Wen-Lian Hsu, Giuseppe F. Italiano, Ming-Yang Kao, Sanjiv Kapoor, Tak-Wah Lam, Erran Li Li, Jing Li, Xiang-Yang Li, Peter Bro Miltersen, Seffi Naor, Chung Keung Poon, Kirk Pruhs, Rajeev Raman, Paul Spirakis, Zheng Sun, Wing Kin Sung, Jan van Leeuwen, Jie Wang, Lusheng Wang, Weizhao Wang, Yu Wang, JinHui Xu, Yinfeng Xu, and Binhai Zhu.

The submitted papers to AAIM 2007 were from Algeria, Canada, Chile, China (mainland and Taiwan), Germany, Hong Kong, India, Italy, Japan, Mexico, Netherlands, Portugal, South Korea, Spain, Sweden, Switzerland, Turkey, Ukraine, UK, and USA.

Each paper was evaluated by at least two Program Committee members and most papers were actually evaluated by at least three Program Committee members, asisted in some cases by external reviews and comments. In addition to these selected papers, the conference also includeed three invited keynote talks by Anna Karlin from the University of Washington, Tuomas Sandholm from CMU, and Shang-Hua Teng from Boston University.

We thank all the people who made this meeting possible: the authors for submitting their papers to AAIM 2007, the Program Committee members and external reviewers (listed on the pages that follow) for their excellent work, and the three invited keynote speakers. Finally, we thank the Washington State University, Vancouver campus, for their support and the local organizers and our colleagues for their assitance.

June 2007

Ming-Yang Kao
Xiang-Yang Li

Organization

Program Committee Chairs

Ming-Yang Kao   Northwestern University, Chicago, USA
Xiang-Yang Li   Illinois Institute of Technology, Chicago, USA,
                       and Microsoft Research Asia, Beijing, China,
                       and Nanjing University, Nanjing, China

Program Committee

Hee-Kap Ahn	Korean Advanced Institute of Science and Technology
Takao Asano	Chuo University
Amotz Bar-Noy	City Uinversity of New York
Hans Bodlaender	University of Utrecht
Peter Brucker	University of Osnabrück
Leizhen Cai	Chinese University of Hong Kong
Gruia Calinescu	Illinois Institute of Technology
Jianer Chen	Texas A&M University
Siu-Wing Cheng	Hong Kong University of Science and Technology
Marek Chrobak	University California at Riverside
Yang Dai	University of Illinois, Chicago
Rudolf Fleischer	Fudan University
Jie Gao	University of SUNY Stony Brook
Joachim Gudmundsson	National ICT Australia
Bhaskar DasGupta	University of Illinois at Chicago
Gregory Gutin	Royal Holloway, University London and University Haifa
Wen-Lian Hsu	Academia Sinica, Taiwan
Giuseppe F. Italiano	University of Rome "Tor Vergata"
Ming-Yang Kao	Northwestern University, Co-chair
Sanjiv Kapoor	Illinois Institute of Technology
Tak-Wah Lam	University of Hong Kong
Erran Li Li	Bell Labs
Jing Li	Case Western Reserve University
Xiang-Yang Li	Illinois Institute of Technology, Co-chair
Peter Bro Miltersen	University of Aarhus
Seffi Naor	Technion and Microsoft Research
Chung Keung Poon	City University of Hong Kong
Kirk Pruhs	University of Pittsburgh
Rajeev Raman	Leicester University

Paul Spirakis   University of Patras and CTI Greece
Zheng Sun   Google Inc.
Wing Kin Sung   National University of Singapore
Jan van Leeuwen University of Utrecht
Jie Wang   University of Massachusetts
Lusheng Wang   City University of Hong Kong
Weizhao Wang   Google Inc.
Yu Wang   University of North Carolina at Charlotte
JinHui Xu   University of SUNY Buffalo
Yinfeng Xu   Xi'an Jiaotong University
Binhai Zhu   Montana State University

Organizing Committee

Yu Wang   University of North Carolina at Charlotte
Wen-Zhan Song   Washington State University
Xiang-Yang Li   Illinois Institute of Technology, Chicago, USA,
                  and Microsoft Research Asia, Beijing, China,
                  and Nanjing University, Nanjing, China

External Referees

Marjan van den Akker Peter Brass   Jia-Ming Chang
Yixuan Chen   Adrian Dumitrescu Thomas Erlebach
Qizhi Fang   Feifeng Zheng   Herman Haverkort
Danny Hermelin   Shudong Jin   Rohit Khandekar
Marc van Kreveld   Lap Chi Lau   Jang-Won Lee
Xin Li   Mingen Lin   Lopamudra Mukherjee
Maurizio Naldi   Tomasz Radzik   Sandeep Sen
Chan-Su Shin   Vikas Singh   Marinus Veldhorst
Thomas Wolle   Shiquan Wu   Mingyu Xiao
Yang Yang   Ei-Wen Yang   Yucheng Dong

Table of Contents

Session 4: Graph Theory

Session 5: Newtork Algorithm

Session 6: Game Theory

Session 7: Option Theory

Session 8: Computational Geometry

Session 9: Graph Theory and Combinatorics

Session 10: Networks and Data

Invited Lecture

Solving Generalized Maximum Dispersion with Linear Programming

Gerold Jäger[1], Anand Srivastav[2], and Katja Wolf[3]

[1] Department of Computer Science
Washington University
Campus Box 1045, One Brookings Drive
St. Louis, Missouri 63130-4899, USA
jaegerg@cse.wustl.edu
[2] Institut für Informatik
Christian-Albrechts-Universität zu Kiel
Christian-Albrechts-Platz 4, D-24118 Kiel, Germany
asr@numerik.uni-kiel.de
[3] Zentrum für Paralleles Rechnen Universität zu Köln
Weyertal 80, D-50931 Köln, Germany
wolf@zpr.uni-koeln.de

Abstract. The GENERALIZED MAXIMUM DISPERSION problem asks for a partition of a given graph into p vertex-disjoint sets, each of them having at most k vertices. The goal is to maximize the total edge-weight of the induced subgraphs. We present the first LP-based approximation algorithm.

Keywords: Approximation Algorithms, Randomized Algorithms, Generalized Maximum Dispersion.

1 Introduction

Let $G = (V, E)$ be an undirected graph ($n = |V|, m = |E|$) with non-negative edge weights w_{ij} for $(i, j) \in E$ (for convenience, we set $w_{ij} = 0$ for $(i, j) \notin E$, implicitly assuming that G is a complete graph). The weight $\omega(S)$ of a subgraph of G induced by $S \subset V$ is the sum $\sum_{i \in S, j \in S} w_{ij}$.

For non-negative integers p, k with $pk \leq n$ the GENERALIZED MAXIMUM DISPERSION problem is to find p disjoint induced subgraphs of G, each with at most k vertices, such that the sum of the edge weights of the subgraphs is maximum.

For a motivation of this problem consider the following problem: A large scale manufacturer wants to expand in a new region and can use n locations. He wants to expand in exactly p business areas (e.g. restaurants, groceries, home-improvement markets etc.). For each such area he is allowed to use up to k locations, where $pk \leq n$. Furthermore assume, that all n locations are ready, i.e. do not cause any extra costs. Shops of the same area are attractive for the manufacturer, if they are as far as possible from each other (which means that for one customer only one shop comes into question). Thus the manufacturer wishes to maximize the sum over all distances between locations of the same area, i.e. he has to solve exactly our GENERALIZED MAXIMUM DISPERSION problem.

GENERALIZED MAXIMUM DISPERSION is $\mathcal{NP}$-hard, which can be easily seen by a reduction from the MAXIMUM CLIQUE problem. As DENSE SUBGRAPH is a special case, it follows from [16] that there is even no PTAS for GENERALIZED MAXIMUM DISPERSION.

A promising and often successful approach to cope with the hardness of a combinatorial optimization problem is to design polynomial-time approximation algorithms. Given an instance I of a maximization problem and an (approximation) algorithm A the *approximation factor* $r_A(I)$ is defined by $r_A(I) = A(I)/OPT(I) \leq 1$.

Previous Work. To the best of our knowledge, the GENERALIZED MAXIMUM DISPERSION problem is only considered by Hassin, Rubinstein and Tamir [14]. They provide a polynomial-time algorithm with approximation factor $\frac{1}{2-1/\lceil k/2 \rceil}$ in graphs where the edge weights satisfy the triangle inequality.

As expressed in the name, the GENERALIZED MAXIMUM DISPERSION is a natural generalization of the MAXIMUM DISPERSION problem (which is also denoted by MAX-k-DENSE-SUBGRAPH problem). For a given weighted graph the MAXIMUM DISPERSION problem chooses one vertex set (i.e. $p = 1$) with exactly k vertices, where the total edge-weight of the induced subgraph is to be maximized. The MAXIMUM DISPERSION problem is also $\mathcal{NP}$-hard and remains $\mathcal{NP}$-hard when the weights satisfy the triangle inequality [18]. For further results about this problem see [3,4,5,6,7,8,12,13,15,20].

A similar problem, called MAX-p-Section, is considered by Andersson [1]. He provides an approximation algorithm based on semidefinite programming, which is a natural generalization of the approximation algorithm for the special MAX-Bisection problem given by Frieze and Jerrum [9]. Many of the techniques used in this paper – as relaxation and randomized rounding – base on the pioneer paper of Goemans and Williamson [11] who applied semidefinite programming on MAX-CUT.

The Results. We present a randomized rounding algorithm for the GENERALIZED MAXIMUM DISPERSION problem which achieves for every $0 < \delta \leq \frac{1}{2}$ and $0 < \varepsilon < 1$ with probability at least $\frac{\varepsilon k}{12n}$ a solution with value at least $\frac{(1-\delta)^2(1-\varepsilon)k}{2n-pk} W$, provided that $k \geq \frac{3(1-\delta)}{\delta^2} \ln(\frac{36np}{\varepsilon k})$, where W is the value of an optimal solution. We also show how this algorithm can be derandomized. By iterating the randomized algorithm we obtain the same approximation guarantee even under the weaker condition $k \geq \frac{3(1-\delta)}{\delta^2} \ln(4p)$. A key point in our algorithm is that it can be viewed as a combination of direct randomized rounding and random sampling in the following sense: Instead of rounding the fractional solution directly by taking the fractional solution as rounding probabilities, we take a convex sum of the fractional solution and a certain probability for uniformly distributing the vertices among the subgraphs.

The paper is organized as follows. In Section 2 we introduce a linear relaxation for the GENERALIZED MAXIMUM DISPERSION problem. Depending on this relaxation, we analyze a randomized algorithm in Section 3 and a deterministic algorithm in Section 4.

2   A Linear Relaxation for Generalized Maximum Dispersion

In GENERALIZED MAXIMUM DISPERSION the task is to construct disjoint subsets $S_1, \ldots, S_p \subset V$ maximizing the total weight of the subgraphs induced by the sets S_ℓ, $\ell = 1, \ldots, p$. Let W denote the value of an optimal solution for the given instance. An integer program is given below. For each vertex $i \in V$ we introduce a p-variate vector $x_{i1}, \ldots, x_{ip}$, where $x_{i\ell} = 1$ is interpreted as $i \in S_\ell$. The constraints (2) enforce that each vertex belongs to at most one of the subgraphs, (3) mirrors the cardinality constraints. $z_{ij\ell} = 1$ and (1) imply that both, i and j, are part of the same subset S_ℓ, and the weight of the corresponding edge contributes to the objective function.

$$\text{maximize} \qquad \sum_{(i,j)\in E} w_{ij} \sum_{\ell=1}^{p} z_{ij\ell}$$

$$\text{subject to } 0 \le z_{ij\ell} \le x_{i\ell},\ 0 \le z_{ij\ell} \le x_{j\ell} \text{ for } (i,j) \in E,\ \ell = 1,\ldots,p \qquad (1)$$

$$\sum_{\ell=1}^{p} x_{i\ell} \le 1 \qquad \text{for } i = 1, \cdots, n \qquad (2)$$

$$\sum_{i=1}^{n} x_{i\ell} \le k \qquad \text{for } \ell = 1, \ldots, p \qquad (3)$$

$$x_{i\ell}, z_{ij\ell} \in \{0, 1\} \qquad \text{for } i, j = 1, \cdots, n,\ \ell = 1, \cdots, p \quad (4)$$

When we relax the integrality constraints (4), an optimal fractional solution $x_{i\ell}^*, z_{ij\ell}^* \in [0, 1]$ can be computed in polynomial time using standard linear programming techniques. We now round the fractional solution to an integer. Our rounding scheme is a mixture of direct LP-based randomized rounding (with the fractional solution $x_{i\ell}^*$) and distributing the vertices among the sets uniformly at random with probability k/n. We define the following probabilities

$$p_{i\ell} := cx_{i\ell}^* + (1 - c)k/n \quad \text{for} \quad i = 1, \ldots, n,\ \ell = 1, \ldots, p.$$

$$p_i := \sum_{\ell=1}^{p} p_{i\ell} \quad \text{for} \quad i = 1, \ldots, n.$$

$c \in [0, 1]$ is a constant, depending on k and n, which we will specify later so as to obtain a good approximation factor.

For $\ell = 1, \ldots, p$ let $e_\ell \in \mathbb{N}^p$ be the 0/1-vector whose ℓ-th component is 1 and the other entries are 0, and let $e_{p+1} \in \mathbb{N}^p$ be the zero vector.

Let $0 < \delta \leq \frac{1}{2}$ and let X_i, $1 \leq i \leq n$, be n mutually independent random variables with values in $\{e_\ell \mid \ell = 1, \ldots, p+1\}$ with

$$\Pr[X_i = e_\ell] = \begin{cases} (1-\delta)p_{i\ell} & \text{for } \ell = 1, \ldots, p \\ 1 - (1-\delta)p_i & \text{for } \ell = p+1 \end{cases}$$

We can think of the X_i's as n mutually independent $(p+1)$-faced dice.

Algorithm RANDDISP

1. For $\ell = 1, \ldots, p$ define

$$x_{i\ell} = \begin{cases} 1 & \text{if } X_i = e_\ell \\ 0 & \text{else} \end{cases}$$

If $X_i = e_{p+1}$, then $x_{i\ell} = 0$ for $\ell = 1, \ldots, p$.

2. For $i, j \in \{1, \ldots, n\}$ and $\ell \in \{1, \ldots, p\}$ define

$$z_{ij\ell} = x_{i\ell} x_{j\ell}$$

RANDDISP generates a random 0/1 assignment for the $x_{i\ell}$'s and the $z_{ij\ell}$'s. We show that this is a feasible solution for GENERALIZED MAXIMUM DISPERSION with non-zero probability.

3 A Randomized Algorithm for Generalized Maximum Dispersion

Let $\omega := \sum_{(i,j) \in E} w_{ij} \sum_{\ell=1}^{p} z_{ij\ell}$ be the weight resulting from the above assignment.

Theorem 1. *Let $0 < \delta \leq \frac{1}{2}$, $0 < \varepsilon < 1$ and $c_\delta = \frac{3(1-\delta)}{\delta^2}$. If $k \geq c_\delta \ln(\frac{36np}{\varepsilon k})$, then with probability at least $\frac{\varepsilon k}{12n}$ the $z_{ij\ell}$, $x_{i\ell}$, $i,j \in \{1, \ldots, n\}$, $\ell \in \{1, \ldots, p\}$ build a feasible solution for GENERALIZED MAXIMUM DISPERSION and*

$$\omega \geq (1-\delta)^2 (1-\varepsilon) \frac{k}{2n - pk} W.$$

PROOF Let $A_0, A_1, \ldots, A_p$ be the following events. A_0 is the event $\omega < (1-\delta)^2 (1-\varepsilon) \frac{k}{2n-pk} W$. For $\ell = 1, \ldots, p$, A_ℓ is the event $\sum_{i=1}^{n} x_{i\ell} > k$.

We will derive upper bounds for $\Pr[A_\ell]$, $\ell = 0, \ldots, p$.
Let $a(n) = 1/(1 + \frac{\varepsilon k}{8n})$ and let $b(n) = 1 - a(n)$.

Claim 1: $\Pr[A_0] \leq a(n)$

Proof of Claim 1:

$$\mathbb{E}[\omega] = (1-\delta)^2 \sum_{(i,j)\in E} w_{ij} \sum_{\ell=1}^{p} p_{i\ell} p_{j\ell}$$

$$= (1-\delta)^2 \sum_{(i,j)\in E} w_{ij} \sum_{\ell=1}^{p} \left[cx_{i\ell}^* + (1-c)k/n \right]\left[cx_{j\ell}^* + (1-c)k/n \right]$$

$$= (1-\delta)^2 \sum_{(i,j)\in E} w_{ij} \sum_{\ell=1}^{p} \left[c^2 x_{i\ell}^* x_{j\ell}^* + c(1-c)(x_{i\ell}^* + x_{j\ell}^*)k/n + (1-c)^2 k^2/n^2 \right]$$

$$\geq (1-\delta)^2 \left[c(1-c)\Big(\sum_{(i,j)\in E} w_{ij} \sum_{\ell=1}^{p} 2z_{ij\ell}^* \Big)k/n \;+\; \omega(V)\, p(1-c)^2 k^2/n^2 \right]$$

$$\geq (1-\delta)^2 \left[2c(1-c)k/n + p(1-c)^2 k^2/n^2 \right] W$$

Now choose c so that the factor $2c(1-c)k/n + p(1-c)^2 k^2/n^2$ is maximum. It is easy to show that this is the case for $c = 1 - n/(2n - pk)$. Thus

$$\mathbb{E}[\omega] \geq (1-\delta)^2 \left(2 \cdot \left(1 - \frac{n}{2n-pk}\right) \cdot \frac{n}{2n-pk} \cdot \frac{k}{n} + p \cdot \left(\frac{n}{2n-pk}\right)^2 \cdot \frac{k^2}{n^2} \right) W$$

$$= (1-\delta)^2 \left(2 \cdot \left(1 - \frac{n}{2n-pk}\right) \cdot \frac{k}{2n-pk} + \frac{pk^2}{(2n-pk)^2} \right) W$$

$$= (1-\delta)^2 \left(\frac{2kn - pk^2}{(2n-pk)^2} \right) W$$

$$= (1-\delta)^2 \frac{k}{2n-pk} W$$

We have

$$\Pr[A_0] = \Pr\left[W - \omega > W - (1-\delta)^2(1-\varepsilon)\frac{k}{2n-pk} W \right]$$

$$\leq \Pr\left[W - \omega > W - (1-\varepsilon)\mathbb{E}[\omega] \right]$$

$$\leq \frac{W - \mathbb{E}[\omega]}{W - (1-\varepsilon)\mathbb{E}[\omega]} \qquad \text{(Markov inequality)} \qquad (5)$$

$$\leq a(n) \qquad\qquad\qquad\qquad\qquad\qquad\qquad\qquad\qquad (6)$$

Claim 1 is proved, if we have shown:

$$\frac{W - \mathbb{E}[\omega]}{W - (1-\varepsilon)\mathbb{E}[\omega]} \leq a(n) \qquad\qquad\qquad (7)$$

As (7) is equivalent to

$$W - W \cdot a(n) \leq (1 - (1-\varepsilon) \cdot a(n)) \cdot \mathbb{E}[\omega]$$

it is sufficient to show (7) for $\mathbb{E}[\omega] = (1-\delta)^2 \frac{k}{2n-pk} W$.

In this case (7) is equivalent to

$$\left(1 - (1-\delta)^2 \cdot \frac{k}{2n - pk}\right) \cdot \left(1 + \frac{\varepsilon k}{8n}\right) \le 1 - (1-\varepsilon)\cdot(1-\delta)^2 \cdot \frac{k}{2n - pk}$$

and to

$$\frac{\varepsilon k}{8n} \le (1-\delta)^2 \cdot \frac{k}{2n - pk} \cdot \left(1 + \frac{\varepsilon k}{8n} - 1 + \varepsilon\right) \tag{8}$$

Because of $\delta \le \frac{1}{2}$, (8) follows from

$$2n - pk \le \frac{1}{4} \cdot (k + 8n)$$

which is true for $p \ge -\frac{1}{4}$.

Claim 2: $\Pr[A_\ell] \le \frac{b(n)}{4p}$ for $\ell = 1, \ldots, p$

Proof of Claim 2:

$$\mathbb{E}\left[\sum_{i=1}^{n} x_{i\ell}\right] = (1-\delta)\sum_{i=1}^{n} p_{i\ell}$$

$$= (1-\delta)\left[c \sum_{i=1}^{n} x_{i\ell}^* + (1-c)k\right]$$

$$\le (1-\delta)\left[ck + (1-c)k\right]$$

$$= (1-\delta)k.$$

With the Angluin-Valiant inequality (see [2])

$$\Pr[A_\ell] = \Pr\left[\sum_{i=1}^{n} x_{i\ell} > k\right]$$

$$= \Pr\left[\sum_{i=1}^{n} x_{i\ell} > (1-\delta) \cdot k \cdot \left(1 + \frac{\delta}{1-\delta}\right)\right]$$

$$\le e^{-\frac{\delta^2}{(1-\delta)^2 \cdot 2\left(1 + \frac{\delta}{3(1-\delta)}\right)} \cdot (1-\delta)k} \qquad \text{Angluin-Valiant inequality}$$

$$\le e^{-\frac{\delta^2 k}{3(1-\delta)}} \qquad \text{using } \frac{\delta}{1-\delta} \le \frac{3}{2}$$

$$\le \frac{\varepsilon k}{36np} \qquad \text{using } k \ge c_\delta \ln\left(\frac{36np}{\varepsilon k}\right)$$

$$\le \frac{b(n)}{4p}$$

The correctness of the last inequality follows from the equivalence to

$$\varepsilon k\left(1 + \frac{\varepsilon k}{8n}\right) \le 9n\left(1 + \frac{\varepsilon k}{8n}\right) - 9n$$

which is equivalent to

$$72n^2 \leq (8n + \varepsilon k)(9n - \varepsilon k)$$

and to $n \geq \varepsilon k$ which is true. So Claim 2 is proved.

Hence

$$\Pr[A_0 \vee A_1 \vee \cdots \vee A_p] \leq \sum_{\ell=0}^{p} \Pr[A_\ell] \leq a(n) + b(n)/4$$

$$\leq \frac{3}{4} a(n) + \frac{1}{4} \leq 1 - \frac{\varepsilon k}{12n} \tag{9}$$

The correctness of the last inequality follows from the equivalence to

$$9n \leq 9n \cdot \left(1 + \frac{\varepsilon k}{8n}\right) - \varepsilon k \cdot \left(1 + \frac{\varepsilon k}{8n}\right)$$

and to $n \geq \varepsilon k$.

Therefore the theorem is proved. □

There are two interesting cases depending on the magnitude of k.

Case 1: $k = o(n)$. Due to (6) the probability that the objective function is smaller than $(1 - \delta)^2 (1 - \varepsilon) \frac{k}{2n - pk} W$ is not larger than $a(n)$. If $k = o(n)$, $a(n)$ tends to 1 as n tends to infinity. In this case the success probability of RANDDISP tends to zero. But if we iterate RANDDISP L times and take the best result, then the probability of the event A_0 is only $a(n)^L$. Let us call the iterated algorithm RANDDISP(L). With $f = \frac{\varepsilon k}{8n}$, the inequality $a(n)^L \leq 1/4$ is equivalent to $L \geq (\ln(1 + f))^{-1} \ln 4$. Since $f \leq 1/8$ because of $\varepsilon k \leq n$, Taylor expansion of $\ln(1 + f)$

$$\ln(1 + f) = f - \frac{f^2}{2} + \frac{f^3}{3} - \frac{f^4}{4} + \cdots$$

with $f - \frac{f^2}{2} \geq \frac{15}{16} f$ shows that $\ln(1 + f) \geq 15f/16$, thus for $L \geq \frac{16 \ln 4}{15 f}$ we get $a(n)^L \leq 1/4$. Now we can argue as in the proof of Theorem 1 and obtain under the weaker condition $k \geq c_\delta \ln(4p)$ the desired approximation:

Theorem 2. Let $0 < \delta \leq \frac{1}{2}$, $0 < \varepsilon < 1$, $c_\delta = \frac{3(1 - \delta)}{\delta^2}$ and let $L = \lceil \frac{12n}{\varepsilon k} \rceil$. If $k \geq c_\delta \ln(4p)$, then with probability at least $\frac{1}{2}$, RANDDISP(L) generates a solution $z_{ij\ell}$, $x_{i\ell}$, $i, j \in \{1, \ldots, n\}$, $\ell \in \{1, \ldots, p\}$ for GENERALIZED MAXIMUM DISPERSION such that

$$\omega \geq (1 - \delta)^2 (1 - \varepsilon) \frac{k}{2n - pk} W$$

Case 2: $k = \Omega(n)$. In this case $a(n)$ is a constant smaller than one, and as above the result can be proved under the weaker condition $k \geq c_\delta \ln(4p)$. But since $k = \Omega(n)$, for every fixed δ this condition is automatically satisfied.

Note that using a non-linear rounding function, Goemans [10] has achieved the better approximation guarantee of k/n for the situation where a *single* subgraph is to be determined. However, his analysis does not appear to generalize to $p \geq 2$.

For the complete graph $G = K_n, k = n/p$ and unit edge weights an optimum partition has weight $n^2/2p - n/2$. On the other hand $x_{i\ell} = x_{j\ell} = z_{ij\ell} = 1/p$, for all i, j, ℓ, is a feasible solution of the fractional linear program with objective function value $(n^2 - n)/2$. So the integrality gap between the integer optimum and the fractional optimum is $\frac{1}{p}(1 - o(1))$. For our approximation factor holds

$$(1 - \delta)^2(1 - \varepsilon)\frac{k}{2n - pk} \leq \frac{k}{2n - pk} \leq \frac{k}{n} \leq \frac{1}{p}$$

where for $\delta = 0$, $\varepsilon = 0$, and $k = n/p$ the equality holds. Thus the approximation factor cannot be improved essentially using the relaxation from this paper.

Of course, Hassin, Rubinstein and Tamir's [14] approximation factor – for the special case that the triangle equality is satisfied – is better in general than our approximation factor because of $\frac{1}{p} \leq \frac{1}{2 - 1/\lceil k/2 \rceil}$ for $p \geq 2$.

4 A Deterministic Algorithm for Generalized Maximum Dispersion

RANDDISP can be derandomized using the method of conditional probabilities and pessimistic estimators (see [17,19] for details):

Theorem 3. *Let T be the time to compute a fractional, optimal solution for* GENERALIZED MAXIMUM DISPERSION. *Let $0 < \delta \leq \frac{1}{2}$, $0 < \varepsilon < 1$ and $c_\delta = \frac{3(1-\delta)}{\delta^2}$. If $k \geq c_\delta \ln(\frac{36np}{\varepsilon k})$, then a feasible solution for* GENERALIZED MAXIMUM DISPERSION *which satisfies*

$$\omega \geq (1 - \delta)^2(1 - \varepsilon)\frac{k}{2n - pk} W$$

can be found in $O(T + pn^2 \log(\frac{pn}{\varepsilon k}))$ time.

PROOF

The proof relies on a slight modification of the algorithmic Angluin-Valiant inequality for multivalued random variables (Theorem 2.13 in [19]). There the probabilities of all events under consideration are estimated by the Angluin-Valiant inequality, whereas here the probability of the event A_0 is bounded by Markov's inequality and the probabilities of all other events A_ℓ, $\ell = 1, \ldots, p$, have been estimated by the Angluin-Valiant inequality. We must ensure that Theorem 2.13 in [19] is valid in this case as well. We follow the notation in the proof of Theorem 1. In the proof of Theorem 2.13 in [19] upper bounds for the conditional probabilities

$$\Pr[A_\ell \mid X_1, \ldots, X_i]$$

must be computed efficiently, for all $i = 1, \ldots, n$ and $\ell = 0, \ldots, p$. For the events A_ℓ, $\ell = 1, \ldots, p$ this can be done as in the proof of Theorem 2.13 in [19]. For A_0 the Markov inequality helps. By (5) we have

$$\Pr[A_0 \mid X_1, \ldots, X_i] \leq \frac{W - \mathbb{E}[\omega \mid X_1, \ldots, X_i]}{W - (1 - \varepsilon)\mathbb{E}[\omega \mid X_1, \ldots, X_i]}$$

The conditional expectations $\mathbb{E}[\omega \mid X_1, \ldots, X_i]$ can easily be computed in $O(pn)$-time evaluating the quadratic function ω at the values determined by the $X_1, \ldots, X_i$ and then taking the expectation. With this observation the proof of Theorem 2.13 in [19] can be lifted to cover all events $A_0, A_1, \ldots, A_p$. We invoke Theorem 2.13 in [19], with n random variables $X_1, \ldots, X_n$ with values in the set

$$\{y \in \{0, 1\}^p \mid \sum_{\ell=1}^{p} y_\ell \in \{0, 1\}\}$$

and $p + 1$ events defined as follows. For $\ell = 0, \ldots, p$ we define

$$\psi_\ell = \begin{cases} \sum_{i=1}^{n} x_{i\ell} & \text{for } \ell = 1, \ldots, p \\ \sum_{(i,j) \in E} w_{ij} \sum_{r=1}^{p} x_{ir}x_{jr} & \text{if } \ell = 0. \end{cases}$$

The corresponding events in Theorem 2.13 in [19], defined via the functions ψ_ℓ are the events A_ℓ, $\ell = 0, \ldots, p$. The upper bounds $f(\ell)$ for $\Pr[A_\ell]$ are according to the proof of Theorem 1: $a(n)$ if $\ell = 0$ and $\frac{b(n)}{4p}$ for $\ell = 1, \ldots, p$. Using (9), condition (27) preceding Theorem 2.13 in [19] reads as

$$\sum_{\ell=0}^{p} f(\ell) \leq 1 - \gamma,$$

with $\gamma = \frac{\varepsilon k}{12n}$. By Theorem 2.13 in [19] we can construct a vector $(X_1, \ldots, X_n)$ with $X_i \in \{0, 1\}^p$ and $\sum_{\ell=1}^{p} x_{i\ell} \in \{0, 1\}$ for all $i = 1, \ldots, n$ such that $(X_1, \ldots, X_n)$ simultaneously satisfies all events $\overline{A_0}, \ldots, \overline{A_p}$, as required. The time needed for the construction of $(X_1, \ldots, X_n)$ is according to Theorem 2.13 in [19]

$$O\left((p+1)n^2 \log\left(\frac{(p+1)n}{\gamma}\right)\right) = O\left(pn^2 \log\left(\frac{pn}{\varepsilon k}\right)\right).$$

Together with the computation time T for the fractional solution we are done. □

Acknowledgement

We would like to thank the anonymous referees for their helpful comments.

The research was funded in part by the United States National Science Foundation grant IIS-053525.

References

1. G. Andersson. An approximation algorithm for Max p-Section. *Proceedings of Symposium on Theoretical Aspects in Computer Science (STACS)*, Lecture Notes in Comput. Sci. 1563: 237-247, 1999.
2. D. Angluin and L.G. Valiant. Fast Probabilistic Algorithms for Hamiltonian Circuits and Matchings. *J. Comput. System Sci.* 18(2): 155-193, 1979.
3. Y. Asahiro, K. Iwama, H. Tamaki, T. Tokuyama. Greedily Finding a Dense Subgraph. *J. Algorithms* 34(2): 203-221, 2000.
4. A. Czygrinow. Maximum dispersion problem in dense graphs. *Oper. Res. Lett.* 27: 223-227, 2000.
5. U. Feige, G. Kortsarz, D. Peleg. The Dense k-Subgraph Problem. *Algorithmica* 29(3): 410-421, 2001.
6. U. Feige, M. Langberg. Approximation algorithms for maximization problems arising in graph partitioning. *J. Algorithms* 41(2): 174-211, 2001.
7. U. Feige, M. Seltser. On the Densest k-Subgraph Problem. Technical report, Department of Applied Mathematics and Computer Science, The Weizmann Institute, Rehovot, September 1997.
8. S.P. Fekete, H. Meijer. Maximum Dispersion and Geometric Maximum Weight Cliques. *Algorithmica* 38(3): 501-511, 2003.
9. A. Frieze, M. Jerrum. Improved approximation algorithms for MAX k-CUT and MAX BISECTION. *Algorithmica* 18: 67-81, 1997.
10. M.X. Goemans. Mathematical programming and approximation algorithms. Lecture given at the Summer School on Approximate Solution of Hard Combinatorial Problems, Udine, September 1996.
11. M.X. Goemans, D.P. Williamson. Improved approximation algorithms for maximum cut and satisfiability problems using semidefinite programming. *J. ACM* 42(6): 1115-1145, 1995.
12. E. Halperin, U. Zwick. A unified framework for obtaining improved approximation algorithms for maximum graph bisection problems. *Random Structures Algorithms* 20(3): 382-402, 2002.
13. Q. Han, Y. Ye, J. Zhang. An improved rounding method and semidefinite programming relaxation for graph partition. *Math. Program.* 92(3): 509-535, 2002.
14. R. Hassin, S. Rubinstein and A. Tamir. Approximation algorithms for maximum dispersion. *Oper. Res. Lett.* 21(3): 133-137, 1997.
15. G. Jäger, A. Srivastav. Improved Approximation Algorithms for Maximum Graph Partitioning Problems. *J. Comb. Optim.* 10(2): 133-167, 2005.
16. S. Khot. Ruling Out PTAS for Graph Min-Bisection, Densest Subgraph and Bipartite Clique. *Proceedings of the 45th Annual IEEE Symposium on Foundations of Computer Science (FOCS)*, IEEE Computer Society, 136-145, 2004.
17. P. Raghavan. Probabilistic construction of deterministic algorithms: approximating packing integer programs. *J. Comput. System Sci.* 37(2): 130-143, 1988.
18. S.S. Ravi, D.J. Rosenkrantz and G.K. Tayi. Heuristic and Special Case Algorithms for Dispersion Problems. *Oper. Res.* 42(2): 299-310, 1994.
19. A. Srivastav and P. Stangier. Algorithmic Chernoff-Hoeffding inequalities in integer programming. *Random Structures Algorithms* 8(1): 27-58, 1996.
20. A. Srivastav and K. Wolf. Finding Dense Subgraphs with Semidefinite Programming. *Proceedings of the 1st International Workshop on Approximation Algorithms for Combinatorial Optimization (WAOA)*, Lecture Notes in Comput. Sci. 1444: 181-191, 1998.

Significance-Driven Graph Clustering*

Marco Gaertler, Robert Görke, and Dorothea Wagner

Faculty of Informatics, Universität Karlsruhe (TH)
{gaertler,rgoerke,wagner}@informatik.uni-karlsruhe.de

Abstract. Modularity, the recently defined quality measure for clusterings, has attained instant popularity in the fields of social and natural sciences. We revisit the rationale behind the definition of modularity and explore the founding paradigm. This paradigm is based on the trade-off between the achieved quality and the expected quality of a clustering with respect to networks with similar intrinsic structure. We experimentally evaluate realizations of this paradigm systematically, including modularity, and describe efficient algorithms for their optimization. We confirm the feasibility of the resulting generality by a first systematic analysis of the behavior of these realizations on both artificial and on real-world data, arriving at remarkably good results of community detection.

1 Introduction

Discovering natural groups and large scale inhomogeneities is a crucial task in the exploration and analysis of large and complex networks. This task is usually realized with clustering methods. The majority of algorithms for graph clustering are based on the paradigm of intra-cluster density versus inter-cluster sparsity. Several formalizations have been proposed and evaluated, an overview of such techniques is given in [1]. Recently, Girvan and Newman [2] proposed the objective function *modularity*, which indirectly incorporates this paradigm. Modularity evaluates a clustering based on the fraction of intra-cluster edges compared to the expected value of this number. Modularity was first introduced as a quality measure of a clustering, however, a simple greedy algorithm, solely based on modularity, has been proposed shortly after by Newman [3]. The definition of modularity and this algorithm evoked a surge of interest, yielding many studies concerning different applications, such as protein interaction dependencies, recommendation systems or social network analysis, and possible adjustments (see e.g. [4,5,6,7]). Moreover, a range of alternative algorithmic approaches has been proposed, based on a greedy agglomeration [8,9], on spectral division [10,11], simulated annealing [12,13] and extremal optimization [14]. Recently, it has been proven, that it is $\mathcal{NP}$-hard to optimize modularity [15],

---

* This work was partially supported by the DFG under grants WA 654/14-3 and by EU under grant DELIS (contract no. 001907) and CREEN project (contract no. 012684).

M.-Y. Kao and X.-Y. Li (Eds.): AAIM 2007, LNCS 4508, pp. 11–26, 2007.

which justifies the need for heuristics and approximations. Note that little is known on the approximability of modularity.

However, to our knowledge, no systematic evaluation of modularity-based clustering has been performed, yet. In this work, we conduct such an evaluation and, additionally, we advance towards a profound understanding of the rationales modularity is based upon. We formally state and investigate the generalized founding paradigm for the *significance* of a clustering as the trade-off between the achieved quality and the expected quality for random networks incorporating the intrinsic properties of the original network. We experimentally evaluate realizations of this paradigm (including modularity itself) and extensively evaluate algorithms that each optimize a realization, followed by a partial discussion of their behavior. Furthermore, we present an algorithm that efficiently optimizes the particularly promising realization *relative performance significance* in $O(n^2 \log n)$ time. We compare the performance of these algorithms in terms of clustering quality to that of other clustering algorithms, on a set of random pre-clustered graphs and complement our findings with results on real data. Our results indicate the feasibility of the paradigm in that, on the whole, our algorithms surpass the benchmark algorithms, and in that the generality of the approach is justified by specific realizations excelling on real-world data.

This paper is organized as follows: After introducing the necessary preliminaries for graph clustering and some quality measures (Sec. 2), we give the formal definition of our significance paradigm and present some realizations (Sec. 3). Section 4 scrutinizes the greedy algorithms which are employed to obtain clusterings with high significance score, including an efficient implementation for a *quick divisive merge*. The setup and the results of the experimental evaluation are described in Section 5 which are followed by a conclusion.

2   Preliminaries

Throughout this paper, we will use the notation of [1]. More precisely, we assume that $G = (V, E, \omega)$ is an undirected, weighted, and simple graph and $\omega : E \rightarrow [0, 1]$. For a node v, we define the node weight $\omega(v)$ as the sum of the weight of its incident edges. Let $|V| =: n, |E| =: m$ and $\mathcal{C} = \{C_1, \ldots, C_k\}$ a partition of V. We call $\mathcal{C}$ a *clustering* of G and the C_i *clusters*; $\mathcal{C}$ is called *trivial* if either $k = 1$, or all clusters C_i contain only one element. In the following, we often identify a cluster C_i with the induced subgraph of G, i.e., the graph $G[C_i] := (C_i, E(C_i), \omega_{|E(C_i)})$, where $E(C_i) := \{\{v, w\} \in E : v, w \in C_i\}$. Then $E(\mathcal{C}) := \bigcup_{i=1}^{k} E(C_i)$ is the set of *intra-cluster edges* and $E \setminus E(\mathcal{C})$ the set of *inter-cluster edges*. The set $E(C_i, C_j)$ denotes the set of edges connecting nodes in C_i to nodes in C_j. The number of intra-cluster edges is denoted by $m(\mathcal{C})$ and the number of inter-cluster edges by $\overline{m}(\mathcal{C})$. We denote the number of non-adjacent pairs of nodes that are in the same cluster as $m(\mathcal{C})^c$, and the number of non-connected pairs of nodes that are not in the same cluster as $\overline{m}(\mathcal{C})^c$. Modularity is defined ([9]) as:

$$\text{mod}\,(\mathcal{C}) := \frac{m(\mathcal{C})}{m} - \frac{1}{4m^2} \sum_{C \in \mathcal{C}} \left(\sum_{v \in C} \deg(v) \right)^2 \tag{1}$$

We measured the quality of clusterings with a range of of quality indices, discussed e.g., in [1], however, we set our focus on the indices *coverage* ([1]) and *performance* ([16]) in this work, since they are the most well studied ones. In brief, coverage is the fraction of intra-cluster edges and performance is the fraction of correctly classified node pairs. For a discussion of these indices we refer the reader to the given references, and simply state their formal definitions in Equations (2) and (3):

$$\text{cov}(\mathcal{C}) := \frac{m(\mathcal{C})}{m} = \frac{m(\mathcal{C})}{m(\mathcal{C}) + \overline{m}(\mathcal{C})} \tag{2} \qquad \text{perf}(\mathcal{C}) := \frac{m(\mathcal{C}) + \overline{m}(\mathcal{C})^c}{\frac{1}{2}n(n-1)} \tag{3}$$

The fact that modularity can be expressed as *coverage* minus the expected value of *coverage* (see Sec. 3.1 and [9]) motivates the general paradigm we state in the next section. These definitions generalize in a natural way as to take edge weights into account. Thus, $\omega(\mathcal{C})$ ($\overline{\omega}(\mathcal{C})$) denotes the sum of the weights of all intra-cluster (inter-cluster) edges and W denotes the sum of all edge weights. In the terms $\omega(\mathcal{C})^c$ and $\overline{\omega}(\mathcal{C})^c$ the weight between a non-adjacent pair of nodes is set to the maximum edge weight in E and the weight between adjacent nodes is the difference of the maximum edge weight and the weight of the edge between them.

3   The Significance Paradigm

In the *significance paradigm* a good clustering is characterized by having a high quality compared to the value obtained for a random network that reflects specific *structural properties* that are expected to be present in the graph, as predefined in an appropriate null hypothesis. The structural properties of a graph can include characteristics such as the sequence of degrees, the number of nodes, the clustering coefficient, the degree distribution etc. These properties need not determine a graph completely but define a family of graphs incorporating them. A configuration is a specific realization of these properties, i.e., a specific graph. Every realization of the significance paradigm requires a quality measure, a null hypothesis, and a mode of comparison of both. Since measuring the quality of a clustering is a well-studied field, we present a way to extend a given quality index $\mathcal{M}$ to our paradigm. As, in this context, modularity (1) extends coverage (2), the concept of significance is a true generalization of modularity.

Definition 1. *Given a quality index $\mathcal{M}$ and a clustering $\mathcal{C}$, we define the significance $S_{\mathcal{M}}^{\odot}$ of a clustering $\mathcal{C}$ as the corresponding quality index respecting our paradigm in the following way:*

$$S_{\mathcal{M}}^{\odot}(\mathcal{C}) := \mathcal{M}(\mathcal{C}) \odot \mathbb{E}_{\Omega}[\mathcal{M}(\mathcal{C})] \ , \qquad (4)$$

where $\mathbb{E}_{\Omega}[\mathcal{M}]$ is the expected value of the quality index $\mathcal{M}$ for the clustering $\mathcal{C}$ with respect to a suitable probability space Ω and $\odot$ is a binary operator on real numbers.

The following example illustrates the intention of this definition. Although, many quality measures $\mathcal{M}$ are normalized to the interval $[0, 1]$, it is often hard to associate a specific value with a meaningful interpretation as showcased in Figures 1 and 2. This can be an intrinsic pattern of the measure and not merely

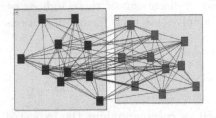

Fig. 1. A random split of a $G(n, p)$ with $n = 20$ and $p = 0.5$, where coverage yields a value of 0.66

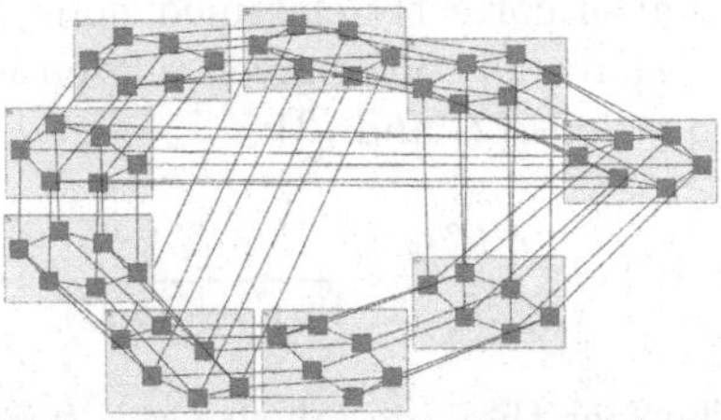

Fig. 2. A meaningful clustering of an interconnected, six-sided tube, yielding a coverage of 0.43

an artifact of the normalization. However, the comparison of the achieved quality with the expected quality (with respect to a suitable model) intuitively yields a 'relative' variant of the measure. Thus, it potentially allows for a better interpretation of the scale (modularity value for Fig. 1: 0.128, for Fig. 2: 0.316).

3.1   Probability Space

In the following we briefly discuss a suitable probability space Ω required for Definition 1, which we use throughout this paper. The importance of the structural properties of a graph may differ depending on the application, we restrict ourselves to a basic setup, where the degrees of the nodes are considered to be a defining property. Thus Ω consists of the family of all graphs having the same sequence of degrees as the input graph, with uniform probability.

However, even for simple quality indices, such as coverage, we are not aware of a closed formulation for the expectation. Instead of designing random processes to obtain expectation values, we consider a heuristic that was already used in [9]. The heuristic assigns to each edge $e = \{v, w\}$ a probability of $\deg(v)\deg(w)/2m$ in the unweighted case, and an expected weight of $\omega(v)\omega(w)/2W$ in the weighted case. While this heuristic does not rigorously respect the expected sum of all node weights, the error term tends to zero for graphs with a sublinear maximum degree, see Appendix A. A degenerate example that clearly illustrates this discrepancy is the unweighted, complete graph K_n. For this graph, the original probability space contains only one element, thus yielding $\mathcal{M}(\mathcal{C}) = \mathbb{E}_{\Omega}[\mathcal{M}(\mathcal{C})]$, which is not true for the heuristic.

The concept of significance is related to the notion of p-values in statistical hypothesis testing. The p-value of a value t observed for a random variable T is the probability that under the assumption of a given null hypothesis, T assumes a value at least as unfavorable to the null hypothesis as the observed value t. In general, the null hypothesis is rejected, if the p-value is smaller than the statistical significance level (see e.g. [17]). However, in our concept we do not reject a null hypothesis, which we assume to reasonably describe observed graphs. Instead, we compare the achieved quality of a clustering to the expected value, in order to judge its relevance.

3.2   Implementations of the Significance Paradigm

The heuristic presented above enables us to study four implementations of the significance paradigm, namely, *coverage* and *performance* as quality indices and subtraction and division as the binary operators. Using *coverage* and subtraction, modularity is one of the implementations. Table 3.2 summarizes the formulas of the resulting four implementations of the significance paradigm. For a discussion of *performance* in weighted graphs see [1], the proof for $\mathbb{E}[performance]$ can be found in the Appendix A.

Table 1. Quality indices and their expected values ($\omega_{\max}$: maximum edge weight)

measure	*coverage*	*performance*
$\mathcal{M}$	$\frac{m(C)}{m}$	$\frac{m(C)+\overline{m}(C)^c}{0.5 \cdot n(n-1)}$
$\mathbb{E}[\mathcal{M}]$	$\sum_{C \in \mathcal{C}} \left(\frac{\sum_{v \in C} \deg(v)}{2m} \right)^2$	$\frac{\sum_{C \in \mathcal{C}}((\sum_{v \in C} \deg(v))^2/m - (\sum_{v \in C} 1)^2) + n^2 - 2m}{n(n-1)}$
$\mathcal{M}_{\text{weighted}}$	$\frac{\omega(C)}{W}$	$\frac{\omega(C)+\omega_{\max}\overline{m}(C)^c+(\omega_{\max}\overline{m}(C)-\overline{\omega}(C))}{0.5 \cdot n(n-1) \cdot \omega_{\max}}$
$\mathbb{E}[\mathcal{M}_{\text{weighted}}]$	$\sum_{C \in \mathcal{C}} \left(\frac{\sum_{v \in C} \omega(v)}{2W} \right)^2$	$\frac{\sum_{C \in \mathcal{C}}(\sum_{v \in C} \omega(v))^2/W + \omega_{\max}(n^2 - \sum_{C \in \mathcal{C}} |C|^2) - 2W}{n(n-1)\omega_{\max}}$

Note that the weighted versions of modularity are true generalizations of the unweighted case, since setting each weight to 1 yields the unweighted formulas. Thus, we restrict our analyses on the weighted case. The weighted variant of modularity has also been defined by Newman in [18]. Based on Table 3.2 we now define the following implementations, with a nomenclature borrowed from approximation theroy:

$$S^-_{\text{cov}} := coverage - \mathbb{E}[coverage] \quad \text{(equals modularity)} \qquad S^{\div}_{\text{cov}} := \frac{coverage}{\mathbb{E}[coverage]}$$

$$S^-_{\text{perf}} := performance - \mathbb{E}[performance] \qquad\qquad S^{\div}_{\text{perf}} := \frac{performance}{\mathbb{E}[performance]}$$

$$\underbrace{\phantom{S^-_{\text{perf}} := performance - \mathbb{E}[performance]}}_{\text{absolute variants (subtractive)}} \qquad \underbrace{\phantom{S^{\div}_{\text{perf}} := \frac{performance}{\mathbb{E}[performance]}}}_{\text{relative variants (divisive)}}$$

As we shall see in the evaluation in Section 5, these implementations differ significantly in their behavior, although they are all derived from the same paradigm.

4   Significance-Clustering Algorithms

In this section, we describe the algorithms that are used to obtain clusterings with high values of significance. As suggested by Newman in [3] we employ a straightforward greedy heuristic. This allows for a fair evaluation for all the variants of significance introduced in Section 3.2. The usage of a heuristic or an approximation is strongly encouraged by the fact that an $\mathcal{NP}$-completeness proof of this optimization problem has recently been presented in [15].

4.1   The Greedy Algorithm

For a given significance measure S the greedy algorithm starts with the singleton clustering and iteratively merges those two clusters that yield largest increase or the smallest decrease in significance. After $n-1$ merges the clustering that achieved the highest significance is returned. The algorithm maintains a symmetric matrix ΔS with entries $\Delta S_{i,j}$ equaling $S(\mathcal{C}_{i,j}) - S(\mathcal{C})$, where $\mathcal{C}$ is the current clustering and $\mathcal{C}_{i,j}$ is obtained from $\mathcal{C}$ by merging clusters C_i and C_j. The pseudo-code for the greedy algorithm is given in Algorithm 1.

Algorithm 1. GREEDY SIGNIFICANCE

   Input: Graph $G = (V, E, \omega)$
   Output: Clustering $\mathcal{C}$ of G
1 $\mathcal{C} \leftarrow$ Singletons
2 Initialize matrix ΔS (as described above)
3 Initialize S
4 **while** $|\mathcal{C}| > 1$ *and* $\exists i, j : \Delta S_{i,j} > 0$ **do**
5   Find $\{i, j\}$ with $\Delta S_{i,j}$ is the maximum entry in the matrix ΔS
6   Merge clusters i and j
7   Update ΔS
8      $S \leftarrow S + \Delta S_{i,j}$
9 Return clustering with highest significance

4.2   Runtime Analysis

Due to the special structure of the significance measures introduced in Section 3.2, a precise analysis can be given. We define the matrices $\Delta \mathcal{M}$ and $\Delta \mathbb{E}[\mathcal{M}]$ analogously to ΔS. First, note that *coverage* and *performance* can be updated locally. More precisely, let $\mathcal{C}$ be a clustering, then $cov(\mathcal{C}_{i,j})$ can be computed using $cov(\mathcal{C})$, $E(C_i)$, $E(C_j)$ and $E(C_i, C_j)$. The same holds for *performance* and for both values of expectation. Thus, we obtain for the absolute variants the update formula: $\Delta S_{k,(ij)} = \Delta \mathcal{M}_{i,k} + \Delta \mathcal{M}_{j,k} - \Delta \mathbb{E}[\mathcal{M}_{k,i}] - \Delta \mathbb{E}[\mathcal{M}_{k,j}]$, where $C_{(ij)}$

corresponds to the merged cluster of C_i and C_j; see [19] for details. Due to this locality, only two rows and columns are affected, namely the ith and the jth, requiring $O(n \log n)$ update time, using heaps to store the rows. The heaps help to quickly find the maximum entry in ΔS, i.e. the best improvement. However, for the relative variants, the whole matrix has to be updated, yielding at least a quadratic runtime for a standard implementation. In Section 4.3, we present a faster method using *kinetic heaps* and geometry.

Employing a simple data structure for clusterings e.g. representing clusters by linked lists and storing a collection of pointers to their head, one observes that Step 1, Step 3 and Step 9 run in $O(n)$ time. The matrix ΔS is initialized in $O(n^2)$ time. The loop at Step 4 is executed $n-1$ times. Step 5 runs in $O(n)$ time, since the rows of ΔS are stored as heaps. The merge of two clusters (step 6) and the update of S (step 8) require at most linear time. Thus, the total runtime is dominated by the the the time required for updating ΔS, which yields the following lemma:

Lemma 1. *Algorithm 1 runs in $O(n^2 \log n)$ time for the absolute variants.*

For the relative variants, an analog to Lemma 1 yields a runtime of $O(n^3)$, however, in Lemma 2 we improve this upper bound for relative variants employing Algorithm 2. It is not hard to see that the first local optimum of S, that the absolute greedy heuristic attains, is its global optimum, since then the matrix ΔS is non-positive, allowing no further increase in S with any future merge. This is due to the linear update of ΔS that consists of simply adding up specific entries, thus keeping the matrix non-positive after future merges. This can result in a substantial decrease in running time if the graph is known to have a fine clustering structure, i.e., if $|\mathcal{C}| \in \omega(1)$ (if the number of clusters is dependent on n). The loop at Step 4 can then be terminated after $o(n)$ merges. A similar observation holds for the relative variants, as described in the next section.

4.3   Quick Divisive Merge

Considering the relative variants, after a merge, both $\mathcal{M}$ and $\mathbb{E}[\mathcal{M}]$ change, and thus all values of $\Delta S_{k,l}$ have to be recomputed after each merge, even though $\Delta \mathcal{M}_{k,l}$ and $\Delta \mathbb{E}[\mathcal{M}]_{k,l}$ might not change. The update of ΔS thus requires $O(n^2)$ time. In Algorithm 2 we show how this update can be performed in $O(n \log n)$ time using a geometric embedding of ΔS. The algorithm uses the fact that if we assign to each entry $\{i,j\}$ of ΔS a point $p_{ij} = (\mathcal{M}(\mathcal{C}_{i,j}), \mathbb{E}[\mathcal{M}(\mathcal{C}_{i,j})]) = (\mathcal{M} + \Delta \mathcal{M}_{ij}, \mathbb{E}[\mathcal{M}] + \Delta \mathbb{E}[\mathcal{M}]_{ij})$, the best merge corresponds to the point $p_{\max}$ that maximizes $y(p)/x(p)$. Since we only regard increases, and $\Delta \mathbb{E}[\mathcal{M}]_{ij} \geq 0$, we only need to consider the points in quadrant one. The point $p_{\max}$ can be found by a tangent query through 0. Thus $p_{\max}$ lies on the convex hull of this set of points, unless we encounter the degenerate case, where $x(p) = 0$, which can easily be detected and handled. After a merge, the points that correspond to rows or columns k or l, are updated as sketched out in Section 4.2. Then, all points are implicitly shifted with respect to the new values of $\mathcal{M}$ and $\mathbb{E}[\mathcal{M}]$ in line 3, which means that R is initialized as 0 before the first merge.

In terms of running time, the crucial steps in Algorithm 2 are lines 1 and 2. Brodal and Jacob [20] introduce kinetic heaps using space only linear in the number of points stored, that support the tangent queries in line 1 in $O(\log n)$ time and insertions and deletions in $O(\log n)$ amortized time per operation. This yields a total running time of $O(n \log n)$ per merge of clusters, since only points assigned to entries of ΔS that correspond to clusters k or l have to be recomputed. The update of the values $\mathcal{M}$ and $\mathbb{E}[\mathcal{M}]$ is realized by simply shifting the origin accordingly, saving $\Omega(n^2)$ updates of points. Thus, we arrive at the following:

Lemma 2. *By employing* quick divisive merge *(Algorithm 2), Algorithm 1 runs in $O(n^2 \log n)$ time for the relative variants.*

Similar to the absolute variants, the relative heuristic finds a single optimum. The update of S and of the entries of ΔS corresponds to a vector addition. The greedy strategy chooses the update that results in the best increase in S. If at any time S cannot be increased, then no entry of ΔS yields a greater gradient in the plane than S. But then in turn, no vector addition can ever produce an updated entry of ΔS that has a higher gradient than that of S. Thus, the first peak of S that the relative algorithm finds is the maximum of the algorithm. It can even be observed that there are no further local optima, as the algorithm always chooses the most gentle decrease in S.

Note that the above observations and lemmas generalize to all implementations of significance, such as *relative performance significance*, where a merge of two clusters entails an addition of corresponding entries of ΔS (or of $\Delta \mathcal{M}_w$ and $\Delta \mathbb{E}[\mathcal{M}_w]$). However, special attention has to be paid to implementations, where points in quadrant three can also lead to an increase. Then, two tangent queries have to be performed and the update has to be adapted.

If a graph is known to have a fine community structure (e.g. $\Omega(\sqrt{n})$ communities), the running time of Algorithm 1 decreases accordingly, if it is reasonable to assume that the algorithm identifies the community structure.

Algorithm 2. Quick Divisive Merge

Input: $\Delta \mathcal{M}, \Delta \mathbb{E}[\mathcal{M}]$, geometric 2-D data structure P of points p_{ij} as described above, reference point R
Output: Best merge, updated matrices $\Delta' \mathcal{M}, \Delta' \mathbb{E}[\mathcal{M}]$
1 Find $p_{\max} = p_{kl}$ with tangent query through R
2 Merge clusters corresponding to $p_{\max}$ and update P, $\Delta \mathcal{M}$ and $\Delta \mathbb{E}[\mathcal{M}]$ accordingly
3 $R \leftarrow R - (\Delta \mathcal{M}_{kl}, \Delta \mathbb{E}[\mathcal{M}]_{kl})$

---

5   Evaluation

In the following we first describe the general model used to generate the instances for the experimental evaluation, then we present and discuss the results. Our setup for the evaluation is an adaption of the benchmark presented in [21].

5.1 Random Uniform Clustered Graphs

We use a random partition generator $\mathcal{P}(n, s, v)$ that creates a partition $(P_1, \ldots, P_k)$ of $\{1, \ldots, n\}$ with $|P_i|$ being a normal random variable with expected value s and standard deviation s/v. Note that k depends on the choice of n, s and v, and that the last element $|P_k|$ of $\mathcal{P}(n, s, v)$ is possibly significantly smaller than the others. In order to settle this, we relaxed the size constraint i.e., if the last cluster size variable $|P_k|$ is too small or too large but the number of unassigned or additional nodes is less than one third of the expected cluster size, we add or delete the corresponding nodes. However, if the gap exceeds one third, we reject the partition and generate a new one. This may bias the generation process, yet we observed only few rejections during our experiments.

Given a partition $\mathcal{P}(n, s, v)$ and probabilities p_{in} and p_{out}, a uniformly random clustered graph $(G, \mathcal{C}, \omega)$ is generated by inserting intra-cluster edges with probability p_{in} and inter-cluster edges with probability p_{out}. In case a graph generated that way is not connected, additional edges combining the components are added. Edge weights ω are then set as to reflect the given partitioning. A weight from $[0, p_{\text{out}}]$ for each inter-cluster edge and from $[p_{\text{in}}, 1]$ each for intra-cluster edge is uniformly at random selected and assigned.

For our experiments we considered the following values of p_{in}, p_{out} and (n, s, v). We set $v = 4$ and choose s uniformly at random from $\{\frac{n}{\ell} \mid 2 \leq \ell \leq \sqrt{n}\}$. For $n = 100$ and $n = 1000$ we consider all combinations of p_{in} and p_{out} at a distance of 0.05. Among these we roughly refer to combinations supporting *dense*, *sparse*, *strong* and *random* community structure by A_{dense}, A_{sparse}, A_{strong} and $A_{\text{rand.}}$, respectively, as sketched out in Figure 3. For each combination and each algorithm, experiments were repeated until statistical significance has been attained with an α-level of 0.95 and a confidence interval of

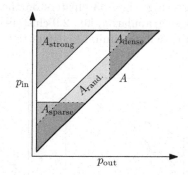

Fig. 3. Combinations of p_{in} and p_{out}

length 0.1 for each quality index measured. With an average of 223 runs per experiment we conducted about one million runs.

5.2 Evaluation Process

We conduct a qualitative comparison of the four implementations of the significance paradigm, as described in Section 3.2. The corresponding clusterings with high significance scores are obtained with the greedy heuristic based on Algorithm 1. Similar to the benchmark presented in [21], we restrict ourselves to qualitative aspects of the resulting clustering, and basic structural aspects, such as the number of clusters. Although this is only peripheral to our work, the effective runtimes of our basic and non-optimized Java 1.5 implementations ranged from a few milliseconds for 100 nodes using absolute variants to several minutes for 1000 nodes using relative variants, on an AMD Opteron 2.2 GHz machine.

5.3   Computational Results

Experiments. As a reference to the benchmark, we compare our findings to an established clustering algorithm, *Markov Clustering* (MCL) [16], The results of the experiments with respect to *performance* and *coverage* are given in Figure 4 and 5, respectively. We omit further plots illustrating the results on other benchmark algorithms (e.g. GMC [1]), other quality indices (e.g. *inter-cluster conductance*, see [21]) and structural observations due to space limitations.

At a first glance, the statistical results of the two absolute variants (S_{cov}^{-} and S_{perf}^{-}) strongly resemble each other, see Figure 4(c) and 4(d). For example their achieved quality does not differ by more than 2.6% with respect to performance, see Figure 4. Similar observations hold for the number of clusters. However, the relative variants ($S_{cov}^{\div}$ and $S_{perf}^{\div}$) essentially differ, see Figure 4(e) and 4(f). Alongside the disagreement on the quality indices, $S_{cov}^{\div}$ tends to identify fine clusterings, i.e., 33 clusters on the average,while $S_{perf}^{\div}$ finds clusterings with a coarse granularity, i.e., 2.9 clusters on the average. The absolute variants exhibit a surprisingly similar behavior to the initial clustering with respect to the quality

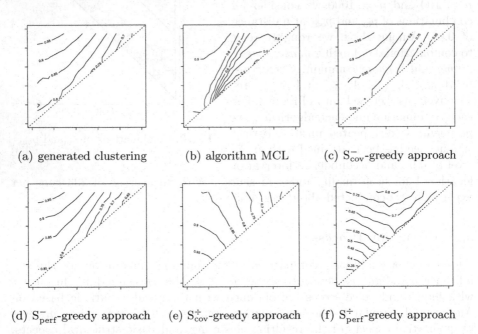

(a) generated clustering   (b) algorithm MCL   (c) S_{cov}^{-}-greedy approach

(d) S_{perf}^{-}-greedy approach   (e) $S_{cov}^{\div}$-greedy approach   (f) $S_{perf}^{\div}$-greedy approach

Fig. 4. Results showing the achieved *performance*. Probability p_{in} is shown on the y-axis, the x-axis holds p_{out}.

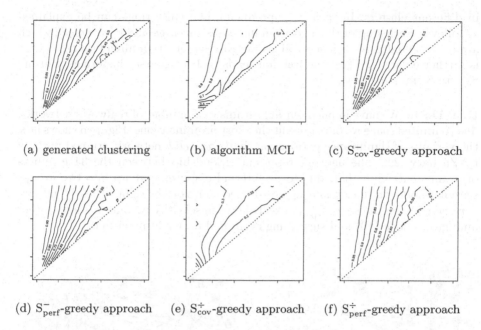

(a) generated clustering   (b) algorithm MCL   (c) S_{cov}^{-}-greedy approach

(d) S_{perf}^{-}-greedy approach   (e) $S_{cov}^{\div}$-greedy approach   (f) $S_{perf}^{\div}$-greedy approach

Fig. 5. Results showing the achieved *coverage*. Probability p_{in} is shown on the y-axis, the x-axis holds p_{out}.

indices. The same holds for $S_{perf}^{\div}$ with respect to *coverage* and *inter-cluster conductance*, however, the behavior is different for *performance*, but still acceptable scores are attained. In contrast, $S_{cov}^{\div}$ clearly fails to achieve high values of *coverage* and *inter-cluster conductance*, while its *performance* score is surprisingly good. As observed in [21], the benchmark algorithms do not substantially surpass the initial clustering in general. The same observation holds for the realizations of significance.

In an overall assessment of the achieved clustering quality, the two absolute variants excel with respect to *performance* for almost all generated instances, with a small advantage of S_{perf}^{-} over S_{cov}^{-} (standard modularity). This is particularly meaningful since both do not yield an inappropriately high number of clusters, which would artificially increase *performance*. With respect to *coverage*, the absolute variants are only surpassed by the few algorithms that produced a substantially coarser clustering, among those $S_{perf}^{\div}$. An interesting observation is that, using the significance measures as quality indices themselves, all four greedy algorithms attain the maximum corresponding score for most testsets. However, in the case of A_{strong}, the obtained differences in the significance measures are small among most algorithms.

Explaining Some Artifacts. The high values of *performance*, attained by $S_{cov}^{\div}$ for A_{sparse} are due to the fact that the large number of clusters identified by this algorithm yields a large fraction of non-connected pairs of nodes that are in different clusters. In turn, $S_{cov}^{\div}$ producing fine clusterings can be explained as follows. Each step of the algorithm increases *coverage* and $\mathbb{E}[coverage]$, which are both bounded by 1. These values increase faster, if an already large cluster is further enlarged. Thus, the fraction tends to 1 for coarse clusterings, causing $S_{cov}^{\div}$ to terminate early.

Real Data. We have applied our algorithms to a number of realworld networks, due to limited space we only present the most prominent one. Figure 6 shows how the variants of significance perform on the *karate club* network, studied initially by Zachary [22]. The network represents friendship between the 34 members of a university club that, due to an internal dispute, split up into two groups (circular nodes on the left and square-shaped nodes on the right). Clearly, *relative performance significance* ($S_{perf}^{\div}$) excels here, misclassifying only a single rather ambiguous node (10) and surpassing even modularity in precision.

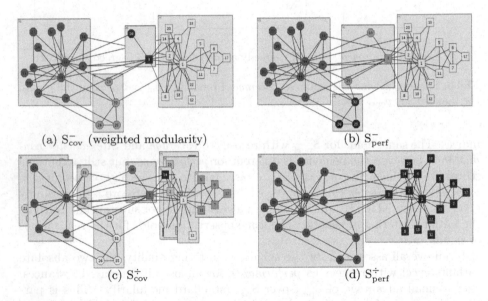

(a) S_{cov}^{-} (weighted modularity)   (b) S_{perf}^{-}

(c) $S_{cov}^{\div}$ (d) $S_{perf}^{\div}$

Fig. 6. The results of the greedy significance algorithms (groupings) on Zachary's karate club [22] are in agreement with our experimental evaluation. The variants S_{cov}^{-} and S_{perf}^{-} produce almost the same clustering. While both clusterings are meaningful and close to the real grouping (node shapes), *relative performance significance* ($S_{perf}^{\div}$) yields a bisection which is almost exactly the real grouping. Obviously, $S_{cov}^{\div}$ fails to find a reasonable clustering for this network.

Figure 7 shows an anonymized graph of the email contacts at our department over a period of three months (approx. 44300 emails). Nodes represent persons and weighted edges represent the number of email contacts between two coworkers. The grouping depicts the department's internal structure while the node colors (gray values) show the findings of community structure of the greedy algorithm based on S_{perf}^-. Since this example is based on the intuition that the graph structure reflects the grouping, we cleaned the network of artifact nodes with no links to other nodes in its reference cluster (approx. 7.5% of the original nodes).

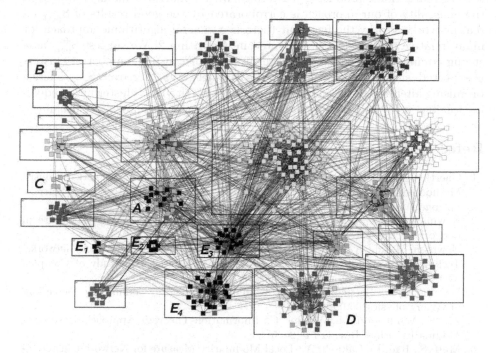

Fig. 7. A network of email contacts at our department. The grouping depicts the department's internal structure as a reference, and the node colors (gray values) are the community detection result of *absolute performance significance* (S_{perf}^-). Inside reference clusters, S_{perf}^- misclassifies only 6.8% of nodes, most of which are due to the highly ambiguous reference cluster A, which is split in half by the algorithm. The clustering of S_{perf}^- yields a noticeably higher ($\approx 6\%$) *coverage*, which is partly due to 9 clusters each being merged into other clusters they are strongly connected with. In terms of *inter-cluster conductance* and all four realizations of significance, S_{perf}^- slightly surpasses the reference. However, the *performance* of the reference clustering is approx. 2.4% higher than that found by S_{perf}^-. On the whole, a closer investigation explains most disagreements between the two clusterings, e.g., note the artifact nodes in clusters B, C, D and the strong connections between clusters $A, E_1, \ldots, E_4$, which account for the aggregation done by the algorithm.

6   Conclusion

Based on modularity [2], a recently introduced quality measure for graph clusterings, we formally stated a new clustering paradigm, significance, that considers the trade-off between the achieved quality and the expected quality with respect to networks with a similar intrinsic structure. The performed experimental study is a systematic evaluation of this paradigm and a substantial advance towards a generalized understanding of the concept modularity is based upon. Summarizing, the evaluation yields that the significance paradigm is highly feasible for several realizations, producing clusterings with remarkable quality. Moreover, the generality of our approach is corroborated by the good results of $S_{perf}^{\div}$ on real networks and by the fact that there is a general algorithmic approach for many relative realizations, as presented in Algorithm 2. We suggest $S_{perf}^{\div}$ as a strong community detection algorithm if a low or constant number of clusters is expected, such as coarsely structured social networks. Moreover, S_{perf}^{-} offers a promising alternative to standard modularity, since it yields clusterings of equaling quality, yet it is based on the more appropriate quality index *performance*.

References

1. Gaertler, M.: Clustering. In Brandes, U., Erlebach, T., eds.: Network Analysis: Methodological Foundations. Volume 3418 of Lecture Notes in Computer Science. Springer-Verlag (2005) 178–215
2. Newman, M.E.J., Girvan, M.: Finding and evaluating community structure in networks. Phys. Rev. E **69** (2004)
3. Newman, M.E.J.: A fast algorithm for detecting community structure in networks. Technical report, Department of Physics and Center for the Study of Complex Systems, University of Michigan (2003)
4. Fortunato, S., Barthelemy, M.: Resolution Limit in Community Detection. arXiv.org physics/0607100 (2006)
5. Ziv, E., Middendorf, M., Wiggins, C.: Information-Theoretic Approach to Network Modularity. Phys. Rev. E **71** (2005)
6. Muff, S., Rao, F., Caflisch, A.: Local Modularity Measure for Network Clusterizations. Phys. Rev. E **72** (2005)
7. Fine, P., Paolo, E.D., Philippides, A.: Spatially Constrained Networks and the Evolution of Modular Control Systems. In: 9th Intl. Conference on the Simulation of Adaptive Behavior (SAB). (2006)
8. Newman, M.E.J.: Fast Algorithm for Detecting Community Structure in Networks. Physical Review E **69** (2004)
9. Clauset, A., Newman, M.E.J., Moore, C.: Finding community structure in very large networks. Phys. Rev. E **70** (2004)
10. Newman, M.: Modularity and Community Structure in Networks. In: Proceedings of the National Academy of Sciences. (2005) 8577–8582
11. White, S., Smyth, P.: A Spectral Clustering Approach to Finding Communities in Graph. In: SIAM Data Mining Conference. (2005)
12. Guimerà, R., Sales-Pardo, M., Amaral, L.A.N.: Modularity from Fluctuations in Random Graphs and Complex Networks. Physical Review E **70** (2004)
13. Reichardt, J., Bornholdt, S.: Statistical Mechanics of Community Detection. arXiv.org cond-mat/0603718 (2006)
14. Duch, J., Arenas, A.: Community Detection in Complex Networks using Extremal Optimization. Physical Review E **72** (2005)
15. Brandes, U., Delling, D., Gaertler, M., Görke, R., Hoefer, M., Nikoloski, Z., Wagner, D.: Maximizing modularity is hard, arxiv preprint. http://arxiv.org/abs/physics/0608255 (2006)

16. van Dongen, S.M.: Graph Clustering by Flow Simulation. PhD thesis, University of Utrecht (2000)
17. Coffin, M., Saltzmann, M.J.: Statistical analysis of computational tests of algorithms and heuristics. **12** (2000)
18. Newman, M.: Analysis of Weighted Networks. Technical report, Cornell University, Santa Fe Institute, University of Michigan (2004)
19. Clauset, A., Newman, M.E.J., Moore, C.: Finding community structure in very large networks. Technical report, University of New Mexico, University of Michigan (2004)
20. Brodal, G.S., Jacob, R.: Dynamic planar convex hull. In: FOCS. (2002) 617–626
21. Brandes, U., Gaertler, M., Wagner, D.: Experiments on Graph Clustering Algorithms. In: Proceedings of the 11th Annual European Symposium on Algorithms (ESA'03). Volume 2832 of Lecture Notes in Computer Science. (2003) 568–579
22. Zachary, W.: An information flow model for conflict and fission in small groups. Journal of Anthropological Research **33** (1977) 452–473

A   Expected *performance* and the Heuristic

Lemma 3. *If the probability of an edge between nodes v and w is defined to be $\frac{\omega(v)\omega(w)}{(2W)}$ the expected value of* performance *is (for unweighted edges set $\omega(e) \equiv 1$ for all edges)*

$$\frac{2\frac{1}{2W}\sum_i(\sum_{v\in C_i}\omega(v))^2 + \omega_{\max}(n^2 - \sum_i|C_i|^2) - 2W}{n(n-1)\omega_{\max}}$$

Proof.

$\mathbb{E}(p)$

$$= \frac{\mathbb{E}(\sum_{intra-e}\omega(e)) + \mathbb{E}(\sum_{inter-e}(\omega_{\max}-\omega(e))) + \mathbb{E}(\sum_{non-inter-e}\omega_{\max})}{\frac{1}{2}n(n-1)\omega_{\max}}$$

$$= \frac{\sum_i\frac{1}{2}\sum_{v\in C_i}\sum_{w\in C_i}2W\frac{\omega(v)\omega(w)}{(2W)^2} + \sum_i\frac{1}{2}\sum_{v\in C_i}\sum_{w\notin C_i}(\omega_{\max} - 2W\frac{\omega(v)\omega(w)}{(2W)^2})}{\frac{1}{2}n(n-1)\omega_{\max}}$$

$$= \frac{\frac{1}{2W}\sum_i(\sum_{v\in C_i}\omega(v))^2 + \sum_i\sum_{v\in C_i}\sum_{w\notin C_i}(\omega_{\max})}{n(n-1)\omega_{\max}}$$

$$- \frac{\frac{1}{2W}\sum_i\sum_{v\in C_i}\sum_{w\notin C_i}(\omega(v)\omega(w))}{n(n-1)\omega_{\max}}$$

$$= \frac{2\frac{1}{2W}\sum_i(\sum_{v\in C_i}\omega(v))^2 + \omega_{\max}(n^2 - \sum_i|C_i|^2) - 2W}{n(n-1)\omega_{\max}}$$

Lemma 4. *The heuristic of setting the expected weight of an edge between nodes v and w to $\frac{\omega(v)\omega(w)}{(2W)}$ does not lead to an expected total edge weight of W.*

Proof.

$$\sum_{\substack{\{v,w\} \\ v,w \in V}} \frac{\omega(v) \cdot \omega(w)}{2W} = \frac{1}{2} \left(\sum_{\substack{(v,w) \\ v,w \in V}} \frac{\omega(v) \cdot \omega(w)}{2W} + \sum_{v \in V} \frac{(\omega(v))^2}{2W} \right) \tag{5}$$

$$= \frac{1}{4W} \left(\sum_{v \in V} \omega(v) \right)^2 + \frac{1}{4W} \sum_{v \in V} (\omega(v))^2 = W + \underbrace{\frac{1}{4W} \sum_{v \in V} (\omega(v))^2}_{>0 \text{ since } W \neq 0} \neq W \tag{6}$$

Thus, the expected sum of edge weights does not equal W.

Note that if we do not allow self-loops we obtain a similar result. However, since in most real-world graphs the maximum node degree is clearly sublinear in the total sum of edge weights, the relative error, caused by the abovementioned edge probabilities, tends to zero. This can be observed in Equation 6.

An Improved Approximation Algorithm for Maximum Edge 2-Coloring in Simple Graphs

Zhi-Zhong Chen and Ruka Tanahashi

Department of Mathematical Sciences, Tokyo Denki University, Hatoyama, Saitama
350-0394, Japan
`chen@r.dendai.ac.jp`

Abstract. We present a polynomial-time approximation algorithm for legally coloring as many edges of a given simple graph as possible using two colors. It achieves an approximation ratio of $\frac{468}{575}$. This improves on the previous best (trivial) ratio of $\frac{4}{5}$.

1   Introduction

Given a graph G and a natural number t, the *maximum edge t-coloring problem* (called MAX EDGE t-COLORING for short) is to find a maximum set F of edges in G such that F can be partitioned into at most t matchings of G. Motivated by call admittance issues in satellite based telecommunication networks, Feige et al. [1] introduced the problem and proved its APX-hardness. Their APX-hardness proof indeed shows that the problem remains APX-hard even if we restrict the input graph to a simple graph and fix the input integer t to 2. We call this restriction (special case) of the problem MAX SIMPLE EDGE 2-COLORING.

Since MAX EDGE t-COLORING and its special cases are hard, it is interesting to design approximation algorithms for them. As observed by Feige et al. [1], MAX EDGE t-COLORING is obviously a special case of the well-known maximum coverage problem (see [3]). Since the maximum coverage problem can be approximated by a greedy algorithm within a ratio of $1 - (1 - \frac{1}{t})^t$ [3], so can be MAX EDGE t-COLORING. In particular, the greedy algorithm achieves an approximation ratio of $\frac{3}{4}$ for MAX EDGE 2-COLORING which is the special case of MAX EDGE t-COLORING where the input number t is fixed to 2. Feige et al. [1] has improved the trivial ratio $\frac{3}{4}$ to $\frac{10}{13}$ by an LP approach. They also pointed out that for MAX SIMPLE EDGE 2-COLORING, the ratio $\frac{10}{13}$ can be further improved to $\frac{4}{5}$ by the following simple algorithm:

Input: A simple graph G.
1. Compute a maximum subgraph H of G such that the degree of each vertex in H is at most 2 and there is no 3-cycle in H. (Comment: This step can be done in $O(n^3)$ time [2].)
2. Remove one edge from each odd cycle of H.
Output: H.

Kosowski et al. [6] also considered MAX SIMPLE EDGE 2-COLORING. They presented an approximation algorithm that achieves a ratio of $\frac{28\Delta - 12}{35\Delta - 21}$, where Δ

M.-Y. Kao and X.-Y. Li (Eds.): AAIM 2007, LNCS 4508, pp. 27–36, 2007.

is the maximum degree of a vertex in the input simple graph. This ratio can be arbitrarily close to the trivial ratio $\frac{4}{5}$ because Δ can be very large.

In this paper, we present a polynomial-time approximation algorithm for MAX SIMPLE EDGE 2-COLORING which achieves a ratio of $\frac{468}{575}$. To achieve this, we first design a randomized algorithm and then derandomize it. The analysis of our algorithm is quite nontrivial.

Kosowski et al. [6] showed that approximation algorithms for MAX SIMPLE EDGE 2-COLORING can be used to obtain approximation algorithms for certain packing problems and fault-tolerant guarding problems. Combining their reductions and our improved approximation algorithm for MAX SIMPLE EDGE 2-COLORING, we can obtain improved approximation algorithms for their packing problems and fault-tolerant guarding problems immediately.

2   Basic Definitions

Throughout the remainder of this paper, a graph means a simple undirected graph (i.e., it has neither parallel edges nor self-loops).

Let G be a graph. We denote the vertex set of G by $V(G)$, and denote the edge set of G by $E(G)$. The *degree* of a vertex v in G, denoted by $d_G(v)$, is the number of edges incident to v in G. A vertex v of G with $d_G(v) = 0$ is called an *isolated vertex*. An *independent set* of G is a set S of vertices of G such that no two vertices of S are adjacent in G.

A *cycle* in G is a connected subgraph of G in which each vertex is of degree 2. A *path* in G is a connected subgraph of G in which exactly two vertices are of degree 1 and the others are of degree 2. The *length* of a cycle or path C is the number of edges in C. A cycle of odd (respectively, even) length is called an *odd* (respectively, *even*) cycle. A *k-cycle* is a cycle of length k. Similarly, a k^+-*cycle* is a cycle of length at least k. A *path component* (respectively, *cycle component* of G is a connected component of G that is a path (respectively, cycle). Note that an isolated vertex of G is not a path component of G.

For a function b mapping each vertex v of G to a nonnegative integer, a *b-matching* of G is a subgraph H of G such that $d_H(v) \le b(v)$ for all vertices v of H. When $b(v) \le 1$ for all vertices v of G, a b-matching of G is called a *matching* of G. The *size* of a b-matching M of G, denoted by $|M|$, is the number of edges in M. Given a graph G and a function b mapping each vertex v to a nonnegative integer, the *maximum b-matching problem* is to find a b-matching of G of maximum size. Similarly, given a graph G, the *maximum matching problem* is to find a matching of G of maximum size. G is *edge-2-colorable* if $E(G)$ can be partitioned into two matchings. Thus, MAX SIMPLE EDGE 2-COLORING is the problem of finding a maximum edge-2-colorable subgraph in a given graph. Note that G is edge 2-colorable if and only if each connected component of G is an isolated vertex, a path, or an even cycle.

For a random event A, $\Pr[A]$ denotes the probability that A occurs. For two random events A and B, $\Pr[A \mid B]$ denotes the probability that A occurs given the occurrence of B. For a random variable X, $\mathcal{E}[X]$ denotes its expected value.

3   The Algorithm

Throughout this section, fix a graph G and a maximum edge-2-colorable subgraph Opt.

The basic ideas behind our algorithm are as follows. Like the simple algorithm described in Section 1, our algorithm first computes a maximum subgraph H of G such that the degree of each vertex in H is at most 2 and there is no 3-cycle in H. Still like the simple algorithm, our algorithm needs to delete one edge from each odd cycle of H so that it becomes edge 2-colorable. After this deletion process, G may have some edges e such that both endpoints of e are of degree at most 1 in H and appear in different connected components of H. For convenience, let us call such an edge of G an *available bridge*. Clearly, an available bridge can be added to H without invalidating the edge-2-colorability of H. Of course, we do not want to add only one available bridge to H but want to add many available bridges to H simultaneously. Unfortunately, two available bridges may conflict each other, i.e., adding them to H simultaneously invalidates the edge-2-colorability of H. In order to find a large set of nonconflicting available bridges, we may first construct an auxiliary graph B whose endpoints are the endpoints of paths of H (after the deletion process) and whose edges are the available bridges, and then find a maximum b-matching N in B where $b(v) = 2 - d_H(v)$ for each vertex v of B. We now add the edges of N to H simultaneously, obtaining a graph H' whose connected components are paths, even cycles, or odd cycles. Each odd cycle of H' must contain at least two edges of N and we can delete exactly one edge of N from each odd cycle of H'. The resulting H' is now edge 2-colorable and contains at least $\frac{1}{2}|N|$ more edges than the output of the simple algorithm.

Unfortunately, $|N|$ may be small and hence the above basic ideas do not work. So, we modify the ideas as follows. First, we split the deletion process into two: *7-arbitrary* and *5-random*. In the 7-arbitrary deletion process, we only delete one (arbitrary) edge from each odd 7^+-cycle of H. This process at most decreases the number of edges in H by a fraction of $\frac{1}{7}$, which is significantly smaller than the fraction of $\frac{1}{5}$ decreased by the original deletion process. After the 7-arbitrary deletion process, we construct an auxiliary graph A (instead of constructing B as above), where $V(A)$ consists of those vertices v in H such that v is of degree at most 1 in H or appears on a 5-cycle in H and $E(A)$ consists of those edges $\{u, v\}$ of G such that u and v belong to different connected components of H. We further construct a maximum b-matching M in A (instead of constructing N as above), where $b(v) = 1$ if v is a vertex of a 5-cycle in H while $b(v) = 2 - d_H(v)$ otherwise. Obviously, $|M|$ may be much larger than $|N|$, especially when there are a lot of 5-cycles in H. In particular, if we let $E_{opt}^{A,ex}$ denote the set of edges contained in both A and Opt, we can easily prove that $|M| \geq \frac{1}{2}|E_{opt}^{A,ex}|$ (see Lemma 4 below). Now, it comes the 5-random deletion process: We select one edge from each 5-cycle uniformly at random and delete it from H. As before, we call an edge e of A an *available bridge* if both endpoints of e become of degree at most 1 in H after the 5-random deletion process. Obviously, each edge of M

becomes an available bridge with probability at least $\frac{2}{5}$. Thus, we can expect at least $\frac{1}{5}|E_{opt}^{A,ex}|$ available bridges in M, which can then be added to H as before to obtain H'.

By the discussion in the last paragraph, if $|E_{opt}^{A,ex}|$ is significantly large, then the above modified ideas lead to a randomized algorithm whose output can be expected to contain significantly more edges than the output of the simple algorithm. So, the problematic case happens when $|E_{opt}^{A,ex}|$ is not significantly large. A large portion of the analysis of our algorithm is devoted to this case. Intuitively speaking, we can prove that in this problematic case, the simple algorithm should achieve an approximation ratio better than $\frac{4}{5}$.

We next give a formal description of our algorithm. Given G, our algorithm finds an edge-2-colorable subgraph of G as follows.

1. Compute a maximum subgraph H of G such that each connected component of H is an isolated vertex, a path, or a 4^+-cycle. (*Comment:* The set of vertices v of H with $d_H(v) \leq 1$ is an independent set of G because of the maximality of H.)
2. Remove one (arbitrary) edge from each odd 7^+-cycle of H.
3. For $i \in \{0,1\}$, let T_i be the set of vertices v of H with $d_H(v) = i$.
4. Let V_{5c} be the set of vertices on 5-cycles of H.
5. Construct an auxiliary graph A, where $V(A) = T_0 \cup T_1 \cup V_{5c}$ and $E(A)$ consists of those $\{u,v\} \in E(G)$ such that no connected component of H contains both u and v.
6. Compute a maximum b-matching M in A, where $b(v) = 2 - d_H(v)$ for each $v \in T_0 \cup T_1$ and $b(v) = 1$ for each $v \in V_{5c}$.
7. Choose one edge from each 5-cycle of H uniformly and independently at random and remove it from H.
8. Let M' be the set of all edges $\{u,v\}$ in M such that $d_H(u) + d_M(u) \leq 2$ and $d_H(v) + d_M(v) \leq 2$.
9. Add the edges in M' to H. (*Comment:* Each connected component of H is an isolated vertex, a path, or a cycle.)
10. For each odd cycle C in H, select one edge in $E(C) \cap M'$ uniformly and independently at random and delete it from H.
11. Output H.

3.1   The First Analysis

For each $i \in \{1,2,7,9,10\}$, let H_i be the content of graph H immediately after Step i of our algorithm. Note that H_{10} is the output of our algorithm. Related to H_1, we define two sets and two numbers as follows:

- E_{5c} is the set of edges on the 5-cycles of H_1.
- $E_{\overline{5c}} = E(H_1) - E_{5c}$.
- n_{7c+} is the number of odd 7^+-cycles of H_1.
- n_{pc} is the number of path components of H_1.

Lemma 1. $|V(H_1) - V(A)| = |E_{\overline{5c}}| - 2n_{7c+} - n_{pc}$.

Proof. For each path component P of H_1, $|E(P)| = |V(P)| - 1$ and two vertices of P are contained in A. So, each path component of H_1 contributes 1 to the value of $|E_{\overline{5c}}| - |V(H_1) - V(A)|$. Similarly, for each odd 7^+-cycle C of H_1, $|E(C)| = |V(C)|$ and two vertices of C are contained in A. Thus, each odd 7^+-cycle of H_1 contributes 2 to the value of $|E_{\overline{5c}}| - |V(H_1) - V(A)|$. This completes the proof of the lemma.

Related to Opt, we define five sets of edges as follows:

- E_{opt}^A is the set of edges $\{u, v\}$ in Opt such that both u and v are vertices of A.
- $E_{opt}^{\overline{A}} = E(Opt) - E_{opt}^A$.
- $E_{opt}^{A,5c}$ (respectively, $E_{opt}^{A,7c+}$) is the set of edges $\{u, v\} \in E_{opt}^A$ such that some 5-cycle (respectively, path component) of H_2 contains both u and v. (*Comment:* By the maximality of $E(H_1)$, there is no edge $\{u, v\} \in E_{opt}^A$ such that some path component of H_1 contains both u and v. So, the endpoints of each edge in $E_{opt}^{A,7c+}$ must appear on the same odd 7^+-cycle of H_1.)
- $E_{opt}^{A,ex} = E_{opt}^A - (E_{opt}^{A,5c} \cup E_{opt}^{A,7c+})$.

Lemma 2. $|E_{opt}^A| \geq |E(Opt)| - 2|E_{\overline{5c}}| + 4n_{7c+} + 2n_{pc}$.

Proof. Since each vertex can be adjacent to at most two edges in Opt, $|E(Opt) - E_{opt}^A| \leq 2|V(H_1) - V(A)|$. So, the lemma follows from Lemma 1 immediately.

Corollary 1. $|E_{opt}^{A,ex}| \geq |E(Opt)| - 2|E_{\overline{5c}}| + 4n_{7c+} + 2n_{pc} - |E_{opt}^{A,5c}| - |E_{opt}^{A,7c+}|$.

Proof. Obviously, $|E_{opt}^{A,ex}| = |E_{opt}^A| - (|E_{opt}^{A,5c}| + |E_{opt}^{A,7c+}|)$. So, the corollary follows from Lemma 2 immediately.

The following key lemma shows that each edge of M is include in the output with a significantly high probability.

Lemma 3. *For each edge $e \in M$, $\Pr[e \in E(H_{10})] \geq \frac{8}{75}$.*

Proof. Fix an arbitrary edge $e = \{u, v\}$ in M. We distinguish three cases as follows:

Case 1: Both $u \in V_{5c}$ and $v \in V_{5c}$. In this case, since $\Pr[d_{H_7}(u) = 1] = \frac{2}{5}$, $\Pr[d_{H_7}(v) = 1] = \frac{2}{5}$, $d_M(u) \leq 1$, and $d_M(v) \leq 1$, we have $\Pr[e \in M'] = \frac{4}{25}$. Thus, it remains to show that $\Pr[e \in E(H_{10}) \mid e \in M'] \geq \frac{2}{3}$. Assume that $e \in M'$. If no odd cycle of H_9 contains e, then we are done. So, further assume that some odd cycle C of H_9 contains e. We claim that C contains at least three edges of M'. For a contradiction, assume that C contains only one edge e' of M' other than e. Obviously, if we delete e and e' from C, we obtain two paths P_1 and P_2 both of which are connected components of H_7. Moreover, one of u and v is an endpoint of P_1 and the other is an endpoint of P_2. Now, since $u \in V_{5c}$ and $v \in V_{5c}$, P_1 and P_2 must have been obtained in Step 7 by deleting one edge from each 5-cycle of H_1. So, both P_1 and P_2 are of length 4. However, this implies that the length of C is 10 which is even, a contradiction. Hence, the claim holds. By the claim, we have $\Pr[e \in E(H_{10}) \mid e \in M'] \geq \frac{2}{3}$ immediately.

Case 2: Exactly one of u and v is contained in V_{5c}. We assume that $u \in V_{5c}$ but $v \notin V_{5c}$; the other case is similar. Then, $\Pr[d_{H_7}(u) = 1] = \frac{2}{5}$ and $d_M(u) \leq 1$. Moreover, $v \in T_0$ or $v \in T_1$. In the former case, $d_{H_7}(v) = 0$ and $d_M(v) \leq 2$. In the latter case, $d_{H_7}(v) = 1$ and $d_M(v) \leq 1$. So, in both cases, $d_{H_7}(v) + d_M(v) \leq 2$. Consequently, $\Pr[e \in M'] \geq \frac{2}{5} \cdot 1 = \frac{2}{5}$. Thus, it suffices to show that $\Pr[e \in E(H_{10}) \mid e \in M'] \geq \frac{1}{2}$. Assume that $e \in M'$. If no odd cycle of H_9 contains e, then we are done. On the other hand, if some odd cycle C of H_9 contains e, then the assumption $u \in V_{5c}$ guarantees that C contains at least two edges of M' and hence $\Pr[e \in E(H_{10}) \mid e \in M'] \geq \frac{1}{2}$.

Case 3: Both $u \notin V_{5c}$ and $v \notin V_{5c}$. Then, as discussed in Case 2 about v, we have $d_{H_7}(u) + d_M(u) \leq 2$ and $d_{H_7}(v) + d_M(v) \leq 2$. So, $\Pr[e \in M'] = 1$. Thus, it suffices to show that $\Pr[e \in E(H_{10}) \mid e \in M'] \geq \frac{1}{2}$. Assume that $e \in M'$. If no odd cycle of H_9 contains e, then we are done. So, further assume that some odd cycle C of H_9 contains e. We claim that C contains at least two edges of M'. For a contradiction, assume that the claim is false. Then, the path obtained from C by deleting e is a connected component of H_2. However, this contradicts the construction of graph A in Step 5. Thus, the claim holds. Consequently, $\Pr[e \in E(H_{10}) \mid e \in M'] \geq \frac{1}{2}$.

By Lemma 3 and the algorithm, we have the following corollary immediately:

Corollary 2. $\mathcal{E}[|E(H_{10})|] \geq |E_{\overline{5c}}| - n_{7c+} + \frac{4}{5}|E_{5c}| + \frac{8}{75}|M|$.

The following lemma proves a simple lower bound on $|M|$.

Lemma 4. $|M| \geq |E_{opt}^{A,ex}|/2$.

Proof. Let M'' be a maximum matching in graph A. Since Opt has no odd cycle, $E_{opt}^{A,ex}$ can be partitioned into two matchings of A. So, $|M''| \geq |E_{opt}^{A,ex}|/2$. On the other hand, since M is a maximum b-matching of A with $b(v) \geq 1$ for each $v \in V(A)$, we have $|M| \geq |M''|$. Thus, the lemma holds.

Theorem 1. $\mathcal{E}[|E(H_{10})|] \geq \frac{146}{175}|E(Opt)| + \frac{2}{105}|E_{5c}| - \frac{4}{75}|E_{opt}^{A,5c}| - \frac{4}{75}|E_{opt}^{A,7c+}|$.

Proof. Combining Corollary 2 and Lemma 4, we have

$$\mathcal{E}[|E(H_{10})|] \geq |E_{\overline{5c}}| - n_{7c+} + \frac{4}{5}|E_{5c}| + \frac{4}{75}|E_{opt}^{A,ex}|.$$

So, by Corollary 1 and a simple calculation, we have

$$\mathcal{E}[|E(H_{10})|] \geq \frac{4}{75}|E(Opt)| + \frac{67}{75}|E_{\overline{5c}}| + \frac{4}{5}|E_{5c}| - \frac{59}{75}n_{7c+} + \frac{8}{75}n_{pc} - \frac{4}{75}|E_{opt}^{A,5c}| - \frac{4}{75}|E_{opt}^{A,7c+}|.$$

Consequently, since $n_{7c+} \leq (|E(H_1)| - |E_{5c}|)/7$ and $|E_{\overline{5c}}| = |E(H_1)| - |E_{5c}|$, we have

$$\mathcal{E}[|E(H_{10})|] \geq \frac{4}{75}|E(Opt)| + \frac{82}{105}|E(H_1)| + \frac{2}{105}|E_{5c}| - \frac{4}{75}|E_{opt}^{A,5c}| - \frac{4}{75}|E_{opt}^{A,7c+}|.$$

Now, since $|E(H_1)|$ is at least as large as $|E(Opt)|$, the theorem follows.

The following corollary shows that our algorithm achieves an expected ratio of $\frac{304}{375}$.

Corollary 3. $\mathcal{E}[|E(H_{10})|] \geq \frac{304}{375}|E(Opt)| + \frac{2}{125}|E_{\overline{5c}}|.$

Proof. Obviously, $|E_{opt}^{A,7c+}| \leq n_{7c+} \leq \frac{1}{7}|E_{\overline{5c}}|$. Moreover, since Opt cannot contain a 5-cycle, $E_{opt}^{A,5c}$ contains at most four edges $\{u, v\}$ with $u \in V(C)$ and $v \in V(C)$ for each 5-cycle C of H_1. Consequently, $|E_{opt}^{A,5c}| \leq \frac{4}{5}|E_{5c}| = \frac{4}{5}|E(H_1)| - \frac{4}{5}|E_{\overline{5c}}|$. Also recall that $|E(H_1)| \geq |E(Opt)|$. Now, by the last inequality in the proof of Theorem 1, the corollary follows. $\qquad\square$

In the next subsection, we will give another analysis of the algorithm and combine it with the analysis in this section to obtain a better ratio.

3.2 The Second Analysis

Let K be the graph with vertex set $V(A)$ and edge set $E_{opt}^A - (E_{opt}^{A,5c} \cup E_{opt}^{A,7c+})$.

Lemma 5. *There are at least $\frac{5}{4}|E_{opt}^{A,5c}|$ vertices $v \in V_{5c}$ with $d_K(v) < 2$.*

Proof. Fix an arbitrary 5-cycle C of H_1. Let F be the number of edges $\{u, v\} \in E_{opt}^{A,5c}$ with $\{u, v\} \subseteq V(C)$. Let W be the set of the endpoints of the edges in F. Obviously, for each $v \in W$, $d_K(v) < 2$. We claim that $|W| \geq \frac{5}{4}|F|$. To see this, first observe that we always have $|W| \geq |F| + 1$. Moreover, $|F| \leq 4$ because Opt cannot contain a 5-cycle. Thus, the claim holds. The claim implies the lemma immediately because summing up $\frac{5}{4}|F|$ over all 5-cycles C of H_1 yields the bound $\frac{5}{4}|E_{opt}^{A,5c}|$. $\qquad\square$

Besides Lemma 4, we have another (less obvious) lower bound on $|M|$.

Lemma 6. $|M| \geq |E_{opt}^A| - |E_{5c}| - 2n_{7c+} - 2n_{pc} + \frac{1}{4}|E_{opt}^{A,5c}| + |E_{opt}^{A,7c+}|.$

Proof. Let h be the number of vertices $v \in V_{5c}$ with $d_K(v) = 2$. By Lemma 5, $h \leq |E_{5c}| - \frac{5}{4}|E_{opt}^{A,5c}|$.

Let ℓ be the number of vertices $v \in T_1$ with $d_K(v) = 2$. We claim that $\ell \leq 2(n_{7c+} + n_{pc}) - 2|E_{opt}^{A,7c+}|$. To see this, first observe that $|T_1| = 2(n_{7c+} + n_{pc})$. Moreover, if $\{u, v\} \in E_{opt}^{A,7c+}$, then both $d_K(u) \leq 1$ and $d_K(v) \leq 1$. Now, since no two edges of $E_{opt}^{A,7c+}$ can share an endpoint, the claim holds.

Obviously, if we modify K by removing one edge from each $v \in V_{5c} \cup T_1$ with $d_K(v) = 2$, we obtain a b-matching of A. So, since M is a maximum b-matching of A, we have

$$|M| \geq |E_{opt}^A| - |E_{opt}^{A,5c}| - |E_{opt}^{A,7c+}| - h - \ell.$$

Thus, by the aforementioned bounds on h and ℓ, the lemma holds. $\qquad\square$

Theorem 2. $\mathcal{E}[|E(H_{10})|] \geq \frac{82}{105}|E(Opt)| + \frac{2}{105}|E_{5c}| + \frac{2}{75}|E_{opt}^{A,5c}| + \frac{8}{75}|E_{opt}^{A,7c+}|.$

Proof. Combining Corollary 2 and Lemma 6, we have

$$\mathcal{E}[|E(H_{10})|] \geq |E_{\overline{5c}}| - \frac{91}{75}n_{7c+} + \frac{52}{75}|E_{5c}| + \frac{8}{75}|E^A_{opt}| - \frac{16}{75}n_{pc} + \frac{2}{75}|E^{A,5c}_{opt}| + \frac{8}{75}|E^{A,7c+}_{opt}|.$$

So, by Lemma 2 and a simple calculation, we have

$$\mathcal{E}[|E(H_{10})|] \geq \frac{8}{75}|E(Opt)| + \frac{59}{75}|E_{\overline{5c}}| + \frac{52}{75}|E_{5c}| - \frac{59}{75}n_{7c+} + \frac{2}{75}|E^{A,5c}_{opt}| + \frac{8}{75}|E^{A,7c+}_{opt}|.$$

Consequently, since $n_{7c+} \leq |E_{\overline{5c}}|/7$, we have

$$\mathcal{E}[|E(H_{10})|] \geq \frac{8}{75}|E(Opt)| + \frac{354}{525}|E_{\overline{5c}}| + \frac{52}{75}|E_{5c}| + \frac{2}{75}|E^{A,5c}_{opt}| + \frac{8}{75}|E^{A,7c+}_{opt}|.$$

Since $|E_{5c}| + |E_{\overline{5c}}| = |E(H_1)|$, we have

$$\mathcal{E}[|E(H_{10})|] \geq \frac{8}{75}|E(Opt)| + \frac{354}{525}|E(H_1)| + \frac{10}{525}|E_{5c}| + \frac{2}{75}|E^{A,5c}_{opt}| + \frac{8}{75}|E^{A,7c+}_{opt}|.$$

Now, since $|E(H_1)| \geq |E(Opt)|$, the theorem follows.

Corollary 4. $\mathcal{E}[|E(H_{10})|] \geq \frac{1258}{1575}|E(Opt)| + \frac{2}{105}|E_{5c}|.$

Proof. By Theorem 1, we have

$$\frac{1}{3}\mathcal{E}[|E(H_{10})|] \geq \frac{146}{525}|E(Opt)| + \frac{2}{315}|E_{5c}| - \frac{4}{225}|E^{A,5c}_{opt}| - \frac{4}{225}|E^{A,7c+}_{opt}|.$$

On the other hand, by Theorem 2, we have

$$\frac{2}{3}\mathcal{E}[|E(H_{10})|] \geq \frac{164}{315}|E(Opt)| + \frac{4}{315}|E_{5c}| + \frac{4}{225}|E^{A,5c}_{opt}| + \frac{16}{225}|E^{A,7c+}_{opt}|.$$

So, summing up the left sides and the right sides of the above two inequalities respectively, we have

$$\mathcal{E}[|E(H_{10})|] \geq \frac{1258}{1575}|E(Opt)| + \frac{2}{105}|E_{5c}|.$$

Theorem 3. $\mathcal{E}[|E(H_{10})|] \geq \frac{468}{575}|E(Opt)|.$

Proof. By Corollary 3, we have

$$\frac{25}{46}\mathcal{E}[|E(H_{10})|] \geq \frac{152}{345}|E(Opt)| + \frac{1}{115}|E_{\overline{5c}}|.$$

By Corollary 4, we have

$$\frac{21}{46}\mathcal{E}[|E(H_{10})|] \geq \frac{629}{1725}|E(Opt)| + \frac{1}{115}|E_{5c}|.$$

So, summing up the left sides and the right sides of the above two inequalities respectively, we have

$$\mathcal{E}[|E(H_{10})|] \geq \frac{463}{575}|E(Opt)| + \frac{1}{115}|E(H_1)| \geq \frac{468}{575}|E(Opt)|.$$

3.3 Derandomization

Our algorithm makes random choices only in Steps 7 and 10. To derandomize Step 10, we just modify it as follows:

10. For each odd cycle C in H, select an arbitrary edge of C and delete it from H.

Because the input graph is unweighted, it does not matter which edge is deleted from each odd cycle in Step 10. So, it should be clear that the above modification of Step 10 does not affect the approximation ratio achieved by the algorithm.

In Step 7, we make a random choice for each 5-cycle. In our above analysis of the algorithm, only the proof of Lemma 3 is based on the mutual independence between these random choices. Indeed, by carefully inspecting the proof, we can see that the proof is still valid even if the random choices made in Step 7 are only pairwise independent. So, we can derandomize it via conventional approaches. Therefore, we have the following theorem:

Theorem 4. *There is an $O(n^2 m^2)$-time approximation algorithm for* MAX SIMPLE EDGE 2-COLORING *achieving a ratio of $\frac{468}{575}$, where n (respectively, m) is the number of vertices (respectively, edges) in the input graph.*

Proof. We estimate the running time of the derandomized algorithm as follows. Step 1 can be done in $O(n^2 m^2)$ time [2]. Obviously, Steps 2 through 4 can be done in $O(n)$ time. Step 5 can be trivially done in $O(n^2)$ time. Since $b(v) \leq 2$ for each vertex v, Step 6 can be done in $O(\sqrt{n}m)$ time. In Step 7, we need to generate $O(n)$ pairwise independent random integers. A conventional way to do this uses two random seeds s_1 and s_2 both of value $O(n)$. So, the sample space of (s_1, s_2) is of size $O(n^2)$. For each sample (s_1, s_2) in the space, we perform Steps 8 through 11 to obtain an output $H(s_1, s_2)$. This takes a total time of (n^3) because Steps 8 through 11 can be done in $O(n)$ time. We then find the sample (s_1, s_2) in $O(n^2)$ time such that $|H(s_1, s_2)|$ is maximized, and further output $H(s_1, s_2)$.

4   An Application

Let G be a graph. An *edge cover* of G is a set F of edges of G such that each vertex of G is incident to at least one edge of F. For a natural number k, a *[1,Δ]-factor k-packing* of G is a collection of k disjoint edge covers of G. The *size* of a [1,Δ]-factor k-packing $\{F_1, \ldots, F_k\}$ of G is $|F_1| + \cdots + |F_k|$. The problem of deciding whether a given graph has a [1,Δ]-factor k-packing was considered in [4,5]. In [6], Kosowski *et al.* defined the *minimum [1,Δ]-factor k-packing problem* (MIN-k-FP) as follows: Given a graph G, find a [1,Δ]-factor k-packing of G of minimum size or decide that G has no [1,Δ]-factor k-packing at all.

According to [6], MIN-2-FP is of special interest because it can be used to solve a fault tolerant variant of the guards problem in grids (which is one of the art gallery problems [7,8]). Indeed, they proved the following:

Lemma 7. *If* MAX SIMPLE EDGE 2-COLORING *admits an approximation algorithm* A *achieving a ratio of* α, *then* MIN-2-FP *admits an approximation algorithm* B *achieving a ratio of* $2 - α$. *Moreover, if the time complexity of* A *is* $T(n)$, *then the time complexity of* B *is* $O(T(n))$.

So, by Theorem 4, we have the following immediately:

Theorem 5. *There is an* $O(n^2 m^2)$-*time approximation algorithm for* MIN-2-FP *achieving a ratio of* $\frac{682}{575}$, *where* n *(respectively,* m*) is the number of vertices (respectively, edges) in the input graph.*

Previously, the best ratio achieved by a polynomial-time approximation algorithm for MIN-2-FP was $\frac{6}{5}$ [6], although MIN-2-FP admits a polynomial-time approximation algorithm achieving a ratio of $\frac{42Δ-30}{35Δ-21}$, where $Δ$ is the maximum degree of a vertex in the input graph [6].

5   Final Remarks

When the input graph is restricted to simple graphs, MAX EDGE t-COLORING is easier to approximate for large values of t as follows: Given G and t, first compute a maximum b-matching M of G where $b(v) = t$ for all vertices v of G, then partition $E(M)$ into $t+1$ matchings $M_1, \ldots, M_{t+1}$, and finally output the largest t matchings among $M_1, \ldots, M_{t+1}$. This algorithm achieves an approximation ratio of $\frac{t}{t+1}$. In particular, this algorithm only achieves a ratio of $\frac{2}{3}$ when $t = 2$. The simple algorithm for MAX SIMPLE EDGE 2-COLORING pointed out by Feige *et al.* (see Section 1) can be viewed as an improvement over the above algorithm. Our new algorithm can be viewed as a further improvement.

References

1. U. Feige, E. Ofek, and U. Wieder. Approximating Maximum Edge Coloring in Multi-graphs. *Proceedings of the 10th International Conference on Integer Programming and Combinatorial Optimization* (IPCO), Lecture Notes in Computer Science, **2462** (2002) 108-121.
2. D. Hartvigsen. *Extensions of Matching Theory.* Ph.D. Thesis, Carnegie-Mellon University, 1984.
3. D. Hochbaum. *Approximation Algorithms for NP-Hard Problems.* PWS Publishing Company, Boston, 1997.
4. D.P. Jacobs and R.E. Jamison. Complexity of Recognizing Equal Unions in Families of Sets. *Journal of Algorithms,* **37** (2000) 495-504.
5. K. Kawarabayashi, H. Matsuda, Y. Oda, and K. Ota. Path Factors in Cubic Graphs. *Journal of Graph Theory,* **39** (2002) 188-193.
6. A. Kosowski, M. Malafiejski, and P. Zylinski. Packing Edge Covers in Graphs of Small Degree. *Manuscript,* 2006.
7. J. O'Rourke. *Art Gallery Theorems and Algorithms.* Oxford University Press, 1987.
8. J. Urrutia. *Art Gallery and Illumination Problems.* Handbook on Computational Geometry, Elsevier Science, Amsterdam, 2000.

Digraph Strong Searching: Monotonicity and Complexity

Boting Yang and Yi Cao

Department of Computer Science, University of Regina
{boting,caoyi200}@cs.uregina.ca

Abstract. Given a digraph, suppose that some intruders hide on vertices or along edges of the digraph. We want to find the minimum number of searchers required to capture all the intruders hiding in the digraph. In this paper, we propose and study two digraph searching models: strong searching and mixed strong searching. In these two search models, searchers can move either from tail to head or from head to tail when they slide along edges, but intruders must follow the edge directions when they move along edges. We prove the monotonicity of each model respectively, and show that both searching problems are NP-complete.

1 Introduction

The graph searching problem was introduced by Parsons [9], which is to find the minimum number of searchers required to capture all the intruders hiding in a graph. Megiddo et al. [8] proposed the edge search problem, in which there are three types of actions for searchers, i.e., placing, removing and sliding, and an edge is cleared only by a sliding action in a proper way. Kirousis and Papadimitriou [6] proposed the node search problem, in which there are only two types of actions for searchers, i.e., placing and removing, and an edge is cleared if both end vertices are occupied by searchers. Kirousis and Papadimitriou showed that the node search number is equal to the pathwidth plus one. Bienstock and Seymour [3] introduced the mixed search problem that combines the edge search and node search problems. Thus, in the mixed searching problem, an edge is cleared if both end vertices are occupied by searchers or cleared by a sliding action in a proper way. In these three graph searching problems, intruders are invisible and they can move at a great speed at any time along a path that contains no searchers.

Monotonicity is a very important issue in graph searching problems. Megiddo et al. [8] showed that the edge search problem is NP-hard. This problem belongs to the NP class follows from the monotonicity result of [7] in which LaPaugh showed that recontamination of edges cannot reduce the number of searchers needed to clear a graph. Bienstock and Seymour [3] proposed a method that gives a succinct proof for the monotonicity of the mixed search problem, which implies the monotonicity of the edge search problem and the node search problem. Fomin and Thilikos [4] provided a general framework that can unify monotonicity results in a unique minmax theorem.

M.-Y. Kao and X.-Y. Li (Eds.): AAIM 2007, LNCS 4508, pp. 37–46, 2007.
© Springer-Verlag Berlin Heidelberg 2007

An undirected graph is not always sufficient in representing all the information of a real-world problem. Johnson et al. [5] generalized the concepts of tree-decomposition and treewidth to digraphs and introduced a cops-and-robber game on digraphs accordingly. Barat [2] introduced another cops-and-robber game on digraphs. He proved that an optimal monotonic search strategy for a digraph needs at most one more cop than the cop number of the digraph and he conjectured that the monotonicity is held for this cops-and-robber game on digraphs. Alspach et al. [1] proposed four digraph search models in which searchers cannot be removed from digraphs. Yang and Cao [10] studied two digraph search models which are different from this paper in that all searchers and intruders must follow the edge directions when they move along edges. Yang and Cao [11] also introduced the directed vertex separation and investigated the relations between different digraph search models, directed vertex separation, and directed pathwidth.

In some applications, searchers may not obey the edge directions. For example, if an intruder hiding in a building with one-way locking doors (people can go out but cannot go in without a key), then the intruder must go one-way through a door (suppose that he has no keys), but the security personnel can go both ways through a door because they have the keys. This motivates us to introduce two search models on digraphs: the strong searching model and the mixed strong searching model, in which intruders must follow the edge directions and searchers need not.

2   Definitions and Notation

All digraphs in this paper contain at least one edge. Throughout this paper, we use D to denote a digraph, and use (u, v) to denote a directed edge with tail u and head v, and $u \rightsquigarrow v$ to denote a directed path from u to v. Initially, all edges of D are *contaminated*.

In the *strong searching model*, each intruder can move at a great speed at any time from vertex u to vertex v along a directed path $u \rightsquigarrow v$ that contains no searchers. There are three types of actions for searchers: (1) placing a searcher on a vertex, (2) removing a searcher from a vertex and (3) sliding a searcher along an edge from one end vertex to the other. A *strong search strategy* is a sequence of actions such that the final action leaves all edges of D *uncontaminated* (or *cleared*). A contaminated edge (u, v) can be cleared in one of three ways by one sliding action: (1) sliding a searcher from u to v along (u, v) while at least one searcher is located on u, (2) sliding a searcher from u to v along (u, v) while all edges with head u are already cleared, and (3) sliding a searcher from the head v to the tail u along the edge (u, v). The digraph D is *cleared* if all of its edges are cleared. The minimum number of searchers needed to clear D in the strong searching model is the *strong search number* of D, denoted by $ss(D)$.

In the strong searching model, a cleared edge (u, v) will be recontaminated if there is a directed path $w \rightsquigarrow v$ containing edge (u, v) and a contaminated edge (w, w') such that there is no searchers stationing on any internal vertex of this directed path. It is easy to see that only removing and sliding actions may cause recontamination in the strong searching model. In order to simplify the case analysis in a monotonicity proof, we want to restrict the cause for recontamination. For this reason, we will define the actions of searchers in the mixed strong searching model in a slightly different way such that only removing actions may cause recontamination. Its precise definition will be given in Definition 1.

We say that a vertex in D is *occupied* at some moment if at least one searcher is located on this vertex at this moment. Any searcher that is not on D at some moment is said to be *free* at this moment.

Let S be a strong search strategy and let A_i be the set of cleared edges immediately after the ith action. S is *monotonic* if $A_i \subseteq A_{i+1}$ for each i. We say that the strong searching model has the property of *monotonicity* (or is *monotonic*) if for any digraph D, there exists a monotonic strong search strategy that can clear D using ss(D) searchers.

3 Monotonicity of Mixed Strong Searching

We will show the monotonicity of the mixed strong searching model in this section, which means that recontamination does not help to reduce the mixed strong search number of a digraph.

Given a digraph in which all edges are contaminated, a *mixed strong search strategy* is a sequence of actions such that all edges of the digraph are cleared after the last action is executed. The *mixed strong searching model* can be obtained by modifying the strong searching model as follows. Recall that there are three ways to clear an edge by a sliding action in the strong searching model. In the mixed strong searching model, we replace the first way by the *node-search-clearing* rule: an edge can be cleared if both of its end vertices are occupied. Another modification is to disallow the recontamination caused by a sliding action. Thus, in a mixed strong search strategy, each sliding along an edge must clear this edge if it is contaminated. More precisely, we define the five types of actions in the mixed strong searching model as follows.

Definition 1. Let $S = (s_1, s_2, \ldots, s_n)$ be a mixed strong search strategy for a digraph D. For each action s_i in S, let A_i be the set of cleared edges and Z_i be the set of occupied vertices immediately after s_i, and let $A_0 = Z_0 = \emptyset$. Each action s_i, $1 \le i \le n$, is one of the following five types:

(a) *(placing a searcher on v)* $Z_i = Z_{i-1} \cup \{v\}$ for some vertex $v \in V(D) - Z_{i-1}$ and $A_i = A_{i-1}$ (note that each vertex in Z_i has exactly one searcher because $v \in V(D) - Z_{i-1}$);

(b) *(removing the searcher from v)* $Z_i = Z_{i-1} - \{v\}$ for some vertex $v \in Z_{i-1}$ and $A_i = \{e \in A_{i-1} \mid$ if there is a directed path $u \leadsto w$ containing e and an edge $e' \in E(D) - A_{i-1}$ such that w is the head of e and u is the tail of e', then $u \leadsto w$ has an internal vertex in $Z_i\}$;

(c) *(node-search-clearing e)* $Z_i = Z_{i-1}$ and $A_i = A_{i-1} \cup \{e\}$ for some edge $e = (u, v) \in E(D) - A_{i-1}$ with both ends u and v in Z_{i-1};

(d) (*forward edge-search-clearing e*) $Z_i = (Z_{i-1} - \{u\}) \cup \{v\}$ and $A_i = A_{i-1} \cup \{e\}$
for some edge $e = (u, v) \in E(D) - A_{i-1}$ with $u \in Z_{i-1}$ and $v \in V(D) - Z_{i-1}$
and every (possibly 0) edge with head u belongs to A_{i-1}.

(e) (*backward edge-search-clearing e*) $Z_i = (Z_{i-1} - \{v\}) \cup \{u\}$ and $A_i = A_{i-1} \cup \{e\}$
for some edge $e = (u, v) \in E(D) - A_{i-1}$ with $v \in Z_{i-1}$ and $u \in V(D) - Z_{i-1}$
and either every edge with head v except e belongs to A_{i-1} or no edge with
tail v belongs to A_{i-1}.

The *mixed strong search number* of a digraph D, denoted by mss(D), is the
minimum number of searchers needed to clear D in the mixed strong searching
model.

A mixed strong search strategy is *monotonic* if any cleared edge cannot be
recontaminated. We say that the mixed strong searching model is *monotonic* if
for any digraph D, there exists a monotonic mixed strong search strategy that
can clear D using mss(D) searchers.

From Definition 1, we know that at most one edge can be cleared in one
action and each vertex is occupied by at most one searcher at any time. Note
that recontamination in the mixed strong searching model is caused only by
removing actions.

Definition 2. Let D be a digraph. For an edge set $X \subseteq E(D)$, a vertex in $V(X)$
is *critical* if it is the tail of an edge in X and the head of an edge in $E(D) - X$.
The set of all critical vertices in $V(X)$ is denoted by $\delta(X)$.

From [2], we know that the parameter $|\delta|$ has the following property.

Lemma 1. *For any* $X, Y \subseteq E(D)$, $|\delta(X \cap Y)| + |\delta(X \cup Y)| \leq |\delta(X)| + |\delta(Y)|$.

Definition 3. For a digraph D, let mss(D) $= k$ and let $S = (s_1, s_2, \ldots, s_n)$ be
a mixed strong search strategy for D using k searchers. If s_i is a removing action
and there are k critical vertices immediately before s_i, then we say that s_i is a
k-critical-removing action. If S contains no k-critical-removing actions, we say
that S is a *standard* mixed strong search strategy.

We now show that k-critical-removing actions can be avoided.

Lemma 2. *Let* D *be a digraph. If* mss(D) $= k$, *then there always exists a standard mixed strong search strategy that clears* D *using* k *searchers.*

Proof. For a mixed strong search strategy S, let $N(S)$ be the number of actions
before the last k-critical-removing action in S. Among all the mixed strong search
strategies of D using k searchers, choose a strategy $S^* = (s_1, s_2, \ldots, s_n)$ such that
$N(S^*)$ is minimum. Clearly, if $N(S^*) = 0$, then there is no k-critical-removing action in S^* and S^* is a standard mixed strong search strategy of D. Suppose that
$N(S^*) = i \geq 1$. Then s_{i+1} is the last k-critical-removing action. For $1 \leq j \leq n$,
let A_j be the set of cleared edges and Z_j be the set of occupied vertices immediately after s_j, and let $A_0 = Z_0 = \emptyset$. Suppose that s_{i+1} removes the searcher
from vertex $w \in Z_i$. Since s_{i+1} is a k-critical-removing action, we know that

$|\delta(A_i)| = k$. Thus, $\delta(A_i) = Z_i$. It follows that all edges with tail w are contaminated immediately after s_{i+1}. We will modify S^*, by replacing s_i and s_{i+1} with a sequence of actions, to obtain a new strategy S' such that the following two conditions are satisfied: (1) the number of actions before the last k-critical-removing action in S' is less than i, i.e., $N(S') < i$; and (2) D has the same cleared edge set just before s_{i+2} in both S' and S^*. From Definition 1, we know that immediately after a placing, removing, or forward edge-search-clearing action, there are at most $k - 1$ critical vertices. Since $|\delta(A_i)| = k$, action s_i can only be a node-search-clearing (case 1) or a backward edge-search-clearing (case 2).

CASE 1. The action s_i is a node-search-clearing action that clears $(u, v) \in E(D) - A_{i-1}$. If $w = u$, then replace s_i and s_{i+1} by one removing action "removing the searcher on w". If $w = v$, let α be the searcher on w at s_i. It is easy to see that $A_{i+1} - A_{i-1} = \{(u, v)\}$. In this case (i.e., $w = v$), we replace s_i and s_{i+1} by four actions consecutively: "removing α from w", "placing α on w", "node-search-clearing (u, w)", and "removing α from w". Notice that the fourth action is not a k-critical-removing action because all out-edges of w are contaminated, that is, w is not critical at this step. Thus, only the first action may be a k-critical-removing action. If $w \neq u$ and $w \neq v$, then replace s_i and s_{i+1} by two actions consecutively: "removing the searcher on w" and "node-search-clearing (u, v)".

CASE 2. The action s_i is a backward edge-search-clearing action that clears $(u, v) \in E(D) - A_{i-1}$. From Definition 1, we know that $w \neq v$ because $v \notin Z_i$. If $w = u$, then replace s_i and s_{i+1} by a removing action "removing the searcher on v". If $w \neq u$, it is easy to see that $w, v \in Z_{i-1}$. Let α be the searcher on w and β be the searcher on v just after s_{i-1}. Since s_i is a backward edge-search-clearing action, we know that immediately before s_i, either every edge with tail v is contaminated or every edge with head v except (u, v) is cleared. In the former case, we replace s_i and s_{i+1} by two actions consecutively: "removing α from w" and "backward edge-search-clearing (u, v)". In the latter case, if there is a directed path from w to v such that this path does not pass through u and no internal vertex in this path is occupied immediately after s_{i-1}, then it is easy to see that just after s_{i+1} is done in S^*, all edges with tail v are contaminated. We replace s_i and s_{i+1} by four actions consecutively: "removing α from w", "removing β from v", "placing β on v", and "backward edge-search-clearing (u, v)". Note that the second action is not a k-critical-removing action because the number of critical vertices is at most $k - 1$ just before this action. Thus, only the first action may be a k-critical-removing action. If there is no such a path, all edges with head v are cleared just after s_{i+1} is done in S^*. We replace s_i and s_{i+1} by two actions consecutively: "removing α from w" and "backward edge-search-clearing (u, v)".

It is easy to see that immediately before s_{i+2}, D has the same cleared edge set in both S^* and S'. Thus, S' can also clear D with k searchers and $N(S') \leq i - 1$. This contradicts the assumption that $N(S^*)$ is minimum. Therefore, $N(S^*) = 0$ and S^* is a standard mixed strong search strategy of D using k searchers.

We now define the k-strong campaign that corresponds to the sequence of the cleared edge sets in a mixed strong search strategy.

Definition 4. Given a digraph D and a nonnegative integer k, a sequence $(X_0, X_1, \ldots, X_n)$ of subsets of $E(D)$ is called a k-*strong campaign* if it satisfies the following three conditions:

  (i) $X_0 = \emptyset$, $X_n = E(D)$, $|\delta(X_i)| \leq k$, and $|X_j - X_{j-1}| \leq 1$, for $1 \leq j \leq n$;
 (ii) if $|\delta(X_i)| = k$ for some i satisfying $0 \leq i \leq n-1$, then $X_{i+1} = X_i \cup \{(u,v)\}$ for some edge $(u,v) \in E(D) - X_i$ such that $v \in \delta(X_i)$ and either $u \in \delta(X_i)$ or $v \notin \delta(X_{i+1})$; and
(iii) if $|\delta(X_i)| = k$ and $|\delta(X_{i+1})| < k$ for some i satisfying $0 \leq i \leq n-2$, then $X_{i+1} \subset X_{i+2}$.

A k-strong campaign is *progressive* if $X_0 \subseteq X_1 \subseteq \cdots \subseteq X_n$ and $|X_j - X_{j-1}| = 1$, for $1 \leq j \leq n$.

Let Y_m be a digraph such that only one vertex in Y_m has indegree m and outdegree 0, which is called the *sink* of Y_m, and all other m vertices of Y_m have indegree 0 and outdegree 1. An *auxiliary digraph* of a digraph D, denoted by D_m^*, is a digraph which contains two disjoint subdigraphs D and Y_m It is easy to see that $\mathrm{mss}(D_m^*) = \mathrm{mss}(D)$.

Lemma 3. *Let D be a digraph. If $\mathrm{mss}(D) = k$, then we can construct an auxiliary digraph D_m^*, $m \geq 0$, such that there exists a k-strong campaign in D_m^*.*

Lemma 4. *Let D be a digraph. If there is a k-strong campaign in D, then there is a progressive k-strong campaign in D.*

Lemma 5. *Let D be a digraph and $(X_0, X_1, \ldots, X_n)$ be a progressive k-strong campaign in D, and let $X_i - X_{i-1} = \{e_i\}$ for $1 \leq i \leq n$. Then there is a monotonic mixed strong search strategy that clears D using k searchers such that the edges of D are cleared in the order $e_1, e_2, \ldots, e_n$.*

Proof. We construct the monotonic mixed strong search strategy inductively. Suppose that we have cleared the edges $e_1, \ldots, e_{j-1}$, $2 \leq j \leq n$, in order and no other edges have been cleared yet. Before we clear e_j, we may remove searchers such that each vertex in $\delta(X_{j-1})$ is occupied by a searcher and all other searchers are free. Let $e_j = (u,v)$. If $|\{u,v\} \cup \delta(X_{j-1})| \leq k$, then we may place free searchers on both ends of e_j and then clear e_j by a node-search-clearing action. Assume $|\{u,v\} \cup \delta(X_{j-1})| > k$. There are only two cases.

CASE 1. If $|\delta(X_{j-1})| = k-1$, then $\{u,v\} \cap \delta(X_{j-1}) = \emptyset$. Since $(u,v) \notin X_{j-1}$ and there is no searcher on v, no edge with tail v belongs to X_{j-1}. We can place a searcher on v and then clear e_j by a backward edge-search-clearing action.

CASE 2. If $|\delta(X_{j-1})| = k$, since the k-strong campaign $(X_0, X_1, \ldots, X_n)$ is progressive, it follows from Definition 4 that $v \in \delta(X_{j-1})$. Since $|\{u,v\} \cup \delta(X_{j-1})| > k$, we know that $u \notin \delta(X_{j-1})$. It follows from Definition 4 that $v \notin \delta(X_j)$. Thus, we can clear e_j by sliding the searcher on v along (u,v) to u.

In all cases we can clear e_j without any recontamination and no more than k searchers are needed.

Now we can prove the monotonicity of the mixed strong searching model.

Theorem 1. *For a digraph D, if $\mathrm{mss}(D) = k$, then there is a monotonic mixed strong search strategy that clears D using k searchers.*

4   Monotonicity of Strong Searching

In this section, we will prove the monotonicity of the strong searching model by using a relation between strong searching and mixed strong searching under certain transformation.

Lemma 6. *For a digraph D, if D' is a digraph obtained from D by replacing each edge $(u, v) \in E(D)$ with a directed path (u, v', v) of length two, then $\text{mss}(D') \leq \text{ss}(D)$.*

Proof. Let $(s_1, s_2, \ldots, s_n)$ be a strong search strategy that clears D using k searchers. We will inductively construct a mixed strong search strategy $(S'_1, S'_2, \ldots, S'_n)$ that clears D' using k searchers, where S'_i is a subsequence of actions corresponding to s_i. Since s_1 is a placing action, let S'_1 be the same placing action. Suppose that we have constructed $S'_1, S'_2, \ldots, S'_{j-1}$ such that the following two conditions are satisfied: (1) the set of occupied vertices immediately after s_{j-1} is the same as the set of occupied vertices immediately after the last action in S'_{j-1}, and (2) if $(u, v) \in E(D)$ is cleared immediately after s_{j-1}, then the corresponding two edges (u, v') and (v', v) are also cleared immediately after the last action in S'_{j-1}. Note that there are three types of actions for s_j. We now construct S'_j in the following cases.

CASE 1. If s_j is a placing action that places a searcher on an unoccupied vertex, S'_j will take the same action. If s_j is a placing action that places a searcher on an occupied vertex, S'_j is empty.

CASE 2. If s_j is a removing action that removes the only searcher from a vertex, S'_j will take the same action. If s_j is a removing action that removes a searcher from a vertex occupied by at least two searchers, S'_j is empty.

CASE 3. If s_j is a sliding action that slides a searcher from vertex u to vertex v along edge (u, v) and clears (u, v), we have two subcases.

CASE 3.1. All edges with head u are cleared in D immediately before s_j. By the hypothesis, the vertex $u \in V(D')$ is also occupied and all edges with head u are also cleared in D' immediately after the last action in S'_{j-1}. If v is not occupied, then we can construct S'_j as follows: "edge-search-clearing (u, v')", and "edge-search-clearing (v', v)". If v is occupied, then we can construct S'_j as follows: "edge-search-clearing (u, v')", "node-search-clearing (v', v)", and "removing the searcher from v'".

CASE 3.2. At least one edge with head u is contaminated in D immediately before s_j. We know that there is at least one searcher on u while performing s_j, which implies that u is occupied by at least two searchers immediately before s_j. By the hypothesis, the vertex $u \in V(D')$ is also occupied and we have at least one free searcher immediately after the last action in S'_{j-1}. If v is not occupied, then we can construct S'_j as follows: "placing a searcher on v'", "node-search-clearing (u, v')", and "edge-search-clearing (v', v)". If v is occupied, then we can construct S'_j as follows: "placing a searcher on v'", "node-search-clearing (u, v')", "node-search-clearing (v', v)", and "removing the searcher from v'".

CASE 4. If s_j is a sliding action that slides a searcher from vertex u to vertex v along edge (u, v) but does not clear (u, v), we know that immediately before

s_j, u is occupied by only one searcher and at least one edge with head u is contaminated. By the hypothesis, the vertex $u \in V(D')$ is also occupied immediately after the last action in S'_{j-1}. If v is occupied, then S'_j consists of only one action: "removing the searcher from u". If v is not occupied, then S'_j consists of two actions: "removing the searcher from u" and "placing it on v".

CASE 5. If s_i is a sliding action that slides a searcher along (u, v) from v to u without any recontamination, then we have two subcases regarding v.

CASE 5.1. There are at least two searchers on v just before s_i. By the hypothesis, the vertex $v \in V(D')$ is also occupied and we have at least one free searcher immediately after the last action in S'_{j-1}. If u is unoccupied, then we can construct S'_i as follows: "placing a searcher on v'", "node-search-clearing (v', v)" and "backward edge-search-clearing (u, v')". If u is occupied, then we can construct S'_i as follows: "placing a searcher on v'", "node-search-clearing (v', v)", "node-search-clearing (u, v')" and "removing the searcher from v'".

CASE 5.2. There is only one searcher on v just before s_i, and either all the edges with tail v are contaminated, or every edge with head v except (u, v) is cleared. If u is unoccupied, then we can construct S'_i as follows: "backward edge-search-clearing (v', v)" and "backward edge-search-clearing (u, v')". If u is occupied, then we can construct S'_i as follows: "backward edge-search-clearing (v', v)", "node-search-clearing (u, v')" and "removing the searcher from v'".

CASE 6. If s_i is a sliding action that slides a searcher along (u, v) from v to u with recontamination, then there is only one searcher λ on v that is the head of at least two contaminated edges and tail of at least one cleared edge just before s_i. If u is unoccupied, then we can construct S'_i as follows: "removing λ from v", "placing λ on v", "backward edge-search-clearing (v', v)" and "backward edge-search-clearing (u, v')". If u is occupied, then we can construct S'_i as follows: "removing λ from v", "placing λ on v", "backward edge-search-clearing (v', v)", "node-search-clearing (u, v') and "removing λ from v'".

It is easy to see that $(S'_1, S'_2, \ldots, S'_n)$ can clear D' using k searchers. Therefore, $\mathrm{mss}(D') \leq \mathrm{ss}(D)$.

Theorem 2. *For a digraph D, if $\mathrm{ss}(D) = k$, then there is a monotonic strong search strategy that clears D using k searchers.*

5 NP-Completeness Results

Kirousis and Papadimitriou [6] proved that the node search problem is NP-complete. In this section, we will establish a relationship between mixed strong searching and node searching. Using this relation, we can prove the mixed strong searching problem is NP-complete. We can then prove the strong searching problem is NP-complete from Theorem 2.

The minimum number of searchers needed to clear a graph G in the node searching model is the *node search number* of G, denoted by $\mathrm{ns}(G)$.

Theorem 3. *Let G be an undirected graph. If $\tilde{G}$ is a digraph obtained from G by replacing each edge $uv \in E(G)$ with two directed edges (u, v) and (v, u) of multiplicity 2, then $\mathrm{mss}(\tilde{G}) = \mathrm{ns}(G)$.*

Proof. In the mixed strong search model, there are five types of actions, placing, removing, node-search-clearing, forward edge-search-clearing and backward edge-search-clearing, and in the node search model, there are only two types of actions, placing and removing. Note that there is no "clearing" action in the node search model corresponding to the node-search-clearing, forward edge-search-clearing or backward edge-search-clearing action. A contaminated edge is cleared in the node search model if both end vertices are occupied, while a contaminated edge is cleared in the mixed strong search model only by a node-search-clearing, forward edge-search-clearing or backward edge-search-clearing action.

We first show that $mss(\tilde{G}) \leq ns(G)$. Let S_n be a monotonic node search strategy that clears G using k searchers. Notice that S_n is a sequence of placing and removing actions. We will construct a mixed strong search strategy S_m by inserting some node-search-clearing actions into S_n as follows. Initially, we set $S_m = S_n$. For each placing action s in S_n, let A_s be the set of cleared edges just after s and B_s be the set of cleared edges just before s. If $A_s - B_s \neq \emptyset$, then for each edge $uv \in A_s - B_s$, we insert four node-search-clearing actions into the current S_m just after s such that they clear all the four edges corresponding to uv. It is easy to see that S_m can clear $\tilde{G}$ using k searchers. Therefore, $mss(\tilde{G}) \leq ns(G)$.

We now show that $ns(G) \leq mss(\tilde{G})$. Let S_m be a monotonic mixed strong search strategy that clears $\tilde{G}$ using k searchers. We first prove that there is no forward edge-search-clearing action in S_m. Suppose that s' is a forward edge-search-clearing action in S_m, which clears an edge (u, v) (i.e., one of the edges with tail u and head v) by sliding a searcher from u to v. From Definition 1, all in-edges of u are cleared. Since (v, u) is cleared but (u, v) is contaminated just before s', the vertex v must contain a searcher to protect (v, u) from recontamination. From Definition 1, (u, v) must be cleared by a node-search-clearing action because both u and v are occupied just before s'. This is a contradiction. Thus, S_m consists of only four types of actions: placing, removing, node-search-clearing and backward edge-search-clearing. Let S_n be a sequence of actions obtained from S_m by replacing each node-search-clearing action and each backward edge-search-clearing with an empty action. We next prove that S_n is a node search strategy that clears G using k searchers.

When an edge (u, v) is cleared by a node-search-clearing action in S_m, the corresponding edge uv in G is also cleared just before the corresponding empty action in S_n because both u and v are occupied. Note that for any edge $uv \in E(G)$, the corresponding four edges in $\tilde{G}$ cannot be cleared just by backward edge-search-clearing actions in S_m. Thus, when one of these four edges is cleared by a node-search-clearing action, uv is also cleared just before the corresponding empty action in S_n. Since S_m is monotonic, it is easy to see that this uv will keep cleared to the end of the search process if the four corresponding edges all are cleared. Thus, S_n can clear G using k searchers, and therefore, $ns(G) \leq mss(\tilde{G})$.

The Strong (resp. Mixed Strong) Searching problem can be described as follows: Given a digraph D and an integer k, can we use k searchers to clear D under the strong (resp. mixed strong) search model? From Theorem 3, we can prove the following result.

Theorem 4. *The Mixed Strong Searching problem is NP-complete.*

From Theorems 2 and 4, we can prove that the Strong Searching problem is NP-hard. From Theorem 2, we can prove that the Strong Searching problem belongs to NP. Therefore, we have the major result of this section.

Theorem 5. *The Strong Searching problem is NP-complete.*

References

1. B. Alspach, D. Dyer, D. Hanson and B. Yang, Some basic results in arc searching, *Technical report CS-2006-10*, University of Regina, 2006.
2. J. Barat, Directed path-width and monotonicity in digraph searching, *Graphs and Combinatorics*, 22 (2006) 161–172.
3. D. Bienstock and P. Seymour, Monotonicity in graph searching, *Journal of Algorithms*, 12 (1991) 239–245.
4. F. Fomin and D. Thilikos, On the monotonicity of games generated by symmetric submodular functions, *Discrete Applied Mathematics*, 131 (2003) 323–335.
5. T. Johnson, N. Robertson, P. Seymour and R. Thomas, Directed tree-width, *Journal of Combinatorial Theory, Series B*, 82 (2001) 138–154.
6. L. Kirousis and C. Papadimitriou, Searching and pebbling, *Theoret. Comput. Sci.*, 47 (1996) 205–218.
7. A. LaPaugh, Recontamination does not help to search a graph. *Journal of ACM*, 40 (1993) 224–245.
8. N. Megiddo, S. Hakimi, M. Garey, D .Johnson and C. Papadimitriou, The complexity of searching a graph, *Journal of ACM*, 35 (1998) 18–44.
9. T. Parsons, Pursuit-evasion in a graph. *Theory and Applications of Graphs*, Lecture Notes in Mathematics, Springer-Verlag, 426–441, 1976.
10. B. Yang and Y. Cao, Monotonicity of Digraph Directed Searching, preprint, 2006.
11. B. Yang and Y. Cao, Digraph searching, directed vertex separation and directed pathwidth, preprint, 2006.

Algorithms for Counting 2-SAT Solutions and Colorings with Applications

Martin Fürer* and Shiva Prasad Kasiviswanathan

Computer Science and Engineering, Pennsylvania State University
{furer,kasivisw}@cse.psu.edu

Abstract. An algorithm is presented for exactly counting the number of maximum weight satisfying assignments of a 2-CNF formula. The worst case running time of $O(1.246^n)$ for formulas with n variables improves on the previous bound of $O(1.256^n)$ by Dahllöf, Jonsson, and Wahlström. The algorithm uses only polynomial space. As a direct consequence we get an $O(1.246^n)$ time algorithm for counting maximum weighted independent sets in a graph.

1 Introduction

There has a been a growing interest in the analysis of algorithms for NP-hard problems, such as satisfiability [7] or graph coloring [2]. Improvements in the exponential bounds are critical, for even a slight improvement from $O(c^n)$ to $O((c - \epsilon)^n)$ can significantly change the range of the problem being tractable. Most of the super-polynomial algorithms known are only for decision problems. As a natural extension we have counting problems, where we wish to not only decide the existence of solution, but to count the number of solutions. Counting problems are not only mathematically interesting, but they also arise in many applications [16,17].

The decision problem of weighted 2-SAT is NP-hard and the corresponding counting problem (#2-SAT) is #P-complete even for the unweighted case [13,19]. The class #P (proposed by Valiant [18]) is defined as $\{f : \exists$ a non deterministic polynomial time Turing Machine M such that on input x, M has exactly $f(x)$ accepting leaves$\}$. We consider the problem of counting the number of maximum weight satisfying assignments of 2-CNF formulas. Earlier works in this area include papers by Dubois [10], Zhang [20], Littman *et al.* [15]. The algorithm by Zhang [20] runs in $O(1.618^n)$ time, whereas the algorithm by Littman *et al.* runs in $O(1.381^n)$ time.

Over a series of work done by Dahllöf *et al.* [4,5,6] presented an $O(1.2561^n)$ time algorithm for #2-SAT. We improve this upper bound. Our algorithm uses polynomial space and counts the number of maximum weighted satisfying assignments of a 2-CNF formula in $O(1.2461^n)$ time. The weighted 2-SAT problem is closely related to the well studied problem of finding (or counting) maximum

---

* This material is based upon work supported by the National Science Foundation under Grant CCR-0209099.

M.-Y. Kao and X.-Y. Li (Eds.): AAIM 2007, LNCS 4508, pp. 47–57, 2007.

weighted independent sets. Using a standard reduction (see for example [6]) we also obtain an $O(1.2461^n)$ time algorithm for the problem of counting maximum weighted independent sets (further discussion of this result is omitted).

Since the preliminary version of the paper appeared as the technical report [11] our results have found applications in a variety of problems. Anglesmark [1] noted that the faster algorithm for the decision 2-SAT leads to a faster algorithm for the problem of MAX Hamming Distance $(2,2)$-CSP. Very recently, Björklund and Husfeldt [3] used our results to obtain a faster algorithm for determining the chromatic number of graph. They also present a $2^n n^{O(1)}$ time and space algorithm for counting k-colorings of a graph. Analogous results were also obtained by Koivisto [12].

Our *main improvement* in the running time for 2-SAT, is by improved handling of the subproblem with a restriction to at most three occurrences of every variable, i.e., the corresponding constraint graph of the formula (formally defined later) is of degree at most three. Here, the decisive parameter determining the running time is the number of degree 3 nodes. However, more progress in eliminating degree 3 nodes is possible when there are many of them. For example, when the average degree is more than $12/5$, we can find a degree 3 vertex with a neighbor of degree 3, and both are eliminated in at least one of the assignments of the first degree-3 vertex. We take advantage of this by choosing a different complexity measure above $12/5$. Our improved time bounds for degree 3 propagate to formulas of higher degrees, because the average degree has a tendency to shrink during the iterative algorithm. This extension to higher degrees is done with the framework of Dahllöf *et al.* [6].

2   Preliminaries and Problem Definitions

We will employ notation similar to that proposed in [6]. A *propositional variable*, or variable takes values *true* or *false*. A *literal* is a variable (x) or its negation $(\neg x)$. A *clause* is a finite non empty collection of literals. A propositional formula in *conjunctive normal form* is a conjunction of disjunction of literals. A k-CNF formula is a propositional formula in conjunctive normal form with the restriction that each clause contains at most k literals.

#2-SAT is the problem of computing the number of *maximum weight models* (a.k.a. satisfying assignments) for a 2-CNF formula. With each literal l, a weight $w(l) \in \mathbb{N}$ and a count $c(l) \geq 1$ are associated; the vectors W and C are the corresponding vectors. For a set of literals L, we define the weight and cardinality of a model M respectively as

$$\mathcal{W}(M) = \sum_{\{l \in L : l \text{ is true in } M\}} w(l),$$
$$\mathcal{C}(M) = \prod_{\{l \in L : l \text{ is true in } M\}} c(l).$$

Given a 2-CNF formula F, let $Var(F)$ denotes the set of variables of F and $n(F) = |Var(F)|$. A variable which occurs only as x or only as $\neg x$ is called *monotone*.

We define the *constraint graph* $G(F) = (Var(F), E)$, as an undirected graph where the vertex set is the set of variables and the edge set E is defined as

$$\{(x, y) \: : \: x, y \text{ appear in the same clause of } F\}.$$

The *degree* $d(x)$ of a variable x is the degree of x in G. We use $d(F)$ to denote the maximum degree of any variable in F and $n_d(F)$ is the number of variables of degree d in F. *Singleton* variables are variables of degree 1. For convenience, we use a slightly altered notion of neighborhood. The *neighborhood* of a vertex x in the graph G, denoted by $N_G(x)$, is the set $\{y \: : \: (x, y) \in E\} \cup \{x\}$. The size of the neighborhood of x is $S(x) = \sum_{y \in N_G(x)} d(y)$.

We define $m(F)$ as $\sum_{x \in Var(F)} d(x)$. Both $n(F)$ and $m(F)$ are used as measures of formula complexity. For $\mathcal{M}$ being the set of all maximum weight models for F and M' being any arbitrary maximum weight model in $\mathcal{M}$ define

$$\#2\text{-Sat}(F, C, W) = \left(\sum_{M \in \mathcal{M}} \mathcal{C}(M), \mathcal{W}(M') \right).$$

Throughout this paper, $\tilde{O}(f(n))$ will denote $n^{O(1)} f(n)$.

2.1 Estimation of Tree Size

Loosely speaking, the idea behind our algorithms is recursive decomposition based on a popular approach that has originated in papers by Davis, Putnam, Logemann and Loveland [9,8]. The recurrent idea behind these algorithms is to choose a variable x and to recursively count the number of satisfying assignments where x is true as well as those where x is false, i.e., we *branch on* x.

For $\#2$-SAT we follow the analysis of Kullmann [14]. In the implicit branching tree constructed, let x be a node with 2 branches labeled with positive real numbers t_1, t_2. The labels are the measures of the reduction in complexity in the respective branch. The *branching number* is the largest real-valued solution of the function $\mathcal{F}(a) = 1 - \sum_{i=1}^{2} a^{-t_i}$. For a branching tuple (t_1, t_2) the branching number is denoted by $\tau(t_1, t_2)$. The branch from F to F_i is labeled by $t_i = \triangle f(F) = f(F) - f(F_i)$, where $f(F)$ is some algorithm specific measure of complexity. Defining $f_{max}(n) = max_{n(F)=n} f(F)$, ensures a running time of $\tilde{O}(\gamma^{f_{max}(n)})$, where γ is the largest branching number occurring in any tuple of the branching tree. We will define the function f such that the worst case branching number is $\tau(1, 1) = 2$. Let l, l' and l'' denote literals over the variable set $Var(F)$. In a step satisfying literal l, we eliminate all the clauses of the form $(l \vee l')$ and replace all clauses of the form $(\neg l \vee l'')$ by l''. We call a branching as *maximally unbalanced* if clauses of only one form occur.

2.2 Helper Functions

This subsection deals with important functions used for reducing the input formula. We use similar functions and structures as in [6], some of which have been reproduced for completeness. The first function called PROPAGATE (Figure 1) simplifies the formula by removing dead variables. The input to the algorithm is the formula, count vector and weight vector. The four steps of the algorithm are performed until not applicable. The function returns the updated formula, the weight of the variables removed, and count for the eliminated variables.

Function PROPAGATE(F, C, W)
(Initialize $w \leftarrow 0$, $c \leftarrow 1$)
1) If there exists a clause $(1 \vee \ldots)$, then it is removed. Any variable a in the clause which gets removed is handled according to following cases
    a) If $w(a) = w(\neg a)$, then $c \leftarrow c \cdot (c(a) + c(\neg a))$; $w \leftarrow w + w(a)$.
    b) If $w(a) < w(\neg a)$, then $c \leftarrow c \cdot (c(\neg a))$; $w \leftarrow w + w(\neg a)$.
    c) If $w(a) > w(\neg a)$, then $c \leftarrow c \cdot c(a)$; $w \leftarrow w + w(a)$.
2) If there is a clause of the form $(0 \vee \ldots)$, then remove 0 from it.
3) If there is a clause of the form (a), then remove the clause and $c \leftarrow c \cdot c(a)$; $w \leftarrow w + w(a)$, and, if a still appears in F then set $F \leftarrow F[a = 1]$.
4) Return(F, c, w).

Fig. 1. Function PROPAGATE

Function REDUCE(F, v, f)
(Assume $F = F_0 \wedge F_1$ with $Var(F_0) \cap Var(F_1) = \{x\}$)
1) Let $f(F_i) \leq f(F_{1-i})$, $i \in \{0, 1\}$.
2) Let $(c_t, w_t) \leftarrow$ #2-SAT($F_i[x = 1], C, W$).
3) Let $(c_f, w_f)bec$ #2-SAT($F_i[x = 0], C, W$).
4) Modify the vectors C and W so that $c(x) \leftarrow c_t \cdot c(x)$, $c(\neg x) \leftarrow c_f \cdot c(\neg x)$, $w(x) \leftarrow w_t + w(x)$, $w(\neg x) \leftarrow w_f + w(\neg x)$.
5) Return #2-SAT(F_{1-i}, C, W).

Fig. 2. Function REDUCE

Another function called REDUCE (Figure 2) reduces the input formula. It takes advantage of the fact that if a formula F can be partitioned into sub-formulas F_0 and F_1 such that each clause of F belongs to either of them, and $|Var(F_0) \cap Var(F_1)| = 1$, then we can remove F_0 or F_1 while appropriately updating count and weight associated with the common variable. Let $Var(F_0) \cap Var(F_1) = \{x\}$. The input to the function is the formula, x, and some algorithm specific measure of complexity f. We say that in such a situation that REDUCE is *applicable*. Among F_0, F_1 we always remove the one having a smaller value under f. Note that REDUCE needn't be a polynomial time operation. We apply the function REDUCE as long as it is applicable. A formula F is called a *maximally reduced* formula if this routine doesn't apply. It can easily be verified that the value of #2-SAT(F, C, W) is preserved under both PROPAGATE and REDUCE routines.

3   Algorithm for #2-SAT

We have a main Algorithm C-2-SAT (Figure 3) which makes use of another function C-2-SAT$_6$ when $d(F) \leq 6$. The algorithms operate on all connected components of the constraint graph. The phrase *branch on* used in the algorithms is a shorthand for technicalities described in Figure 5. Let $F[x = 0]$ be the result of assigning $x = 0$ in F. $F[x = 1]$ is defined accordingly. The proof of correctness of the Algorithm C-2-SAT is straightforward and can be shown as in [6].

Algorithm C-2-SAT(F, C, W)
1) If F contains no clauses, then return (1,0).
2) If F contains an empty clause, then return (0,0).
3) If $G(F)$ is disconnected, then return (c, w) where $c \leftarrow \prod_{i=0}^{j} c_i$, $w \leftarrow \prod_{i=0}^{j} w_i$, and $(c_i, w_i) \leftarrow$ C-2-SAT(F_i, C, W) for connected components $G(F_0), \ldots, G(F_j)$.
4) If there exists a non-monotone variable x with $d(x) \geq 6$, then branch on x.
5) If $d(F) \leq 6$, then return C-2-SAT$_6(F, C, W)$.
6) Pick a variable x of maximum degree and branch on it.

Fig. 3. Algorithm C-2-SAT

We now concentrate on the analysis of the running time for C-2-SAT$_6$. For the analysis we use a continuous and piecewise linear function $f(n, m)$ similar to the one introduced by [6] as a measure of complexity where $n = n(F)$ and $m = m(F)$. A branching variable is chosen to optimize the progress in the next step. There is a worst case branching associated with each value of m/n. Using a classical model of complexity, such as $n(F)$, means that the worst case branching numbers are smaller near the top of the branching tree and increase as we go down. Informally, this is because the maximum degree $d(F)$ is smaller at the bottom of the branching tree, hence smaller pieces of the formula are removed. The estimation of the running time as $\tilde{O}(\gamma^{f_{max}(n)})$ (with $f_{max}(n) = \max\{f(n, m) : m \in \mathbb{N}\}$) is best when the branching numbers are uniform throughout.

Algorithm C-2-SAT$_6(F, C, W)$
(Assume $d(F) \leq 6$)
1) If F contains no clauses, then return (1,0).
2) If F contains an empty clause, then return (0,0).
3) If $G(F)$ is disconnected, then return (c, w) where $c \leftarrow \prod_{i=0}^{j} c_i$, $w \leftarrow \prod_{i=0}^{j} w_i$, and $(c_i, w_i) \leftarrow$ C-2-SAT$_6(F_i, C, W)$ for connected components $G(F_0), \ldots, G(F_j)$.
4) If REDUCE is applicable, then apply it.
5) Pick a variable x of maximum degree, with the maximum $S(x)$. There are two subcases:

      a) If $N_{G(F)}(x)$ is connected to the rest of the graph[1] using only two external[2] vertices (say) p and q, such that $d(p) \geq d(q)$, then branch on p.
      b) Else branch on x.

Fig. 4. Algorithm C-2-SAT$_6$

The complexity measure introduced by Dahllöf *et al.* [6] incorporates the effects of the decreasing m/n quotient in the upper time bound, leading to a

---

[1] Subgraph induced by the vertex set $Var(F) \setminus N_{G(F)}(x)$.
[2] $\{p, q\} \cap N_{G(F)}(x) = \varnothing$.

better worst case running time estimates. We will find a sequence of worst cases as the m/n quotient increases. Each worst case i is associated with a piecewise linear function $f_i(n,m) = a_i n + b_i m$, a lower limit k_i for m/n below which it is possible that worse cases appear, and an upper limit k_{i+1} for the m/n above which that worst case can't occur. We define the coefficient χ_i by $\chi_i = a_i + k_i b_i$ implying that $f(n,m) = f_i(n,m) = f_{i-1}(n,m) = \chi_i n$ for $m/n = k_i$. Now $f_i(n,m)$ can also be expressed as $f_i(n,m) = \chi_i n + (m - k_i n)b_i$. We define a *Interval* i as the range k_i to k_{i+1}. The formal definitions of the functions are (similar to [6]):

1) Let $(F_t, c_t, w_t) \leftarrow$ PROPAGATE$(F[x = 1], C, W)$.
2) Let $(F_f, c_f, w_f) \leftarrow$ PROPAGATE$(F[x = 0], C, W)$.
3) Let $(c'_t, w'_t) \leftarrow$ C-2-SAT(F_t, C, W) and $(c'_f, w'_f) =$ C-2-SAT(F_f, C, W).
4) Let $W_{true} \leftarrow w(x) + w_t + w'_t$, $W_{false} \leftarrow w(\neg x) + w_f + w'_f$, $C_{true} \leftarrow c(x) \cdot c_t \cdot c'_t$, and $C_{false} \leftarrow c(\neg x) \cdot c_f \cdot c'_f$. There are 3 cases:
    a) If $W_{true} = W_{false}$, then return$(C_{true} + C_{false}, W_{true})$.
    b) Else if $W_{true} > W_{false}$, then return(C_{true}, W_{true}).
    c) Else if $W_{true} < W_{false}$, then return(C_{false}, W_{false}).

Fig. 5. Shorthand for the phrase branch on x

$$f(n,m) = f_i(n,m) \text{ where } k_i < m/n \le k_{i+1},\ 0 \le i \le 18,$$
$$f_i(n,m) = a_i n + b_i m = \chi_i n + (m - k_i n)b_i,\ 0 \le i \le 18,$$
$$\chi_i = \chi_{i-1} + (k_i - k_{i-1})b_{i-1},\ 1 \le i \le 19, \chi_0 = 0.$$

The values of k_i, χ_i, a_i, b_i are in Figure 6. Also provided are the worst case recurrences and the corresponding running times. $\tilde{O}(2^{\chi_i n})$ is the upper limit on the running time for a formula F with $m(F)/n(F) \le k_i$. Following are some interesting properties of $f(n,m)$ used in the analysis. The first property can be observed from the Figure 6, and the second is derived in [6].
1) $f(n,m) > f(n-1,m)$ if $m > 3.75n$.
2) $f(n,m) \ge f(n_1, m_1) + f(n - n_1, m - m_1)$ if $0 \le n_1 \le n, 0 \le m_1 \le m$.

3.1   Worst Case Branching

In this subsection we discuss the worst case branching situation for C-2-SAT$_6$. The following lemma shows that if we change Interval during branching it is even better for our measure f. Therefore, the worst case occurs only when both m/n and m_1/n_1 are in the same *Interval*. All proofs omitted in this section can be found in the full version [11].

Lemma 1. *Let $f(n,m)$ and a_i, b_i, m_1, n_1 be defined as above. Then $\triangle f(n,m) = f(n,m) - f(n_1, m_1) \ge \triangle f_i(n,m) = f_i(n,m) - f_i(n_1, m_1)$ if $k_i < m/n \le k_{i+1}$.*

A worst case branching occurs if the branching is maximally unbalanced. If $d(F) = 2$ in Step 5a, then F is a cycle and we are done after one branching.

Interval i	k_i, k_{i+1}	Worst case	χ_i	b_i	a_i	Time
0	0, 2		0	0	0	$\mathrm{poly}(n)$
1	2, 2.4	$\tau(4a_1 + 12b_1, 4a_1 + 12b_1)$	0	1/4	-1/2	$2^{n/9.99}$
2	2.4, 2+2/3	$\tau(2a_2 + 8b_2, 4a_2 + 14b_2)$	0.1	0.188	-0.352	$2^{n/6.65}$
3	2+2/3, 3	$\tau(a_3 + 6b_3, 4a_3 + 16b_3)$	0.150	0.155	-0.265	$2^{n/4.94}$
4	3, 3.2	$\tau(2a_5 + 10b_5, 5a_5 + 18b_5)$	0.202	0.090	-0.068	$2^{n/4.54}$
5	3.2, 3.5	$\tau(a_5 + 8b_5, 5a_5 + 20b_5)$	0.220	0.089	-0.067	$2^{n/4.04}$
6	3.5, 3.75	$\tau(a_6 + 8b_6, 5a_6 + 22b_6)$	0.247	0.076	-0.018	$2^{n/3.75}$
7	3.75, 4	$\tau(a_7 + 8b_7, 5a_7 + 24b_7)$	0.266	0.065	0.021	$2^{n/3.54}$
8	4, 4+4/29	$\tau(a_8 + 10b_8, 6a_8 + 26b_8)$	0.282	0.036	0.136	$2^{n/3.47}$
9	4+4/29, 4+4/9	$\tau(a_9 + 10b_9, 6a_9 + 28b_9)$	0.287	0.032	0.153	$2^{n/3.36}$
10	4+4/9, 4+4/7	$\tau(a_{10} + 10b_{10}, 6a_{10} + 30b_{10})$	0.297	0.028	0.169	$2^{n/3.33}$
11	4+4/7, 4.8	$\tau(a_{11} + 10b_{11}, 6a_{11} + 32b_{11})$	0.301	0.026	0.182	$2^{n/3.25}$
12	4.8, 5	$\tau(a_{12} + 10b_{12}, 6a_{12} + 34b_{12})$	0.307	0.023	0.195	$2^{n/3.21}$
13	5, 5+5/47	$\tau(a_{13} + 12b_{13}, 7a_{13} + 36b_{13})$	0.311	0.006	0.278	$2^{n/3.20}$
14	5+5/47, 5+1/3	$\tau(a_{14} + 12b_{14}, 7a_{14} + 38b_{14})$	0.312	0.006	0.281	$2^{n/3.18}$
15	5+1/3, 5.5	$\tau(a_{15} + 12b_{15}, 7a_{15} + 40b_{15})$	0.313	0.005	0.283	$2^{n/3.17}$
16	5.5, 5+5/8	$\tau(a_{16} + 12b_{16}, 7a_{16} + 42b_{16})$	0.314	0.005	0.286	$2^{n/3.16}$
17	5+5/8, 5+5/6	$\tau(a_{17} + 12b_{17}, 7a_{17} + 44b_{17})$	0.315	0.004	0.289	$2^{n/3.15}$
18	5+5/6, 6	$\tau(a_{18} + 12b_{18}, 7a_{18} + 46b_{18})$	0.316	0.004	0.291	$2^{n/3.15}$

Fig. 6. Parameter table for $a_i n + b_i m$ $(k_i < m/n \leq k_{i+1})$

Thus, assume that we are in Step 5a with $d(F) \geq 3$, and after branching on p we obtain the maximally reduced formulas F_1 and F_2. In both branches, we eliminate at least p by assignment and $N_{G(F)}(x)$ by REDUCE. It can be seen that in the worst case, in one branch we have $\triangle n_1 = d(x) + 2$, $\triangle m_1 \geq S(x) + 6$. Also in the worst case the other branch will have $\triangle n_2 = d(x) + 4$, $\triangle m_2 \geq S(x) + 10$.

The Step 5b of the Algorithm C-2-SAT$_6$ is the more interesting case and requires special attention. Let $x \in F$ be the variable we branch on to get maximally reduced formulas F_1 and F_2. In the worst case, in one branch we have $\triangle n_1 = 1 +$ #degree 2 nodes in $N_{G(F)}(x) \setminus \{x\}$, $\triangle m_1 = 2 \cdot (d(x) +$ #degree 2 nodes in $N_{G(F)}(x) \setminus \{x\})$. Also in the worst case the other branch will have $\triangle n_2 = d(x) + 1$, $\triangle m_2 \geq 2\lceil \frac{S(x)+3}{2} \rceil$ (Lemma 2). We use these bounds as our worst cases throughout the paper. If $m > 3.75n$ (when both a_i and b_i are positive) it is obvious from the Property 1 of f that Step 5b is harder than Step 5a. For any Interval i with $m \leq 3.75n$ one can easily verify with the given a_i and b_i that Step 5a always has a worst case branching number less than 2. Also long chains of degree 2 nodes don't hurt when $m \leq 2.4n$ because $2b_i + a_i = 0$, $i \in \{0, 1\}$ and long chains are beneficial if $m > 2.4n$ because $2b_i + a_i > 0$, $i \in \{2, 3, \ldots, 18\}$. So, from now on we will only be focusing on Step 5b.

Lemma 2. *Let $x \in F$ be the variable we branch on in Step 5b of the Algorithm C-2-SAT$_6$. In a worst case branching, i.e., the branching is maximally unbalanced, we decrease $m(F)$ by at least $2\lceil \frac{S(x)+3}{2} \rceil$ in at least one of the branchings.*

We also use the following lemma (stated without proof) from [6] that makes a connection between the values of $m(F)/n(F)$ and worst case branchings.

Lemma 3. *(Dahllöf et al. [6]) Let F be a non-empty formula with $m(F)/n(F) = k \in \mathbb{Q}$, and define $\alpha(x)$ and $\beta(x)$ such that*

$$\alpha(x) = d(x) + |\{y \in N_{G(F)}(x) : d(y) < k\}|,$$
$$\beta(x) = 1 + \sum_{\{y \in N_{G(F)}(x) : d(y) < k\}} 1/d(y).$$

There exists some variable $x \in Var(F)$ such that $d(x) \geq k$ and $\alpha(x)/\beta(x) \geq k$.

3.2   Performance of C-2-SAT$_6$

The proof will be divided according to the values of $m(F)/n(F)$. We branch on some variable $x \in F$, eventually resulting in two maximally reduced formulas F_1 and F_2. It is possible to end up with more than two maximally reduced formulas, then the above applies to all connected components $G(F_i)$ and by Property 2 the total work is smaller. It is shown that in each Interval i the worst case branching number is 2. The case where $m \leq 2n$ is a straight forward consequence of applying REDUCE.

We now handle denser cases. First the case where $m \leq 3n$. As mentioned earlier, this case is one of the our main sources of improvement from Dahllöf *et al.* [6]. The benefits of this propagate through to the later cases.

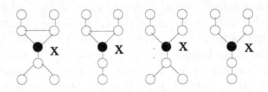

Fig. 7. All possible neighborhood configurations with $d(F) = d(x) = 3$

Lemma 4. *Let F be a maximally reduced formula with $m \leq 3n$ and $d(F) = 3$, then C-2-SAT$_6(F, C, W)$ runs in $\tilde{O}(2^{\chi_4 n})$ time.*

Proof. We divide this case into worst cases depending on the number of degree 3 nodes (0,1,2 or 3) adjacent to x (Figure 7). k_i to k_{i+1} captures the range of m/n where each worst case can appear. For example, if no degree 3 nodes are adjacent to one another, then the worst case is a bipartite graph with $2n/5$ degree 3 nodes on one side and $3n/5$ degree 2 nodes on the other side. This leads to a value of $(3 \cdot 2/5 + 2 \cdot 3/5) = 12/5$ for k_2. This lemma uses $f_1(n, m)$ to $f_3(n, m)$ as measures.

Interval 1: $m/n \in (2, 12/5]$, $d(F) = 3$. In this case we focus on the number of degree 3 variables $n_3(F)$. We will decrease this number by 4 in one step (in the worst case). Therefore, we actually use $n_3(F)/4$ as a measure, i.e., $b_1 = 1/4$ and

$a_1 = -1/2$. If F is maximally reduced then $n_3(F) = m(F) - 2n(F)$. We can show that $\triangle n_3(F) \geq 4$ along any branch by proving that $\triangle m \geq 2\triangle n + 4$. Let V and $V_1 \subseteq V$ be the set of variables in F and F_1 respectively. The reduction in m is

$$\triangle m = \sum_{v \in V - V_1} d(v) + 2|\{\text{clauses } C' \text{ in } F : C' \text{ has variables from both } V \setminus V_1, V_1\}|.$$

Since there are no singleton variables in F, $d(v) = 3$, the first term is at least $2\triangle n + 1$, and since REDUCE does not apply, the second term is at least 2. Taking them together and also noting the fact that $m(F)$ is even, we have $\triangle m \geq 2\triangle n + 4$, so $\triangle n_3(F) \geq 4$. The same argument also works for F_2. Furthermore, $\chi_2 = \chi_1 + (12/5 - 2)b_1 = 0.1$.

Interval 2: $m/n \in (12/5, 2 + 2/3]$, $d(F) = 3$. In this case x has at least one degree 3 node as its neighbor. There are two worst case recursions to be considered in this case. Here we use the estimates for $\triangle n_1, \triangle m_1, \triangle n_2, \triangle m_2$ from Section 3.1.

Case 1: $S(x) = 10$. In this case x has exactly one degree 3 node as its neighbor. The worst case branching is when the branching is maximally unbalanced, with a branching number of $\tau(3a_2 + 10b_2, 4a_2 + 14b_2) < 2$.

Case 2: $S(x) = 11$. In this case x has exactly two degree 3 nodes as its neighbors. The worst case branching is when the branching is maximally unbalanced, with a branching number of $\tau(2a_2 + 8b_2, 4a_2 + 14b_2) = 2$. As given in Figure 6 we have $b_2 \approx 0.1884$ and $a_2 \approx -0.3520$ and $\chi_3 = \chi_2 + (8/3 - 12/5)b_2 \approx 0.1502$.

Interval 3: $m/n \in (2 + 2/3, 3]$, $d(F) = 3$.

In this case x has three degree 3 nodes as its neighbors. The worst case branching number is $\tau(a_3 + 6b_3, 4a_3 + 16b_3) = 2$. As given in Figure 6 we have $b_3 \approx 0.1557$ and $a_3 \approx -0.2650$ and $\chi_4 = \chi_3 + (3 - 8/3)b_3 \approx 0.2021$.

As we see the worst case branching number is 2 and the worst case running time for C-2-SAT$_6$ with maximum degree 3 and $m \leq 3n$ is $\tilde{O}(2^{\chi_4 n})$.   ☐

Depending on the range of m, we are allowed to have different values of $\triangle(F)$ and the proof of the following lemma proceeds as in the previous lemma.

Lemma 5. *For a maximally reduced formula F, the Algorithm C-2-SAT$_6$ (F, C, W) runs in time, $\tilde{O}(2^{\chi_8 n})$ if $3n < m \leq 4n$, $\tilde{O}(2^{\chi_{13} n})$ if $4n < m \leq 5n$, $\tilde{O}(2^{\chi_{19} n})$ if $5n < m \leq 6n$.*

Putting together Lemmas 4, 5, we get that C-2-SAT$_6(F, C, W)$ has a worst case running time of $O(2^{n/3.15})$. The main Algorithm C-2-SAT (Figure 3) checks for three different cases.

Theorem 1. *Algorithm C-2-SAT(F, C, W) runs in $O(1.2461^n)$ time.*

Proof. The runtime bound $T(n)$ of C-2-SAT(F, C, W) is of the form:

$$T(n) \leq \max\{T(n - 2) + T(n - 6), 1.2461^n, T(n - 1) + T(n - 8),$$

where the first term is from Step 4, the second is from Step 5, and the last is from Step 6. Note in Step 6 we branch on a variable with degree greater than 6. The recurrence $T(n) = T(n-2)+T(n-6)$ solves to $T(n) = \tilde{O}(\tau(2,6)^n) \approx O(1.2106^n)$, and the recurrence $T(n) = T(n-1) + T(n-8)$ solves to $T(n) = \tilde{O}(\tau(1,8)^n) \approx O(1.2320^n)$. Therefore, the Algorithm C-2-SAT runs in $O(1.2461^n)$ time.   □

References

1. ANGELSMARK, O. Constructing algorithms for constraint satisfaction and related problems : Methods and applications. *PhD. thesis, Linköping University* (2005).
2. BEIGEL, R., AND EPPSTEIN, D. 3-coloring in time $O(1.3289^n)$. *Journal of Algorithms 54*, 2 (2005), 168–204.
3. BJÖRKLUND, A., AND HUSFELDT, T. Inclusion–exclusion algorithms for counting set partitions. In *Foundations of Computer Science* (2006), IEEE Computer Society, pp. 575–582.
4. DAHLLÖF, V., AND JONSSON, P. An algorithm for counting maximum weighted independent sets and its applications. In *Symposium on Discrete Algorithms* (2002), SIAM, pp. 292–298.
5. DAHLLÖF, V., JONSSON, P., AND WAHLSTRÖM, M. Counting satisfying assignments in 2-SAT and 3-SAT. In *Annual International Computing and Combinatorics Conference (COCOON)* (2002), vol. 2387, Springer, pp. 535–543.
6. DAHLLÖF, V., JONSSON, P., AND WAHLSTRÖM, M. Counting models for 2SAT and 3SAT formulae. *Theoretical Computer Science 332*, 1-3 (2005), 265–291.
7. DANTSIN, E., GAVRILOVICH, M., HIRSCH, E., AND KONEV, B. MAX SAT approximation beyond the limits of polynomial-time approximation. *Annals of Pure and Applied Logic 113* (2002), 81–94.
8. DAVIS, M., LOGEMANN, G., AND LOVELAND, D. A machine program for theorem-proving. *Communications of the ACM 5*, 7 (1962), 394–397.
9. DAVIS, M., AND PUTNAM, H. A computing procedure for quantification theory. *Journal of Association Computer Machinery 7* (1960), 201–215.
10. DUBOIS, O. Counting the number of solutions for instances of satisfiability. *Theoretical Computer Science 81*, 1 (1991), 49–64.
11. FÜRER, M., AND KASIVISWANATHAN, S. P. Algorithms for counting 2-SAT solutions and colorings with applications. TR05-033, Electronic Colloquium on Computational Complexity, 2005.
12. KOIVISTO, M. An $O^*(2^n)$ algorithm for graph coloring and other partitioning problems via inclusion–exclusion. In *Foundations of Computer Science* (2006), IEEE Computer Society, pp. 583–590.
13. KOZEN, D. C. *The Design and Analysis of Algorithms*. Springer, Berlin, 1992.
14. KULLMANN, O. New methods for 3-SAT decision and worst-case analysis. *Theoretical Computer Science 223* (1999), 1–72.
15. LITTMAN, M. L., PITASSI, T., AND IMPAGLIAZZO, R. On the complexity of counting satisfying assignments. In *The Working notes of LICS 2001 Workshop on Satisfiability* (2001).
16. ROTH, D. On the hardness of approximate reasoning. *Artificial Intelligence 82*, 1–2 (1996), 273–302.
17. VADHAN, S. P. The complexity of counting in sparse, regular, and planar graphs. *SIAM Journal on Computing 31*, 2 (2002), 398–427.

18. VALIANT, L. G. The complexity of computing the permanent. *Theoretical Computer Science 8*, 2 (1979), 189–201.
19. VALIANT, L. G. The complexity of enumeration and reliability problems. *SIAM Journal of Computing 8*, 3 (1979), 410–421.
20. ZHANG, W. Number of models and satisfiability of sets of clauses. *Theoretical Computer Science 155*, 1 (1996), 277–288.

Collaborative Ranking: An Aggregation Algorithm for Individuals' Preference Estimation*

Joachim Giesen[1], Dieter Mitsche[2], and Eva Schuberth[2]

[1] Max-Planck Institut für Informatik, Saarbrücken, Germany
[2] Department of Computer Science, ETH Zürich, Switzerland

Abstract. We consider the problem of estimating an individual's product preferences for substitute goods or services. The preferences are elicited by questionnaires that pose a few choice tasks to individuals from the population (respondents). The simplest choice task is a pairwise comparison. To elicit a respondent's ranking of n products completely $\Omega(n \log n)$ pairwise comparisons are necessary. These are easily too many in settings where the incentive for the respondent is not high though he might be willing to answer a few questions truthfully. One approach to cope with this complexity is to aggregate the answers of several respondents in order to estimate an individual's complete preference ranking. Here we describe such an aggregation mechanism based on spectral clustering and prove its validity in statistical model of population and respondents.

1 Introduction

Preference elicitation is daily practice in market research. Its goal is to assess a person's preferences concerning a set of products or a distribution of preferences over a population. The products in the set are usually substitute goods or services, i.e., products that serve the same purpose and can replace each other.

In applications like market share prediction estimating the distribution of the population's preferences is enough whereas in other applications like recommendation systems, see for example [2, 3, 4], one needs to have a good estimate of an individual's preferences. Preferences can be captured by a value function that assigns to every product a value. Every value function induces a (partial) ranking of the products. The higher the value of a product, the higher is the product's position in the ranking. A ranking in general contains less information than a value function. But the direct assessment of a person's value function is difficult and often leads to unreliable results. Ranking products is an easier task, especially if the ranking is obtained from pairwise comparisons (or more general choice tasks), which are popular in market research because they often simulate

---

* This reseach was partially funded by the Swiss National Science Foundation under the project "Algorithms for Robust Conjoint Analysis" and partly supported by EC, COMBSTRU project (HPRN-CT-2002-00278).

M.-Y. Kao and X.-Y. Li (Eds.): AAIM 2007, LNCS 4508, pp. 58–67, 2007.

real buying situations. Fortunately, in many applications a ranking of the products is enough. Sometimes it is even enough to determine the highest ranking product. But often one is also interested in the ordering of the in-between products. For example the highest ranking product for a group of persons might be costly to produce. If it turns out that there is a product which is much cheaper to produce but is still high up in the ranking of most of the persons, then it might be reasonable to produce the latter product.

Here we study the problem of eliciting product preferences from respondents based on pairwise comparison questionnaires. Eliciting a respondent's ranking would require him to perform $\Theta(n \log n)$ comparisons, where n is the number of products. This might be too much if n exceeds a certain value. For example in web based questionnaires respondents get worn out already after only very few comparisons. If the number of comparisons they have to perform exceeds their "tolerance threshold" they either cancel the whole interview or stop to do the comparisons carefully. The tolerance threshold does not scale with n but is usually a small constant. Therefore we want to ask only a constant number of questions but increase the number of respondents for larger product sets. According to their answers we segment the respondents to consumer types. In the end we compute the rankings for the consumer types from the sparse input. Ideally the latter ranking approximates the rankings of the respondents of this consumer type well. Our contribution is a spectral segmentation algorithm that we analyze in a reasonable statistical model of population and respondents, but the formulation of the algorithm is independent of the model.

2 Ranking Algorithm

Given a set of products, we want persons (respondents) to rank the products by pairwise comparisons. Note that other choice tasks, e.g., a one out of three choice task, can be interpreted as multiple paired comparisons. In practice it is infeasible to pose enough choice tasks to deduce the ranking. Our approach is to aggregate the answers obtained from the respondents in order to obtain a few rankings that are typical for the population of respondents and approximate the individual's rankings well. That is, we assume that the population can be segmented into a small number of (consumer) types, i.e., persons that belong to the same consumer type have similar preferences and thus rank the products similarly whereas the rankings of two persons that belong to different types differ substantially. At first we describe how we want to elicit preferences.

Elicitation procedure. Let X be a set of n products. We want to infer a ranking of X for consumers who have to answer l different paired comparison questions (one out of two choice tasks) chosen independently at random. We refer to the consumers who have to perform the choice tasks as respondents. Let m be the number of respondents. Note that here l is a constant independent of n whereas m is dependent on n (m being the only nonconstant parameter, n is also considered as a constant).

Next we give an outline of our ranking algorithm. We refer to any $L \subseteq \binom{X}{2}$ of l product pairs shortly as l-tuple. In the following it turns out to be convenient to consider each product pair as an ordered pair (x, y) such that we can refer to x as the first product and y as the second product of this pair. Our algorithm has four phases. In the first phase we choose a random l-tuple L and then segment all respondents that did all comparisons corresponding to the pairs in L into types of similar preference. In the second phase we use the segmentation of this subset of respondents to compute typical partial rankings of the products covered by the l-tuple L for each segment. Then we extend the partial rankings to complete rankings of all products for all the consumer types that we determined in the second phase. Finally in the last phase, we also segment all the respondents into their respective consumer types that have not been segmented before.

(1) Segmenting the respondents for a given l-tuple. Given an l-tuple L the algorithm SEGMENTRESPONDENTS segments all respondents that did all comparisons corresponding to the pairs in L into types of similar preference. The algorithm has two parameters:

(1) A parameter $0 < \alpha < 1$ that imposes a lower bound of αm on the size of the smallest consumer type that it can identify. Respondents from smaller types will be scattered among other types.
(2) A symmetric $(m_L + l) \times (m_L + l)$-matrix $B = (b_{ij})$ that contains the data collected from the m_L respondents, where m_L is the number of respondents that did all comparisons corresponding to the pairs in L. The column and row indices $1, \ldots, m_L$ of B are indexed by the corresponding respondents and the column and row indices $m_L + 1, \ldots, m_L + l$ are indexed by the l ordered product pairs in L. For $i \in \{1, \ldots, m_L\}$, $j \in \{m_L + 1, \ldots, m_L + l\}$, we set $b_{ji} := b_{ij} := -1$, if respondent i prefers in the $(j - m_L)$'th product comparison the second product over the first one, and $b_{ji} := b_{ij} := 1$, if he prefers the first product over the second. All other entries b_{ij} are set to 0.

SEGMENTRESPONDENTS(B, α)
1   $k :=$ number of eigenvalues of B that are larger than some threshold
         depending on m, n, and l.
2   $P_B :=$ projector onto the eigenspace corresponding to the k most positive
         eigenvalues and the k most negative eigenvalues of B.
3   $P'_B :=$ restriction of P_B onto its first m_L columns and m_L rows.
4   **for** $r := 1$ **to** m_L **do**
5   **for** $s := 1$ **to** m_L **do**
6           $(c_r)_s := \begin{cases} 1 & : & (P'_B)_{rs} \geq 0.49\binom{n}{l}/m \\ 0 & : & \text{otherwise} \end{cases}$
7   **end for**
8   **if** $0.99\alpha m \leq \|c_r\|^2 \binom{n}{2}$ **do**
9   mark r.
10   **end if**
11 **end for**

12   $I := \{1, \ldots, m_L\}, \mathcal{C} := \emptyset$
13   **while** $\{j \in I : j \text{ marked}\} \neq \emptyset$ **do**
14   unmark arbitrarily chosen $i \in \{j \in I : j \text{ marked}\}$
15       $C := \left\{ j \in I : j \text{ marked}, \frac{\langle c_i, c_j \rangle}{\|c_i\| \|c_j\|} \geq 0.97 \right\}$
16       $C' := \left\{ j \in I : j \text{ marked}, \frac{\langle c_i, c_j \rangle}{\|c_i\| \|c_j\|} \geq 0.8 \right\}$
17   **if** $|C| \binom{\binom{n}{2}}{l} \geq 0.9 \alpha m$ **and** $|C'| \binom{\binom{n}{2}}{l} \leq |C| \binom{\binom{n}{2}}{l} + 0.02 \alpha m$ **do**
18           $I := I \setminus C$
19           $\mathcal{C} := \mathcal{C} \cup \{C\}$
20   **end if**
21   **end while**
22   **return** $\mathcal{C}$

The threshold in line 1 of the algorithm SEGMENTRESPONDENTS is used to estimate the number of different consumer types using a carefully chosen threshold that only depends on the known quantities m, n and l. In line 3 the projector is restricted to its first m_L rows and first m_L columns, since this allows us to identify the columns of P'_B with respondents. From each column i of P'_B we compute a vector $c_i \in \{0, 1\}^{m_L}$ whose j'th entry is 1 if and only if the corresponding entry b_{ij} in the matrix B is not less than $0.49 \binom{\binom{n}{2}}{l} / m$, see line 6. We also check whether c_i does not contain a too small number of 1 entries. The intuition is that c_i provides the characteristic vector of the "typical consumer type" which the i'th respondent belongs to. This idea is exploited in the **while**-loop enclosed by lines 13 and 21, where two respondents are grouped together if their corresponding vectors c_i and c_j make a small angle, see line 15 and 16. The use of two sets C and C' is there to avoid taking a vector c_i that is not the characteristic vector of a a typical type in order to compute such a type. All vectors which are never put into any set C will be discarded. We interpret the computed segments as typical consumer types, i.e., each segment represents a different consumer type.

(2) Computing partial rankings for the consumer types. For each consumer type computed by the algorithm SEGMENTRESPONDENTS we determine a typical partial ranking of the products covered by the l-tuple L simply by majority vote. For every ordered product pair $(x, y) \in L$ we say that a type prefers x over y if more than half of the respondents of this type have stated that they prefer x over y, otherwise we say the consumer type prefers y over x.

(3) Extending the partial rankings. For a consumer type to extend the partial ranking of the products covered by L to all products in X, we proceed as follows: we replace an arbitrary element $j \in L$ by an element in $X \setminus \bigcup_{Y \in L} Y$. Let L' be the resulting set of pairs. We run the algorithm SEGMENTRESPONDENTS on L' to segment all the respondents that did all comparisons corresponding to the pairs in L'. The segments of respondents computed from L and from L' will then be merged and the replacement process is repeated until the typical consumer preferences for all $\binom{n}{2}$ product pairs are determined.

(4) Classifying all respondents into the typical consumer types. Finally, all not yet classified respondents, i.e., those respondents corresponding to an l-tuple not used to determine the typical consumer preferences, get also classified. Assume that such a respondent did pairwise product comparisons for pairs in the l-tuple L^*. Such a respondent is classified to be of that consumer type whose ranking restricted to L^* best matches the answers he provided (ties broken arbitrarily).

Let us summarize all parameters of our ranking algorithm in the following table.

n	number of products
m	number of respondents (dependent on n)
l	number of comparisons performed by each respondent (independent of n)
α	parameter in $(0,1)$ that poses a lower bound of αm on the size of the smallest consumer type that it can be identified.

3   Statistical Model

The ranking algorithm that we presented in the last section is formulated independent of a model of population and respondents. But in order to theoretically analyze any procedure that computes typical rankings from the input data a model of the population and a model of the respondents is necessary. Providing such a model gives rise to a *reconstruction problem*, namely, given data obeying the model reconstruct the model parameters. Here we want to turn the preference elicitation problem into a reconstruction problem by providing a reasonable model.

Population model. We assume that the population can be partitioned into k typical consumer types, or short types. Let $\alpha_i \in (0,1)$ be the fraction of the i'th type in the whole population. For each type there is a ranking σ_i, $i = 1, \ldots, k$, i.e., a permutation, of the n products. It will become convenient to encode a ranking as a vector u with $\binom{n}{2}$ entries in $\{\pm 1\}$, one entry for each product pair (x, y). In the vector the entry at position (x, y) is 1 if x is preferred over y and -1 otherwise. We will refer to the vector u also simply as ranking. As a measure of separation of two permutations we use the Hamming distance which is the number of inverted pairs, i.e., the number of pairs (x, y), $x, y \in X$, where $x \prec y$ in the one permutation and $y \prec x$ in the other permutation. Here $x \prec y$ means that y is ranked higher than x. Obviously the maximum separation of two permutations is $\binom{n}{2}$.

Respondent model. We assume that the set of respondents faithfully represents the population, i.e., α_i is also (roughly) the fraction of respondents of type

i among all respondents. Given a respondent let σ be the ranking that corresponds to the type of this respondent. For any comparison of products x and y, with $x \prec y$ according to σ, we assume that the respondent states his preference of y over x with probability $p > 1/2$. In our model each comparison is a random experiment with success probability p, where success means that a respondent answers according to his type. Note that this allows the respondents' answers to violate transitivity, i.e., we still ask comparisons whose outcome could have been derived already from transitivity. From a practical perspective this seems meaningful since stated preferences are often not transitive and the amount of "non-transitivity" is interesting extra information that could be exploited otherwise.

Let us summarize all parameters of our models in the following table.

k	number of types
p	probability for a respondent not to deviate from its type when performing a comparison
$\delta(i,j)$	separation of typical ranking σ_i and σ_j
α_i	fraction of respondents of type i among all respondents
m_i	$\alpha_i m$, i.e., number of respondents of the i'th type

This model allows to analyze procedures that compute typical rankings from the input data. It leads to the ranking reconstruction problem.

Ranking reconstruction problem. Given the data obtained by our elicitation procedure (see previous section) from a population that follows the model described above, the ranking reconstruction problem asks to reconstruct the number k of consumer types, their corresponding typical rankings and to associate every (many) respondent(s) with his (their) correct type.

4 Analysis

The reconstruction problem becomes harder if the typical rankings are not well separated. To make this more precise we introduce the following notions of well-separation.

Well-separation. For $0 < \epsilon < 1$ we say that the typical consumer types are ϵ-well-separated if for any two different consumer type rankings u_i and u_j, we have

$$(1 - \epsilon)\binom{n}{2} \leq 2\,\delta(u_i, u_j) \leq (1 + \epsilon)\binom{n}{2}.$$

Given a ranking (vector) u and an l-tuple L, we denote the projection of u onto L by $\pi_L(u)$, i.e., $\pi_L(u)$ is a vector in $\{\pm 1\}^l$. We say that an l-tuple L is ϵ-well-separating, if for any two different consumer types with associated rankings u_i and u_j we have

$$|\langle \pi_L(u_i), \pi_L(u_j)\rangle| \leq 3\epsilon l.$$

In this case, we also say that consumer types i and j are ϵ-well-separated by L.

Even for well-separated typical rankings it will not always be possible to solve the ranking reconstruction problem, especially if the number of respondents m is too small compared to the number of products n. In the following we outline a proof[1] that with high probability our ranking algorithm solves the ranking reconstruction problem approximately if:

(1) All parameters besides m are considered constant and the number m of respondents is large enough as some function of n.
(2) The number of l of comparisons depends on α, p and ϵ, i.e., it is larger when α (fraction of smallest type) becomes smaller, or when the non-error probability p gets closer to $1/2$, or when ϵ (well-separation of the typical types) becomes smaller. Also l is assumed to much smaller than $\binom{n}{2}$.

We start by showing that a randomly chosen l-tuple is well-separating with high probability.

Lemma 1. *Suppose that the typical consumer type rankings are ϵ-well-separated and suppose that $l = l(p, \alpha, \epsilon) \ll \binom{n}{2}$ is a large constant but independent of n. Then with probability at least $1 - e^{-100}$, a randomly chosen l-tuple is ϵ-well-separating.*

Unless stated differently we will assume in the following that L is an ϵ-well-separating l-tuple. For some l-tuple L let m_L^i denote the number of consumers of type $i \in \{1, \ldots, k\}$ that compared exactly the l product pairs in L.

In order to estimate the number of consumer types k in line 1 of the algorithm SEGMENTRESPONDENTS we want to compute the largest eigenvalues in absolute value of B (see Section 2 for the definition of B). We want to exploit the block structure of B, which can be written as

$$B = \begin{pmatrix} 0 & A \\ A^T & 0 \end{pmatrix},$$

where A is an $m_L \times l$ matrix whose rows are indexed by respondents and whose columns are indexed by product pairs. We are not going to compute the eigenvalues of B directly but use a perturbation argument to estimate them. Therefore we compare B to $\hat{B}$, which we define as the matrix of expected values for the entries in B, i.e., $\hat{b}_{ij} = E[b_{ij}]$. In particular that means that for $i \in \{1, \ldots, m_L\}$ and $j \in \{m_L+1, \ldots, m_L+l\}$ we have $\hat{b}_{ji} = \hat{b}_{ij} = 2p-1$, if respondent i prefers in the j'th product comparison the first product over the second with probability p, $\hat{b}_{ji} = \hat{b}_{ij} = 1-2p$, if he prefers the second product over the first with probability p, and $\hat{b}_{ij} = 0$ otherwise. Note that $\hat{B}$ has a similar form as B, namely,

$$\hat{B} = \begin{pmatrix} 0 & \hat{A} \\ \hat{A}^T & 0 \end{pmatrix}.$$

---

[1] See the full version of this paper for the actual proofs.

The eigenvalues of $\hat{B}$ can be computed from their squared values, i.e., from the eigenvalues of $\hat{B}^2$. The matrix $\hat{B}^2$ has the form

$$\hat{B}^2 = \begin{pmatrix} \hat{A}\hat{A}^T & 0 \\ 0 & \hat{A}^T\hat{A} \end{pmatrix},$$

Thus the eigenvalues of $\hat{B}^2$ can be computed as the eigenvalues of the $m_L \times m_L$ matrix $\hat{A}\hat{A}^T$ and the $l \times l$ matrix $\hat{A}^T\hat{A}$, respectively. Furthermore, $\hat{A}\hat{A}^T$ and $\hat{A}^T\hat{A}$ have the same non-zero eigenvalues. The eigenvalues of $\hat{A}^T\hat{A}$ can be estimated as follows.

Lemma 2. *For $i = 1, \ldots, k$ it holds that*

$$\lambda_i(\hat{A}\hat{A}^T)\binom{\binom{n}{2}}{l} \geq 9lm(2p-1)^2\alpha/10,$$

with probability at least $1 - e^{-cm}$ for some $c > 0$, if $\epsilon \leq (2p-1)^2\alpha^3/90000$.

Corollary 1. *The matrix $\hat{B}$ has rank $2k$ with probability at least $1 - e^{-cm}$.*

Observe now that if λ or $-\lambda$ is a non-zero eigenvalue of $\hat{B}$, then λ^2 is an eigenvalue of $\hat{B}^2$. Hence the absolute value of any of the $2k$ non-zero eigenvalues of $\hat{B}$ is at least $\sqrt{lm(2p-1)^2\alpha 0.9/\binom{\binom{n}{2}}{l}}$ with probability at least $1 - e^{-cm}$. Now we can provide a good value for the threshold that we use in line 1 of the algorithm SEGMENTRESPONDENTS to estimate the number of consumer types k (the constants of this theorem are not needed elsewhere). This shows that with high probability we can reconstruct the number of different consumer types correctly.

Theorem 1. *Let L be a randomly chosen l-tuple for $l = l(p, \alpha, \epsilon)$ sufficiently large but independent of n. With probability at least $1 - 2e^{-100}$, B has exactly k positive eigenvalues and exactly k negative eigenvalues whose absolute value is larger than*

$$10\sqrt{m/\binom{\binom{n}{2}}{l}}.$$

In the following let k denote the number of eigenvalues of B that are larger than $10\sqrt{m/\binom{\binom{n}{2}}{l}}$. We will show next that we do not just know bounds for the eigenvalues of $\hat{B}$ that hold with high probability, but we can also say something about the structure of the projectors onto the corresponding eigenspaces. Later we will use these projectors to cluster the respondents. Denote in the following by $P_A^{(k)}$ the projector onto the space spanned by eigenvectors corresponding to the k largest eigenvalues of the symmetric matrix A. Let P_B' be the restriction of $P_B^{(k)} + (I - P_B^{(m_L+l-k)})$ onto its first m_L rows and first m_L columns, i.e., P_B' is an $m_L \times m_L$ matrix. In the full version of this paper we introduce another $m_L \times m_L$ matrix C to which we compare P_B'. The matrix C is block diagonal and we can characterize P_C^k explicitly: we have $(P_C)_{rs}^{(k)} = 1/m_L^i$, if both respondents r and s belong to type i (and compare all the product pairs in the l-tuple L), and $(P_C)_{rs}^{(k)} = 0$ otherwise.

Lemma 3. *With the probability at least* $1 - 2e^{-100}$*, we have*

$$\|P_B' - P_C^{(k)}\|_2 < \alpha^2/1000.$$

From this lemma we get the following theorem.

Theorem 2. *With probability at least* $1 - 3e^{-100}$*, for a randomly chosen l-tuple L, for every consumer type i, the algorithm* SEGMENTRESPONDENTS *misclassifies at most 3% of the m_L^i respondents.*

The theorem tells us that with high probability for all consumer types at most a 3%-fraction of the respondents gets misclassified. Conditioned under this event, we can now easily, by majority vote, extract the rankings of the k consumer types: for each reconstructed consumer type and for each product pair $(x, y) \in L$ we say that the consumer type prefers x over y if more than half of the respondents of this type have stated that they prefer x over y, otherwise we say the consumer type prefers y over x.

Lemma 4. *With probability at least* $1 - 4e^{-100}$*, for every consumer type i and its typical ranking u_i, $\pi_L(u_i)$ is reconstructed perfectly.*

Next we use that if L is well-separating and we exchange one pair in L with a pair from $X \setminus \bigcup_{Y \in L} Y$ to get an l-tuple L', then also L' is almost well-separating and basically everything we proved for L remains valid for L'. That is, with high probability for both L and L', respectively, exactly the same number k of typical consumer types will be computed whose rankings all agree on all the $l-1$ product comparisons in $L \cap L'$. Thus the segments computed for L and L' can be easily merged.

Theorem 3. *Suppose that $l = l(\alpha, p, \epsilon)$ is a sufficiently large constant (but independent of n). Then, with probability at least* $1 - 5e^{-100}$*, all consumer type rankings can be reconstructed perfectly.*

Finally, we show that also most of the respondents get classified correctly.

Theorem 4. *Suppose that $l = l(p, \alpha, \epsilon)$ is a sufficiently large constant (but independent of n). Then with probability at least* $1 - 6e^{-100}$*, for each consumer type i, $i = 1, \ldots, k$, at most a e^{-97}-fraction of the respondents of that type is misclassified.*

5   Conclusion

We have studied the problem to elicit product preferences of a population, where the preferences are represented as a ranking of the products. During elicitation we ask many respondents to perform a few pairwise comparisons. We provide an algorithm to process the elicited data and introduce models of population and respondents and analyze our algorithm in their context. The following theorem summarizes the analysis of our collaborative ranking algorithm.

Theorem 5. *For any $\xi > 0$ and any $\gamma > 0$, we can choose $l = l(p, \alpha, \epsilon)$ to be a sufficiently large constant (but independent of n) such that for our model of population and respondents and a sufficiently large number of respondents we can with probability $1 - \xi$ correctly*

(1) infer the number k of different consumer types, and
(2) segment a $(1 - \gamma)$-fraction of the respondents into correct types, and
(3) reconstruct the k typical consumer rankings.

Acknowledgment. We thank Michael Krivelevich and Philipp Zumstein for helpful discussions and an anonymous reviewer who pointed out a crucial mistake in an earlier version of this paper.

References

[1] T.K. Dey. Curve and surface reconstruction. *Handbook of Discrete and Computational Geometry*, 2004.
[2] P. Drineas, I. Kerenidis, and P. Raghavan. Competitive recommendation systems. *Proceedings of 32nd IEEE Symposium on Theory of Computing*, pages 82–90, 2002.
[3] S. R. Kumar, P. Raghavan, S. Rajagopalan, and A. Tomkins. *Recommendation systems: A probabilistic analysis. Proc. 39th IEEE Symposium on Foundations of Computer Science*, pages 664-673, 1998.
[4] P. Resnick and H.R. Varian. CACM special issue on recommender systems. *Communications of the ACM*, 40(3), 1997.

A Compact Encoding of Rectangular Drawings with Efficient Query Supports

Katsuhisa Yamanaka and Shin-Ichi Nakano

Department of Computer Science, Gunma University,
1-5-1 Tenjin-cho Kiryu, Gunma, 376-8515 Japan
{yamanaka,nakano}@msc.cs.gunma-u.ac.jp

Abstract. A rectangular drawing is a plane drawing in which every face is a rectangle. In this paper we give a simple encoding scheme for rectangular drawings. Given a rectangular drawing R with maximum degree 3, our scheme encodes R with $\frac{5}{3}m + o(n)$ bits for each n-vertex rectangular drawing R, where m is the number of edges of R, and supports a rich set of queries, including adjacency and degree queries on the faces, in constant time.

1 Introduction

We wish to encode a given graph G into 0-1 binary string S satisfying the following three conditions.
(1) S can be decoded to reconstruct G.
(2) The length of S is as short as possible.
(3) A variety of queries for G can be computed from S efficiently.

For ordered trees, the following string S_T of length $2(n-1)$ bits has been known for a long time. Given an ordered tree T we traverse T starting at the root with depth first manner. If we go down an edge then we code it with '0', and if we go up an edge then we code it with '1'. Thus, any n-vertex ordered tree T has a string S_T of length $2(n-1)$ bits, Then, (1) S_T can be decoded to reconstruct T, and (2) the length of S_T is $2(n-1)$ bits. However it seems to be impossible to compute efficiently, say in $O(1)$ time, whether two designated vertices of T are adjacent or not. However, by appending a string of length $o(n)$ bits to S_T, one can compute the adjacency in $O(1)$ time [8,9].

On the other hand, the number of ordered trees with n vertices is known as the Catalan number C_{n-1}, and it is defined as $C_n = \frac{1}{(n+1)} \frac{(2n)!}{n!n!}$ in [13, p.145]. Since each ordered tree with n vertices must correspond to a distinct string, the average length of S_T is at least $\log C_{n-1} = 2n - o(n)$ bits for S_T [2,8,9].

Thus, the length of S_T above is asymptotically optimal.

Also, for plane graphs many encoding schemes are known [1,2,7,8,9,12,14]. Among them [2] gives a scheme to encode a maximal planar graph into a string of length $2m + n + o(n)$ bits with supporting adjacency and degree queries in $O(1)$ time, where m is the number of edges.

In this paper we design an encoding scheme for rectangular drawings. A rectangular drawing is a plane drawing in which every face is a rectangle. A based

M.-Y. Kao and X.-Y. Li (Eds.): AAIM 2007, LNCS 4508, pp. 68–81, 2007.

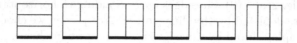

Fig. 1. Based rectangular drawings with exactly three inner faces

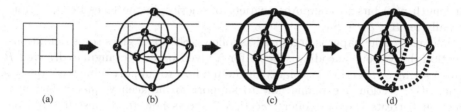

Fig. 2. The encoding scheme

rectangular drawing is a rectangular drawing with one designated base line segment on the outer face. For instance, there are six based rectangular drawings with three faces, as shown in Fig. 1. Each base line segment is depicted by a thick line. Such based rectangular drawings play an important role in many applications, including VLSI floorplanning [5]. Since the size of modern VLSI is extremely huge, a compact representation for VLSI is needed. Note that, if a graph has a vertex with degree five or more, then the graph has no rectangular drawing. Also, if a graph has a degree 4 vertex v, we can replace v by two degree 3 vertices connected by an edge of zero length(although it introduces more vertices). Also note that the degree of each of the four corner vertices on the outer face is always two. We assume there are no other vertices with degree two, since they are redundant. Thus we can assume that the degree of every inner vertex is three, as in most literatures, [5,10,15]. This convention simplifies discussions.

The main result of the paper is as follows. Given a based rectangular drawing R with n vertices and m edges, we give a scheme to encode R into a string S satisfying the following three conditions (1)–(3).

(1) S can be decoded to reconstruct R,

(2) the length of S is $\frac{5}{3}m + o(n)$, and

(3) a variety of queries on the faces for R can be computed from S in $O(1)$ time.

Our strategy is as follows.

Given a based rectangular drawing R, we first compute the planar dual D_R of R (See Fig. 2(b)). Note that each edge of D_R corresponds to either vertical or horizontal adjacency between a pair of adjacent faces. Since every inner vertex of R has degree three, every inner face of D_R is a triangle.

If we use an encoding scheme for maximal planar graphs in [2], we can encode D_R into a string of length $2m + n + o(n) = \frac{7}{3}m + o(n)$ bits. However we lose the information that each edge corresponds to either vertical or horizontal adjacency, so we cannot reconstruct R.

In our strategy we classify the edges of D into the following two types. (1) The edges corresponding to vertical adjacency of faces. (These are depicted by thick lines in Fig. 2(c).) (2) The edges corresponding to horizontal adjacency of faces. (These are depicted by thin lines in Fig. 2(c).) Then we choose a spanning tree T of the graph induced by the edges in (1), and we further classify the edges of (1) into either (1a) edges included in T or (1b) others (See Fig. 2(d)). Finally we only store the edges of (1a) and (2) into a string S_R of length $\frac{5}{3}m$ bits, and we show that S_R is enough to reconstruct R. Then we also design a string S_A of length $o(n)$ bits to compute a variety of queries on the faces in $O(1)$ time. Finally let $S = S_R + S_A$.

The rest of the paper is organized as follows. Section 2 gives some definitions. Section 3 gives our encoding scheme for a given based rectangular drawing R into a string S_R. Section 4 explains for S_A and how to support some basic queries efficiently. Section 5 explains how to support an adjacency query efficiently. Section 6 treats a degree query. Section 7 shows how to reconstruct R from S. Finally Section 8 is a conclusion.

2   Preliminaries

In this section we give some definitions.

Let G be a connected graph. A *tree* is a connected graph with no cycle. A *rooted* tree is a tree with one vertex r chosen as its *root*. An *ordered* tree is a rooted tree with fixed orderings for siblings.

A drawing of a graph is *plane* if it has no two edges intersect geometrically except at a vertex to which they are both incident. A plane drawing divides the plane into connected regions called *faces*. The unbounded face is called *the outer face*, and other faces are called *inner faces*. We regard *the contour* of a face as the clockwise cycle formed by the vertices and edges on the boundary of the face. A graph is *planar* if it has a plane drawing. A *plane graph* is a planar graph with a fixed planar drawing. Let n be the number of vertices of a graph, m be the number of edges, and f be the number of faces. Since every face is enclosed by at least three edges and every edge is on the contours of exactly two faces, we have $3f \leq 2m$ for any plane graph. The inequality and Euler's formula $n - m + f = 2$ means that $m \leq 3n - 6$ holds for any plane graph.

A *rectangular drawing* is a plane drawing in which every face (including the outer face) is a rectangle. See some examples in Fig. 1. Note that a graph with maximum degree five or more does not have any rectangular drawing. A *based rectangular drawing* is a rectangular drawing with one designated base line segment on the contour of the outer face. The designated base line segment is called *the base*, and we always draw the base as the lowermost horizontal line segment of the drawing. For examples, all based rectangular drawings with three inner faces are shown in Fig. 1, in which each base is depicted by a thick line.

Two faces F_1 and F_2 are *ns-adjacent* (north-south adjacent) if they share a horizontal line segment on their contours. Two faces F_1 and F_2 are *ew-adjacent* (east-west adjacent) if they share a vertical line segment on their contours. If two based rectangular drawings P_1 and P_2 have a one-to-one correspondence between faces preserving ns- and ew-adjacency, and in which each base corresponding to the other, then we say that P_1 and P_2 are isomorphic.

In rectangular drawing, vertex v with degree three is *w-missing* (west missing) if v has edges to top, bottom, and right. Similarly we define *e-missing*, *n-missing*, and *s-missing*. Note that every vertex with degree three is either w-missing, e-missing, n-missing, or s-missing.

3   The String of a Rectangular Drawing

In this section, given a based rectangular drawing R, we design a string S_R for R such that (1) S_R can be decoded to reconstruct R, and (2) the length of S_R is $\frac{5}{3}m + \frac{22}{3}$ bits, where m is the number of edges of R. In the next section, by using an additional string S_A of length $o(n)$ bits, we explain how to compute some queries for R in $O(1)$ time. The details of S_A are shown in the next section.

As we mentioned in Section 1, the idea of our encoding scheme is shown in Fig. 2.

We now need some definitions.

Given a based rectangular drawing R, by rotating either 0, 90, 180, or 270 degrees in clockwise, respectively, one can have four based rectangular drawings. Without loss of generality, we can assume that R is the based rectangular drawing with the maximum number of s-missing vertices among the four rotated drawings. Because, otherwise, we can first rotate R so that it has the maximum number of s-missing vertices, and store the amount of rotation in two bits, and preprocess each query with this rotation information.

We have the following lemma.

Lemma 1. *Let n_S be the number of s-missing vertices. Then, $n_S \geq \frac{n-4}{4}$ holds, where n is the number of vertices of R.*

Proof. The four corners of the outer rectangle of R have degree two, and all other vertices have degree three. Each vertex with degree three is either w-, e-, n-, or s-missing. Since n_S is the largest, $n_S \geq \frac{n-4}{4}$ holds. □

Given a based rectangular drawing R, we compute the planar dual D_R of R as follows. First, by extending the uppermost and lowermost horizontal line segments, we divide the outer face of R into four faces, those are west, east, north, and south faces. Then we put a vertex in each face of R, and we connect two vertices by an edge if the corresponding two faces are adjacent. See Fig. 2(b). Note that each of the divided four outer faces is also replaced with a vertex.

Then, we classify the edges of D_R into two subsets of edges as follows. Note that each edge of D_R corresponds to some adjacency of two faces of R, and it is either ns-adjacency or ew-adjacency. Let E_{NS} be the set of edges of D_R corresponding to some ns-adjacency, and E_{EW} be the set of edges corresponding to some ew-adjacency. (See Fig. 2(c).) Let D_{NS} be the subgraph of D_R induced by the edges in E_{NS}, and D_{EW} be the subgraph of D_R induced by the edges in E_{EW}.

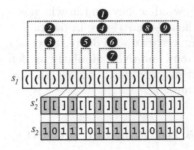

Fig. 3. The string $S_R = S_1 + S_2$ for a rectangular drawing in Fig. 2

Next, we further classify the edges in E_{NS} into two subsets of edges as follows. For each face f of R let $p(f)$ be the westmost face among the faces locating on the north side of f. We call $p(f)$ *the parent face* of f. Especially for the (outer) north face f_N, $p(f_N)$ is not defined. Let E_{NS}^T be the set of edges corresponding to the adjacency between some face f and $p(f)$. Now E_{NS}^T induces a spanning tree of D_R. Let T_{NS} be the spanning tree. An example is shown in Fig. 2(d), in which T_{NS} is drawn by thick lines, while the edges in $E_{NS} - E_{NS}^T$ is drawn by dashed lines.

We are going to construct the string S_1 from T_{NS}, then the string S_2 from T_{NS} and D_{EW}, then finally construct the string $S_R = S_1 + S_2$ for R.

First we construct S_1 as follows. (It is identical string for an ordered tree mentioned in Section 1.) Starting the vertex corresponding to the north face, we traverse T_{NS} with depth first manner. If we go down an edge then we code it with '(', and if we go up an edge then we code it with ')'. Let S_1' be the resulting string. Let $S_1 = (S_1')$, that is the string obtained from S_1' by adding '(' at the head and ')' at the end. (Actually, we encode each '(' to '0' and each ')' to '1', respectively.) Fig. 3 shows a string S_1 of Fig. 2.

The i-th '(' and its matching parenthesis ')' in S_1 correspond to the i-th vertex of T_{NS} in preorder. (See Figs. 2 and 3.)

Next we construct S_2 as follows.

We first construct S_2' from S_1 as follows. We replace each i-th '(' with $|west(i)|$ of consecutive ']'s, and its matching parenthesis ')' with $|east(i)|$ of consecutive '['s, where $west(i)$ is the set of faces locating on the west side of f_i corresponding to the i-th vertex of T_{NS} in preorder, and $east(i)$ is the set of faces locating on the east side of f_i. For example, in Figs. 2, $west(5) = \{f_2\}$ and $east(5) = \{f_6, f_7\}$. Since $west(1) = west(2) = west(3) = \phi$ and $east(1) = east(f+3) = east(3) = \phi$ hold, they have no corresponding string, where f is the number of faces of R. Note that every inner face f_i of R satisfies $|west(i)| \geq 1$ and $|east(i)| \geq 1$. The $|west(i)|$ of consecutive ']'s in S_2' correspond to the $|west(i)|$ of ew-adjacencies between faces in $west(i)$ and f_i. Also the k-th ']' among the $|west(i)|$ of consecutive ']'s corresponds to the k-th ew-adjacency of f_i to west from north. Similarly, the $|east(i)|$ of consecutive '['s correspond to the $|east(i)|$ of ew-adjacencies between faces in $east(i)$ and f_i. Also the k-th '[' among $|east(i)|$ of consecutive '['s corresponds to the k-th ew-adjacency of f_i to east from south. Note that, since T_{NS} is a tree, S_2' has a nested structure. (We will explain a reason in Section 5.)

Next we construct S_2 from S_2' as follows. We replace each consecutive $|west(i)|$ of ']'s with one '1' followed by $(|west(i)| - 1)$ of consecutive '0's. Similarly, we replace each consecutive $|east(i)|$ of '['s with one '1' followed by $(|east(i)| - 1)$ of consecutive '0's. An example is shown in Fig. 3. Note that each '1' in S_2 corresponds to the border of some $west(i)$ or $east(i)$. Also note that no information of edges in $E_{NS} - E_{NS}^T$ is stored.

Now we estimate the length of $S_R = S_1 + S_2$. Let n be the number of vertices of R, m be the number of edges of R, and f be the number of faces of R. Then D_R has $m + 4$ edges, because we extend the uppermost and lowermost horizontal line segments of R. Since the four corners on the outer face of R have degree two, and all other vertices have degree three, the following equation holds.

$$2m = 3(n - 4) + 2 \cdot 4 \tag{1}$$

First we estimate the length of S_1. The string S_1' stores the pair of '(' and ')' for each edge of T_{NS}, then by adding two more bits we have S_1. Thus $|S_1| = 2|E_{NS}^T| + 2$.

Next we estimate the length of S_2. Since the pair of '[' and ']' is stored for each edge corresponding to some ew-adjacency, $|S_2| = 2|E_{EW}| = 2((m + 4) - |E_{NS}|)$ holds.

Therefore,

$$|S_R| = |S_1| + |S_2|$$
$$= (2|E_{NS}^T| + 2) + (2m + 8 - 2|E_{NS}|) = 2m + 10 - 2|E_{NS} - E_{NS}^T|$$

We have the following lemma.

Lemma 2. $|E_{NS} - E_{NS}^T| = n_S + 2$, where n_S is the number of s-missing vertices of R.

Proof. We show that there is a one-to-one mapping between s-missing vertices and edges in $E_{NS} - E_{NS}^T$. Assume that face f_u is located on the north side of face f_d, and $f_u \neq p(f_d)$ holds. Let u and d be the vertices corresponding to f_u and f_d. Then we can assign the lower-left corner x of the face f_u, which is s-missing, to the edge (u, d). Since the definition of E_{NS}^T implies that each vertex has at most one 'downward' edge in $E_{NS} - E_{NS}^T$, any duplication does not happen in the assignment above. Since by the division of the outer face we have created two of s-missing vertices to R, the claim holds. □

By Lemma 2,
$$|S_R| = 2m + 10 - 2|E_{NS} - E_{NS}^T| = 2m + 10 - 2(n_S + 2)$$
By Lemma 1,
$$|S_R| \leq 2m + 10 - 2\left(\frac{n - 4}{4} + 2\right) - 2m - \frac{n}{2} + 8$$
By equation (1), $|S_R| \leq \frac{5}{3}m + \frac{22}{3}$ holds.

We will show that S_R can be decoded to reconstruct R in Section 7.

4 The Basic Queries

In this section we survey some basic queries for strings.

Let $S[i,j]$ be the substring of S from position i to position j. We denote $S[i,i] = S[i]$.

Let S_1 be the string in Section 3 consisting of '(' and ')'. Operation $rank(S_1, i, ()$ computes the number of '('s up to and including the position i in S_1. Operation $select(S_1, k, ()$ computes the position of the k-th '(' in S_1. Note that, if $j = select(S_1, k, ()$, then $k = rank(S_1, j, ()$ holds. Similaly we define $rank(S_1, i,))$ and $select(S_1, k,))$. For example, if $S_1 = ((()) (()(())) ()())$, as shown in Fig. 3, then $rank(S_1, 6, () = 4$ and $select(S_1, 3,)) = 8$.

Next, we define the *balanced* parentheses. The string $S_a = ()$ is balanced. Assume that two strings S_b and S_c are balanced, then two strings $S_d = S_b S_c$ and $S_e = (S_b)$ are also balanced. In a balanced string there is a natural one-to-one correspondence between '(' and ')'. If $S_1[i] = $ '(' corresponds to $S_1[j] = $ ')', we say that $S_1[i]$ *matches* $S_1[j]$, and we write $match(S_1, i) = j$ and $match(S_1, j) = i$. If $match(S_1, i) = j$, then $match(S_1, j) = i$ holds.

Next we define $enclose(S_1, i)$ in the following two cases. If $S_1[i] = $ '(', then, $enclose(S_1, i)$ is the position of '(' which immediately encloses the pair i and $match(S_1, i)$. Otherwise, $S_1[i] = $ ')', then $enclose(S_1, i)$ is the position of '(' which immediately encloses the pair $match(S_1, i)$ and i. For example, $enclose(S_1, 9) = 6$ in Fig. 3.

Next we define $wrapped(S_1, i)$ in the following two cases. If $S_1[i] = $ '(', then let c be the number of positions k in S_1 such that $enclose(S_1, k) = i$. Otherwise, $S_1[i] = $ ')', then let c be the number of position k in S_1 such that $enclose(S_1, k) = j$, where $j = match(S_1, i)$. In both cases we define $wrapped(S_1, i) = c$.

The following lemma is known.

Lemma 3. ([1,3,6,8,9])

Given a balanced string S_1 of length $2n$, using an additional string S_1^A of $o(n)$ bits, one can compute the following operations in $O(1)$ time for each. One can construct S_1^A in $O(n)$ time.

  (1) $rank(S_1, i, ()$ and $rank(S_1, i,))$ [3,6].
  (2) $select(S_1, i, ()$ and $select(S_1, i,))$ [3].
  (3) $match(S_1, i)$ [8,9].
  (4) $enclose(S_1, i)$ [8,9].
  (5) $wrapped(S_1, i)$ [1].

Each pair of '(' and ')' in S_1 corresponds to a vertex of D_R. The i-th vertex of T_{NS} in preorder corresponds to the i-th '(' and its matching ')'.

Let S_2 be the string in Section 3. Then S_2 has a nested structure. Similarly, we can define operations $rank$, $select$, $match$, $enclose$, and $wrapped$ for S_2.

Also we add some basic operations for S_2.

By definition, S_2 can be divided into $(|S_1| - 6)$ substrings. (See Fig 3.) Let L_i be the substring in S_2 corresponding to $west(i)$. We can find L_i in S_2 as follows. Assume that $select(S_1, i, () = a$. Then, $L_i = S_2[select(S_2, a-4, 1), select(S_2, a-3, 1)) - 1]$ holds. For example, in Fig. 3, if $i = 5$, then $a = select(S_1, 5, () = 7$.

Since the first four characters in S_1 have no corresponding substring in S_2 (See Fig. 3), L_5 begin at the position of $(7-4)$-th '1' in S_2. Thus, L_5 begin at position $select(S_2, 7 - 4, 1) = 4$. Also, L_5 ends at position $select(S_2, 7 - 4 + 1, 1) - 1$, because the next substring begin at position $select(S_2, 7 - 4 + 1, 1)$ in S_2. Hence $L_5 = S_2[4, 4]$. Note that each L_i in S_2 start with '1', and each '1' is a starting position for some L_i. L_i corresponds to the $|west(i)| = |L_i|$ of ew-adjacency between each face in $west(i)$ and the f_i corresponding to the i-th vertex. Also the k-th ']' in L_i corresponds to the k-th ew-adjacency of f_i to west from north. In particular, the pair of the first ']', at position $p_1 = select(S_2, a - 4, 1)$, and its matching '[', at position $p_2 = match(S_2, p_1)$, corresponds to the ew-adjacency between face f_i and face f_{WN}, where f_{WN} is the northmost face among the faces locating on the west side of f_i. Let v_j be the vertex corresponding to face f_{WN}. Then, we write $f_{WN}(i) = j$. Thus, f_{WN} is the northmost $(= N)$ face among the faces locating on the west $(= W)$ side of f_i. Then,

$$f_{WN}(i) = rank(S_1, match(S_1, rank(S_2, p_2, 1) + 4), ()$$

holds. Similarly, let f_{WS} be the southmost face among the faces locating on the west side of f_i. Then,

$$f_{WS}(i) = rank(S_1, match(S_1, rank(S_2, match(S_2, select(S_2, a - 3, 1) - 1), 1) + 4), ()$$

holds.

Similarly, let R_i be the substring in S_2 corresponding to $east(i)$. We can find R_i in S_2 as follows. Assume $select(S_1, i, () = a$ and $match(S_1, a) = b$. Then $R_i = S_2[select(S_2, b - 4, 1), select(S_2, b - 3, 1) - 1]$ holds. R_i corresponds to the $|east(i)| = |R_i|$ of ew-adjacency between each face in $east(i)$ and the f_i corresponding to the i-th vertex. Also the k-th '[' in R_i corresponds to the k-th ew-adjacency of f_i to east from south. Let f_{EN} and f_{ES} be the northmost and southmost faces among the faces locating on the east side of f_i, respectively. Then similarly,

$$f_{EN}(i) = rank(S_1, rank(S_2, match(S_2, select(S_2, b - 3, 1) - 1), 1) + 4, (),$$

and

$$f_{ES}(i) = rank(S_1, rank(S_2, match(S_2, select(S_2, b - 4, 1)), 1) + 4, ()$$

hold.

Lemma 4. *Given a balanced string S_R of length $2n$, using an additional stirng S_A of $o(n)$ bits, one can compute L_i, R_i, $f_{WN}(i)$, $f_{WS}(i)$, $f_{EN}(i)$, and $f_{ES}(i)$ in $O(1)$ time. One can construct S_A in $O(n)$ time.*

Proof. It is trivial by Lemma 3.   □

Let v be a lower-left corner of face f_i corresponding to the i-th vertex. Then, v is either s-missing or w-missing. If v is s-missing, we write $s\text{-}miss(i, WS) = True$. Otherwise, $s\text{-}miss(i, WS) = False$. We can compute $s\text{-}miss(i, WS)$ as follows. If $f_{ES}(f_{WS}(i)) = i$, then $s\text{-}miss(i, WS) = True$, otherwise, $s\text{-}miss(i, WS) = False$. We have the following lemma.

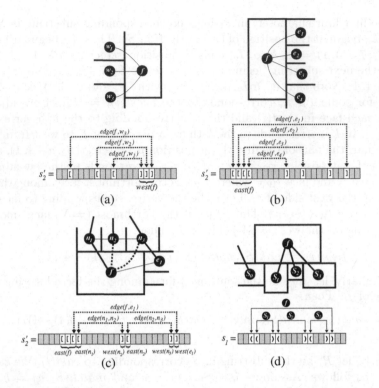

Fig. 4. Illustration for how to compute the four adjacency queries

Lemma 5. *For each four corner of a face, one can compute which direction is missing in $O(1)$ time.*

5   The Adjacency Query

In this section we explain how we can compute the adjacency query in $O(1)$ time. We can compute such query with a help of the additional string S_A of $o(n)$ bits.

Given two faces f_i and f_j of based rectangular drawing R, we consider the following four adjacency queries.

(wa) $f_j \in west(i)$: Is face f_j located on the west side of face f_i? (See Fig. 4(a).)
(ea) $f_j \in east(i)$: Is face f_j located on the east side of face f_i? (See Fig. 4(b).)
(na) $f_j \in north(i)$: Is face f_j located on the north side of face f_i? (See Fig. 4(c).)
(sa) $f_j \in south(i)$: Is face f_j located on the south side of face f_i? (See Fig. 4(d).)

Let L_i be the substring in S_2 corresponding to $west(i)$, and R_i be the substring in S_2 corresponding to $east(i)$. Let D_R be the planar dual of R.

We first consider query (wa).

We can observe that $f_j \in west(i)$ if and only if some '[' in R_j matches some ']' in L_i. Such a pair can be efficiently computed as follows.

Assume that $R_j = S_2[a, b]$ and $L_i = S_2[c, d]$. Without loss of generality one can assume that $a < b < c < d$. We have the following three cases.

Case 1: $match(S_2, d) > b$.

None ']' in L_i matches any '[' in R_j. Hence $f_j \notin west(i)$.

Case 2: $a \le match(S_2, d) \le b$.

In this case $S_2[d]$ matches some '[' in R_j. Hence $f_j \in west(i)$.

Case 3: $match(S_2, d) < a$.

We have two subcases.

Case 3(a): $match(S_2, a) < c$.

None '[' in R_j matches any ']' in L_i. Thus $f_j \notin west(i)$.

Case 3(b): $c \le match(S_2, a) < d$.

In this case $S_2[a]$ matches some ']' in L_i. Hence $f_j \in west(i)$.

Therefore, we can compute query (wa) in $O(1)$ time.

The computation for query (ea) is symmetric to query (wa). So we omit the detail for (ea).

Next we consider for (na). If f_j is the parent face of f_i, $f_j \in north(i)$. Otherwise, we have to compute whether f_j is located on the north side of f_i by the edge in $E_{NS} - E_{NS}^T$. Note that no information for the edges in $E_{NS} - E_{NS}^T$ is stored (directly) in S_1 or S_2. However, we can compute the adjacency query (na) in the following way.

First we show that the nested structure of S_2 corresponds to an inclusion structure of regions of planar dual D_R. We need some definitions. Let $\{v_1, v_2, \ldots, v_{f+3}\}$ be the set of vertices of D_R. The root v_1 of T_{NS} corresponds to the north (outer) face. For each edge $e = (v_i, v_k) \in E_{EW}$ we assign the following region $R(v_i, v_k)$ of D_R as follows. Let P_i be the path v_i to v_1 and P_k be the path v_k to v_1. Then $R(v_i, v_k)$ is the region enclosed by two pathes P_i, P_k, and edge e. No region $R(v_i, v_k)$ can include only a proper part of the other. (Since, otherwise, assume that $R(v_i, v_k)$ includes a proper part of $R(v_j, v_l)$, now there is an edge $(v_j, v_l) \in E_{EW}$ such that $v_j \in R(v_i, v_k)$ and $v_l \notin R(v_i, v_k)$, however then edge (v_j, v_l) intersect some edge on the boundary of $R(v_i, v_k)$, a contradiction.) By the planarity of D_R, we have the following lemma.

Lemma 6. $R(v_i, v_k)$ *properly includes* $R(v_j, v_l)$ *if and only if the pair of '[' and ']' for edge* (v_i, v_k) *encloses the pair of '[' and ']' for edge* (v_j, v_l).

Now let consider face f_j which is locating on the north side of f_i but $f_j \ne p(f_i)$. Assume $k = f_{EN}(i)$, and its corresponding vertex is v_k. Let v be the upper-right corner of f_i. If v is n-missing, we assume $l = f_{WS}(j)$. (See Fig. 5(a).) Otherwise we assume $l = f_{ES}(j)$. (See Fig. 5(b).) Let v_l be the l-th vertex of T_{NS} in preorder. Each of the two edges (v_i, v_k) and (v_j, v_l) represents some ew-adjacency. Let $R'(v_i, v_k)$ be the region obtained from region $R(v_i, v_k)$ by eliminating the face with edge (v_i, v_k) on its contour. Then the pair of '[' and ']' for edge (v_i, v_k) immediately encloses the pair of '[' and ']' for edge (v_j, v_l) if and only if the region $R'(v_i, v_k)$ has edge (v_j, v_l) on the contour of $R'(v_i, v_k)$.

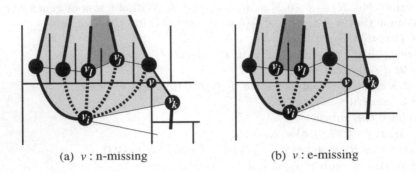

(a) v : n-missing   (b) v : e-missing

Fig. 5. Illustration for query (na)

We have the following lemma.

Lemma 7. *Assume that face f_j is not the parent face of f_i. Face f_j is located on the north side of f_i if and only if the pair of '[' and ']' for edge (v_i, v_k) immediately encloses the pair of '[' and ']' for edge (v_j, v_l). (See Fig. 5.)*

Proof. ($\Rightarrow$) Since f_j is located on the north side of f_i, region $R(v_i, v_k)$ includes region $R(v_j, v_l)$. (See Fig. 5.) Then the pair of '[' and ']' for (v_i, v_k) encloses the pair of '[' and ']' for (v_j, v_l). In addition, when the face having edge (v_i, v_k) on its contour is eliminated from $R(v_i, v_k)$, (v_j, v_l) appears on the contour of the resulting region. Thus the pair of '[' and ']' for (v_i, v_k) immediately encloses the pair of '[' and ']' for (v_j, v_l).

($\Leftarrow$) Since the pair of '[' and ']' for (v_i, v_k) immediately encloses the pair of '[' and ']' for (v_j, v_l), when the face having edge (v_i, v_k) on its contour is eliminated from $R(v_i, v_k)$, (v_j, v_l) appears on the contour. Thus f_j is located on the north side of f_i. as shown in Fig. 5. Note that each of the two edges (v_i, v_k) and (v_j, v_l) corresponds to some ew-adjacency.   □

By Lemma 7, we can compute the query (na) in the following way. We have the following two cases.

Case 1: f_j is the parent face of f_i.

$f_j \in north(i)$ holds. Note that, by using the enclose operation in S_1, we can compute whether f_j is the parent face of f_i or not in $O(1)$ time.

Case 2: Otherwise.

By Lemma 7, the pair of '[' and ']' for edge (v_i, v_k) immediately encloses the pair of '[' and ']' for edge (v_j, v_l) if and only if $f_j \in north(i)$. Let v be the upper-right corner of f_i. We have the following two subcases. Assume that $a = match(S_1, select(S_1, i, ()))$, that is a is the position of ')' for v_i in S_1.

Case 2(a): v is n-missing. (See Fig. 5(a).)

In this case $l = f_{WS}(j)$ holds. Assume $b = select(S_1, j, ()$. Then

$$enclose(S_2, select(S_2, b - 3, 1) - 1) = select(S_2, a - 3, 1) - 1$$

if and only if $f_j \in north(i)$. Since $select(S_2, b - 4, 1)$ in above equation indicates the first ']' among $|west(j)|$ of consecutive ']'s, $select(S_2, b - 3, 1)$ indicates the *next* character to $|west(j)|$ of consecutive ']'s. Thus $select(S_2, b - 3, 1) - 1$

indicates the last ']' among $|west(j)|$ of consecutive ']'s. Note that the indicated ']' above corresponds to edge (v_j, v_l).

Case 2(b): v is e-missing. (See Fig. 5(b).)

In this case $l = f_{ES}(j)$ holds. Similar to Case 2(a), assume $c = match(S_1, select(S_1, j, ())$. Then

$$enclose(S_2, select(S_2, c - 4, 1)) = select(S_2, a - 3, 1) - 1$$

if and only if $f_j \in north(i)$.

Therefore we can compute query (na) in constant time.

Finally we consider for query (sa). Face f_i is located on the north side of f_j if and only if f_j is located on the south side of f_i. Thus we can compute the query (sa) by query (na).

6   The Degree Query

In this section we explain how to compute the number of neighbour faces in the designated direction. We can compute such a query in $O(1)$ time with a help of the additional string S_A of $o(n)$ bits. We consider the following four degree queries (wd), (ed), (nd), and (sd) for each direction.

(wd) $|west(i)|$: How many faces are located on the west side of face f_i?

In Section 3 we explained how to find L_i in S_2. Since $|west(i)| = |L_i|$, we can compute the query (wd) in $O(1)$ time.

(ed) $|east(i)|$: How many faces are located on the east side of face f_i?

The computation of query (ed) is symmetric to (wd). So we omit the detail.

(nd) $|north(i)|$: How many faces are located on the north side of face f_i?

Let v_i be the vertex corresponding to f_i. Let $f_{EN}(i) = k$ be the northmost face among faces locating on east side of f_i, and v_k be its correponding vertex. Let $g_1, g_2, \ldots, g_{|north(i)|}$ be the faces locating on the north side of f_i from west to east order. Let $v_1^g, v_2^g, \ldots, v_{|north(i)|}^g$ be the vertices corresponding to $g_1, g_2, \ldots, g_{|north(i)|}$, respectively. Note that face g_1 is the parent face of f_i. Let v be the upper-right corner of f_i. We have the following two cases.

Case 1: v is n-missing

Assume that the pair of '[' and ']' in S_2' for edge (v_i, v_k) is at position a and b in S_2', respectively. Note that the values of a and b can be computed in $O(1)$ time by basic queries. Then the pairs immediately enclosed by the pair of $S_2'[a]$ and $S_2'[b]$ are only pairs of '[' and ']' for edges $(v_1^g, v_2^g), (v_2^g, v_3^g), \ldots, (v_{|north(i)|-1}^g, v_{|north(i)|}^g)$. Thus $|north(i)| = (wrapped(S_2, a)/2) + 1$.

Case 2: v is e-missing

Similar to Case 1, assume that the pair of '[' and ']' in S_2' for edge (v_i, v_k) is at position a and b in S_2', respectively. Then the pairs immediately enclosed by the pair of $S_2'[a]$ and $S_2'[b]$ are only pairs of '[' and ']' for edges $(v_1^g, v_2^g), (v_2^g, v_3^g)$, $\ldots, (v_{|north(i)|-1}^g, v_{|north(i)|}^g), (v_{|north(i)|}^g, v_k)$. Thus $|north(i)| = wrapped(S_2, a)/2$.

Therefore we can compute query (nd) in $O(1)$ time.

(sd) $|south(i)|$: How many faces are locating on the south side of face f_i?

Let v be the lower-left corner of f_i. We have the following two cases.

Case 1: v is w-missing

In this case face f is located on the south side of f_i if and only if its parent face is f_i. Thus we need to compute the number of faces with the parent face f_i. Hence $|south(i)| = wrapped(S_1, select(S_1, i, ()))/2$. Therefore we can compute $|south(i)|$ in $O(1)$ time.

Case 2: v is s-missing

In this case there is exactly one face on the south side of f_i such that its parent face is not f_i. Such a face is the westmost face among the faces locating on the south side of f_i. For other faces the condition is similar to Case 1. Thus $|south(i)| = 1 + wrapped(S_1, select(S_1, i, ()))/2$. Hence we can compute $|south(i)|$ in $O(1)$ time.

Therefore we can compute the number of neighbour faces of f_i in the designated direction in $O(1)$ time. Thus we can compute the degree of f_i in $O(1)$ time.

7   Reconstruction of R

Given the string $S = S_R + S_A$ for a based rectangular drawing R, we can reconstruct R by using the queries in Section 4 and Section 5. We have the following theorem.

Theorem 1. *Given the string S_R of a based rectangular drawing R, one can construct an additional string S_A of $o(n)$ bits in $O(n)$ time. Then one can reconstruct R from S in $O(n)$ time.*

Proof. Proof by induction. If we know about all west and north adjacencies for each $1, 2, \ldots, (i-1)$-th face, we can also compute west and north adjacencies for the i-th face in $O(|north(i)| + |west(i)|)$ time by the queries. Thus the running time of the algorithm is $O(m) = O(n)$ in total.   □

8   Conclusion

In this paper we designed a compact string S for a based rectangular drawing. Given the string, we can (1) compute the adjacency and degree queries on the faces in $O(1)$ time and (2) reconstruct the original based rectangular drawing R from S in $O(n)$ time. Also S supports a variety of queries for R.

Let N_k be the number of based rectangular drawings with k inner faces. By implementing an efficient enumeration algorithm for based rectangular drawings, [10,11] gave $N_{11} = 10948768$. Thus at least $24 > \log N_{11} = 23.5$ bits are needed to encode a based rectangular drawing with 11 inner faces, while the total length of our string is $5 \cdot 11 - 6 = 49$ bits without S_A. Thus we conjecture that there are still many chances to reduce the length of the string. For $N_{12} = 89346128$, at least $27 > \log N_{12} = 26.4$ bits are needed for a based rectangular drawing with 12 inner faces, while our encoding needs $5 \cdot 12 - 6 = 54$ bits.

Our future work is to give more compact strings for based rectangular drawings with efficient query supports. A lower bound of the length of the strings for based rectangular drawings is also needed to evaluate "the compactness" of the strings.

References

1. Y.-T. Chiang, C.-C. Lin and H.-I Lu, *Orderly spanning trees with applications to graph encoding and graph drawing,* Proc. of SODA 2001, pp.506–515, 2001.
2. R.C.-N Chuang, A. Garg, X. He, M.-Y. Kao, and H.-I Lu, *Compact encodings of planar graphs via canonical orderings and multiple parentheses,* Proc. of ICALP 1998, LNCS 1443, pp.118–129, 1998.
3. D. Clark, *Compact pat trees,* Ph. D, thesis, Department of Computer Science, University of Waterloo, 1998.
4. R.L. Graham, D.E. Knuth and O. Patashnik, *Concrete mathematics, 2nd ed.,* Addison-Wesley, 1994.
5. X. He, *On floor-plan of plane graph,* SIAM J. on Comput., vol.28, no.6, pp.2150–2167, 1999.
6. G. Jacobson, *Succinct static data structures,* Ph. D, thesis CMU-CS-89-112, Carnegie Mellon University, 1989.
7. K. Keeler and J. Westbrook, *Short encodings of planar graphs and maps,* Discrete Applied Math., 58, no.3, pp.239–252, 1995.
8. J.I. Munro and V. Raman, *Succinct representation of balanced parentheses, static trees and planar graphs,* Proc. of FOCS 1997, pp.118–126, 1997.
9. J.I. Munro and V. Raman, *Succinct representation of balanced parentheses and static trees,* SIAM J. Comput., vol.31, no.3, pp.762–776, 2001.
10. S. Nakano, *Enumerating floorplans with n rooms,* Proc. of ISAAC 2001, LNCS 2223, pp.107–115, 2001.
11. S. Nakano, *Enumerating floorplans with some properties,* Interdisciplinary Information Sciences, vol.8, no.2, pp.199–206, 2002.
12. C. Papadimitriou and M. Yannakakis, *A note on succinct representations of graphs,* Information and Control, vol.71, pp.181–185, 1986.
13. K. H. Rosen (Eds.), *Handbook of discrete and combinatorial mathematics,* CRC Press, Boca Raton, 2000.
14. G. Turan, *Succinct representations of graphs,* Discrete Applied Math., vol.8, pp.289–294, 1984.
15. K.-H. Yeap and M. Sarrafzadeh, *Floor-planning by graph dualization: 2-concave rectilinear modules,* SIAM J. Comput., vol.22, pp.500–526, 1993.

A New Efficient Algorithm for Computing the Longest Common Subsequence

M. Sohel Rahman[*,**] and Costas S. Iliopoulos[* * *]

Algorithm Design Group
Department of Computer Science, King's College London,
Strand, London WC2R 2LS, England
{sohel,csi}@dcs.kcl.ac.uk
http://www.dcs.kcl.ac.uk/adg

Abstract. In this paper, we present a new and efficient algorithm for solving the LCS problem for two strings. Our algorithm runs in $O(\mathcal{R} \log \log n + n)$ time, where $\mathcal{R}$ is the total number of ordered pairs of positions at which the two strings match.

1 Introduction

The *longest common subsequence*(LCS) problem is a classic and well-studied problem in computer science with extensive applications in diverse areas ranging from spelling error corrections to molecular biology. A subsequence of a string is obtained by deleting zero or more symbols of that string. The longest common subsequence problem for two strings, is to find a common subsequence in both strings, having maximum possible length. More formally, suppose we are given two strings $X[1..n] = X[1]X[2] \ldots X[n]$ and $Y[1..n] = Y[1]Y[2] \ldots Y[n]$. A subsequence $S[1..r] = S[1]S[2] \ldots S[r]$, $0 < r \leq n$ of X is obtained by deleting $n - r$ symbols from X. A common subsequence of two strings X and Y, denoted $cs(X,Y)$, is a subsequence common to both X and Y. The longest common subsequence of X and Y, denoted $lcs(X,Y)$ or $LCS(X,Y)$, is a common subsequence of maximum length. We denote the length of $lcs(X,Y)$ by $r(X,Y)$. In this paper, we assume that the two given strings are of equal length. But our results can be easily extended to handle two strings of different length.

Problem "LCS". LCS Problem for 2 Strings. *Given strings X and Y, compute the Longest Common Subsequence of X and Y.*

The longest common subsequence problem for k strings $(k > 2)$ was first shown to be NP-hard [13] and later proved to be hard to be approximated [10]. The restricted but, probably, the more studied problem that deals with two

---

[*] Supported by the Commonwealth Scholarship Commission in the UK under the Commonwealth Scholarship and Fellowship Plan (CSFP).
[**] On Leave from Department of CSE, BUET, Dhaka-1000, Bangladesh.
[* * *] Supported by EPSRC and Royal Society grants.

M.-Y. Kao and X.-Y. Li (Eds.): AAIM 2007, LNCS 4508, pp. 82–90, 2007.

strings has been studied extensively [20,16,15,14,9,8,7]. The classic dynamic programming solution to LCS problem, invented by Wagner and Fischer [20], has $O(n^2)$ worst case running time. Masek and Paterson [14] improved this algorithm using the "Four-Russians" technique [1] to reduce the worst case running time to $O(n^2/\log n)^1$. Since then not much improvement in terms of n can be found in the literature. However, several algorithms exist with complexities depending on other parameters. For example, Myers in [15] and Nakatsu et al. in [16] presented an $O(nD)$ algorithm, where the parameter D is the simple Levenshtein distance between the two given strings [11]. Another interesting and perhaps more relevant parameter for this problem is $\mathcal{R}$, which is the total number of ordered pairs of positions at which the two strings match. More formally, we say a pair (i,j), $1 \leq i,j \leq n$, defines a match, if $X[i] = Y[j]$. The set of all matches, M, is defined as follows:

$$M = \{(i,j) \mid X[i] = Y[j], 1 \leq i,j \leq n\}$$

Observe that $|M| = \mathcal{R}$. Hunt and Szymanski [9] presented an algorithm to solve Problem LCS in $O((\mathcal{R}+n)\log n)$ time. They also cited applications, where $\mathcal{R} \sim n$ and thereby claimed that for these applications the algorithm would run in $O(n \log n)$ time. For a comprehensive comparison of the well-known algorithms for LCS problem and study of their behaviour in various application environments the readers are referred to [4].

In this paper, we revisit the much studied LCS problem for two strings and present new algorithms using some novel ideas and interesting observations. Our main result is an $O(\mathcal{R} \log \log n + n)$ algorithm for Problem LCS. The rest of the paper is organized as follows. In Sections 2 we present an $O(n^2 + \mathcal{R} \log \log n)$ algorithm, namely LCS-I, to solve Problem LCS, which is an easy extension of the algorithms and techniques of [17]. LCS-I provides the base of our new improved algorithm, LCS-II, described in Section 3. Using some novel techniques, we achieve $O(\mathcal{R} \log \log n + n)$ running time for LCS-II. Finally, we briefly conclude in Section 4.

2   A New Algorithm

In this section, we present Algorithm LCS-I which works in $O(n^2 + \mathcal{R} \log \log n)$ time. Note that, LCS-I is an easy extension of the algorithms presented in [17]^2 and the main contribution of this paper is an improved algorithm, namely LCS-II, with $O(\mathcal{R} \log \log n + n)$ running time, to be presented in Section 3.

From the definition of LCS it is clear that, if $(i,j) \in M$, then we can calculate $\mathcal{T}[i,j], 1 \leq i,j \leq n$ by employing the following equation [17]:

---

[1] Employing different techniques, the same worst case bound was achieved in [5].

[2] In [17], only the running time of LCS-I was mentioned without explicitly devising the algorithm. We revisit LCS-I here for the sake of completeness because LCS-II heavlily depends on the concept used in LCS-I.

$$T[i,j] = \begin{cases} \text{Undefined} & \text{if } (i,j) \notin M, \\ 1 & \text{if } (i = 1 \text{ or } j = 1) \text{ and } (i,j) \in M \\ \max_{\substack{1 \le \ell_i < i \\ 1 \le \ell_j < j \\ (\ell_i, \ell_j) \in M}} \{(T[\ell_i, \ell_j])\} + 1 & \text{if } (i, j \ne 1) \text{ and } (i,j) \in M. \end{cases}$$

$$(1)$$

Here we have used the tabular notion $T[i,j]$ to denote $r(X[1..i], Y[1..j])$. From Equation 1, it follows that only the entries $T[i,j]$ such that $(i,j) \in M$ are useful. Therefore, we can ignore all $T[i,j]$ with $(i,j) \notin M$ from the calculation. In order to do that, we need a preprocessing step to construct the set M in sorted order according to their position they would be considered in the algorithm (we consider a row by row operation). Such a preprocessing algorithm, referred to as "Algorithm Pre" in the rest of this paper, was presented in [17] which runs in $O(\mathcal{R} \log \log n + n)$ time[3]. After we have computed the set M (using Algorithm Pre), we can start computing the entries $T[i,j], (i,j) \in M$ according to the Equation 1. Since we are not calculating all the entries of the table, we need to use a global variable and appropriate pointers to keep track of the actual LCS. For the efficient implementation of the computation of Equation 1, we utilize the following facts observed in [17].

Fact 1. ([17]) *Suppose $(i,j) \in M$. Then for all $(i',j), i' > i$ $((i,j'), j' > j)$, we must have $T[i',j] \ge T[i,j]$ $(T[i,j'] \ge T[i,j])$.* $\square$

Fact 2. ([17]) *The calculation of the entry $T[i,j], (i,j) \in M, 1 \le i,j \le n$ is independent of any $T[\ell, q], (\ell, q) \in M, \ell = i, 1 \le q \le n$.* $\square$

Following the techniques of [17], we also use the following problem and relevant result.

Problem "RMAX". Range Maxima Query Problem. *Suppose we are given a sequence $A = a_1 a_2 ... a_n$. A Range Maxima (Minima) Query specifies an interval $I = (i_s, i_e), 1 \le i_s \le i_e \le n$ and the goal is to find the index ℓ with maximum (minimum) value a_ℓ for $\ell \in I$.*

Theorem 1. ([6,3]) *The RMAX problem can be solved in $O(n)$ preprocessing time and $O(1)$ time per query.* $\square$

The algorithm LCS-I proceeds as follows. We maintain an array H of length n where, for $T[i,j]$ we have, $H[\ell] = \max_{1 < k < i, (k,\ell) \in M}(T[k, \ell]), 1 \le \ell \le n$. The 'max' operation, here, returns 0, if there does not exist any $(k, \ell) \in M$ within the range. Given the updated array H, we can easily perform the task by using the constant time range maxima query (Theorem 1). And Fact 1 makes it easy to maintain the array H on the fly, as we proceed as follows. As usual, we proceed in a row by row manner. We use another array S, of length n, as a temporary storage. When we find an $(i,j) \in M$, after calculating $T[i,j]$, we store $S[j] = T[i,j]$. We continue to store in this way as long as we are in the same

---

[3] In [17], the running time of Algorithm Pre is reported as $O(\mathcal{R} \log \log n)$ under the usual assumption that $\mathcal{R} \ge n$.

row. As soon as we find an $(i', j) \in M, i' > i$, i.e. we start processing a new row, we update H with new values from S. The correctness of the above procedure follows from Fact 1 and 2. However, for the constant time range maxima query we need to do a $O(n)$ preprocessing as soon as H is updated. But due to Fact 2, it is sufficient to perform this preprocessing once per row. So, the computational effort added for this preprocessing is $O(n^2)$ in total. Therefore we get the following theorem.

Theorem 2. *LCS-I solves Problem LCS in $O(n^2 + \mathcal{R} \log \log n)$ time using $\theta(max(\mathcal{R}, n))$ space.* □

The outline of LCS-I is presented formally in the form of Algorithm 1. Note that we can shave off the $\log \log n$ term from the running time reported in Theorem 2 as follows. Since we have an n^2 term anyway in the running time, we do not need to compute the set M in the preprocessing step using Algorithm Pre. Instead, we consider each $\mathcal{T}[i, j], 1 \le i, j \le n$ and perform useful computation only when $(i, j) \in M$.

Algorithm 1. Outline of LCS-I

1: Construct the set M using Algorithm Pre. Let $M_i = (i, j) \in M, 1 \le j \le n$.
2: globalLCS.Instance $= \epsilon$
3: globalLCS.Value $= 0$
4: **for** $i = 1$ *to* n **do**
5:    $S[i].Value = 0$ {Initialize the temporary array S}
6:    $S[i].Instance = \epsilon$
7: **end for**
8: **for** $i = 1$ *to* n **do**
9:    $H = S${Update H for the next row}
10:   Preprocess $H.Value$ for Range Maxima Query
11:   **for** each $(i, j) \in M_i$ **do**
12:     $maxindex = RMQ_H(1, j - 1)${Range Maxima Query on Array H}
13:     $\mathcal{T}.Value[i, j] = H[maxindex].Value + 1$
14:     $\mathcal{T}.Prev[i, j] = H[maxindex].Instance$
15:     $S[j].Value = \mathcal{T}.Value[i, j]$
16:     $S[j].Instance = (i, j)$
17:     **if** globalLCS.value $< \mathcal{T}.Value[i, j]$ **then**
18:       globalLCS.Value $= \mathcal{T}.Value[i, j]$
19:       globalLCS.Instance $= (i, j)$
20:     **end if**
21:   **end for**
22: **end for**
23: **return** globalLCS

---

3   The Improved Algorithm

In this section, we present the main result of this paper. In particular, we improve the running time of LCS-I, as reported in Theorem 2, with some nontrivial modifications. The resulting Algorithm, LCS-II, will eventually run in $O(\mathcal{R} \log \log n + n)$ time. As is explained in the previous section, LCS-I exploits the constant time query operation (Theorem 1) of Problem RMAX. However, due to the $O(n)$ preprocessing step of RMAX, we can't eliminate the n^2 term from the running time of LCS-I. But a very important, albeit easily observable, fact is that the range maxima queries made in LCS-I is always of a special form.

Fact 3. *All the range maxima queries in Algorithm LCS-I are of the form* $RMQ(1, j)$, $0 \le j \le n$. □

From Fact 3, it seems that Problem RMAX may be too general a tool to solve LCS and it seems to be worthwhile to look for a better solution exploiting the special query structure reported in Fact 3. Indeed, as we shall show that we can exploit this special structure in the query to avoid the $O(n)$ preprocessing step and hence the n^2 term from the running time reported in Theorem 2. However the price we pay is that the query time increases to $O(\log \log n)$. We present the idea as follows.

$$... \to (\alpha_i, x_i) \to (\alpha_j, x_j) \to (\alpha_k, x_k) \to ...$$

Fig. 1. Partial E_A with e_i, e_j, and e_k

Assume that we have an array $A[1..n]$ on which we want to apply the range maxima queries. We now use an elegant data structure (referred to as vEB data structure henceforth) invented by van Emde Boas [19] that allows us to maintain a sorted list of integers in the range $[1..n]$ in $O(\log \log n)$ time per insertion and deletion. In addition to that it can return $next(i)$ (successor element of i in the list) and $prev(i)$ (predecessor element of i in the list) in constant time. We maintain a vEB data structure E_A, where each element $e_i \in E_A, 1 \le i \le |E_A|$ is a 2-tuple $(Value, Pos)$. The order in E_A is maintained according to $e_i.Pos, 1 \le i \le |E_A|$. Now consider 3 entries $e_i, e_j, e_k \in E_A$ such that $e_j = next(e_i), e_k = next(e_j)$. Let $e_i = (\alpha_i, x_i), e_j = (\alpha_j, x_j)$ and $e_k = (\alpha_k, x_k)$ (Figure 1). The invariant we maintain is as follows:

$$RMQ(1..x) = \alpha_i, \ prev(e_i).Pos < x \le x_i$$

$$RMQ(1..x) = \alpha_j, \ x_i < x \le x_j$$

$$RMQ(1..x) = \alpha_k, \ x_j < x \le x_k$$

Assuming that we have the above data structure at our disposal, answering a query is easy as follows. Consider a query $RMQ(1..x')$. To answer this query, we just need to return the 'Value' of the element, which would be next in order, if a new element with $Pos = x'$ were inserted in E_A. So, we create an entry $e' = (null, x')$ and insert it in E_A and get the 'Value' of the $Next(e')$. Finally, we delete $e' = (null, x')$ from E_A. The only thing we need to ensure is that if there is already an entry e in E_A such that $e.Pos = x'$, e' must be placed before e in E_A. This is to ensure that $Next(e') = e$, as required. This can be easily achieved if we take 'Value' into account while preserving the order in E_A for equal values of 'Pos' and assume 'null' to be a lesser value than any other 'Value'. Note, however, that, by definition, in 'normal' state, there can be no two elements in E_A having same value for 'Pos'. The steps are formally presented in Algorithm 2.

Algorithm 2. Steps to answer the query $RMQ(1..x')$ on array A

1: Insert $e' = (null, x')$ in E_A
2: $Result = Next(e').Value$
3: Delete e' from E_A
4: **return** Result

The correctness of Algorithm 2 follows from the invariants maintained for E_A. Now it remains to show how we can maintain that invariant under update operations in the context of the Algorithm LCS-I. Recall that our goal is to get the answer of appropriate range maxima queries on the array H in Algorithm LCS-I and we operate in a row by row basis. For the sake of convenience, we use the following notation.

$$M_i = \{(i,j)|X[i] = Y[j], 1 \leq j \leq n\}$$

We start with reporting the following fact.

Fact 4. $T[i,j] = 1$, for all $(i,j) \in M_1$.   □

In cases, where $M_1 = \emptyset$ or a number of subsequent $M_i = \emptyset, i > 1$, we have the following fact.

Fact 5. $T[i,j] = 1$, for all $(i,j) \in M_i$ such that $M_k = \emptyset$, for all $1 \leq k < i$.   □

$$(0, j' - 1) \rightarrow (1, n) \rightarrow (\infty, \infty)$$

Fig. 2. Initial E_H

We initialize E_H with three elements, $e_s = (0, j' - 1)$, $e_e = (1, n)$ and $e_\infty = (\infty, \infty)$, where $(1, j') \in M_1$ and there exists no $j < j'$ such that $(1, j) \in M_1$ (Figure 2). Note that, if $M_1 = \emptyset$ then, for initialization, we have to use M_i instead of M_1 such that $M_k = \emptyset$, for all $1 \leq k < i$ (Fact 5). This initialization of E_H correctly maintains the invariants for the processing of the next row. Indeed, for the next row, we must have $RMQ(1..x) = 0$ if $x \leq j' - 1$ ($j' - 1$ is defined as above) and $RMQ(1..x) = 1$ otherwise. The last element, e_∞, is required to tackle the general cases and here we assume ∞ to be greater than any number. Now let us consider the case, where we process the subsequent rows. It is important to note that as we process a particular row i, for each $(i, j) \in M_i$, we need to update E_H; but this update is effective only for the next row, i.e. row $i + 1$. So, as we process row i we perform the update on a temporary copy and as soon as

$$\ldots \to (\alpha_i, x_i) \to (\alpha_j, x_j) \to (\alpha_k, x_k) \to \ldots$$

Fig. 3. Partial E_H^{i+1} with $e_i, e_j,$ and e_k

row i is completely processed we actually change the E_H to make it ready for row $i + 1$. In what follows, for the sake of convenience, we denote by E_H^i the 'state' of E_H which is used to process row i.

Now consider the case that we are in row i and processing the match $(i, x' + 1) \in M_i$. It is easy to see that we need the answer of the query $RMQ(1..x')$, which can be obtained easily applying Algorithm 2 on E_H^i. So, according to LCS-I, we would compute $T[i, x' + 1] = RMQ(1..x') + 1 = \alpha' (let)$. Now we need to consider the updating of E_H^i to get the E_H^{i+1} to be used when processing row $i + 1$. We initialize E_H^{i+1} with E_H^i and for each match $(i, j) \in M_i$ we continue to update E_H^{i+1} so that we get the 'correct' E_H^{i+1} as soon as the processing of row i is finished. The update process is as follows. In what follows, we assume (without the loss of generality) that we have $e_i, e_j, e_k \in E_H^{i+1}$ such that $e_j = next(e_i), e_k = next(e_j)$ (Figure 3). Let $e_i = (\alpha_i, x_i), e_j = (\alpha_j, x_j)$ and $e_k = (\alpha_k, x_k)$. Assume, without the loss of generality, that $x_i < x' + 1 \le x_j$. Since we have the value α' at position $x' + 1$, the query $RMQ(1..x' + 1)$ should return $\zeta \ge \alpha'$ when we are processing subsequent rows. So, first we check whether $RMQ(1..x' + 1) \ge \alpha'$ on the current E_H^{i+1}. It is clear that if the answer is positive, we don't need to do any update at all. Otherwise, we have, at position x_j or before it (off course after x_i) a higher value α'. So we insert a new element (α_j, x') to E_H^{i+1}, because up to x' we have no change in the RMQ answers. Now we have to change $e_j = (\alpha_j, x_j)$. But this change may be influenced by $e_k = (\alpha_k, x_k)$ as follows. We have two cases.

Case 1.a: $\alpha_k = \alpha'$. In this case, we just need to delete e_j because $e_k = (\alpha_k = \alpha', x_k)$ has already taken into account the updated value α' at position $x' + 1 \le x_k$ (Figure 4).

$$\ldots \to (\alpha_i, x_i) \to (\alpha_j, x') \to (\alpha_k = \alpha', x_k) \to \ldots$$

Fig. 4. Updated E_H for Case 1.a

Case 1.b: $\alpha_k > \alpha'$. In this case, $e_k.Value$ is greater than the updated value at position $x' + 1 \le x_j$. So it is clear that up to position x_j, we have α' as the highest value and hence we need to update e_j such that $e_j.value = \alpha'$ (Figure 5).

$$\ldots \to (\alpha_i, x_i) \to (\alpha_j, x') \to (\alpha', x_j) \to (\alpha_k, x_k) \to \ldots$$

Fig. 5. Updated E_H for Case 1.b

So far we have discussed the algorithm without analyzing the running time. Theorem 3 below reports the running time of the Algorithm LCS-II.

Theorem 3. *LCS-II solves Problem LCS in $O(\mathcal{R}\log\log n + n)$ time.*

Proof. It is clear that for each $(i,j) \in M$ we do the following 3 steps.
1: Perform appropriate range maxima query using Algorithm 2
2: Compute $T[i,j]$ from the result of Step 1
3: Update E_H based on $T[i,j]$

It is easy to see that Step 1, i.e., Algorithm 2 requires $O(\log\log n)$ time. Step 2 requires $O(1)$ time. In Step 3 we perform the update. Here, we first check, using Algorithm 2, whether any update is indeed required. This, again, requires $O(\log\log n)$ time. Finally, if update is required, then we need to perform constant number of insertion and/or deletion requiring, again, $O(\log\log n)$ time. So, for each $(i,j) \in M$ the computation effort spent is $O(\log\log n)$. Therefore, Algorithm LCS-II requires $O(\mathcal{R}\log\log n + n)$ time to compute LCS. □

4 Conclusion

In this paper, we have studied the classic and much studied LCS problem for two strings. Using some new ideas, we have presented an $O(\mathcal{R}\log\log n + n)$ time algorithm to solve the problem where $\mathcal{R}$ is the total number of ordered pairs of positions at which the two strings match. Although, $\mathcal{R} = O(n^2)$, there are large number of applications for which we have $R \sim n$. Typical of such applications include finding the longest ascending subsequence of a permutation of integers from 1 to n, finding a maximum cardinality linearly ordered subset of some finite collection of vectors in 2-space etc (for more details see [9] and references therein). So in these situations our algorithm would exhibit an almost linear $O(n\log\log n)$ behavior. The techniques we have used to develop our algorithm are new and, we believe, of independent interest. It would be interesting to see whether our techniques could be extended for variants of LCS problems (e.g. constrained LCS [18,2], rigid LCS [12]). Furthermore, in this paper, we have implicitly presented an algorithm for the range maxima query problem. Our algorithm allows restricted dynamic updates and considers a restricted sets of queries. It would be interesting to see whether we can lift the restrictions and/or improve the query and pre-processing time. Moreover, we believe that this algorithm could be used in many other problems requiring similar sort of restricted updates and queries.

References

1. V.L. Arlazarov, E.A. Dinic, M.A. Kronrod, and I.A. Faradzev. On economic construction of the transitive closure of a directed graph (english translation). *Soviet Math. Dokl.*, 11:1209–1210, 1975.
2. Abdullah N. Arslan and Ömer Eğecioğlu. Algorithms for the constrained longest common subsequence problems. *Int. J. Found. Comput. Sci.*, 16(6):1099–1109, 2005.
3. Michael A. Bender and Martin Farach-Colton. The lca problem revisited. In *LATIN*, pages 88–94, 2000.

4. Lasse Bergroth, Harri Hakonen, and Timo Raita. A survey of longest common subsequence algorithms. In *SPIRE*, pages 39–48, 2000.
5. Maxime Crochemore, Gad M. Landau, and Michal Ziv-Ukelson. A sub-quadratic sequence alignment algorithm for unrestricted cost matrices. In *Symposium of Discrete Algorithms (SODA)*, pages 679–688, 2002.
6. H. Gabow, J. Bentley, and R. Tarjan. Scaling and related techniques for geometry problems. In *STOC*, pages 135–143, 1984.
7. F. Hadlock. Minimum detour methods for string or sequence comparison. *Congressus Numerantium*, 61:263–274, 1988.
8. Daniel S. Hirschberg. Algorithms for the longest common subsequence problem. *J. ACM*, 24(4):664–675, 1977.
9. James W. Hunt and Thomas G. Szymanski. A fast algorithm for computing longest subsequences. *Commun. ACM*, 20(5):350–353, 1977.
10. Tao Jiang and Ming Li. On the approximation of shortest common supersequences and longest common subsequences. *SIAM Journal of Computing*, 24(5):1122–1139, 1995.
11. V.I. Levenshtein. Binary codes capable of correcting deletions, insertions, and reversals. *Problems in Information Transmission*, 1:8–17, 1965.
12. Bin Ma and Kaizhong Zhang. On the longest common rigid subsequence problem. In *CPM*, pages 11–20, 2005.
13. David Maier. The complexity of some problems on subsequences and supersequences. *Journal of the ACM*, 25(2):322–336, 1978.
14. William J. Masek and Mike Paterson. A faster algorithm computing string edit distances. *J. Comput. Syst. Sci.*, 20(1):18–31, 1980.
15. Eugene W. Myers. An o(nd) difference algorithm and its variations. *Algorithmica*, 1(2):251–266, 1986.
16. Narao Nakatsu, Yahiko Kambayashi, and Shuzo Yajima. A longest common subsequence algorithm suitable for similar text strings. *Acta Inf.*, 18:171–179, 1982.
17. M. Sohel Rahman and Costas S. Iliopoulos. Algorithms for computing variants of the longest common subsequence problem. In *ISAAC*, pages 399–408, 2006.
18. Yin-Te Tsai. The constrained longest common subsequence problem. *Inf. Process. Lett.*, 88(4):173–176, 2003.
19. P. van Emde Boas. Preserving order in a forest in less than logarithmic time and linear space. *Information Processing Letters*, 6:80–82, 1977.
20. Robert A. Wagner and Michael J. Fischer. The string-to-string correction problem. *J. ACM*, 21(1):168–173, 1974.

Scheduling a Flexible Batching Machine[*]

Baoqiang Fan[1,**], Jianzhong Gu[1], and Guochun Tang[2]

[1] Department of Mathematics and Information, Ludong University, Yantai 264025,
People's Republic of China
baoqiangfan@163.com
[2] Institute of Management Engineering, Shanghai Second Polytechnic University,
Shanghai 201209, People's Republic of China
gtang@sh163.net

Abstract. Minimizing total completion time $\sum C_j$ on normal batching machine is solvable in polynomial time for fixed $B(B > 1)$, while Minimizing total completion time $\sum C_j$ for arbitrary B and minimizing total weighted completion time $\sum W_j C_j$ are open problems. In this paper, we consider the problem of scheduling jobs on a flexible batching machine in order to minimizing the total completion time. We prove that the problem is strongly NP-hard. Then the problem with agreeable is NP-hard even if there have three fixed capacities all the time.

Keywords: Scheduling; Batching machine; Complexity.

1   Introduction

A batch processing machine is one that can handle up to B jobs simultaneously all the time. The jobs that are processed together form a batch, and all jobs contained in the same batch start and complete at the same time since the completion time of a job is equal to the completion time of the batch to which it belongs. The processing time of a batch is equal to the largest processing time of any job in the batch (denoted by $p - batch$) or the sum of processing times of all jobs in the batch (denoted by $s - batch$). The scheduling models for batching machine are motivated by burn-in operations in semiconductor manufacturing [11]. There are two variants: the unbounded model, where $B \geq n$ and the bounded model, where $B < n$ [16]. In this paper, we address bounded problems of scheduling a p-batch processing machine, i.e. we assume that $B < n$ and the processing time of a batch is equal to the largest processing time of any job in the batch.

A number of researchers have directed their attention toward batching problems. Santos and Magazine [14], and Tang [15] present integer programming formulations and several procedures to determine optimal batches of jobs for a single-stage production system. Ikura and Gimple [9] are the first researchers

[*] Supported by the National Natural Science Foundation of China (No. 10371071 and 70618001) and Shanghai Municipal Education Commission Address (No. 07zz178).
[**] Corresponding author.

to address the problem of scheduling batch processing machines from a deterministic scheduling perspective. J.Ahmadi, al et. [1] examine a class of problems defined by a two or three machine flowshop where one of the machines is a batch processing machine. More work on batching and scheduling includes Coffman, Nozari and Yannakakis [5], Julien and Magazine [10], Vickson, Magazine and Santos [17], and Potts and Kovalyov [12]. Webster and Baker [18], and P.Brucker, et al. [2] present overviews of algorithms and complexity results for scheduling batch processing machines.

In the bounded problems of scheduling a batching machine, the capacity B of the machine, the maximum number of jobs that the machine can process simultaneously, is fixed. But a burn-in oven in semiconductor manufacturing has different capacities for different sizes of wafer. So the capacity B of the machine is not constant. This kind machine is called a flexible batching machine. To be able to refer to the problems under study in a concise manner, the problem of minimizing total completed time on a single flexible batching machine is represented by $1|d-Batch|\sum C_j$, where d stands for different capacities of the flexible batching machine, and the bounded problems of scheduling a normal(not flexible)batching machine is denoted by $1|p-Batch|\sum C_j$ or $1|B|\sum C_j$, where B means a batching machine.

In this paper, we deal with the problem $1|d-Batch|\sum C_j$ described as follows. There are n independent jobs to be processed on a flexible batching machine which can handle up to B^i jobs simultaneously in time $[t_{i-1},t_i)$, where B^i is the capacity of the machine in $[t_{i-1},t_i)$ and $t_0 = 0, i = 1,2,\cdots$. For $j = 1,2,\cdots,n$, each job J_j requires processing during a given non-negative uninterrupted time p_j, and is available for processing from time zero onwards. All data are assumed to be deterministic. The objective is to determine a schedule π so that the total completion time $\sum C_j$ is minimized. When capacity B^i and t_i are agreeable(i.e. $t_i \le t_j$ implies $B^i \le B^j$), we represent the problem by $1|inc-d-Batch|\sum C_j$, where inc stands for increasing.

If all the capacities are identical, such as $B^i \equiv B(i = 1,2,\cdots)$, our problem $1|d-Batch|\sum C_j$ becomes a normal batching scheduling problem $1|B|\sum C_j$. Minimizing total completion time $\sum C_j$ is undoubtedly the most vexing bounded problem. Chandru et al. [3,4] present heuristics and a branch-and-bound algorithm as well as an $O(m^3 b^{m+1})$ time dynamic programming algorithm for the case of m different job processing times($m < n$). Hochbaum and Landy [8] present a faster algorithm that requires $O(m^2 3^m)$ time. P. Brucker al et. [2] prove that $1|B|\sum C_j$ is solvable in polynomial time for fixed B, where $B > 1$, by deriving an $O(n^{B(B-1)})$ time dynamic programming algorithm for its solution. C. K. Poon and W. Yu [13] present another algorithm that runs in $O(n^{6B})$ time for $1|B|\sum C_j$. B. Fan and G. Tang [6] are the first researchers to address the problem of scheduling a flexible batch machine. They proved that $1|d-Batch|C_{max}$ is Strongly NP-hard and its two agreeble cases are NP-hard. As to the unbounded model, P.Brucker al et. [2] given a characterization of a class of optimal schedules, which leads to a generic dynamic programming algorithm for minimizing any regular cost function $\sum_{j=1}^{n} f_j$ and an $O(n \log n)$ time algorithm for minimizing $\sum W_j C_j$.

Minimizing total completion time $\sum C_j$ on normal batching machine is solvable in polynomial time for fixed $B(B > 1)$, while Minimizing total completion time $\sum C_j$ for arbitrary B and minimizing total weighted completion time $\sum W_j C_j$ are open problems. In this paper, we prove that the flexible batching problem $1|d - Batch| \sum C_j$ is strongly NP-hard. Then we show that $1|inc - d - Batch|C_{max}$ is NP-hard even if there have three fixed capacities all the time in Section 2. Section 3 is a brief conclusion.

2 NP-Hardness of Total Completion Time Problem

In this section, firstly, we show that the problem of minimizing total completion time on a flexible batching machine is strongly NP-hard. This is done by reducing the strongly NP-hard 3-Partition [7] to the decision version of our problem $1|d - Batch| \sum C_j$.

3-Partition. Given positive integers t, A and a set of integers $S = \{a_1, \cdots, a_{3t}\}$ with $\sum_{j=1}^{3t} a_j = tA$ and $A/4 < a_j < A/2$ for $1 \leq j \leq 3t$, does there exit a partition $\langle S_1, S_2, \cdots, S_t \rangle$ of S into 3-element sets such that

$$\sum_{a_j \in S_i} a_j = A,$$

for each j?

Given an instance $\mathscr{P}$ of 3-Partition, we first describe the details of the corresponding flexible batching problem $\mathscr{L}$. There are basically three classes of jobs in $\mathscr{L}$. The first class, $J^1 = \{J_{ij}^1 | 1 \leq i \leq t, j = 1, 2\}$, where job lengths are specified as follows:

$$p_{ij}^1 = \frac{1}{4}A, i = 1, 2, \cdots, t - 1, j = 1, 2.$$

The second class, $J^2 = \{J_i^2 | 1 \leq i \leq 3t\}$, with job lengths specified as follows:

$$p_i^2 = a_i, i = 1, 2, \cdots, 3t.$$

The finally class, $J^3 = \{J_i^3 | 1 \leq i \leq \bar{n}\}$, with job lengths specified as follows:

$$p_i^3 = t^2 A - \frac{5}{4} tA, i = 1, 2, 3.$$

We define the machine can handle up to $B^1 = 2$ jobs simultaneously in time $[t_{2i-1}, t_{2i}), i = 1, 2, \cdots, t-1, B^2 = \bar{n}$ in time $[t_{2t-1}, t_{2t})$, and $B^3 = 1$ in other time, where $t_{2(i-1)} = \frac{5}{4}(i-1)A, t_{2i-1} = t_{2(i-1)} + A, i = 1, 2, \cdots, t,$ and $t_{2t} = t^2 A - \frac{1}{4}A$. The bound is given by $\delta = \frac{39}{8} t^2 A - \frac{15}{8} tA + \frac{3}{2}A$. We are asked to answer whether there exists a schedule π for instance $\mathscr{L}$ such that its total completion time less than or equal to δ.

Clearly, the construction of $\mathscr{L}$ takes a polynomial time under the binary coding. In the following, we will show that $\mathscr{L}$ has a schedule π such that $\sum C_j \leq \delta$ if and only if the 3-Partition instance $\mathscr{P}$ has a partition $\langle S_1, S_2, \cdots, S_t \rangle$ exists which has the desired form. We first introduce some useful properties associated with feasible schedules and optimal schedules of $\mathscr{L}$.

Lemma 1. *For any feasible schedule π of jobs $\{J_{ij}^1 | 1 \leq i \leq t, j = 1, 2\}$ and $\{J_i^2 | 1 \leq i \leq 3t\}$ in $\mathscr{L}$, the total completed time of such schedule satisfies*

$$\sum C_j(\pi) > \frac{5}{4}t(t-1)A + \frac{5}{3}A.$$

Proof. The result is established by scheduling the jobs $J^1 \cup J^2$ and $J^1 \cup \hat{J}^2$ respectively, where the job lengths $\hat{p}_i^2 = \frac{1}{4}A, i = 1, 2, \cdots, 3t$ in $\hat{J}^2$ and others are same as that in $\mathscr{L}$. $\qquad\qquad\square$

Suppose π is an optimal schedule of $\mathscr{L}$. There have a conclusion about the structure of π as following.

Lemma 2. *If π satisfy $\sum C_j(\pi) \leq \delta$, then such schedule has the following two properties:*

1. *The third class jobs are processed as a batch with the start processing time t_{2t-1} in schedule π.*
2. *The jobs of $J^1 \cup J^2$ are completed no later than the time t_{2t-1} in schedule π.*

Proof. Notice the construction of $\mathscr{L}$ as the Fig.1, since $p_i^3 = t^2 A - \frac{5}{4}tA > t_{2t-1}, i = 1, 2, 3$ and $\sum C_j(\pi) \leq \delta$, we can easily show that the first property holds by a standard interchange argument.

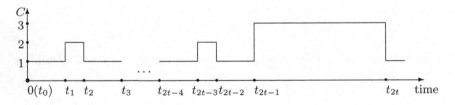

Fig. 1. The construction about capacities of $\mathscr{L}$, where C denotes the capacity of machine

From the Lemma 1, if $\sum C_j(\pi) \leq \delta$, the jobs of $J^1 \cup J^2$ must be processed before the jobs of J^3 in schedule π. Summarizing previous discussions, we have the second property. $\qquad\qquad\square$

Based on the Lemma 1 and Lemma 2, we now prove the following theorem.

Theorem 1. *The problem $1|d - Batch| \sum C_j$ is strongly NP-hard.*

Proof. Suppose a partition $\langle S_1, S_2, \cdots, S_t \rangle$ exists which has the desired form. That is, each set S_i consists of three elements a_{i1}, a_{i2} and a_{i3} such that for all $1 \leq i \leq t, \sum_{j=1}^{3} a_{ij} = A$. Then the following schedule π has total completion time $\sum C_j \leq \delta = \frac{39}{8} t^2 A - \frac{15}{8} tA + \frac{3}{2} A$. About the first class jobs in such schedule, the jobs $\{J_{i1}^1, J_{i2}^1\}$ are processed as a batch with the start processing time $S(\{J_{i1}, J_{i2}\}) = t_{2i-1}, i = 1, 2, \cdots, t-1$, and the third class jobs J^3 processed as a batch in the time interval $[t_{2t-1}, t_{2t})$.

See Fig.2. Note that this basic framework leaves a series of t *"time slots"* open, each of length exactly A and in which the machine can handle up to one job simultaneously. These are precisely tailored so that we can fit in the second class jobs as follows. For each $i = 1, 2, \cdots, t.$

$$S(J_{i1}^2) = t_{2(i-1)}$$
$$S(J_{i2}^2) = t_{2(i-1)} + a_{i1}$$
$$S(J_{i3}^2) = t_{2(i-1)} + a_{i1} + a_{i2}$$

Since $\sum_{j=1}^{3} a_{ij} = A, i = 1, 2, \cdots, t$, this yields a valid schedule with $\sum C_j(\pi) \leq \delta$.

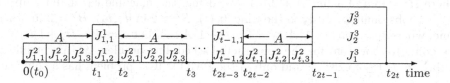

Fig. 2. Where the jobs in the same column are processed as a batch

Conversely, suppose a schedule π with $\sum C_j(\pi) \leq \delta$ does exist. From Lemma 2, the third class jobs must be processed as a batch with the start processing time t_{2t-1} and $J^1 \cup J^2$ are completed before the time t_{2t-1} in schedule π. Since

$$\sum_{j \in J^1 \cup J^2} p_j = \frac{1}{4} A(2t - 2) + tA = \frac{3}{2} tA - \frac{1}{4} A,$$

which is equal to the total length that machine can handle up before t_{2t-1}, it is easy to see that the first class jobs must be scheduled the same way as they are in Fig.2. Thus there are again t slots of length A into which the second class jobs must be placed.

Since the total length of the second class jobs is $\sum_{i=1}^{3t} a_i = tA$, every one of these t slots must be filled completely, and hence must contain a set of the second class jobs whose total length is exactly A. Now since every $a_i > A/4$, no such set can contain more than three jobs. Similarly, since every $a_i < A/2$, no such set can contain less than three jobs. Thus each set contains exactly three jobs of the second class. Hence, by setting $S_i = \{a_i | t_{2(i-1)} < S(p_i^2) \leq t_{2i-1}\}$, $i = 1, 2, \cdots, 3t$, we obtain our desired partition. $\qquad\Box$

We next establish the NP-hardness of the problem $1|inc - d - Batch| \sum C_j$ by a reduction from the NP-complete PARTITION problem.

PARTITION. Given m positive integers $a_1, a_2, \cdots, a_m$, with $\sum_{j=1}^{m} a_j = 2A$, do there exit two disjoint subsets $S_1, S_2 \in I = \{1, 2, \cdots, m\}$ such that

$$\sum_{j \in S_i} a_j = A,$$

for $i = 1, 2$?

Without loss generality, we assume that $m > 2$ throughout the section. To any instance of the PARTITION problem, we construct an instance $\mathscr{L}$ of $1|inc - d - Batch| \sum C_j$ as follows:

There are two classes of jobs in $\mathscr{I}$. The first class, $\{J_{ij}^1 | 1 \leq i \leq m, j = 1, 2, 3\}$, which are classified into m types. For each $i(1 \leq i \leq m)$, define three jobs of type i: J_{i1}, J_{i2}, and J_{i3}. Their processing times are given by

$$p_{i1}^1 = 4iA + a_i, \quad p_{i2} = 4iA - a_i, \quad p_{i3} = 4iA.$$

The second class, $\{J_i^2 | i = 1, 2, \cdots, \mathcal{U}\}$, which processing times are given by

$$p_i^2 = \mathcal{U},$$

where $\mathcal{U} = 16m^3 A + 6m^2 A + 2mA$. We define the machine can handle up to $B^1 = 1$ jobs simultaneously in the time $[0, t_1)$, $B^2 = 2$ in $[t_1, t_2)$, $B^3 = 3$ in other time, where $t_1 = 2m(m+1)A, t_2 = 4m(m+1)A + A$. Let $\delta = \mathcal{U}^2 + (t_2 + 1)\mathcal{U}$. We are going to show that for the constructed scheduling problem $\mathscr{L}$, a schedule π with $\sum C_j(\pi) \leq \delta$ exists if and only if the PARTITION problem has a solution.

Lemma 3. *Suppose π is an optimal schedule of $\mathscr{I}$. If $\sum C_j(\pi) \leq \delta$, then second class jobs are processed as a batch with the start processing time t_2 in π.*

Proof. The result is established by a standard job interchange argument. $\square$

As a result of Lemma 3, the first class jobs must be completed before the time t_2 in one optimal schedule π with $\sum C_j(\pi) \leq \delta$. Otherwise, there exist at least a job which is processed after the third class jobs. Without loss generality, we assume J_i^1 is such job, which start processing time is not less than $t_2 + \mathcal{U}$. Then the total completion time of such optimal schedule π has a lower bound as following.

$$\sum C_j(\pi) > \mathcal{U}^2 + \mathcal{U} + t_2. \tag{1}$$

The inequality (1) is contrary to $\sum C_j(\pi) \leq \delta$. Thus, we have the following conclusion.

Lemma 4. *Suppose π is an optimal schedule of $\mathscr{I}$. If $\sum C_j(\pi) \leq \delta$, then first class jobs are processed before the time t_2.*

Suppose π is a feasible schedule of $1|inc - d - Batch| \sum C_j$ and the capacities B^i are indexed according to the order $0 < B^i < B^{i+1}$, $i = 1, 2, \cdots$. If a batch B_j of π meets $B^i < | B_j | \leq B^{i+1}$, we denote the B_j as B^{i+1} type batch. In the remainder of the proof, we show that PARTITION has a solution if and only if there exists a schedule for the corresponding instance $\mathscr{I}$, and the total completed time of such schedule can not exceed δ.

Theorem 2. *The problem* $1|inc - d - Batch| \sum C_j$ *is NP-hard, even if there have only three capacities all the time.*

Proof. First, suppose that the instance of PARTITION has a solution. Without loss of generality, we assume that $S_1 = \{1, 2, \cdots, k\}$, and $S_2 = \{k + 1, k + 2, \cdots, m\}$. Now, construct schedule π as the following Fig.3. It is easy to check that $\sum C_j(\pi) \leq \delta$.

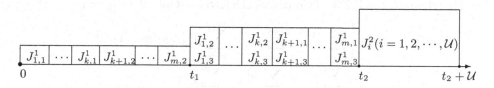

Fig. 3. Illustration of the scheduling π, in which the jobs in the same column are processed as a batch

Conversely, suppose that there exists a schedule π with $\sum C_j(\pi) \leq \delta$. From the Lemma 4, the first class jobs must be completed before t_2. Let σ denotes the partial schedule of the first class jobs in π, then $C_{max}(\sigma) \leq t_2$.

By a standard interchange argument of jobs and batches, we can get a new partial schedule σ' from σ, all the B^1 type batches are processed before the B^2 type batches in such new partial schedule and $C_{max}(\sigma') \leq C_{max}(\sigma) \leq t_2$. About σ', suppose the sum of the time, in which the machine is idle, is $a(a \geq 0)$ before time t_1. Let $|B^1|$ is the sum of the processing time of B^1 type batches in σ'. Because there are only B^1 type batches processed before the time t_1, $a + |B^1| \geq t_1$. Let $d(F)$ be defined for each batch $F \in B^2$ in σ' as follows. $d(F)$ is equal to the difference of the processing times of its two jobs. Then the processing time of batch F is

$$\frac{1}{2} \left(\sum \{p_{ij}^1 | J_{ij}^1 \in F\} + d(F) \right),$$

where $d(F)$ acts as the wasted time during the processing of batch F. From the above expression of the F, we obtain that

$$C_{max}(\sigma') = |B^1| + \frac{1}{2} \left(\sum_{F \in B^2} \sum \{p_{ij}^1 | J_{ij}^1 \in F\} + \sum_{F \in B^2} d(F) \right) + a$$

$$= \frac{1}{2} \sum_{j=1}^{3} \sum_{i=1}^{m} p_{ij}^1 + \frac{1}{2} \left(|B^1| + \sum_{F \in B^2} d(F) \right) + a$$

$$= 3m(m + 1)A + \frac{1}{2} \left(|B^1| + \sum_{F \in B^2} d(F) \right) + a.$$

Since $C_{max}(\sigma') \leq t_2$, it follows that

$$|B^1| + d(F) + 2a \leq 2m(m+1)A + 2A.$$

Due to $a + |B^1| \geq t_1$, $d(F) + a \leq 2A$, hence $d(F) \leq 2A$.

If there exists a batch F in σ' contains two jobs J_i^1, J_j^1 of distinct types, from the constructor of the jobs, $d(F) \geq 4A - a_i - a_j > 2A$. Thus every batch, which includes two jobs in σ', contains two jobs of the same type in the first class.

Let $S_j \subseteq I$ be the subset of $\{J_{1,j}^1, J_{2,j}^1, \cdots, J_{m,j}^1\}$, in which every job is processed as a batch itself, $j = 1, 2, 3$. obviously, $S_1 \cup S_2 \cup S_3 = I$ and $S_1 \cap S_2 \cap S_3 = \emptyset$. Then

$$C_{max}(\sigma') = \max\left\{ \sum_{i \in S_1} p_{i,1}^1 + \sum_{i \in S_2} p_{i,2}^2 + \sum_{i \in S_3} p_{i,3}^3, 2m(m+1)A \right\}$$

$$+ \sum_{i \in S_1} p_{i,3}^1 + \sum_{i \in S_2} p_{i,1}^1 + \sum_{i \in S_3} p_{i,1}^1$$

$$= 4m(m+1)A + \max\left\{ \sum_{i \in S_1} a_i - \sum_{i \in S_2} a_i, 0 \right\} + \sum_{i \in S_2} a_i + \sum_{i \in S_3} a_i$$

$$= 4m(m+1)A + \max\left\{ 2A - \sum_{i \in S_2} a_i, \sum_{i \in S_2} a_i + \sum_{i \in S_3} a_i \right\}$$

$$\leq 4m(m+1)A + A.$$

Accordingly, we can obtain two inequations as the following.

$$2A - \sum_{i \in S_2} a_i \leq A, \tag{2}$$

$$\sum_{i \in S_2} a_i + \sum_{i \in S_3} a_i \leq A. \tag{3}$$

Thus, due to the above inequalities (1) and (2), we have $\sum_{i \in S_2} a_i \geq A$ and $\sum_{i \in S_2} a_i \leq A$. So it follows that $\sum_{i \in S_2} a_i = A$.

Then the PARTITION instance has a solution $X = S_1$.   □

3   Concluding Remarks

In this paper, we address the problem of scheduling n jobs on a flexible batching professor to minimize the total completion time. We will go to searching ecient algorithms for this problem and researching the problem with other objective (i.e. $T_{max}, \sum U_j$). Another extension to the model involves scheduling jobs on m identical parallel machines.

References

1. J. h. Ahmadi, R. h. Almadi, S. Dasu and C. S. Tang. Batching and Jobs on Batch and Discrete Processors. Operations Research, **40** (1992) 750-763.
2. P. Brucker, S. Gladky, H. Hoogeveen, M. Kovalyov, C. Potts, T. Tantenhahn and S. van de Velde. Scheduling a batching machine. Journal of Scheduling, **1** (1998) 1 31-54.

3. V. Chandru, C.-Y. Lee and R. Uzsoy. Minimizing total completion time on batch processing machines. International Journal of Production Research, **31** (1993) 2097-2122.
4. V. Chandru, C.-Y. Lee and R. Uzsoy. Minimizing total completion time on batch processing machines. Operations Research Letter, **13** (1993) 61-65.
5. E. Coffman Jr, A. Nozari and M. Yannakakis. Optimal Scheduling of Products with Two Subassemblies of a Single Machine. Operations Research, **37** (1989) 426-436.
6. B. Fan and G. Tang. Scheduling Jobs on a Flexible Batching Machine: Model, Complexity and Algorithms. TAMC 2006, Lecture Notes in Computer Science, **3959** (2006) 118-127.
7. M.R. Garey and D.S. Johnson. Computers and intactability: A guide to the theory of NP-completeness. Freeman, New York, 1979.
8. D.S. Hochbaum and D. Landy. Scheduling semiconductor burn-in operations to minimize total flowtime. Operations Research, **45** (1997) 874-885.
9. Y. Ikura and M. Gimple. Efficient scheduling algorithms for a single batch processing machine. Operations Research Letter, **5** (1986) 61-65.
10. F. Julien and M. Magazine. Batching and Scheduling Multi-Job Orders. CORS/TIMS/ORSA Vancouver Bulletin, 1989.
11. C.-Y Lee, Reha Uzsoy and Louis A. Martin-Vega. Efficient Algorithms for Scheduling Semiconductor Burn-in Operations. Operations Research, **140** (1992) 40 764-775.
12. C.N. Potts and M.Y. Kovalyov. Scheduling with batching: a review. European Journal of Operational Research, **120** (2000) 228-249.
13. C.K. Poon and W. Yu. On minimizing total completion time in batch machine scheduling. International Journal of Foundations of Computer Science, **15(4)** (2004) 593-604.
14. C. Santos and M. Magazine. Batching in Single Operation Manufacturing Syatem. Operations Research Letter, **4** (1985) 99-103.
15. C.S. Tang. Scheduling Batches on Parallel Machines with Major and Mi- nor Setups, European Journal of Operational Research, **46** (1990) 28-37.
16. G. Tang, F. Zhang, S. Luo and L. Liu. Theory of Modern Scheduling. Shanghai Popular Science Press, 2003.
17. R.G. Vickson, M.J. Magazine and C.A. Santos. Batching and Scheduling of Components at a Single Facility. Working Paper 185-MS-1989, University of Waterloo, Canada.
18. S. Webster and K.R. Baker. Scheduling Groups of Jobs on a Single Machine. Operations Research, **43** (1995) 692-703.

Global Search Method for Parallel Machine Scheduling

Hyun Joon Shin

Department of Industrial Information and Systems Engineering
Sangmyung University
Cheonan-si, Choongnam, 330-720, Korea
hjshin @smu.ac.kr

Abstract. This paper presents a guided multi-restart search (GMRS) algorithm for scheduling parallel machines in terms of global optimum. GMRS consists of a strategic guided local search phase and a phase that generates a beneficial restart point using the information acquired during the local search. The experimental results show that the proposed algorithm considerably improves the solution within a reasonable time.

1 Introduction

This paper considers the problem of scheduling jobs on parallel machines where the release times, due dates, and sequence-dependent setup times of jobs are taken into account. The objective is to minimize the maximum lateness of the jobs (L_{max}), which can give the customers the information on the upper limit of their due date lateness. To state the scheduling problem precisely, suppose that there are N jobs to be scheduled in front of M machines. When a total schedule is expressed as a set Π, the objective function of the scheduling problem is defined as

$$Minimize \quad L_{max}(\Pi) = Max\{L_j \mid j = 1,2,...,N\} \tag{1}$$

where $L_j = C_j - d_j$, and C_j, d_j are the completion time and due date of job j ($j = 1,2,...,N$), respectively [1]. This scheduling problem is known as a strongly NP-hard problem [2].

In order to generate an efficient schedule within a very reasonable time, this paper presents a scheduling algorithm called guided multi-restart search (GMRS), which is characterized by global optimization. GMRS does not provide only means of escaping from local optima, but also methods of breaking fresh solution space.

In general, the performance of global search is largely affected by two factors. One is the intensification strategy - efficiency of the scheme to generate neighborhood schedules from the current schedule. The other is the diversification strategy - efficiency of the scheme to explore unvisited regions throughout the solution space. In this respect, GMRS consists of two main phases: The first phase is an efficient local search heuristic that may employ tabu search, simulated annealing, or genetic algorithm. This paper selects tabu search as a local search heuristic. The second phase on

M.-Y. Kao and X.-Y. Li (Eds.): AAIM 2007, LNCS 4508, pp. 100–107, 2007.

which this study focuses is related to the diversification strategies that prevent searching process from cycling, i.e., endlessly and exclusively revisiting the same set of solutions and guide the search to unexplored regions of the solution space.

In case of tabu search, for example, diversification can be incorporated to some extent into long-term memory, which stores frequency information on the moves throughout the search. This measure, together with residence or transition frequency-based memory structure, is used to avoid moves with high frequency and stimulate moves with low frequency [3,4]. In addition, frequency information within long-term memory may be used in multi-restarting mechanisms. However, due to the frequency-based memory structure that stores only the information on the particular jobs related to a move regardless of whole solution status at the moment, those existing diversification strategies are not strong enough to evade cycling during entire search process [3].

This study proposes a new diversification strategy that employs both a cycling protection technique and a guided restart technique. The cycling protection technique adopts a distance metric for measuring the dissimilarity between schedules and prohibits moves from coming into the protection area set up by given dissimilarity, which is distance. In addition, the guided restart technique leads restart point to the attractive solution space based on previous trajectories and uses genetic operators (mutation, crossover) as a restart point generation scheme.

Schoen [5] proposed global optimization methods consists of global phase and local phase and showed how stochastic techniques coupled with deterministic local search methods can be successfully applied to solve moderately sized multimodal optimization problems. Merkle and Middendorf [6] presented an ant colony optimization (ACO) algorithm for the single machine total weighted tardiness problem and compare it to an existing ACO algorithm. They showed that the ants are guided on their way through the decision space by global pheromone information instead of using only local pheromone information. Ding et al. [7] tackled an optimal process sequence problem with a global optimization strategy based on multi-objective fitness: minimum manufacturing cost, shortest manufacturing time and best satisfaction of manufacturing sequence rules. The hybrid approach proposed by [8] incorporated a genetic algorithm, neural network and analytical hierarchical process, however, had a lack of adaptability. Yang et al. [8] examined evolutionary computations for continuous global optimization. They presented an evolutionary programming algorithm combined with macro mutation, local linear bisection search and crossover operators for global search. Simulated annealing was adopted to prevent premature convergence.

To show the effectiveness of GMRS, the performance of GMRS is compared with that of RHP (rolling horizon procedure) proposed by Ovacik and Uzsoy [2] for the exactly same problem as this study using the benchmarking data given by Uzsoy [9]. The experimental results show that the proposed algorithm considerably improves the solution with reasonable CPU times.

The remainder of this paper is organized as follows. Section 2 presents a description of the proposed algorithm with empirical results shown in Section 3. Section 4 contains a conclusion.

2 Guided Multi-Restart Search (GMRS)

This section introduces the working principle of the GMRS. Virtually, GMRS is composed of a strategic guided local search phase and a phase that generates a beneficial restart point in terms of global optimum using useful information acquired during the local search.

Let protection area (PA) denote the memory stack for forbidding jumping to regions from which the searching may result in the previously visited local optima, and $D(\Pi,\Pi^*)$ denote the distance between the two solutions Π and Π^*, then the algorithm for GMRS can be summarized as:

Step 1: *(Initialization)* Choose an initial solution Π. *current iteration number* = 0. *current optimal* = $+\infty$. $PA = \varnothing$ (Note: this means empty queue).

Step 2: *(Guided Local Search)* Start local search from Π. Let Π^* be the local optimum, however, Π^* must not be within the protection area. Insert $<\Pi^*,d>$ in the protection area memory stack PA, where $d = D(\Pi,\Pi^*)\times\alpha$, where α is a distance factor, $\alpha > 0$, and usually set by experimental analysis. If *current optimal* > Π^* then *current optimal* = Π^*. *current iteration number* = *current iteration number* + 1. If *current iteration number* $\geq$ *stopping criterion*, then stop.

Step 3: *(Restart Point Selection)* From Π^*, generate a new starting schedule Π' such that Π' is not in the regions specified in PA using random restart method or gene generation scheme. Let $\Pi = \Pi'$ and go to Step 2.

In Step 1, a start point (initial schedule), Π is chosen by earliest due date (EDD) rule and the protection area memory stack, PA is initialized. Step 2 finds local optimum Π^* through local search and sets up another new protection area where is distanced d from Π^* by storing PA with the attributes of the local optimum Π^* and the distance d, that is $<\Pi^*,d>$. This means that every point whose distance from each local optimum Π^* in PA is within d must not be the target for local search. Therefore, the point to keep in mind here is that the local search should guide itself not to enter PA during entire local search process. This study employs tabu search as a local search heuristic designated to use insert move operator and tabu list of size 7 according to the suggestion by Laguna et al. [10]. Step 3 generates a restarting point that is not located in PA.

There is a multi-start algorithm, consisting of repeatedly starting local searches from randomly chosen starting point, which is one of global search algorithms. Possibly the algorithm, if correctly tuned, will lead to global optimum. However, its computational performance seems to be quite low [11]. It is a trade-off between random restart and guided restart using previous search history that should be considered here. In order to reflect the trade-off on the restarting point selection mechanism, this study adopts a gene generation scheme, which makes use of specially devised forms of mutation and crossover to assure that restarting point (offspring) will receive information shared by good previous (parent) solutions. This gene generation scheme not only incorporates a type of intensification based on consistency and provides an interesting area for investigation, but also guarantee a global view of search. A detailed description of the gene generation scheme for parallel machines scheduling is given in the literatures [12, 13].

2.1 Protection Area (PA)

In order to prevent searching process from endlessly executing the same sequence of moves efficiently, we need a criterion to decide whether or not every solution obtained during a cycle of local search is out of *PA* and can be selected a local optimum. In this respect, this paper presents dissimilarity measure that is an extension of Tanimoto distance [14] between two schedules.

Given two schedules Π and Π', let n_{Π_m} denote the number of adjacent pairs in the m th machine of the schedule Π and let $n_{\Pi_m,\Pi_m'}$ denote the number of shared adjacent pairs in the m th machine of the schedule Π and the schedule Π'. The distance between Π and Π', $D(\Pi,\Pi')$ is defined as:

$$D(\Pi,\Pi') = \sum_{m=1}^{M}(1-\frac{n_{\Pi_m,\Pi_m'}}{n_{\Pi_m}+n_{\Pi_m'}-n_{\Pi_m,\Pi_m'}}) = \sum_{m=1}^{M}(\frac{n_{\Pi_m}+n_{\Pi_m'}-2n_{\Pi_m,\Pi_m'}}{n_{\Pi_m}+n_{\Pi_m'}-n_{\Pi_m,\Pi_m'}}) \qquad (2)$$

For example, lets consider two cases of single machine schedule ($M = 1$).

Case 1) $\Pi = \{1, 2, 3, 4\}$ and $\Pi' = \{1, 3, 2, 4\}$

Case 2) $\Pi = \{1, 2, 3, 4\}$ and $\Pi' = \{1, 2, 4, 3\}$

Intuitively, the similarity between two schedules in case 2 is greater than that between two schedules in case 1. We can prove this using the equation (2). For case 1, $n_{\Pi_1} = 3$ since there are 3 adjacent pairs (1, 2), (2, 3), and (3, 4) in Π . $n_{\Pi_1'} = 3$ since there are 3 adjacent pairs (1, 3), (3, 2), and (2, 4) in Π'. $n_{\Pi_1,\Pi_1'} = 1$ because there is 1 shared adjacent pair (2, 3) between Π and Π'. The distance $D(\Pi,\Pi') = (3+3-2)/(3+3-1) = 0.8$. In the same manner as case 1, for case 2, $n_{\Pi_1} = 3$, $n_{\Pi_1'} = 3$, and $n_{\Pi_1,\Pi_1'} = 2$ because there are 2 shared adjacent pairs (1, 2) and (3, 4) between Π and Π'. The distance $D(\Pi,\Pi') = (3+3-4)/(3+3-2) = 0.5$. Therefore, the similarity in case 2 is greater than that in case 1 since the nearer distance, the more similar.

3 Computational Results

The RHP proposed by Ovacik and Uzsoy [2] mentioned in Section 1 is used as a comparative algorithm and the scheduling data for this experiment are the same data that Ovacik and Uzsoy created [9]. The benchmarking data are composed of 1800 test problems, each of which is characterized as different R value, the number of jobs, and the number of machines. Here, R is the release time range parameter that decides the release times of the jobs. Of the test problems, those with 150 jobs and 4 machines were used for this experiment. The algorithms proposed in this paper including RHP are all coded using the C++ language and run on PentiumIV 2.54 GHz machine.

Table 1. Experimental design

Design parameter	Values
Distance factor (α)	0.2, 0.6, 1.0, 1.4
Restart method	RS1, RS2, RS3
Restart number	20, 40, 60, 80, 100

As a yardstick for estimating the quality of the solutions, this study defines a comparison value as follows:

$$Comparison \ value = \frac{L_{max} \ obtained \ by \ GMRS}{L_{max} \ obtained \ by \ RHP} \tag{3}$$

Experimental design is given as Table 1, where distance factor (α) has 5 different values, restart method 3 ways, and restart number 5 values, respectively. Restart method RS1 means random restart, RS2 is a method that generates a new start position with 1% mutation (random change) of current local optimum, and RS3 decides a new start position with crossover between recent and current local optimum. RS2 and RS3 could be regarded as a kind of guided method.

The performance of GMRS coupled with the three restart methods was evaluated and shown in Table 2, where the restart number is 100. In this table, GMRS-restart method (e.g., GMRS-RS1) refers to the GMRS that employs the restart method to generate a start position. In addition, $Avg(R)$ means the average value of comparison values when R is given. Table 2 informs that the solution quality of GMRS-RS2 and GMRS-RS3 is much better than that of GMRS-RS1 and Table 3 shows that GMRS methods considerably improve solution compared with RHP heuristics within reasonable computation time.

Table 2. Average comparison values for performance comparison with RS1, RS2, and RS3 (restart number = 100)

α	GMRS-RS1				GMRS-RS2				GMRS-RS3			
	0.2	0.6	1.0	1.4	0.2	0.6	1.0	1.4	0.2	0.6	1.0	1.4
Avg(0.2)	0.892	0.914	0.877	0.892	0.779	0.738	0.678	0.684	0.758	0.725	0.701	0.682
Avg(0.6)	0.885	0.887	0.995	0.886	0.776	0.730	0.698	0.683	0.752	0.720	0.718	0.676
Avg(0.10)	0.896	0.886	0.881	0.886	0.784	0.737	0.690	0.687	0.762	0.721	0.688	0.677
Avg(0.14)	0.888	0.905	0.925	1.000	0.776	0.740	0.692	0.703	0.754	0.727	0.691	0.693
Avg(0.18)	0.894	0.876	0.896	0.997	0.781	0.720	0.692	0.693	0.760	0.712	0.696	0.688
Avg(*)	0.891	0.894	0.915	0.932	0.779	0.733	0.690	0.690	0.757	0.721	0.699	0.683

Table 3. Improvement rate and average CPU times (seconds/problem)

	Improvement rate	Average CPU times
RHP	-	59.62
GMRS-RS1	10.15 %	241.72
GMRS-RS2	38.30 %	237.91
GMRS-RS3	38.85 %	242.52

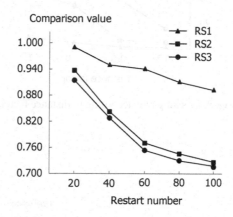

Fig. 1. Effect of restart number on the solution quality

From Tables 2 and 3, we can observe two facts. One is that the restart method plays an important role in improving the solution. This fact gives us useful information that the guided restart methods (RS2 and RS3) generate time-effective solutions. The other is that solution quality depends on the distance factor. Therefore, it is clear that the distance factor is an important decision parameter to be found.

The relationship between the restart number and the quality of the solution is shown in Fig. 1, where each comparison value corresponds to the values in Table 2. The results show that, particularly in RS2 and RS3, the comparison values are enhanced significantly up to a certain value from which the improvement is hardly noticeable. The blocking rate of GMRS with different restart methods, as shown in Fig. 2, increases in proportion along with the values of the distance factor. From this figure and Table 2, we can see that the blocking rate is not related to the improvement rate directly. As a consequence, GMRS-RS2/RS3 has been proved, through the experiment, to be an effective combination in terms of the quality of solution.

Finally, as illustrated Fig. 3, this experiment attempted to trace all moves during whole search, analyze the characteristics of the solution space, and inspect the relative location of initial, first restarting, and best solutions with respect to distance. The distance rate of x axis of Fig.3 means a percentage ratio of distance between *a solution* and the best solution to distance between the best solution and farthest solution from the best solution. From Fig. 3, we can see the problem space tackled in this paper has very rugged surface since there are a lot of good solutions in every distance

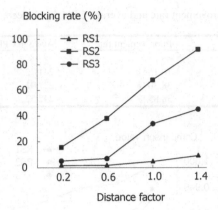

Fig. 2. Blocking rate according to distance factor

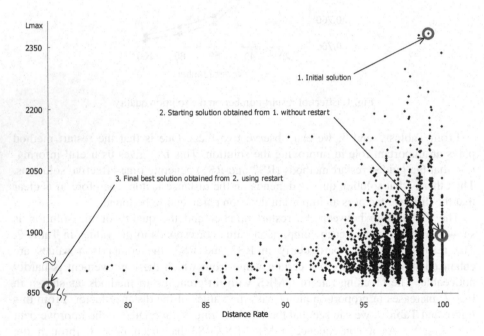

Fig. 3. Global search trajectory and relationship between distance rate and Lmax

rate, which makes the problem difficult to be optimally solved. We can also observe GMRS method went through very long distance to get the best solution via 1. the initial solution and 2. the restarting solution, which means the protection area (PA) incorporated in GMRS method plays a key role in exploring the whole space without trapped in local optima.

4 Conclusions

This paper proposed a guided multi-restart search (GMRS) algorithm, which is characterized by global optimization. GMRS is composed of a strategic guided local search phase and a phase that generates a beneficial restart point in terms of global optimum, using the information acquired during the local search. To show the effectiveness of the proposed algorithm, this study conducted an experiment with benchmarking data. The experimental results showed that the proposed algorithm considerably improved the solution with reasonable CPU times. An application of the algorithm to other practical scheduling problems such as flexible flow line and job shop remains to be further investigated.

References

1. Kim, C.O., Shin., H.J.: Scheduling jobs on parallel machines: a restricted tabu search approach. International Journal of Advanced Manufacturing Technology. Vol.22. (2003) 278-287
2. Ovacik, I.M., Uzsoy, R.: Rolling Horizon Procedures for Dynamic Parallel Machine Scheduling with Sequence Dependent Setup Times. International Journal of Production Research. Vol.33. (1995) 3173-3192
3. Fleurent, C., Glover, F.: Improved Constructive Multistart Strategies for the Quadratic Assignment Problem using Adaptive Memory. INFORMS Journal on Computing. Vol.11, No.2. (1999) 189-203
4. Boese, K.D., Kahng, A.B., and Muddu, S.: A New Adaptive Multi-Start Technique for Combinatorial Global Optimizations. Operations Research Letters. Vol.16, No.2. (1994) 101-113
5. Schoen, F.: Global Optimization Methods for High-Dimensional Problems. European Journal of Operational Research. Vol.119. (1999) 345-352
6. Merkle, D., Middendorf, M.: Ant Colony Optimization with Global Pheromone Evaluation for Scheduling a Single Machine. Applied Intelligence. Vol.18, No.1. (2003) 105-111
7. Ding, L., Yue, Y., Ahmet, K., Jackson, M., and Parkin, R.: Global Optimization of a Feature-based Process Sequence Using GA and ANN Techniques. International Journal of Production Research. Vol.43, No.15. (2005) 3247-3272
8. Yang, Y.W., Xu, J.F., and Soh, C.K.: An Evolutionary Programming Algorithm for Continuous Global Optimization. European Journal of Operational Research. Vol.168, No.2. (2005) 354-369
9. Uzsoy, R (1998) Parallel Machine Scheduling Problem Data Sets.
   http://palette.ecn.purdue.edu/~uzsoy2/Problems/parallel/parameters.html
10. Laguna, M., Barnes, J.W., and Glover, F.: Tabu Search Methods for Single Machine Scheduling Problems. Journal of Intelligent Manufacturing. Vol.2 (1991) 63-74
11. Locatelli, M., Schoen, F.: Fast Global Optimization of Difficult Lennard-Jones Clusters. Computational Optimization and Applications. Vol.21. (2002) 55-70
12. Bean, J.: Genetic Algorithms and Random Keys for Sequencing and Optimization. ORSA Journal on Computing, Vol.6. (1994) 154-160
13. Spears, W , Dejong, K.: On the Virtues of Parameterized Uniform Crossover, Proceedings of the 4th International Conference on Genetic Algorithms. (1991) 230-236
14. Sloan Jr., K.R., Tanimoto, S.L.: Progressive Refinement of Raster Images. IEEE Transactions on Computers. Vol.28, No.11 (1979) 871-874

Releasing and Scheduling of Lots in a Wafer Fab

Subhash C. Sarin[*], Vinod D. Shenai, and Lixin Wang

Grado Department of Industrial and Systems Engineering
Virginia Tech, Blacksburg, VA 24061
{sarins,lxwang}@vt.edu

Abstract. In this paper, we address the problem of both releasing and scheduling of lots in a wafer fab. In the lot release problem, we determine the number
of lots of different products to be released in each period of a planning horizon
in order to minimize total tardiness. The problem of the scheduling of lots at
various workstations is modeled as a mathematical program for the objective of
minimizing the cycle times of the lots and is solved by the Lagrangian relaxation method. Computational results are presented that exhibit that our methodology constantly generates better solutions compared to those obtained by
commonly-used dispatching rules.

Keywords: wafer fab, lot releasing, scheduling, Lagragian relaxation.

1 Introduction and Problem Statement

Wafer fabrication is the most technologically complex and capital intensive phase in
semiconductor manufacturing. The high cost of wafer fabs (over \$3 billion for the
new 300-mm factories) and the need to pay them off quickly means that effective
strategies to operate these facilities are absolutely essential. The operational control
problem that we address in this paper can be stated as follows: Given a set of lots
waiting to be released into a fab or for loading in a processing area, determine the
number of lots to release into the fab as well as the order in which to process the lots
at the processing areas so as to minimize the performance-related measures of average
cycle time and job tardiness.

There are the following key features which need to be incorporated in the problem
formulation and solution methodology. A lot consists of a certain number of wafers
held in a cassette. We define a job in a wafer fab to be a lot of a product. A work order consists of a number of wafers ordered by a customer. The number of lots in each
work order is assumed known which, in turn, is determined by taking into consideration the yield rate. The due date for a work order is the date by which it should be
completed. The planning horizon is the time duration that is sufficient to complete all
the work orders in the fab and its calculation is based upon the capacity of the fab.
The entire planning horizon is divided into planning periods. The duration of each
planning period is assumed to be twenty-four hours.

---

[*] Corresponding author.

M.-Y. Kao and X.-Y. Li (Eds.): AAIM 2007, LNCS 4508, pp. 108–119, 2007.

A station family consists of a number of workstations. The processing step of a lot can be performed at any of the workstations (machines) of a station family. Some processing areas in a wafer fab, such as etching, consists of batch processing machines, where a number of lots are processed simultaneously, as a batch. A wafer has to visit the photolithography processing area numerous times to have all the layers of circuitry fabricated. The fact that a lot visits a processing area more than once, is what is called the reentrant product-flow in a wafer fab. Lot-dedication scheme refers to a policy adopted in the photolithography processing area of certain wafer fabs, where the photolithography processing operations on all (or critical) layers of a wafer are processed on the same workstation where the first layer of that wafer was processed. Preventive maintenance schedules are drawn up months in advance, and a workstation is not available during that time. But, then, there are unscheduled downtimes as well during which a workstation breaks down.

In this paper, we approach this problem in two steps. First, the number of lots of different products to be released into the system during each planning period is determined such that the total tardiness of the product orders is minimized. Second, the schedules of these lots for use in the processing areas are determined so that the cycle time of each lot released into the system is minimized. An integer linear programming model is formulated for the lot release, tardiness problem and an algorithm is presented for obtaining good solutions. The scheduling problem is formulated as a 0-1 integer linear model. We successfully apply Lagrangian relaxation on a carefully chosen set of constraints and present a Lagrangian heuristic for the scheduling of the lots in each period of the planning horizon.

The interest in the production planning problem of semiconductor manufacturing began in the later 1980's. Various techniques and models have been proposed for this problem in the literature (see Uzsoy, Lee and Martin-Vega [1][2]). These include: a linear programming (LP) and simulation-based planner (Hung and Leachman [3]); a two-step LP-based approach (Lee, Yea and Kim [4]); and a hierarchical framework (Grovin [5]). There have been several approaches proposed in the literature for the scheduling problems encountered in a wafer fab (see Uzsoy, Lee and Martin-Vega [1][2]). The most common among these is the use of dispatching rules (Wein [6], Kim et al. [7], and Lu, Ramaswamy and Kumar [8]). However, mathematical programming-based approaches have not been that common even though they have the capability of providing optimal or near-optimal solutions. Some efforts in this regard have been reported by: Sarin et al. [9], Graves et al. [10], and Mehta and Uzsoy [11]. In this paper, we attempt to use such an approach for the problem on hand. Effective algorithms are designed to circumvent their computational intractability.

The rest of this paper is organized as follows. In section 2, we present the lot release problem to minimize tardiness of job orders, and a two-phase algorithm to obtain good solutions is developed. This is followed by the presentation of a model for the scheduling problem in section 3. A Lagrangian relaxation-based algorithm for the solution of this problem is also described in this section. In section 4, results of the experimentation conducted on the proposed solution methodologies for both of these problems are presented. Lastly, concluding remarks are made in section 5.

Before proceeding further, we present the following assumptions employed throughout this paper: (1) the wafer fab produces multiple products; (2) a few station families consist of parallel machines; (3) the parallel machines in a station family are identical in all aspects; (4) the processing times for all processing steps are deterministic; (5) the setup time of each processing step is included in its processing time; (6) the buffer size at any workstation is finite; (7) a due date is provided for every order of a product; (8) all parameters are measured after the system has reached steady state; and (9) the products are compatible at a batching station, and hence, can be processed together.

2   The Lot Release Problem to Minimize Tardiness of Job Orders

2.1   Mathematical Model for the Lot Release Problem

We formulate this problem as an integer programming model, which determines a number of lots for each work order that is to be released into the system, in each time period of the entire planning horizon, so that the total tardiness of the work orders is minimized.

Let f, g, t and k be the subscripts representing a product, the work order number of a product, a single time period, and a station family, respectively; NL_{fg} denote the total number of lots in the work order g of product f; N_{fgt} be the number of lots of work order g of product f released into the system during time period t; BN_{ft} represent the maximum number of lots belonging to product f that can be released into the system during each time period because of a constraint on system capacity; F be the number of products; W_f be the number of work orders of product f; M be the number of station families; T be an upper bound on time required to complete all the work orders; b_k denote the number of parallel machines in the station family k; d_{fg} designate the due date of work order g of product f; d'_{fg} represent the modified due date of work order g of product f; C_{fg} be the completion time of work order g of product f; TD_{fg} designate the tardiness of work order g of product f; $X_{fgt} = 1$, if work order g of product f is released into the system during time period t, and $= 0$ otherwise; r_{fk} represent the number of re-entries of product f into station family k; s_{fk} be the processing time of each step of product f on station family k (in hours); BLT_f denote the base lead time of a lot of product f (the sum of its processing times over all the steps); h_t be the number of hours in time period t; UT_{fg} be an upper bound on the time required to complete work order g of product f; and LT_{fg} be a lower bound on the time required to complete a work order g of product f. We assume each time period t to be of equal time duration, representing one day. The wafer fab is assumed to run continuously for three shifts in a day, and hence, the time available in a single time period is 24 hours. Therefore, $h_t = 24$ hours for all periods. BN_{ft} is the same for all t and, thus, can be rewritten as BN_f. Also, modified due date, $d'_{fg} = d_{fg} - \rho * BLT_f$ where ρ is the lead time factor for the last lot of a work order g of product f.

Problem P1:   Min $\displaystyle\sum_{f=1}^{F}\sum_{g=1}^{W_f} TD_{fg}$

(1) Tardiness definition.

$$TD_{fg} \geq C_{fg} - d'_{fg}, \quad f=1, 2, \dots F, g = 1, 2, \dots W_f \tag{1}$$

(2) The completion time of a work order g of product f is equal to the time period t in which the last lot of the work order g is released into the system.

$$C_{fg} \geq t * X_{fgt}, \quad f=1, 2, \ldots F, \, g = 1, 2, \ldots W_f, \, t=1, 2, \ldots T \qquad (2)$$

(3) The number of lots of a work order g of product f released into the system in time period t is less than or equal to the maximum number of lots of product f, which can be processed by the system in time period t.

$$N_{fgt} \leq BN_{ft} * X_{fgt}, \qquad f=1, 2, \ldots F, \, g = 1, 2, \ldots W_f, \, t=1, 2, \ldots T \qquad (3)$$

(4) The sum of all the lots of a work order belonging to a product that are released into the system, over the planning horizon, is equal to the total number of lots in the work order.

$$\sum_{t=1}^{T} N_{fgt} = NL_{fg}, \quad f=1, 2, \ldots F, \, g = 1, 2, \ldots W_f \qquad (4)$$

(5) The total number of lots, over all the work orders, released into the system should satisfy the capacity of each station family in each period.

$$\sum_{f=1}^{F} \sum_{g=1}^{W_f} N_{fgt} * r_{fk} * s_{fk} \leq h_t * b_k, \quad k=1,\ldots, M \text{ and } t=1,\ldots,T, \qquad (5)$$

$TD_{fg} \geq 0$, C_{fg}, $N_{fgt} \geq 0$ and integer; and $X_{fgt} = 0,1$, $f=1,\ldots,F$, $g=1,\ldots,W_f$, and $t=1,\ldots,T$.

Note that the maximum number of lots of a product f that can be processed in a single time period, t, is equal to the maximum number of lots processed by the bottleneck station of the product, and can be computed as follows. The maximum number of lots of a product f that can be processed in a time period on station family k,

$$M_{kft} = \left\lfloor \frac{h_t * b_k}{s_{fk} * r_{fk}} \right\rfloor, \qquad f=1,\ldots,F, \, k=1,\ldots, M \text{ and } t=1,\ldots,T. \qquad (6)$$

Then,

$$BN_f = \min_k M_{kft}, \qquad f=1,\ldots,F, \, k=1,\ldots, M \text{ and } t=1,\ldots,T \qquad (7)$$

Therefore, an upper bound and a lower bound on the time required for completing a work order g of product f are as follows:

$$UT_{fg} = \left\lceil \frac{NL_{fg}}{BN_f} \right\rceil, \quad LT_{fg} = \left\lfloor \frac{NL_{fg}}{BN_f} \right\rfloor, \quad f=1, 2, \ldots F, \, g = 1, 2, \ldots W_f \qquad (8)$$

Note that $\lceil x \rceil$ and $\lfloor x \rfloor$ represent the smallest integer greater than x and the largest integer smaller than x, respectively.

2.2 A Two-Phase Algorithm for the Lot Release Problem

Problem P1 can be viewed as a k station-family tardiness problem. Since the single machine tardiness problem is NP-hard (Pinedo [12]), so is Problem P1. Therefore, to curtail the computational effort, we determine an initial solution via an extension of the single machine tardiness problem, and then, use it in the above model to generate good solutions in a reasonable amount of time using CPLEX. The initial solution is generated in two phases as follows.

In Phase I, we apply the Wilkerson and Irvine's (W-I) algorithm (as described in Baker [13]) of the single machine tardiness problem to an aggregated version of the parallel machine environment of a wafer fab (call it a virtual single machine) and release the work orders one-at-a-time as follows:

Step 1. Consider the entire manufacturing system as a single machine.
Step 2. Consider each work order as a single job represented by fg.
Step 3. Calculate the modified due date d'_{fg} and the upper bound UT_{fg} on the time required to complete work on work order g of product f.
Step 4. Consider UT_{fg} as the processing time of a job on the virtual single machine.
Step 5. Apply the W-I algorithm to this environment to minimize the tardiness of the aggregated problem and determine the sequence in which to process the jobs (work orders) on the virtual single machine.

The work orders are ranked according to the sequence generated by the W-I algorithm.

In the solution generated by Phase I, each job $f'g'$ will begin processing after the completion of the previous job fg. The completion time t' of the job fg is the time period in which the last lots of the work order fg are released into the system. The next job $f'g'$ will start processing only in the time period $t'+1$. As a result, there may be capacity available in time period $t \le t'$ for processing lots from the work order $f'g'$, but this capacity is not utilized. Hence, the solution obtained in Phase I may involve many capacity "gaps." We remove these capacity gaps in Phase II by releasing the lots from more than one work order at the same time as follows.

Step 1. (Initialization) Rank the work orders (in ascending order) according to the schedule generated by the W-I algorithm (in Phase I).
Step 2. In the first time period, take the first ranked job (work order g of product f) and release the maximum number of lots (BN_f) of product f for the work order g into the system.
Step 3. Go through the schedule, and select the least ranked work order g' of product f' over all the remaining work orders.
Step 4. Release as many lots of the work order g' of product f' as possible so that the system stability is not violated (that is, the system capacity is not exceeded).

Step 5. Repeat Steps 3 and 4 until all the work orders of every product have been evaluated for their release into the system such that the system capacity is not exceeded.

Step 6. Set $t = 2$.

Step 7. For the tth time period ($t=2,3,4,....,T$), release as many lots as possible of the least ranked work order g'' of product f'' into the system.

Step 8. Repeat Steps 3, 4, and 5.

Step 9. Set $t = t+1$.

Step 10. Repeat Step 7, 8, and 9 until time period T or when no lot of any work order remains to be released in the system.

Besides the initial solution determined above, we also use valid inequalities (see Shenai [14]), based on a dominance property among the jobs, in Problem P1. Both of these enhance the effectiveness of the solution of P1 by the CPLEX optimization software.

Above, we have assumed a fixed set of work orders on-hand (static case). However, if new orders are accepted in the middle of the planning period (dynamic case), then P1 needs to be solved frequently, taking into consideration the current status of the existing work orders and information about the new work orders.

3 Scheduling of Lots in a Wafer Fab

3.1 Mathematical Model for the Lot Scheduling Problem

Next, we present a mathematical model that is based on the flow approach (Weiss [15]) and is implemented stage by stage. Each stage is a planning period for which we completely solve the scheduling problem. The duration of this planning period can be taken to be a day while the duration of a time unit in the model is taken as a fraction of an hour (say $1/10^{th}$ of an hour). This is dictated by the fact that supervisors need daily schedules to run the fab, and also, daily information about the availability of resources is known with certainty (except, of course, for unscheduled downtime). We use the following notation in addition to the one presented before. Let i be the subscript representing the lot number; I_f represent the total number of lots of product f in the system; j be the subscript representing the processing step of the lot; J_{fi} be the set of processing steps in the route of the lot i of product f that will be processed (under static lot release policy) or may be processed (under dynamic lot release policy) in a single planning period; b_k be the number of workstations in station family k; t be a subscript representing an individual time unit at which the lot is to be scheduled; $X_{fijkbt} = 1$, if the processing step j of lot i of product f is scheduled on station b of station family k at time unit t, and $= 0$, otherwise; $I(k)$ be the set of operations (combinations of f, i and j) that can be processed on station family k; p_{fijk} be the processing time of step j of lot i of product f on station family k; T represent an upper bound on the time horizon. $T = 240$ time units; B represents the set of batching machines; U_k be the maximum batch size at station family k; L_k be the minimum batch size at station family k; $K(f,i,j)$ represent the set of station families on which operation j of lot i of product f can be processed; $Y_{fij} = 1$, if processing step j of lot i of product f is scheduled in the planning period, and $= 0$, otherwise. (Note: The value of every Y_{fij} is known a priori for a planning period in the static lot release case while it is not known for the dynamic lot release case and is one of the variables to be determined.)

We consider the objective of minimizing, over all the lots, the completion time of those processing steps that are scheduled in the corresponding time period.

Problem P2: Min $\displaystyle\sum_{f=1}^{F}\sum_{i=1}^{I_f}\sum_{j\in J_{fi}}\sum_{k\in K(f,i,j)}\sum_{b=1}^{b_k}\sum_{t=0}^{T} t*X_{fijkbt} - \mu*\sum_{f=1}^{F}\sum_{i=1}^{I_f}\sum_{j\in J_{fi}} Y_{fij}$

where μ is a penalty for not scheduling the job as early as possible; $\mu \geq T$.

(1) Each step of a job must be processed only once or not at all.

$$\sum_{k\in K(f,i,j)}\sum_{b=1}^{b_k}\sum_{t=0}^{T} X_{fijkbt} = Y_{fij}, \quad f = 1, 2, ..., F, \quad i = 1, 2, ..., I_f, \quad \forall j \in J_{fi} \qquad (9)$$

(2) Step $j+1$ of any lot must start only after the completion of step j of that lot.

$$\sum_{k\in K(f,i,j)}\sum_{b=1}^{b_k}\sum_{t=0}^{T}(T-t-p_{fijk})*X_{fijkbt} \geq \sum_{k'\in K(f,i,j+1)}\sum_{b'=1}^{b'_k}\sum_{t=0}^{T}(T-t)*X_{fij+1k'b't} \qquad (10)$$

$$f = 1, 2, ..., F, \quad i = 1, 2, ..., I_f, \quad \forall j \in J_{fi}$$

Also, for the Y variables,

$$Y_{fij} \leq Y_{fij+1}, \qquad f = 1, 2, ..., F, \quad i = 1, 2, ..., I_f, \quad \forall j \in J_{fi} \qquad (11)$$

(3) Lot i of a product should be scheduled before lot $i+1$ of that product at any step j. The lots are numbered according to the chronological order of their release into the wafer fab.

$$\sum_{k\in K(f,i,j)}\sum_{b=1}^{b_k}\sum_{t=0}^{T}(T-t)*X_{fijkbt} \geq \sum_{k'\in K(f,i,j+1)}\sum_{b'=1}^{b'_k}\sum_{t=0}^{T}(T-t)*X_{fij+1k'b't} \qquad (12)$$

$$f = 1, 2, ..., F, \quad i = 1, 2, ..., I_f, \quad \forall j \in J_{fi}$$

(4) Lot $i+1$ of a product can be scheduled at any step j only if lot i is scheduled at that step.

$$Y_{fij} \leq Y_{fi+1j}, \qquad f = 1, 2, ..., F, \quad i = 1, 2, ..., I_f-1, \quad \forall j \in J_{fi} \qquad (13)$$

(5) All stations, except for the batching stations, have a capacity of one. At any given time, the number of lots at a non-batching machine must be less than or equal to one.

$$\sum_{t=0}^{T}\sum_{f,i,j\in I(k)} y_{fijkbt}*X_{fijkbt} \leq 1, \quad \forall k \notin B, \quad \forall b \in b_k, \qquad (14)$$

where, y_{fijkbt} is a column vector with $T+1$ elements and has zero entries except from the tth to the $(t+p_{fijk}-1)$ element, for which the entry is one, and $\mathbf{1}$ is a column vector with $T+1$ elements consisting of ones.

(6) At batching stations, a number of jobs can be processed simultaneously. A batching machine can start processing the lots once the minimum number of lots required to start its processing is reached. This capacity constraint can be captured by a constraint that is identical to constraint (5).

$$\sum_{t=0}^{T} \sum_{f,i,j\in I(k)} y_{fijkbt} * X_{fijkbt} \geq LC_k * D , \quad \forall k \notin B, \ \forall b \in b_k, \quad (15)$$

where, LC_k is a column vector with $T+1$ elements with each element equal to L_k, where L_k is the minimum batch size at the batching machine; and $D = 1$, if a batch of size at least L_k is being processed at the batching station k, and $= 0$, otherwise.

(7) A batching station can only process as many lots simultaneously as the maximum batch size.

$$\sum_{t=0}^{T} \sum_{f,i,j\in I(k)} y_{fijkbt} * X_{fijkbt} \leq UC_k, \quad \forall k \notin B, \ \forall b \in b_k, \quad (16)$$

where, UC_k is a column vector of $T+1$ elements with each element equal to U_k, where U_k is the maximum batch size at the batching machine k.

(8) At a batching machine, a subsequent operation must precede the previous operation by at least the processing time of the previous operation.

$$\sum_{t'=0}^{T} t' X_{f'i'j'k'b't'} + \alpha * p_{f'i'j'k'} + K_2 + T * \left(1 - \sum_{t'=0}^{T} t' * X_{f'i'j'k'b't'}\right) = \quad (17)$$

$$\sum_{t=0}^{T} t * X_{fijkbt} + \beta * p_{fijk} + K_1 + T * \left(1 - \sum_{t=0}^{T} t * X_{fijkbt}\right)$$

$\forall k \in B, \ \forall b, \ b=1, ..., b_k,$ and $\forall f, f' \in F, \ \forall i, i' \in I_f$ and $\forall j, j' \in J_{fi}$

where, α and β are binary variables, K_1 and K_2 are non-negative real variables, and $\alpha + \beta \leq 1, K_1 \leq \beta * T, K_2 \leq \alpha * T$

(9) Other constraints, such as lot dedication, preventive maintenance, handling of hot lots and due date for the last lot of a work order, are also incorporated in this model. More details about these constraints can be found in Shenai [14].

3.2 A Lagrangian Relaxation-Based Algorithm

We apply the Lagrangian relaxation-based approach that relies on subgradient optimization (see Fisher [16]) to the problem on hand. The best results were obtained by relaxing constraint set (2). However, the solution obtained as a result needs to be adjusted in that the operations of a job should be arranged in a sequential manner. However, the sequence of processing steps for a lot is known from the route of that lot, and thus, this adjustment can be made easily. Due to the early start time constraints, it can easily be shown that, at every stage of the Lagrangian heuristic, at least one processing step is always feasible. Consequently, we can generate a feasible solution to the problem by fixing the current feasible variables and repeatedly solving the remaining truncated Lagrangian problem.

4 Computational Results

Consider a wafer fab consisting of 6 products and 10 station families and 25 workstations, which is slightly bigger than the Hewlett-Packard Technology Research Center fab (TRC fab) as presented in [5]. Each product consists of 2 to 4 work orders and the number of lots in each order ranges from 100 to 250. The lot release problem is solved for several combinations of the number of products and the number of machines. The results are presented in Table 1. Note that the solutions generated by the proposed solution methodology are close to optimal while requiring only a fraction of the computational time taken to obtain the optimal solutions. As indicated, for one of the instances, the optimal solution could not be obtained within 10 hours of computational time.

Table 1. Comparison of solution quality and computational time between the proposed solution methodology and the optimal solution

(Number of products)/ (Number of machines)/ (Number of work orders)	2-Phase approach (Tardiness per Work Order) (days)	Optimal solution (Tardiness per Work Order) (days)	Computational time for 2-phase Approach (a) (sec)	Computational time for optimal Solution (b) (sec)	Ratio of solution times (b/a)
6/25/19	5.2	5.0	386.97	106309.34	274
5/27/17	19.2	*	1648.85	–	–
4/24/13	1.0	0.8	103.33	110646.26	1074
3/21/11	2.5	0.5	88.13	5494.55	62
2/18/ 9	11.6	11.6	5.47	123.46	22

* Optimal solution could not be obtained within 36,000 seconds (10 hours) of computation time.

Three experiments were conducted to study the performance of the Lagrangian heuristic for the scheduling of jobs in a wafer fab. All the experiments are performed on systems that have reached a steady state. A system has a starting WIP, and hence, is said to be "full".

The performance measures used to evaluate the solutions are the average and standard deviation of the cycle time. As our experiments involve multiple products, the cycle time of each product type is computed separately. The solution obtained by the Lagrangian heuristic is compared with the solution obtained by using the following six standard dispatching rules available in the literature: First In First Out (FIFO),

Least Balance Ahead (LBA), Least Lots Ahead (LLA), Least percentage of processing time remaining (LPR), Least Slack (LS), and Least Time Remaining (LTR). The Lagrangian heuristic is denoted as LAG.

Experiments 1 and 2 use the same wafer fab system, and are conducted to study the performance of the Lagrangian heuristic in a wafer fab in which different products in the system have different routes and processing times, each product passes through the same bottleneck station, and the product mix that are loaded into the system is different for each planning period. Experiments 1 and 2 differ in the ways the lots are released into the system. Experiment 3 is conducted to study the performance of the Lagrangian heuristic in a wafer fab in which different products pass through different bottlenecks. These experiments generate lot schedules for a time period of 3 or 4 days. This is due to the fact that fab supervisors need daily lot schedules to run the fab and lot schedules for a longer time period are not necessary. The problem data that we used was obtained from a small wafer fab located in Roanoke, Virginia, and is presented in Shenai [14].

In Table 2, the average and the standard deviation of the cycle times given by the Lagrangian heuristic is compared with the corresponding values obtained by using the

Table 2. Summary of results of Experiment 1(cycle time: hours)

Product #	Cycle time	FIFO	LBA	LLA	LPR	LS	LTR	LAG
1	AVG	39.97	38.15	38.15	37.92	37.89	37.90	37.75
	STD DEV	0.12	0.13	0.13	0.15	0.15	0.16	0.15
2	AVG	40.23	38.15	38.15	37.37	39.33	37.37	37.48
	STD DEV	0.34	0.32	0.32	0.33	0.31	0.33	0.32
Lot completed		26	27	27	27	27	28	28

Table 3. Summary of results of Experiment 2

Product #	Cycle time	FIFO	LBA	LLA	LPR	LS	LTR	LAG
1	AVG	41.32	41.47	40.58	40.68	39.07	43.51	40.4
	STDDEV	4.45	6.58	5.63	6.28	5.25	6.22	5.55
2	AVG	35.33	33.75	33.97	36.2	37.99	29.88	30.92
	STDDEV	4.43	4.01	4.0	6.90	6.27	1.4	2.33
Lots completed		24	24	24	25	26	26	26

Table 4. Summary of results of Experiment 3

Product #	Cycle time	FIFO	LBA	LLA	LPR	LS	LTR	LAG
1	AVG	40.71	41.16	42.67	38.45	43.08	38.5	39.61
	STD DEV	6.01	5.93	7.03	5.37	3.57	6.12	6.41
2	AVG	47.05	42.70	44.55	46.62	47.5	46.1	43.38
	STD DEV	6.33	4.78	3.73	5.37	4.99	3.99	4:11
Lots completed		25	26	26	26	21	25	25

standard dispatching rules. The best results corresponding to each performance measure for each product are shaded. Similarly, the summary of results for Experiments 2
and 3 are presented in Tables 3 and 4.

It is clear from the above results that none of the scheduling rules dominates the
others. However, the cycle times obtained by the Lagrangian heuristic are quite close
to the least cycle time for each product. In all cases, the Lagrangian heuristic is not
biased towards a product and provides a good solution (cycle time) for each product.
This is shown in Table 5 where the ratios of the cycle times obtained for each product
and for each scheduling rule in each experiment to the best value among them are
shown. The bottom row provides the average value over all the experiments. This
value is the smallest for the Lagrangian heuristic.

Table 5. Comparison of ratios of the cycle time obtained for each product and for each scheduling rule in each experiment to the best cycle time among them in each experiment

Experiment	Product	FIFO	LBA	LLA	LPR	LS	LTR	LAG
1	1	106%	101%	101%	100%	100%	100%	100%
1	2	108%	102%	102%	100%	105%	100%	100%
2	1	106%	106%	104%	104%	100%	111%	103%
2	2	118%	113%	114%	121%	127%	100%	104%
3	1	106%	107%	111%	100%	112%	100%	103%
3	2	110%	100%	104%	109%	111%	108%	102%
Average		109%	105%	106%	106%	109%	103%	102%

5 Concluding Remarks

In this paper, we have addressed both the lot release and scheduling problems of a
wafer fab. The lot release problem determines the work orders for release into the
wafer fab in each period of a planning horizon in order to minimize the total tardiness.
The proposed solution methodology provides good solutions within a reasonable
amount of computation time.

The scheduling problem for minimizing the cycle time of the lots released into the
wafer fab is formulated as an integer program to capture specific features of a wafer
fab. A Lagrangian relaxation-based method is developed for solving this problem.
Experiments are conducted under different sets of conditions that are encountered in a
wafer fab. The Lagrangian heuristic consistently generates good solutions for the
scheduling problem. The cycle time for each product generated by the Lagrangian
heuristic is closer to the minimum cycle time of that product than that generated by
the dispatching rules.

References

1. Uzsoy R., Lee C. Y., and Martin-Vega L.: A review of production planning and scheduling models in semiconductor industry – part I: system characteristics, performance evaluation and production planning. IIE Transactions, Scheduling and Logistics, vol. 24 (1992)
   47-61.

2. Uzsoy R., Lee C. Y., and Martin-Vega L.: A review of production planning and scheduling models in semiconductor industry – part II: shop-floor control. IIE Transactions, Scheduling and Logistics, vol. 26 (1994) 44-55

3. Hung Y. F. and Leachman R.C.: A production planning methodology for semiconductor manufacturing based on iterative simulation and linear programming calculations. IEEE Transactions on Semiconductor Manufacturing, vol. 9 no. 2 (1996 May) 257-269.

4. Lee Y., Kim S., Yea S., and Kim B.: Production planning in a semiconductor wafer fab considering variable cycle times. Computers in Industrial Engineering, vol. 33, no. 3 (1997) 713-716.

5. Golovin J. J.: A total framework for semiconductor production planning and scheduling. Solid State Technology (1986) 167-170.

6. Wein L.: Scheduling semiconductor wafer fabrication. IEEE Transactions on Semiconductor Manufacturing. vol. 1, no. 3 (1998) 115-129.

7. Kim Y., Kim J., Lim S., and Jun H.: Due date based scheduling and control policies in a multi-product semiconductor wafer fabrication facility. IEEE Transactions on Semiconductor Manufacturing. vol. 11, no. 1 (1998 February) 155-164.

8. Lu S., Ramaswamy D., and Kumar P. R.: Efficient scheduling policies to reduce mean and variance of cycle time in semiconductor manufacturing plants. IEEE Transactions on Semiconductor Manufacturing, vol. 7, no. 3 (1994 August) 374-388.

9. Sarin C. S., Shikalgar S., and Shenai V.: Modeling and analysis for the reduction of cycle time at a wafer fab. Proceedings of the International Conference on Semiconductor Manufacturing Operational Modeling and Simulation (2001).

10. Graves S., Meal H., Stefek D., and Zeghmi A.: Scheduling of re-entrant flow shops. Journal of Operations Management, vol. 3, no. 4 (1983) 197-207.

11. Mehta S. V., and Uzsoy R.: Minimizing total tardiness on a batch processing machine with incompatible job families. IIE Transactions, vol. 30 (1998) 165-178.

12. Pinedo M.: Scheduling: Theory, Algorithms and Systems. Prentice Hall (1995).

13. Baker K. R.: Introduction to Sequencing and Scheduling, John Wiley & Sons (1974).

14. Shenai V.: Releasing and scheduling of jobs in a wafer fab. Master thesis. Virginia Polytechnic Institute and State University (2003).

15. Weiss G.: On optimal draining of re-entrant fluid lines. The IMA Volumes in Mathematics and its Applications, Stochastic Newtorks, vol. 71 (1995) 91-103.

16. Fisher M. L.: The Lagrangian relaxation method for solving integer programming problems. Management Science, vol. 27, no. 1 (1981) 1-18.

Mixed Criteria Packet Scheduling

Chad R. Meiners and Eric Torng

meinersc@cse.msu.edu, torng@cse.msu.edu

Abstract. Packet scheduling in networks with quality of service constraints has been extensively studied as a single criterion scheduling problem. The assumption underlying single criterion packet scheduling is that the value of all packets can be normalized to a single scale, even in cases when packets have different requirements. We demonstrate that this approach can lead to inefficient utilization of network resources.

To improve network efficiency, we model packet scheduling as a *mixed criteria* scheduling problem where there are two distinct sets of jobs: deadline jobs which represent real-time packets in a network and flow jobs which represent other packets in the network. As the names imply, the jobs in these two sets differ by the criteria associated with them. For this problem, the flow jobs are scheduled to minimize the sum of their flow times, and the deadline jobs are scheduled to maximize the value of jobs that complete by their deadlines.

We demonstrate that even when there is only a single deadline job, this mixed criteria scheduling problem is NP-Complete. We give a polynomial time optimal algorithm Slacker for the variant where all jobs have unit size and the value of deadline jobs processed by the deadline must be maximized. Given this constraint Slacker minimizes the total flow time. Furthermore, we show that online Slacker is optimal for flow time while being 2-competitive with respect to the deadline jobs when compared to an optimal algorithm like Slacker that maximizes the value of deadline jobs.

1   Introduction

We investigate the *packet scheduling* problem in a network with quality of service (QoS) considerations. Packet scheduling in QoS networks has been extensively studied as a single criterion scheduling problem [2,3,6,8,10,13,16,17,19]. Typically, each packet has a value or weight, and the goal is to maximize the value of successfully transmitted packets. Constraints on successful packet transmission include a finite buffer size, packets may need to be transmitted in first in first out order, and/or deadlines on packets. The assumption underlying single criterion packet scheduling is that the value of all packets can be normalized to a single scale.

We suggest that this assumption about packet values is not always valid. Networks such as the internet support a variety of applications, each with their own requirements. A video conferencing application requires real-time delivery guarantees for each of its packets while a file download application only seeks to

M.-Y. Kao and X.-Y. Li (Eds.): AAIM 2007, LNCS 4508, pp. 120–133, 2007.

minimize the arrival time of the entire file. It is not obvious that fixed values or weights can be assigned to packets with different requirements.

For example, consider the following two scenarios. In scenario 1, there is one flow job of size x available at time 0, and there is a stream of y unit size video packets arriving one per time unit starting at time 0 with relative deadlines of $x + 1$. Scenario 2 is identical except there is a second video stream of x unit size video packets arriving one per time unit starting at time 0 where each packet must be delivered immediately upon arrival. In scenario 1, the optimal solution is to process the flow job first for a total flow time of x and then process the packets of video stream 1 delivering all packets by their deadlines. In scenario 2, assuming that completing jobs by their deadlines is more important than minimizing flow time when there is a conflict, the optimal solution is to process the second video stream first, then process the first video stream, and finally process the flow job for a total flow time of $y + 2x$. However, given a fixed priority scheme, the flow job packets will either always be scheduled before or after those of video stream 1. The typical assumption is that the flow job packets would have low priority. In scenario 1, this leads to a flow time of $y + x$ which is unnecessarily high.

To gain more insight into the packet scheduling problem, we will examine *mixed criteria* formulations of the packet scheduling problem. In a mixed criteria scheduling problem, there is a mixture of job types. All jobs of the same type should be evaluated using the same criterion such as total flow time, but different job types might use different criteria. In the case of packet scheduling, jobs with real-time constraints will be evaluated based on whether or not they are serviced by their deadlines while jobs with flow time constraints will be evaluated using total flow time. By directly measuring the appropriate criteria for each job type, the mixed criteria model will hopefully lead to more effective packet scheduling algorithms.

If we restrict all jobs to be unit-sized, there is a polynomial-time optimal algorithm that we call *Slacker*, which minimizes total flow time after maximizing the total value of the deadline jobs, and we show that an online variant of Slacker produces an optimal (or better) schedule for the flow jobs while being 2-competitive for maximizing the value of deadline jobs that complete by their deadlines. Furthermore, we show the perhaps surprising result that the introduction of a single deadline job makes this problem NP-complete. Finally, we provide additional results that further delineate the P-NP boundary for this problem.

2   Problem Definition and Variants

An input instance for a scheduling problem is a set of jobs I. For any scheduling algorithm A and any input instance I, $\mathbf{A}(\mathbf{I})$ is the schedule produced by A when given I.

The *DeadFlow* problem is a mixed criteria problem for an input instance $I = I_f \cup I_d$ where $\mathbf{I_f}$ (the flow jobs) and $\mathbf{I_d}$ (the deadline jobs) are non-intersecting sets of jobs. For the flow jobs I_f, each job has a processing time $\mathbf{p_j}$ and a release time r_j. The deadline jobs also have deadlines d_j and weights or values w_j. Given a single processor, the algorithm must schedule both sets of jobs where each job set has its own criterion for optimality. For the flow jobs I_f, the goal is to minimize their total flow time. For the deadline jobs I_d, the goal is to maximize the value of jobs completed by their deadlines. We assume that preemption is allowed. Generalizing standard scheduling notation we represent DeadFlow as

$$1|(r_j, p_j, \text{pmpt}|\sum F_j), (r_k, p_k, d_k, w_k, \text{pmpt}|\sum w_k U_k)$$

Given an algorithm A for DeadFlow and any input instance $I = I_f \cup I_d$, we use $\mathbf{F_A(I)}$ to denote the total flow time of the flow jobs I_f in schedule $A(I)$, $\mathbf{V_A(I)}$ to denote the total weight of jobs in I_d completed by their deadlines in schedule $A(I)$, and we use $\mathbf{M_A(I)}$ to denote the makespan or maximum completion time of schedule $A(I)$.

We say that a schedule is *value-optimal* if it maximizes $V_A(I)$. We define an *optimal schedule* $\mathbf{O(I)}$ as a value-optimal schedule for input instance I that minimizes the total flow time of the flow jobs. It is true that for many input instances I, there are non-value-optimal schedules $A(I)$ such that $F_A(I) < F_O(I)$. We use $O(I)$ to denote an optimal schedule in order to reflect the typical assumption that flow jobs have a lower priority than deadline jobs. This assumption is commonly used because failing to complete a deadline job has more permanent consequences than delaying a flow job. What is different about our approach is that we prioritize deadline jobs at the criteria level rather than the individual job level. This allows our algorithm to schedule flow jobs ahead of deadline jobs when the jobs are not truly in conflict as is the case in Scenario 1 in the introduction. We say that an algorithm A for the DeadFlow problem is a c_1-*value* c_2-*flow approximation algorithm* if $V_A(I) \geq 1/c_1 V_O(I)$ and $F_A(I) \leq c_2 F_O(I)$ for all input instances I.

We consider two special cases of the DeadFlow problem. The *Single DeadFlow scheduling problem* (SDF) is the special case of the DeadFlow problem where the number of the deadline jobs is exactly one. In this case, the single deadline job must be completed by its deadline, so its weight is unimportant. We use this variant to prove hardness results by reducing knapsack variants to SDF.

The *Unit DeadFlow scheduling problem* (UDF) is the special case of the DeadFlow problem when every job is of unit size. With generalized scheduling notation, we represent UDF as

$$1|(r_j, p_j = 1, \text{pmpt}|\sum F_j), (r_k, p_k = 1, d_k, w_k, \text{pmpt}|\sum w_k U_k)$$

We give an optimal polynomial-time algorithm Slacker for UDF in Section 4. We also show that an online variant of Slacker is a 2-value 1-flow approximation algorithm for UDF.

3 Background and Related Work

Scheduling only flow jobs and only deadline jobs have both been studied extensively. Preemptive flow job scheduling is solvable via the shortest remaining processing time algorithm (SRPT). Scheduling to maximize the throughput or value of jobs completed by their deadlines on a single machine is NP-hard [12]. If all jobs have the same size, there is a polynomial time algorithm [7].

However, if there exists a feasible schedule that satisfies every deadline jobs's deadline, several algorithms including Earliest Deadline First (EDF) find a feasible schedule in polynomial time. If all the deadline jobs have unit size, this corresponds to one variant of the single criterion packet scheduling problem [6,8,10,13,17,19] called *buffer management with bounded delay*. The greedy strategy of always scheduling an job with highest weight without consideration of the job's deadline has been shown to be 2-competitive [13]; the best known algorithm is 1.939-competitive [10].

Single criterion packet scheduling has also been studied where jobs do not have deadlines but must be delivered in First In First Out (FIFO) order [2,3,16,17]. The greedy strategy is also 2-competitive for this problem [16]. The best known algorithm is 1.983-competitive [17].

A problem similar to the DeadFlow problem has been studied in the real-time systems community where they must schedule a mixture of hard periodic tasks and soft aperiodic tasks [9,18,23]. In this work, the goal is to minimize the flow time of the aperiodic tasks given that the periodic tasks must complete by their deadlines. Most work makes the simplifying assumption that the aperiodic tasks will be served in FCFS order. Under this assumption, there are simple policies that can optimally schedule each aperiodic job as it arrives. The philosophy here is the same as our Slacker algorithm where the aperiodic tasks are prioritized except that no periodic deadline task is allowed to miss its deadline.

DeadFlow is distinct from traditional multiple criteria scheduling problems which have one set of jobs that must satisfy every criteria. Most research on such problems focuses on designing algorithms that optimize for a particular problem [4,21,22,24] or on proving the existence of good schedules for a variety of environments [14,15,20,25].

We are aware of two previous papers that study mixed criteria scheduling problems [1,5]. This work is similar to ours in that they deal with mixed criterion scheduling problems; however, this work differs from ours by not allowing job preemptions. Agnetis, Mirchandani, and Pacciarelli [1] consider a selection of mixed criteria problems in which each set of jobs has its own representative agent that specifies the set's criterion. Their paper analyzes two competing agents where each agent utilizes one of the following criteria: maximum of regular functions over job completion time, number of late jobs, and total weighted completion time. Baker and Smith [5] examine the group of mixed criteria scheduling problems that result from the combination of the following three criteria: minimizing makespan, minimizing maximum lateness, and minimizing total weighted completion time. They demonstrate that combining weighted completion time with either of the other two criteria results in an NP Complete problem.

4   Unit DeadFlow Problem

The Unit DeadFlow problem, where all jobs are of unit size, is solvable in polynomial time. Furthermore, we give an online algorithm that is a 2-value 1-flow approximation algorithm for UDF.

4.1   Offline Unit DeadFlow

Slacker is our name for the offline algorithm. Below is a detailed description of the Slacker algorithm and a proof of its optimality for UDF. First deadline jobs I_d must be processed to find the maximum weight feasible set of deadline jobs. Initially, our set of jobs M is empty. We consider deadline jobs one by one starting with the heaviest jobs first. A job j is added to M if $M \cup \{j\}$ is feasible. The resulting M has maximum possible weight of all feasible sets of deadline jobs because jobs have unit size so it is impossible that accepting one heavy job will prevent two slightly less heavy jobs from being scheduled. Therefore, we can assume that I_d is feasible and that the goal is to determine when to schedule the deadline jobs. The flow jobs will be scheduled in FIFO order.

This portion of Slacker is composed of two major subroutines. The first, which we name *Inverted EDF* or EDF^{-1}, takes as input a collection of deadline jobs I_d and produces a schedule $EDF^{-1}(I_d)$ that schedules the deadline jobs as late as possible. The second, *FirstIdle*, takes a given schedule $S(I)$ and finds the first idle interval $[t, t+1)$ that can be filled by a later executed deadline job.

Inverted EDF schedules deadline jobs as late as possible by viewing time in reverse with time 0 as time E, the maximum deadline of all jobs in the instance, and time E as time 0. For each deadline job j, EDF^{-1} views its release time as $E - d_j$ and its deadline as $E - r_j$. Inverted EDF then schedules the modified jobs using the earliest deadline first (EDF) algorithm. This guarantees that all feasible deadline jobs will meet their deadlines and will be scheduled as late as possible.

Algorithm 1 (Inverted EDF or EDF^{-1}). *Given a set of deadline jobs I_d, let E denote the largest deadline in I_d, and create $\bar{I}_d$ such that for every $j_k = (r_k, p_k, d_k, w_k) \in I_d$ there exists a $\bar{j}_k = (\bar{r}_k, p_k, \bar{d}_k, w_k) \in \bar{I}_d$ such that $\bar{r}_k = E - r_k$, $\bar{p}_k = p_k$, and $\bar{d}_k = E - d_k$. Let $EDF(\bar{I}_d)$ order jobs by increasing job weight. Construct $EDF^{-1}(I)$ by converting each job interval $(\bar{j}_x, s_x, e_x)$ in $EDF(\bar{I}_d)$ into the job interval $(j_x, E - s_x, E - e_x)$.*

Observation 1. *For any set of deadline jobs I_d and any time t where $EDF^{-1}(I_d)$ is not idle, the deadline job scheduled is an available job with the earliest deadline with ties broken by the FIFO policy.*

Lemma 1. *Let I_d be any set of deadline jobs and let $j = (r_j, 1, d_j)$ be any deadline job in I_d. If j is scheduled in interval $[i, i+1)$ in $EDF^{-1}(I_d)$, then $EDF^{-1}(I_d)$ is busy in the interval $[i+1, d_j)$.*

Proof. This follows given that EDF is a non-idling algorithm; therefore, the only reason that job j is not scheduled after time $i+1$ is that $[i+1, d_j)$ is filled with jobs (or $i+1 = d_j$).

Corollary 1. *For any set of deadline jobs I_d and for any time t, $EDF^{-1}(I_d)$ minimizes the number of deadline jobs scheduled in the interval $[0, t)$.*

Algorithm 2 (FirstIdle). *Given an input $I = I_f \cup I_d$ and a schedule $S(I)$,find the first idle interval $[t, t + 1)$ in $S(I)$ such that there exists a deadline job $(r_k, 1, d_k)$ in I_d such that $r_k \leq t$ and $t + 1 \leq d_k$. If no such idle interval exists, return "null". Otherwise, return (t, j) where $j = (r, 1, d)$ when j is the deadline job that is scheduled the earliest after time $t + 1$ in $S(I)$.*

Given these two pieces, we define Slacker as follows. Note that we define Slacker to take 2 inputs, a set of jobs I and a start time s. We use **Slacker(I)** to stand for $Slacker(I, 0)$.

Algorithm 3 (Slacker). *Given an input $I = I_f \cup I_d$ and a start time s Modify all jobs in I so that their release time is at least s. Let $S(I)$ be the schedule resulting from using FCFS to fill $EDF^{-1}(I_d)$ with the jobs I_f. Apply FirstIdle to $S(I)$. If FirstIdle returns "null", return $S(I)$. Otherwise, FirstIdle returns time t and deadline job j. Let $S'(I)$ be the interval $[s, t)$ on $S(I)$ with j in $[t, t + 1)$, and let I' be the set of all jobs in $S'(I)$ and return $S'(I)$ concatenated with $Slacker(I \setminus I', t + 1)$.*

We next define the concepts of *free jobs* and *delaying jobs*. Intuitively, free jobs are deadline jobs that do not delay flow jobs while delaying jobs are deadline jobs that do delay flow jobs. To formally define these two terms, we use the notation **$S(I_d \setminus \{j\})$** to represent the schedule $S(I_d)$ without the job j, and we use the notation **$I_f(S(I_d))$** to represent the schedule produced by using FIFO given the schedule of deadline jobs, $S(I_d)$.

Definition 1. *For any input instance I and any schedule $S(I)$, deadline job j is a **free job** if and only if $I_f(S(I_d \setminus \{j\}))$ has the same flow time cost as $I_f(S(I_d))$. Any deadline job that is not a free job is a **delaying job**.*

Definition 2. *For any input instance I, any schedule $S(I)$, and any time t, let **$D(S(I), t)$** be the number of delaying jobs in the interval $[0, t]$.*

Slacker is optimal because it satisfies the following property with respect to $D(S(I), t)$.

Lemma 2. *For any problem instance I, any schedule $S(I)$, and any time t, $D(Slacker(I), t) \leq D(S(I), t)$.*

Proof (Proof Sketch). For the sake of contradiction, let S(I) be a better schedule that agrees with Slacker as much as possible; i.e. there is a time x such that $D(S(I), x) < D(Slacker(I), x)$. Without loss of generality, we assume that $S(I)$ uses FIFO ordered flow jobs and EDF ordered deadline jobs (breaking ties using FIFO). Furthermore, we assume that $S(I)$ does not idle when jobs are available. Let t be the largest integer such that $S(I)$ and $Slacker(I)$ are identical on the interval $[0, t)$; that is, $S(I)$ and $Slacker(I)$ first differ in interval $[t, t + 1)$. We have four cases to consider.

The first case is when $S(I)$ has a free deadline job scheduled in $[t, t+1)$. By definition of a free job and the assumption that $S(I)$ and $Slacker(I)$ are identical up to time t, $Slacker(I)$ cannot schedule a flow job in $[t, t+1)$. Thus it must either idle or schedule a deadline job. If we assume that $Slacker(I)$ idles, we come to a contradiction. That is, if $Slacker(I)$ idles on $[t, t+1)$, Slacker did not find a deadline job in I_d that has not been scheduled before t but can be schedule in $[t, t+1)$; therefore there does not exists such a job. However, $S(I)$ demonstrates that such a job exists. Ergo, we know that when $S(I)$ has a free deadline job scheduled in $[t, t+1)$, $Slacker(I)$ has the same deadline job scheduled in $[t, t+1)$, and thus this case is impossible.

In the second case, $Slacker(I)$ has scheduled a delaying deadline job j in $[t, t+1)$ and $S(I)$ has scheduled a flow job in $[t, t+1)$. Let $[h, h+1)$ denote the last interval before time t such that a free deadline job was scheduled in both $S(I)$ and $Slacker(I)$. If no such time exists, let $h + 1 = 0$. $Slacker(I)$ scheduled all the deadline jobs executed after time $h + 1$ using Inverted EDF. Given Corollary 1, Slacker minimizes the number of deadline jobs scheduled in the interval $[h+1, t+1)$. Thus, this case is impossible.

The third case is when $Slacker(I)$ has scheduled a free job in $[t, t+1)$. This case is impossible because we assume $S(I)$ is non-idling.

The fourth and final case is when $S(I)$ has scheduled a delaying deadline job j in $[t, t+1)$ and $Slacker(I)$ has scheduled a flow job j' in $[t, t+1)$. It can be shown that $S(I)$ can be improved making this case impossible as well.

Since $Slacker(I)$ and $S(I)$ cannot differ, $S(I)$ does not exist, and the result follows.

Lemma 3. *Given a problem instance $I = I_f \cup I_d$, any $S(I)$ that minimizes $D(S(I), t)$ for all times t is optimal.*

Proof. For the sake of contradiction assume that $S(I)$ is not optimal. Let $O(I)$ be an optimal schedule. Without loss of generality, we assume that both $S(I)$ and $O(I)$ use FIFO to order the flow jobs and EDF to order the deadlines jobs (breaking ties using FIFO), and that neither schedule idles when jobs are available.

Let t be the largest integer such that $S(I)$ and $O(I)$ are identical on the interval $[0, t)$; that is, $S(I)$ and $O(I)$ first differ in interval $[t, t+1]$. It cannot be the case that one schedule processes a job in this interval and the other does not given our assumption that neither schedule idles the machine unnecessarily. Thus, one schedule executes a delaying deadline job j in the interval $[t, t+1]$ while the other executes a flow job j' in the interval. Given the assumption that $S(I)$ minimizes $D(S(I), q)$ for all times q, it must be the case that $O(I)$ has a delaying deadline job scheduled in the interval $[t, t+1)$ while $S(I)$ has a flow job j' scheduled in the interval $[t, t+1)$.

Since both schedules order flow jobs with FIFO, we know that j' is the first flow job scheduled after time t in $O(I)$. Let $[t', t'+1)$ be the point in time that j' is scheduled in $O(I)$. We know that $t' - t$ deadline jobs are scheduled in the interval $[t, t')$ in $O(I)$. Since $S(I)$ executes job j' in the interval $[t, t')$, there must be a legal way to schedule these $t' - t$ deadline jobs in the interval $[t+1, t'+1)$.

Thus, we can modify $O(I)$ by scheduling job j' in the interval $[t, t+1)$ and these $t' - t$ deadline jobs in the interval $[t + 1, t' + 1]$. This modification produces a new schedule with a better flow time than that of $O(I)$ which is a contradiction. Ergo, $S(I)$ is optimal.

Corollary 2. *Slacker is an optimal algorithm for the Unit DeadFlow problem*

The following observation will be useful for analyzing the online variant of Slacker.

Observation 2. *Given an input $I = I_f \cup I_d$ and any deadline job j, let $I' = I \cup \{j\}$. $F_{Slacker}(I) \leq F_{Slacker}(I')$, and $V_{Slacker}(I) \leq V_{Slacker}(I')$.*

Alternatives and Extensions to Slacker. Slacker is not the only way to solve the offline Unit DeadFlow problem. An alternative approach is to use min-cost flow and min-cost matching techniques. That is, we can view each problem instance as an instance of a transshipment problem or as an instance of an assignment program, both of which are solvable in polynomial time[11].

However, Slacker has two advantages. First, Slacker is more efficient. Second, Slacker remains polynomial when extended to the case where deadline jobs have arbitrary length, something not true of the linear programming approaches of solving this problem.

4.2   Online Unit DeadFlow

We present a 2-value 1-flow algorithm for online UDF, which we call *Online Slacker*.

Definition 3. *For any input instance I and any $t \geq 0$, t is a **release time** if some job in I has $r_j = t$. Let $\mathbf{J(I, t)}$ be the set of jobs with release time exactly t, and let $\mathbf{I_t}$ denote the set of jobs with release time of at most t.*

Algorithm 4 (Online Slacker). *Given an input instance $I = I_f \cup I_d$, at each release time t, Online Slacker applies Slacker to the jobs in I_t that have not yet been scheduled to create a new schedule we denote as $OnlineSlacker(I_t)$. Let s and t be two consecutive release times in I. Then Online Slacker schedules jobs in interval $[s, t)$ according to $Slacker(I_s)$. For the maximum release time t, Online Slacker schedules jobs in interval $[t, \infty)$ according to $OnlineSlacker(I_t)$.*

Definition 4. *For any input instance I and any release time t, let $\mathbf{A(I_t)}$ denote the deadline jobs that are scheduled at some time $t' \geq t$ in $OnlineSlacker(I_t)$. We say that $A(I_t)$ are the jobs **accepted by Online Slacker at time** t. Let $\mathbf{A(I)}$ denote the set of deadline jobs that are in $A(I_t)$ for some time $t \geq 0$. We say that $A(I)$ are the set of deadline jobs accepted by Online Slacker.*

Definition 5. *For any input instance I, let $\mathbf{R(I)}$ denote the set of deadline jobs that are not scheduled by Online Slacker. We say that $R(I)$ is the set of jobs rejected by Online Slacker. We say that a job j is **rejected at release time** t if time t is the first release time after r_j where job j is not in $A(I_t)$. We use $\mathbf{R(I_t)}$ to denote the set of jobs rejected at release time t.*

Once a job is rejected, it is never accepted.

Definition 6. *For any input instance I, let $\mathbf{S(I)}$ denote the set of deadline jobs that are scheduled by Online Slacker.*

Every job scheduled by Online Slacker must be accepted by Online Slacker. However, a job may be accepted by Online Slacker but still get rejected at a later time.

Observation 3. *For any input instance I, any times $t_2 > t_1 \geq 0$ where I_{t_1} and I_{t_2} are both not empty, and any time z such that some delaying deadline job is executed in $[z, z+1)$ in $OnlineSlacker(I_{t_1})$, a delaying deadline job is executed in $[z, z+1)$ in $OnlineSlacker(I_{t_2})$.*

Lemma 4. *For any two input instances I and I' that differ only in the weights of the deadline jobs, the set of times that deadline jobs are scheduled by Online Slacker for I is identical to the set of times that deadline jobs are scheduled by Online Slacker for I'.*

In the uniform weight case, we may assume without loss of generality that once Online Slacker accepts a job, it will eventually schedule the job. The only reason to reject the job would be to schedule some other deadline job, but since all jobs have uniform weight, we may simply reject the other deadline job instead.

The first step is to demonstrate that given a feasible set of deadline jobs, Online Slacker completes at least half of them.

Lemma 5. *Given a feasible input $I = I_f \cup I_d$, Online Slacker legally schedules at least $\left\lceil \frac{|I_d|}{2} \right\rceil$ of the deadline jobs.*

Proof. Given Lemma 4, we assume that all jobs in I_d have uniform weight. This allows us to assume that once a job is accepted, it will be scheduled. We call any time t where $OnlineSlacker(I)$ schedules a deadline job on $[t, t+1)$ a hit. We call any time t where $Slacker(I)$ schedules a deadline job j_1 on $[t, t+1)$ while $OnlineSlacker(I)$ does not schedules a deadline job on $[t, t+1)$ a miss. The number of deadline jobs that Online Slacker fails to schedule is at most the number of misses. We will show that the number of misses is at most the number of hits and the result follows.

Consider any miss at time t where $Slacker(I)$ schedules deadline job j_1. We now argue that job j_1 must either have been already scheduled by Online Slacker or is part of $A(I)$. Job j_1 must have been released by time t or else Slacker could not schedule it at time t. It must have a deadline of at least $t+1$ or else Slacker could not have scheduled it at time t. It must have been feasible when it arrived because it could have been scheduled at time t as Online Slacker's set of times devoted to deadline jobs is monotonically increasing over time by Observation 3. Thus, job j_1 is part of Online Slacker's currently accepted job set and will be scheduled at some later time or it was scheduled in some hole earlier in the schedule. Thus, we can associate each miss with a unique hit, and it follows that the number of misses is at most the number hits.

Lemma 6. *Given any input instance I and any release time t, if $j \in R(I_t)$, then in the interval $[t, d_j)$ in $OnlineSlacker(I_t)$, only deadline jobs with weight at least w_j are scheduled.*

This follows from the optimality of Slacker for the offline version of this problem. We now show that the sum of weighted deadline jobs scheduled by Online Slacker is at least half of the offline optimal.

Lemma 7. *Given a feasible input $I = I_f \cup I_d$, $V_{OnlineSlacker}(I) \geq \frac{V_{Slacker}(I)}{2}$.*

Proof (proof sketch). Given Lemma 4, we know that Online Slacker schedules at least half the deadline jobs in a feasible input instance I. We now show that the weights of those jobs scheduled by Online Slacker must be at least half the weight of all jobs.

To do this, we define a mapping f between rejected jobs and scheduled jobs. We consider rejected jobs in order of rejection time. For jobs with the same rejection time, we consider them in order of deadline breaking ties arbitrarily. For a job j rejected at time t, we define $f(t)$ to be the first unmapped job scheduled at or after time t.

It can be shown that the job $f(j)$ must be scheduled in time $[t, d_j)$. By Lemma 6, the jobs scheduled in $[t, d_j)$ in $OnlineSlacker(I_t)$ must have weight at least w_j. If any of these are later rejected, they must be replaced by jobs of at least the same weight. Thus, the weight of job $f(j)$ must be at least w_j. The lemma then follows.

Finally, we observe that adding a deadline job to any input I cannot decrease the resulting value of the Online Slacker's schedule.

Lemma 8. *Given an input $I = I_f \cup I_d$ and any deadline job j, let $I' = I \cup \{j\}$. $F_{OnlineSlacker}(I) \leq F_{OnlineSlacker}(I')$, and $V_{OnlineSlacker}(I) \leq V_{OnlineSlacker}(I')$.*

Lemma 9. *Given an input $I = I_f \cup I_d$, $V_{OnlineSlacker}(I) \geq \frac{1}{2}V_{Slacker}(I)$.*

Proof. First let I'_d denote the subset of deadline jobs scheduled by Slacker and $I' = I_f \cup I'_d$. From Lemma 7, we see that $V_{OnlineSlacker}(I') \geq \frac{1}{2}V_{Slacker}(I')$. From Lemma 8, it follows that $V_{OnlineSlacker}(I) \geq V_{OnlineSlacker}(I') \geq \frac{1}{2}V_{Slacker}(I') = \frac{1}{2}V_{Slacker}(I)$.

Lastly, we show that the flow time of Online Slacker is at most that of Slacker.

Lemma 10. *Given an input $I = I_f \cup I_d$, $F_{OnlineSlacker}(I) \leq F_{Slacker}(I)$.*

Proof. Given Lemma 4, we can assume that all deadline jobs have the same weight. We can then assume that once a job is accepted, it will eventually be scheduled. $F_{OnlineSlacker}(I) = F_{Slacker}(A(I))$, and since the set of deadline jobs that are in $A(I)$ is a subset of I_d, it follows from Observation 2 that $F_{Slacker}(A(I_t)) \leq F_{Slacker}(I)$. Ergo, $F_{OnlineSlacker}(I) \leq F_{Slacker}(I)$.

Corollary 3. *Online Slacker is a 2-value 1-flow algorithm for the Unit Dead-Flow problem.*

5 NP-Hardness Result

The previous section demonstrates that Slacker produces optimal schedules when jobs are of unit size. However, when flow jobs sizes are unbounded, Slacker does not always minimize total flow time.In fact, the DeadFlow problem with flow jobs of unbounded size is NP-Complete, even if there is a single deadline job. To prove this observation, we present a summary of the reduction process from Partition to SDF.

SDF becomes a knapsack problem for a very restricted class of SDF instances that we call *selection instances*.

Definition 7. *An instance $I = I_f \cup I_d$ is a **selection instance** if I_d and I_f satisfy the following properties: $I_d = \{(r' = 0, p', d' = M_{SRPT}(I_f)\}$, and all the jobs in I_f have unique release times r_i. If we order the jobs in increasing order of release time, $r_i + p_i \leq r_{i+1}$.*

In a selection instance, if we ignore the deadline job, the flow jobs can be scheduled trivially since each job completes by the next job's release time. The main issue for selection instances is determining which flow jobs should be moved to the deadline d' to create space for the deadline job. The cost of moving a flow job j can be broken into two components: the fixed cost of moving a job beyond the deadline d' and the variable cost of rescheduling j after d'. The benefit gained by moving flow job j is that we can schedule up to p_j units of the deadline job in the interval $[r_j, r_j + p_j)$. In knapsack terms, our goal is to move p' units of flow jobs while minimizing the cost incurred in doing so. There are some special properties in these knapsack instances that we must account for in our reductions as is illustrated in Figure 1. These include the observation that no two jobs will have the same fixed cost and the fixed cost for a job j is at least the sum of the sizes of all flow jobs that arrive after job j.

To show that SDF is NP-hard, we will start our reduction from the Partition problem. Note that this problem reduces to standard knapsack in one step. We reduce Partition to *Unique Number Partition* or UNP, the variant of Partition such that every integer in the input is unique. This ensures that each item in the partition input is unique, and we use this property in the next reduction to guarantee a unique fixed cost for each item in the knapsack problem. We then reduce UNP to a constrained variant of the minimization 0-1 Knapsack problem where we ensure that the cost of each flow job is at least the sum of all later arriving flow jobs. This second reduction is very similar to the standard Partition to Knapsack reduction, but it demonstrates that simply choosing which flow jobs must be displaced by the single deadline job, while ignoring the cost of rescheduling them, is NP-hard. The third intermediate problem is a variant of the second problem where we account for the cost of rescheduling the displaced jobs by scaling the fixed costs so that the cost of rescheduling them is sufficiently small. The third problem reduces to SDF by transforming the Knapsack instance into a selection instance.

Selection instance:

$I = I_f \cup I_d$ where $I_f = \{j_1 = (0,3), j_2 = (3,3), j_3 = (6,2), j_4 = (8,1)\}$ and $I_d = \{j_5 = (0,p',9)\}$

Corresponding knapsack instance:
Find the minimum cost set of items whose sizes sum up to at least p' from the set $\{(6,3),(3,3),(1,2),(0,1)\}$.

$SRPT(I_f)$

j_1	j_2	j_3	j_4
0	3	6	8 9

$i =$	1	2	3	4
c_i	6	3	1	0
s_i	3	3	2	1

Fig. 1. Illustration of how an SDF selection instance can be modeled as a knapsack instance if we only consider the fixed cost of displacing a job

Delineating the P-NP Boundary. We identify three factors that contribute to the complexity of any DeadFlow problem: flow job size, deadline job size, and deadline job weight. Of these factors, flow job size is the most significant. When flow jobs have unrestricted size, most natural DeadFlow variants are NP-Complete. When flow jobs size are of unit length, the DeadFlow problem is in P if one of the following two conditions holds: the deadline jobs do not have weights or the deadline jobs have unit length. If the deadline jobs have both weights and arbitrary length, the problem is NP-Complete. Furthermore, our current reduction from Partition to SDF only demonstrates that the problem is weakly NP-hard. It is an open problem whether or not DeadFlow is strongly NP-hard.

6   Conclusions

In this paper, we introduced the mixed criterion DeadFlow scheduling problem that models providing quality of service in networks. Unfortunately, even the simple variant SDF is NP-Complete, which is interesting since both total flow time scheduling and finding a legal schedule for a feasible set of deadline jobs can both be solved in polynomial time using simple online algorithms. We give a polynomial time algorithm Slacker that solves the special case where all jobs have unit size as well as some simple extensions of this problem. Furthermore, we show that an online version of Slacker is a 2-value 1-flow approximation algorithm for the online Unit DeadFlow problem. This result is interesting as our algorithm achieves an almost optimal competitive ratio for the single criterion packet scheduling problem while it maximizes the amount of time devoted to flow jobs. This highlights the value of the mixed criteria approach to the packet scheduling problem.

There are many open questions. For instance, can we find an online algorithm for the Unit DeadFlow problem that is better than 2-competitive for value while being constant competitive for flow time? Furthermore, are there good offline and online algorithms for the DeadFlow problem?

Acknowledgements. The authors thank Laura Dillon for many helpful discussions. This work was supported in part by the National Science Foundation under grant CCR 0105283 and the U.S. Department of the Navy, Office of Naval Research under Grant No. N00014-01-1-0744.

References

1. A. Agnetis, P. B. Mirchandani, D. Pacciarelli, and A. Pacifici. Scheduling problems with two competing agents. *Operations Research*, 52(2):229–242, November 2004.
2. W. A. Aiello, Y. Mansour, S. Rajagopolan, and A. Rosen. Competitive queue policies for differentiated services. *Journal of Algorithms*, 55:113–141, 2005.
3. N. Andelman, Y. Mansour, and A. Zhu. Competitive queuing policies for QoS switches. In *Proc. 14th Symp. on Discrete Algorithms (SODA)*, pages 761–770, 2003.
4. J. Aslam, A. Rasala, C. Stein, and N. Young. Improved bicriteria existence theorems for scheduling. In *Proceedings of the tenth annual ACM-SIAM symposium on Discrete algorithms*, pages 846–847, 1999.
5. K. R. Baker and J. C. Smith. A multiple-criterion model for machine scheduling. *Journal of Scheduling*, 6(1):7 – 16, January-February 2003.
6. N. Bansal, L. Fleischer, T. Kimbrel, M. Mahdian, B. Schieber, and M. Sviridenko. Further improvements in competitive guarantees for QoS buffering. In *Proc. 31st International Colloquium on Automata, Languages, and Programming (ICALP)*, pages 196–207, 2004.
7. P. Baptiste, M. Chrobak, C. Durr, W. Jawor, and N. Vakhania. Preemptive scheduling of equal-length jobs to maximize weighted throughput. http://www.citebase.org/cgi-bin/citations?id=oai:arXiv.org:cs/0209033, 2002.
8. Y. Bartal, F. Y. L. Chin, M. Chrobak, S. P. Y. Fung, W. Jawor, R. Lavi, J. Sgall, and T. Tichy. Online competitive algorithms for maximizing weighted throughput of unit jobs. In *Proc. 21st Symp. on Theoretical Aspects of Computer Science (STACS)*, pages 190–210, 2004.
9. H. Chetto and M. Chetto. Some results of the earliest deadline scheduling algorithm. *IEEE Transactions on Software Engineering*, 15:1261–1269, 1989.
10. M. Chrobak, W. Jawor, J. Sgall, and T. Tichy. Improved online algorithms for better management in QoS switches. In *Proc. 12th European Symp. on Algorithms (ESA)*, pages 204–215, 2004.
11. V. Chvátel. *Linear Programming*. W. H. Freeman and Company, 1983.
12. M. R. Garey and D. S. Johnson. *Computers and Intractability, A Guide to the Theory of NP-Completeness*. Freeman, 1979.
13. B. Hajek. On the competitiveness of online scheduling of unit-length packets with hard deadlines in slotted time. In *Proceedings of 2001 Conference on Information Sciences and Systems*, pages 434–438, 2001.
14. J. A. Hoogeveen. Single-machine scheduling to minimize a function of two or three maximum cost criteria. *Journal of Algorithms*, 21(2):415–433, September 1996.
15. J. A. Hoogeveen and S. L. van de Velde. Minimizing total completion time and maximum cost simulatneously is solvable in polynomial time. *Operations Research Letters*, 17:205–208, 1995.
16. A. Kesselman, Z. Lotker, Y. Mansour, B. Patt-Shamir, B. Schieber, and M. Sviridenko. Buffer overflow management in QoS switches. *SIAM Journal on Computing*, 33:563–583, 2004.

17. A. Kesselman, Y. Mansour, and R. van Stee. Improved competitive guarantees for QoS buffering. *Algorithmica*, 43:63–80, 2005.
18. J. P. Lehoczky and S. Ramos-Thuel. An optimal algorithm for scheduling soft-aperiodic tasks in fixed-priority preemptive systems. In *IEEE Real-Time Systems Symposium*, pages 110–123, 1992.
19. F. Li, J. Sethuraman, and C. Stein. An optimal online algorithm for packet scheduling with agreeable deadlines. In *Proc. 16th Symp. on Discrete Algorithms (SODA)*, pages 801–802, 2005.
20. S. T. McCormick and M. L. Pinedo. Scheduling n independent jobs on m uniform machines with both flow time and makespan objectives: A parametric approach. *ORSA Journal of Computing*, pages 63–77, 1992.
21. K. Pruhs, P. Uthaisombut, and G. Woeginger. Getting the best response for your erg. In *Scandanavian Workshop on Algorithms and Theory*, 2004.
22. A. Rasala, C. Stein, E. Torng, and P. Uthaisombut. Existence theorems, lower bounds and algorithms for scheduling to meet two objectives. In *Proceedings of the thirteenth annual ACM-SIAM symposium on Discrete algorithms*, pages 723 – 731, 2002.
23. M. Spuri and G. C. Buttazzo. Scheduling aperiodic tasks in dynamic priority systems. *Real Time Systems*, 10:179–210, 1996.
24. C. Stein and J. Wein. On the existence of schedules that are near-optimal for both makespan and total weighted completion time. *Operations Research Letters*, 21, 1980.
25. L. N. V. Wassenhove and F. Gelders. Solving a bicriterion scheduling problem. *European Journal of Operations Research*, 4:42–48, 1980.

Efficient Algorithms for k-Disjoint Paths Problems on DAGs*

Rudolf Fleischer, Qi Ge, Jian Li, and Hong Zhu**

Shanghai Key Laboratory of Intelligent Information Processing,
Department of Computer Science and Engineering,
Fudan University, Shanghai, China
{rudolf,qge,lijian83,hzhu}@fudan.edu.cn

Abstract. Given an acyclic directed graph and two distinct nodes s and t, we consider the problem of finding k disjoint paths from s to t satisfying some objective. We consider four objectives, MinMax, Balanced, MinSum-MinMin, and MinSum-MinMax. We use the algorithm by Perl-Shiloach and labelling and scaling techniques to devise an FPTAS for the first three objectives. For the forth one, we propose a general and efficient polynomial-time algorithm.

1 Introduction

In communication networks, one way of providing reliable communication is to find several disjoint paths, either node disjoint or edge disjoint. The advantage is that, if some links are broken, there are still other routing paths.

Different objectives may be used to measure the quality (or usefulness) of the disjoint paths. For example, we may require that the total weight of the disjoint paths to be minimized, the so-called *MinSum objective*. This problem can be solved in polynomial time by standard network flow methods [1,9]. However, for many other objectives, the problems are hard to solve. Li *et al.* [7] proposed the MinMax objective and showed that the problem is strong NP-complete. Yang *et al.* [11] proposed the MinMin objective and proved that the problem is also strong NP-complete.

For acyclic directed graphs (DAGs), the problem seems to be easier. In this paper, we focus on finding disjoint paths on DAGs. We propose efficient algorithms for four different objectives that are practically motivated, MinMax k-DP, Balanced k-DP, MinSum-MinMax k-DP, and MinSum-MinMin k-DP.

Let $G = (V, E)$ be a directed graph, s and t two distinct nodes in V, and $\mathcal{F} : E \mapsto \mathbb{N}$ a positive integral weight function on the edges. Li *et al.* [7] had

---

* This work is supported by National Natural Science Fund (grants #60573025, #60496321, #60373021) and Shanghai Science and Technology Development Fund (grant #03JC14014).
** The order of authors follows the international standard of alphabetic order of the last name. In China, where first-authorship is the only important aspect of a publication, the order of authors should be Qi Ge, Jian Li, Rudolf Fleischer, Hong Zhu.

M.-Y. Kao and X.-Y. Li (Eds.): AAIM 2007, LNCS 4508, pp. 134–143, 2007.
© Springer-Verlag Berlin Heidelberg 2007

proposed the `MinMax` k-`DP` problem, where we want to find k disjoint paths $P_1, \ldots, P_k$ from s to t such that the cost of the most expensive path is minimized, i.e., $\max_{1 \leq i \leq k} \mathcal{F}(P_i)$ is minimized, where $\mathcal{F}(P_i) = \sum_{e \in P_i} \mathcal{F}(e)$ is the weight of P_i. They proved that the problem is strong NP-complete, for directed and undirected graphs, and for edge-disjoint and node-disjoint paths. On DAGs, the problem is NP-complete but has a pseudo-polynomial-time algorithm. We will give an FPTAS for this problem on DAGs.

A variant of this problem is the `Balanced` k-`DP` problem where we want to find k disjoint paths $P_1, \ldots, P_k$ from s to t such that the costs of the cheapest and most expensive path are close together, i.e., $\max_{1 \leq i \leq k} \mathcal{F}(P_i) / \min_{1 \leq i \leq k} \mathcal{F}(P_i)$ is minimized. We can show by reduction from the Hamiltonian Path problem that this problem is strong NP-complete for directed and undirected graphs, and for edge-disjoint and node-disjoint paths. For DAGs, the problem is NP-complete by reduction from the Partition problem. We will give an FPTAS for this problem on DAGs.

In the `MinSum-MinMax` k-`DP` problem we want to find k disjoint $s - t$ paths $P_1, \ldots, P_k$ such that $\sum_{1 \leq i \leq k} \mathcal{F}(P_i)$ is minimized. Among all such paths, we want to find those minimizing $\max_{1 \leq i \leq k} \mathcal{F}(P_i)$. Note that we can show by reduction from the Disjoint Paths problem (see [3]) that this problem is strong NP-complete for directed graphs, and for edge-disjoint and node-disjoint paths. For undirected graphs and DAGs, the problem is NP-complete by reduction from the Partition problem. We will give an FPTAS for this problem on DAGs.

Similarly, in the `MinSum-MinMin` k-`DP` problem we want to minimize $\min_{1 \leq i \leq k} \mathcal{F}(P_i)$ among all k disjoint paths of minimum total length. This problem was proposed by Yang et al. [12]. They showed that, for $k = 2$, the problem is strong NP-complete for directed graphs and has a polynomial-time algorithm for DAGs. The latter algorithm reduces the `MinSum-MinMin` 2DP problem to the `Normalized` α^+-`MinSum` 2DP problem [13] which can be solved in polynomial time. However, the algorithm uses many expensive arithmetic operations like multiplications and divisions, and it is not very intuitive. Moreover, it cannot be generalized to arbitrary constant k. We will propose a more efficient algorithm for arbitrary constant k.

Due to space constraints we state all algorithms for $k = 2$. We shortly sketch how to generalize them to arbitrary constant $k \geq 2$.

2 Preliminaries

2.1 The Perl-Shiloach Algorithm

In this subsection, we introduce the algorithm PSA to find k node-disjoint paths on a DAG by Perl-Shiloach [8], which is a key subroutine in our algorithms.

In the *Disjoint Paths Problem* (DPP) we are given a directed graph $G = (V, E)$ and k pairs of distinct nodes $(s_1, t_1), \ldots, (s_k, t_k)$. We want to find k node- or edge-disjoint paths $P_1, \ldots, P_k$, where P_i is a path from s_i to t_i, for $1 \leq i \leq k$. The decision version of this problem, for node-disjoint and edge-disjoint paths, was shown to be NP-complete by Fortune *et al.* [3], even if $k = 2$. For DAGs, Perl and Shiloach gave a polynomial time algorithm, *PSA* [8].

PSA is actually a reduction from DPP to the *Connectivity problem*. Given a DAG $G = (V, E)$ and k pairs of distinct nodes $(s_1, t_1), \ldots, (s_k, t_k)$, let $v_1, \ldots, v_n$ be a topological order of V, i.e., there are only edges from nodes with lower indices to nodes with higher indices. We construct a graph $G_k = (V_k, E_k)$ as follows:

$$V_k = \{\langle j_1, \ldots, j_k \rangle \mid 1 \leq j_i \leq n \text{ for } 1 \leq i \leq k, j_i \neq j_l \text{ for } 1 \leq i \neq l \leq k\}, \quad (1)$$

$$E_k = \bigcup_{d=1}^{k} \{(\langle j_1, \ldots, j_{d-1}, j_d, j_{d+1}, \ldots, j_k \rangle, \langle j_1, \ldots, j_{d-1}, j_d', j_{d+1}, \ldots, j_k \rangle)$$
$$\mid (v_{j_d}, v_{j_d'}) \in E \text{ and } j_d = \min_{1 \leq l \leq k} j_l\}. \quad (2)$$

For simplicity, we will only describe the algorithms for the case of $k = 2$. For $k \geq 2$, see [2]. To find two disjoint paths from $s = v_1$ to $t = v_n$, we add two nodes $\langle s, s \rangle = \langle 1, 1 \rangle$ and $\langle t, t \rangle = \langle n, n \rangle$ to V_2 to obtain graph G_2':

$$V_2' = \{\langle i, j \rangle \mid 1 \leq i, j \leq n \text{ and } i \neq j\} \cup \{\langle s, s \rangle, \langle t, t \rangle\}, \quad (3)$$

$$E_2' = \{(\langle i, j \rangle, \langle i, k \rangle) \mid (v_j, v_k) \in E, j < i\}$$
$$\cup \{(\langle i, j \rangle, \langle k, j \rangle) \mid (v_i, v_k) \in E, i < j\}. \quad (4)$$

We call the edges in the first set of Eq. (4) *horizontal* edges and the edges in the second set *vertical* edges.

Lemma 1 [8]. *There are two node disjoint paths P_1, P_2 from s to t in G if and only if there is a directed path P from $\langle s, s \rangle$ to $\langle t, t \rangle$ in G_2', and P_1 (P_2) consists of the horizontal (vertical) edges of P.* □

2.2 Edge Disjoint Versus Node Disjoint Paths

We can transform the acyclic edge disjoint case to the acyclic node disjoint case using the method by Li *et al.* [7]. Given a DAG G and two distinct nodes s, t, add new nodes u, v and edges $(u, s), (t, v)$. Form the directed line graph (see [4]), and let s', t' denote the nodes corresponding to $(u, s), (t, v)$ respectively. Replace each node w (except s', t') in the line graph with two nodes w_1, w_2 and an edge (w_1, w_2) such that all edges into (out of) w are now into w_1 (out of w_2). The weight of (w_1, w_2) is the weight of the edge in G corresponding to w. Other edges have weight 0. This weight-preserving transformation gives a one-to-one correspondence between edge disjoint paths in the original graph and node disjoint paths in the new graph. Thus, for all the problems investigated in this paper, we only give algorithms for the node disjoint case.

3   The MinMax 2DP Problem

Our FPTAS for MinMax 2DP on DAGs is based on the pseudo-polynomial-time algorithm by Li *et al.* [7] and the weight scaling technique (cf. [5,6,10]).

First, we use PSA to construct the graph G_2' as in Eqs. (3) and (4). Then MinMax 2DP is equivalent to finding a directed path P from $\langle s, s \rangle$ to $\langle t, t \rangle$ in G_2' minimizing $\max\{\mathcal{F}(P_H), \mathcal{F}(P_V)\}$, where P_H (P_V) denotes the horizontal (vertical) edges of P.

The pseudo-polynomial-time algorithm uses a standard labeling method. If there is a directed path P from $\langle s, s \rangle$ to a node $\langle i, j \rangle \in V_2'$, we label it by $(X, Y, Pred)$, where X (Y) is a positive integer denoting the total weight of all horizontal (vertical) edges in P and $Pred$ is the index of the predecessor of $\langle i, j \rangle$ in P. We compute for each node in topological order a set of labels for all the paths from $\langle s, s \rangle$ to that node. When the algorithm terminates, the label in the label set of $\langle t, t \rangle$ minimizing $\max\{X, Y\}$ is the solution.

Unfortunately, the number of labels lab^2 for one node may be exponentially large, where $lab = (n - 1) \cdot \max_{e \in E} \mathcal{F}(e)$. In order to obtain a polynomial-time algorithm, we must somehow compress the label set. We use the scaling technique known from the Subset Sum FPTAS.

We store the labels of each node $\langle i, j \rangle$ in a 2-dimensional array $L_{i,j}[1 \ldots \ell, 1 \ldots \ell]$, where $\ell = \lfloor \log_{1+\delta} lab \rfloor + 1$. Label $(X, Y, Pred)$ will be stored in $L[\lfloor \log_{1+\delta} X \rfloor + 1, \lfloor \log_{1+\delta} Y \rfloor + 1]$. Each cell of $L_{i,j}$ will store at most one label. We use a set $I_{i,j}$ to keep track of all the entries of $L_{i,j}$ that actually store a label. Then, for each node in the topological order, we compute the label set of the node. If a new label should be stored in an array cell which already contains another label, then we discard the new label. We let $\mathcal{F}_{i,j}$ denote the cost of edge (v_i, v_j) in E. The third component $Pred$ of the label $(X, Y, Pred)$ is now of the form $(\langle i', j' \rangle, (a, b))$ and is used to reconstruct the path P corresponding to the label, where $\langle i', j' \rangle$ is the index of the predecessor of $\langle i, j \rangle$ in P and (a, b) is the index of the cell of $L_{i',j'}$ from which $(X, Y, Pred)$ is computed.

The subroutine LABELSCALING computes for each node a scaled set of labels, while the main algorithm FPTAS-DAG-MinMax-2DP returns the approximate solution.

LABELSCALING($G = (V, E), s, t, \mathcal{F}, \delta$)

1   Construct $G_2' = (V_2', E_2')$ as in Eqs. (3) and (4);
2   Initialize matrices $L_{i,j}[k, l] = NULL$, **for** $1 \leq i, j \leq n$ and
          $1 \leq k, l \leq \lfloor \log_{1+\delta} lab \rfloor + 1$;
3    $I_{i,j} = \emptyset$, **for** $1 \leq i, j \leq n$;
4    $I_{s,s} = \{(1, 1)\}$;
5    $L_{s,s}[1, 1] = (0, 0, NULL)$;
6   **for** $i \leftarrow 1$ **to** n **do**
7   **for** $j \leftarrow 1$ **to** n **do**
8   **for** each $(\langle i, k \rangle, \langle i, j \rangle) \in E'$
9   **for** each $(a, b) \in I_{i,k}$
10   let $(X, Y, Pred)$ be the label in $L_{i,k}[a, b]$;

11   let $c = \lfloor \log_{1+\delta}(X + \mathcal{F}_{k,j}) \rfloor + 1$;
12   if $L_{i,j}[c,b] == NULL$
13   then $L_{i,j}[c,b] = (X + \mathcal{F}_{k,j}, Y, (\langle i,k \rangle, (a,b)))$;
14              $I_{i,j} = I_{i,j} \cup \{(c,b)\}$;
15   for each $(\langle k,j \rangle, \langle i,j \rangle) \in E'_2$
16   for each $(a,b) \in I_{k,j}$
17   let $(X, Y, Pred)$ be the label in $L_{k,j}[a,b]$;
18   let $d = \lfloor \log_{1+\delta}(Y + \mathcal{F}_{k,i}) \rfloor + 1$;
19   if $L_{i,j}[a,d] == NULL$
20   then $L_{i,j}[a,d] = (X, Y + \mathcal{F}_{k,i}, (\langle k,j \rangle, (a,b)))$;
21                  $I_{i,j} = I_{i,j} \cup \{(a,d)\}$;

FPTAS-DAG-MinMax-2DP $(G = (V,E), s, t, \mathcal{F}, \epsilon)$

1    $\delta = (1 + \epsilon)^{\frac{1}{2n-2}} - 1$;
2   LABELSCALING $(G = (V,E), s, t, \mathcal{F}, \delta)$
3   for each index (a,b) in $I_{t,t}$
4   let $L_{t,t}[a,b] = (X, Y, Pred)$;
5   find the (a^*, b^*) minimizing $\max\{X, Y\}$;
6   Reconstruct the two disjoint paths from the third components of the labels;

Lemma 2. *Let $\langle i,j \rangle$ be a node in V'_2 such that there is a directed path P from $\langle s,s \rangle$ to $\langle i,j \rangle$. Let $X = \mathcal{F}(P_H)$ and $Y = \mathcal{F}(P_V)$. When the algorithm LABELSCALING terminates, then there exists an index pair (a,b) such that the label $(\tilde{X}, \tilde{Y}, Pred)$ in $L_{i,j}[a,b]$ satisfies $X/(1+\delta)^{i+j-2} \leq \tilde{X} \leq (1+\delta)^{i+j-2}X$ and $Y/(1+\delta)^{i+j-2} \leq \tilde{Y} \leq (1+\delta)^{i+j-2}Y$.*

Proof. By induction on $i + j$. If $i + j = 2$, the node is $\langle 1,1 \rangle = \langle s,s \rangle$, and $L_{s,s}[1,1] = (0, 0, NULL)$ is the correct result.

Let $i+j = l$ and $3 \leq l \leq 2n$. Suppose there is a path P from $\langle s,s \rangle$ to $\langle i,j \rangle$ with total weight of the horizontal (vertical) edges X (Y). Without loss of generality, assume the predecessor of $\langle i,j \rangle$ in P is $\langle i,j' \rangle$; the vertical case has a similar proof. Let the path from $\langle s,s \rangle$ to $\langle i,j' \rangle$ in P be P^+ , and let X^+ (Y^+) be the total weight of the horizontal (vertical) edges of P^+ . Then, $X = X^+ + \mathcal{F}_{j',j}$ and $Y = Y^+$. Since $j' < j$, $i + j' < i + j$, by the induction assumption there exists a label $(\tilde{X}^+, \tilde{Y}^+, Pred)$ such that $X^+/(1+\delta)^{i+j'-2} \leq \tilde{X}^+ \leq (1+\delta)^{i+j'-2}X^+$ and $Y^+/(1+\delta)^{i+j'-2} \leq \tilde{Y}^+ \leq (1+\delta)^{i+j'-2}Y^+$. So, when LABELSCALING computes the label set of the node $\langle i,j \rangle$ and enters steps 8 and 9, it will compute a new label $(\tilde{X}^+ + \mathcal{F}_{j',j}, \tilde{Y}^+, (\langle i,j' \rangle, (a,b)))$.

If the algorithm enters steps 12 to 14, the label $(\tilde{X}^+ + \mathcal{F}_{j',j}, \tilde{Y}^+, (\langle i,j' \rangle, (a,b)))$ will be stored. Let $\tilde{X} = \tilde{X}^+ + \mathcal{F}_{+j',j}, \tilde{Y} = \tilde{Y}^+$. Then, $\tilde{X} = \tilde{X}^+ + \mathcal{F}_{j',j} \leq (1+\delta)^{i+j'-2}X^+ + \mathcal{F}_{j',j} \leq (1+\delta)^{i+j'-2}(X^+ + \mathcal{F}_{j',j}) \leq (1+\delta)^{i+j-2}X \leq (1+\delta)^{i+j-2}X$, and $\tilde{Y} = \tilde{Y}^+ \leq (1+\delta)^{i+j'-2}Y^+ \leq (1+\delta)^{i+j-2}Y$. Similarly, we have $\tilde{X} \geq X/(1+\delta)^{i+j-2}$ and $\tilde{Y} \geq Y/(1+\delta)^{i+j-2}$.

If the algorithm skips steps 12 to 14, then there exists a label $(\tilde{X}, \tilde{Y}, Pred)$. By the fact that $\lfloor \log_{1+\delta} \tilde{X} \rfloor + 1 = \lfloor \log_{1+\delta}(\tilde{X}^+ + \mathcal{F}_{j',j}) \rfloor + 1$ and $\lfloor \log_{1+\delta} \tilde{Y} \rfloor + 1 = \lfloor \log_{1+\delta} \tilde{Y}^+ \rfloor + 1$, we have $\tilde{X} < (1+\delta)(\tilde{X}^+ + \mathcal{F}_{j',j}) \leq (1+\delta)((1+\delta)^{i+j'-2}X^+ +$

$\mathcal{F}_{j',j} \leq (1+\delta)^{i+j-2}(X^+ + \mathcal{F}_{j',j}) \leq (1+\delta)^{i+j-2}X$, and $\tilde{Y} \leq (1+\delta)\tilde{Y}^+ \leq (1+\delta)^{i+j-2}Y^+ = (1+\delta)^{i+j-2}Y$. Similarly, we have $\tilde{X} \geq X/(1+\delta)^{i+j-2}$ and $\tilde{Y} \geq Y/(1+\delta)^{i+j-2}$. $\qquad\square$

Theorem 1. *The algorithm FPTAS-DAG-MinMax-2DP is an FPTAS for the* MinMax 2DP *problem on DAGs.*

Proof. First, it is easy to verify that the time complexity of the algorithm is $O(n^3(\log_{1+\delta} lab)^2) = O(n^5\epsilon^{-2}(\ln lab)^2)$, where $lab = O(n \cdot \max_{e \in E} \mathcal{F}(e))$.

Second, we will show that the approximation ratio is $1 + \epsilon$. Suppose the optimum solution is P_1, P_2. By Lemma 1, there is a path P in G_2' corresponding to P_1, P_2. Let $X = \mathcal{F}(P_H) = \mathcal{F}(P_1)$ and $Y = \mathcal{F}(P_V) = \mathcal{F}(P_2)$. By Lemma 2, there is a label $(\tilde{X}, \tilde{Y}, Pred)$ in $L_{t,t}$ such that $\tilde{X} \leq (1+\delta)^{2n-2}X = (1+\epsilon)X$ and $\tilde{Y} \leq (1+\delta)^{2n-2}Y = (1+\epsilon)Y$. Let $(X', Y', Pred')$ be the solution returned by the algorithm and $Opt = \max\{X, Y\}$. Then, $\max\{X', Y'\} \leq \max\{\tilde{X}, \tilde{Y}\} \leq \max\{(1+\epsilon)X, (1+\epsilon)Y\} = (1+\epsilon) \cdot \max\{X, Y\} = (1+\epsilon)OPT$. Thus, FPTAS-DAG-MinMax-2DP is an FPTAS for the MinMax 2DP problem on DAGs. $\qquad\square$

For arbitrary constant $k \geq 2$, we can generalize the subroutine LABELSCALING by first constructing $G_k = (V_k, E_k)$ as in Eqs. (1) and (2) from G, using a k-dimensional array $L_{d_1, \ldots, d_k}[1 \ldots \ell, \ldots, 1 \ldots \ell]$, where $\ell = \lfloor \log_{1+\delta} lab \rfloor + 1$, and then updating L the same way as in LABELSCALING. By induction on k, we can get the following result similar to Lemma 2.

Lemma 3. *Let $\langle d_1, \ldots, d_k \rangle$ be a node in V_k such that there is a directed path P from $\langle s, \ldots, s \rangle$ to $\langle d_1, \ldots, d_k \rangle$. Let $X_i = \mathcal{F}(P_{d_i})$, $1 \leq i \leq k$, where P_{d_i} denotes the edges in dimension i of P in G_k. When the algorithm LABELSCALING terminates, there exists an index pair $(l_1, \ldots, l_k)$ such that the label $(\tilde{X}_1, \ldots, \tilde{X}_k, Pred)$ in $L_{d_1, \ldots, d_k}[l_1, \ldots, l_k]$ satisfies $X_i/(1+\delta)^{d_1 + \ldots + d_k - k} \leq \tilde{X}_i \leq (1+\delta)^{d_1 + \ldots + d_k - k}X_i$ for $1 \leq i \leq k$.*

Then we can modify the algorithm FPTAS-DAG-MinMax-2DP by setting $\delta = (1+\epsilon)^{\frac{1}{kn-k}} - 1$ to get an FPTAS for the MinMax k-DP problem.

4 The Balanced 2DP Problem

The FPTAS for Balanced 2DP on DAGS is similar to the one described in Section 3, using the same subroutine LABELSCALING.

FPTAS-DAG-Balanced-2DP$(G = (V, E), s, t, \mathcal{F}, \epsilon)$

1    $\delta = (1+\epsilon)^{\frac{1}{4n-4}} - 1$;
2   LABELSCALING$(G = (V, E), s, t, \mathcal{F}, \delta)$
3   **for** each index (a, b) in $I_{t,t}$
4   lot $L_{t,t}[a, b] - (X, Y, Pred)$;
5   find (a^*, b^*) minimizing $\max\{X, Y\}/\min\{X, Y\}$;
6   Reconstruct the 2 disjoint paths from the third entry of the labels;

Theorem 2. *FPTAS-DAG-Balanced-2DP is an FPTAS for the* Balanced *2DP problem on DAGs.*

Proof. First, it can easily be seen that the time complexity of FPTAS-DAG-Balanced-2DP is $O(n^3(\log_{1+\delta} lab)^2) = O(n^5\epsilon^{-2}(\ln lab)^2)$, where $lab = O(n \cdot \max_{e \in E} \mathcal{F}(e))$.

Next, we prove that the approximation ratio is $1 + \epsilon$. Suppose the optimum solution is P_1, P_2. By Lemma 1, there is a path P in G_2' corresponding to P_1, P_2. Let $X = \mathcal{F}(P_H) = \mathcal{F}(P_1)$ and $Y = \mathcal{F}(P_V) = \mathcal{F}(P_2)$. By Lemma 2, there is a label $(\tilde{X}, \tilde{Y}, Pred)$ in $L_{t,t}$ such that $X/\sqrt{1+\epsilon} = X/(1+\delta)^{2n-2}X \leq \tilde{X} \leq (1+\delta)^{2n-2}X = X\sqrt{1+\epsilon}$ and $Y/\sqrt{1+\epsilon} = Y/(1+\delta)^{2n-2}Y \leq \tilde{Y} \leq (1+\delta)^{2n-2}Y = Y\sqrt{1+\epsilon}$. Let $(X', Y', Pred')$ be the solution returned by the algorithm and $Opt = \max\{X,Y\}/\min\{X,Y\} = \max\{X/Y, Y/X\}$. Then, $\max\{X'/Y', Y'/X'\} \leq \max\{\tilde{X}/\tilde{Y}, \tilde{Y}/\tilde{X}\} \leq \max\{(1+\epsilon)X/Y, (1+\epsilon)Y/X\} = (1+\epsilon) \cdot \max\{X/Y, Y/X\} = (1+\epsilon)OPT$. Thus, FPTAS-DAG-Balanced-2DP is an FPTAS for the Balanced 2DP problem on DAGs. □

We can generalize the algorithm FPTAS-DAG-Balanced-2DP for arbitrary constant $k \geq 2$ by setting $\delta = (1+\epsilon)^{\frac{1}{2kn-2k}} - 1$ and then using a similar way to update L as in FPTAS-DAG-Balanced-2DP. The correctness proof follows from Lemma 3 and a generalization of Theorem 2.

5   The MinSum-MinMax 2DP Problem

In this section, we present an FPTAS for the MinSum-MinMax 2DP problem on DAGs. The difference between this problem and the MinMax 2DP problem is that in this problem we should find two disjoint paths with MinMax objective among the set of two disjoint paths whose total weight is minimized.

In the pseudo-polynomial-time algorithm for the MinMax 2DP problem on DAGs, for each node $\langle i, j \rangle$ in G_2', if there is a path from $\langle s, s \rangle$ to $\langle i, j \rangle$, then we will compute a label to store the information of this path. Now, for the MinSum-MinMax 2DP problem, instead of keeping information for all paths, we only store the information of the shortest paths. This can be done by scanning each node in topological order, and then computing for each node a set of labels corresponding to the shortest paths. We then can use the scaling method to convert the pseudo-polynomial-time algorithm to an FPTAS.

SHORTESTLABELSCALING$(G = (V, E), s, t, \mathcal{F}, \delta)$

1   Construct $G_2' = (V_2', E_2')$ as in Eqs. (3) and (4);
2   Initialize matrices $L_{i,j}[k,l] = NULL$, **for** $1 \leq i, j \leq n$ and
          $1 \leq k, l \leq \lfloor \log_{1+\delta} lab \rfloor + 1$;
3    $I_{i,j} = \emptyset$, **for** $1 \leq i, j \leq n$;
4    $I_{s,s} = \{(1,1)\}$;
5    $L_{s,s}[1,1] = (0, 0, NULL)$;
6   **for** $i \leftarrow 1$ to n **do**
7   **for** $j \leftarrow 1$ to n **do**

```
8        currentmin = +∞;
9        for each (⟨i,k⟩, ⟨i,j⟩) ∈ E'₂
10           for each (a,b) ∈ I_{i,k}
11              let (X, Y, Pred) be the label in L_{i,k}[a,b];
12              if X + F_{k,j} + Y < currentmin
13                 then I_{i,j} = ∅;
14                    set all entries of L_{i,j} to NULL;
15                    currentmin = X + F_{k,j} + Y;
16              if X + F_{k,j} + Y == currentmin
17                 let c = ⌊log_{1+δ}(X + F_{k,j})⌋ + 1;
18                    if L_{i,j}[c,b] == NULL
19                       then L_{i,j}[c,b] = (X + F_{k,j}, Y, (⟨i,k⟩, (a,b)));
20                          I_{i,j} = I_{i,j} ∪ {(c,b)};
21        for each (⟨k,j⟩, ⟨i,j⟩) ∈ E'₂
22           for each (a,b) ∈ I_{k,j}
23              let (X, Y, Pred) be the label in L_{k,j}[a,b];
24              if X + Y + F_{k,i} < currentmin
25                 then I_{i,j} = ∅;
26                    set all entries of L_{i,j} to NULL;
27                    currentmin = X + Y + F_{k,i};
28              if X + Y + F_{k,i} == currentmin
29                 let d = ⌊log_{1+δ}(Y + F_{k,i})⌋ + 1;
30                    if L_{i,j}[a,d] == NULL
31                       then L_{i,j}[a,d] = (X, Y + F_{k,i}, (⟨k,j⟩, (a,b)));
32                          I_{i,j} = I_{i,j} ∪ {(a,d)};
```

FPTAS-DAG-MinSum-MinMax-2DP$(G = (V,E), s, t, \mathcal{F}, \epsilon)$

```
1    δ = (1 + ε)^{1/(2n-2)} − 1;
2    SHORTESTLABELSCALING(G = (V, E), s, t, F, δ);
3    for each index (a,b) in I_{t,t}
4       let L_{t,t}[a,b] = (X, Y, Pred);
5       find (a*, b*) minimizing max{X, Y};
6    Reconstruct the two disjoint paths from the third entry of the labels;
```

Lemma 4. *Let $\langle i,j \rangle$ be a node in V'_2 and P a shortest path from $\langle s,s \rangle$ to $\langle i,j \rangle$. Let $X = \mathcal{F}(P_H)$ and $Y = \mathcal{F}(P_V)$. When the algorithm SHORTEST LABELSCALING terminates, then there exists an index pair (a,b) such that the label $(\tilde{X}, \tilde{Y}, Pred)$ in $L_{i,j}[a,b]$ satisfies $X/(1+\delta)^{i+j-2} \leq \tilde{X} \leq (1+\delta)^{i+j-2}X$ and $Y/(1+\delta)^{i+j-2} \leq \tilde{Y} \leq (1+\delta)^{i+j-2}Y$.*

Proof. First, a simple induction on the topological order of the nodes shows that when SHORTESTLABELSCALING terminates, the labels of each node correspond to shortest paths to the nodes.

Second, similar to the proof of Lemma 2, we can show there is a label $(\tilde{X}, \tilde{Y}, Pred)$ satisfying $X/(1+\delta)^{i+j-2} \leq \tilde{X} \leq (1+\delta)^{i+j-2}X$ and $Y/(1+\delta)^{i+j-2} \leq \tilde{Y} \leq (1+\delta)^{i+j-2}Y$. □

From Lemma 4 and an analysis similar to the proof of Theorem 1, we obtain the following result.

Theorem 3. *FPTAS-DAG-MinSum-MinMax-2DP is an FPTAS for the* MinSum-MinMax 2DP *problem on DAGs.*   □

We can generalize the subroutine SHORTESTLABELSCALING for arbitrary constant $k \geq 2$ in a similar way as in the generalization of subroutine LABELSCALING in Section 3, and thus give an FPTAS for the the MinSum-MinMax k-DP problem on DAGs.

6 The MinSum-MinMin 2DP Problem

In this section, we present an efficient polynomial-time algorithm for the MinSum-MinMin 2DP problem on DAGs.

We introduce some notions. Let $\mathbb{Z}^2 = \{(X, Y) \mid X, Y \in \mathbb{Z}\}$ be the set of all pairs of positive integers. We define the relationship '$<$' on $\mathbb{Z}^2$ as: $(X, Y) < (X', Y')$ if and only if $X < X'$ or ($X = X'$ and $Y < Y'$). The operation '$+$' on two elements of $\mathbb{Z}^2$ is defined as $(X, Y) + (X', Y') = (X + X', Y + Y')$.

We first use PSA to transform the given graph G into G'_2, but with different edge weights than before. We let the weight of an edge in G'_2 be an element of $\mathbb{Z}^2$. Let $\mathcal{F}' : E'_2 \mapsto \mathbb{N}$ be the new weight function on G'_2. For a horizontal edge $(\langle i, j \rangle, \langle i, j' \rangle)$, $\mathcal{F}'((\langle v_i, v_j \rangle, \langle v_i, v_{j'} \rangle)) = (\mathcal{F}_{j,j'}, \mathcal{F}_{j,j'})$, and for a vertical edge $(\langle i, j \rangle, \langle i', j \rangle)$, $\mathcal{F}'((\langle i, j \rangle, \langle i', j \rangle)) = (\mathcal{F}_{i,i'}, 0)$. For G'_2 and weight function $\mathcal{F}'$, we compute the shortest path P from $\langle s, s \rangle$ to $\langle t, t \rangle$. It can be shown that the two disjoint paths from s to t in G corresponding to P in G' are an optimum solution to the MinSum-MinMin 2DP problem.

DAG-MinSum-MinMin-2DP$(G = (V, E), s, t, \mathcal{F})$

1   Construct $G'_2 = (V'_2, E'_2)$ as in Eq. (3) and (4);
2   **for** each $\langle i, j \rangle \in V'_2$
3   let $d_{i,j} = (+\infty, +\infty)$;
4            $p_{i,j} = NULL$;
5    $d_{s,s} = (0, 0)$;
6   **for** $i \leftarrow 1$ to n **do**
7   **for** $j \leftarrow 1$ to n **do**
8   **for** each $(\langle i, k \rangle, \langle i, j \rangle) \in E'_2$
9   **if** $d_{i,k} + (\mathcal{F}_{k,j}, \mathcal{F}_{k,j}) < d_{i,j}$
10   **then** $d_{i,j} = d_{i,k} + (\mathcal{F}_{k,j}, \mathcal{F}_{k,j})$;
11                        $p_{i,j} = \langle i, k \rangle$;
12   **for** each $(\langle k, j \rangle, \langle i, j \rangle) \in E'_2$
13   **if** $d_{k,j} + (\mathcal{F}_{k,i}, 0) < d_{i,j}$
14   **then** $d_{i,j} = d_{k,j} + (\mathcal{F}_{k,i}, 0)$;
15                        $p_{i,j} = \langle k, j \rangle$;
16   Reconstruct the two disjoint paths from $p_{t,t}$;

It can easily be shown that $d_{i,j}$ is the value of the shortest path from $\langle s, s \rangle$ to $\langle i, j \rangle$ with respect to the weight function $\mathcal{F}'$.

When the algorithm DAG-MinSum-MinMin-2DP terminates, for any node $\langle i, j \rangle$ in G'_2, let P be the path from $\langle s, s \rangle$ to $\langle i, j \rangle$ constructed by tracing backwards from $p_{i,j}$ to $\langle s, s \rangle$. Let $\mathcal{F}'(P) = (X, Y)$, then by the definition of $\mathcal{F}'$, we have $\mathcal{F}(P) = X$ and $\mathcal{F}(P_H) = Y$. Since (X, Y) is minimized, X is also minimized, and for any path P' from $\langle s, s \rangle$ to $\langle i, j \rangle$ such that $\mathcal{F}(P') = X$, we have $\mathcal{F}(P_H) \leq \mathcal{F}(P'_H)$. We also have $Y \leq X - Y$, that is, $\mathcal{F}(P_H) \leq \mathcal{F}(P_V)$. Suppose for contradiction, $\mathcal{F}(P_H) > \mathcal{F}(P_V)$, then by the symmetry of the construction of G'_2, there is another path P' that $P_H = P'_V$ and $P_V = P'_H$. Thus, $\mathcal{F}(P') = X$ and $\mathcal{F}(P'_V) = \mathcal{F}(P_H) = Y > P_V = \mathcal{F}(P'_H)$, contradicting the fact that (X, Y) is minimal.

The above result is also true for $\langle t, t \rangle$. This proves the correctness of the algorithm. The running time of the algorithm is $O(|E'|) = O(n^3)$.

We note that our algorithm can easily be generalized to the case of $k > 2$, in contrast to the algorithm by Yang et $al.$ [12]. When $k > 2$, we construct G'_k as in Eqs. (1) and (2), and again set the weight of each edge in G'_k to be an element of $\mathbb{Z}^2$. The first integer of the weight is the sum of the weights of all k paths, and the second integer is the weight of the minimum weight path. Then, we use a standard shortest path algorithm to compute the shortest path in G' under the new weight function.

References

1. P. K. Ahuja, T. L. Magnanti, and J. B. Orlin. *Network Flows, Theory, Algorithms, and Applications.* Prentice Hall, Englewood Cliffs, 1993.
2. J. Bang-Jensen and G. Gutin. *Digraphs: Theory, Algorithms and Applications.* Springer, 2000.
3. S. Fortune, J. Hopcroft, and J. Wyllie. The directed subgraph homemorphism problem. *Theoretical Computer Science*, 10:111–121, 1980.
4. F. Harary. *Graph Theory.* Addison-Wesley, Reading, MA, 1972.
5. R. Hassin. Approximation schemes for the restricted shortest path problem. *Mathematics of Operations Research*, 17:237–260, 1992.
6. O. H. Ibarra and C. E. Kim. Fast approximation for the knapsack and sum of subset problems. *JACM*, 22:463–468, 1975.
7. C. L. Li, S. T. McCormick, and D. Simchi-Levi. The complexity of finding two disjoint paths with Min-Max objective function. *Discrete Applied Mathematics*, 26(1):105–115, 1990.
8. Y. Perl and Y. Shiloach. Finding two disjoint paths between two pairs of vertices in a graph. *J. ACM*, 25(1):1–9, 1978.
9. J. W. Suurballe and R. Tarjan. A quick method for finding shortest pairs of disjoint paths. *Networks*, 14:325–336, 1984.
10. G. Tsaggouris and C. Zaroliagis. Improved FPTAS for Multiobjective Shortest Paths with Applications. CTI Technical Report TR 2005/07/03, July 2005.
11. B. Yang, S. Q. Zheng, and S. Katukam. Finding two disjoint paths in a network with Min-Min objective function. In *Proc. 15th IASTED International Conference on Parallel and Distributed Computing and Systems*, pp. 75–80, 2003.
12. B. Yang, S. Q. Zheng, and E. Lu. Finding two disjoint paths in a network with MinSum-MinMin objective function. Technical Report, Dept. of Computer Science, Univ. of Texas at Dallas, April, 2005.
13. B. Yang, S. Q. Zheng, and E. Lu. Finding two disjoint paths in a network with normalized $\alpha^+ - MIN - SUM$ objective function. In *Proc. 16th International Symp. on Algorithms and Computation*, pp. 954–963, 2005.

Acyclic Edge Colouring of Outerplanar Graphs

Rahul Muthu, N. Narayanan, and C.R. Subramanian

The Institute of Mathematical Sciences, Chennai, India
{rahulm,narayan,crs}@imsc.res.in

Abstract. An *acyclic* edge colouring of a graph is a proper edge colouring having no 2-coloured cycle, that is, a colouring in which the union of any two colour classes forms a linear forest. The *acyclic chromatic index* of a graph is the minimum number k such that there is an acyclic edge colouring using k colours and is usually denoted by $a'(G)$. Determining $a'(G)$ exactly is a very hard problem (both theoretically and algorithmically) and is not determined even for complete graphs. We show that $a'(G) \leq \Delta(G) + 1$, if G is an outerplanar graph. This bound is tight within an additive factor of 1 from optimality. Our proof is constructive leading to an $O(n \log \Delta)$ time algorithm. Here, $\Delta = \Delta(G)$ denotes the maximum degree of the input graph.

1 Introduction

Graph colouring is one of the well-studied areas of graph theory, having many practical applications. Vertex and edge colouring problems have many variants. In this paper, we consider a variant of edge colouring known as the *acyclic edge colouring*. An edge colouring of a graph is *proper* if no pair of incident edges receive the same colour. A proper colouring $\mathcal{C}$ of the edges of a graph G is *acyclic* if there is no 2-coloured (bichromatic) cycle in G with respect to $\mathcal{C}$. To put it differently, the subgraph induced by the union of any two colour classes of $\mathcal{C}$ is a linear forest. We are interested in finding the minimum number k such that there exists an acyclic edge colouring of G using k colours. The number k, known as the *acyclic chromatic index* of G is generally denoted $a'(G)$. The notion of acyclic colouring was introduced by Grünbaum [5].

There is also a vertex analogue of acyclic edge colouring and any acyclic edge colouring of a graph G is an acyclic vertex colouring of its line graph $L(G)$ and the converse is also true. The acyclic chromatic index and its vertex analogue can be used to bound other parameters like *oriented chromatic number* and *star chromatic number* of a graph G, both of which have many practical applications such as in wavelength routing in optical networks [4,7].

All graphs we consider are simple and finite. Throughout this paper, $\Delta = \Delta(G)$ denotes the maximum degree of a graph G. It is quite easy to see that $a'(G) \geq \chi'(G) \geq \Delta$ for any graph G. Here $\chi'(G)$ denotes the chromatic index of G (the minimum number of colours used in any proper edge colouring of G). Determining $a'(G)$ exactly is a very hard problem (both theoretically and algorithmically) and is not solved even for complete graphs.

M.-Y. Kao and X.-Y. Li (Eds.): AAIM 2007, LNCS 4508, pp. 144–152, 2007.

Alon, McDiarmid and Reed [1] first obtained a linear upper bound of 64Δ on $a'(G)$. Molloy and Reed [8] improved this by a refined analysis to $a'(G) \leq 16\Delta$. Muthu, Narayanan and Subramanian [9] improved the bound to $a'(G) \leq 5.91\Delta$ for graphs G with girth (the length of the shortest cycle) at least 9.

Alon, Sudakov and Zaks conjectured ([2]) that $a'(G) \leq \Delta + 2$ for every graph G. The conjecture was shown [2] to be true for almost every d-regular (d fixed) graph. Later, Nesetril and Wormald [11] strengthened the latter result by showing that $a'(G) \leq d + 1$ for almost every d-regular graph.

The bounds mentioned before are based on existential arguments. There are very few constructive results which actually produce an acyclic edge colouring. Skulrattankulchai [12] gives a linear time algorithm for obtaining an acyclic 5-edge colouring of subcubic graphs. Recently, Muthu, Narayanan and Subramanian obtained an efficient (almost linear time in the size of the graph) constructive acyclic $(\Delta + 1)$-edge colouring of grid-like graphs [10]. Subramanian [13] showed by analyzing a simple polynomial time greedy heuristic that one can obtain an acyclic $O(\Delta \log \Delta)$ colouring of any graph G. We are not aware of any other constructive result on acyclic edge colouring.

Note that it is NP-complete to determine if $a'(G) \leq 3$ for an arbitrary subcubic graph G, as shown in [3]. It follows from the reduction used in [3] that it is in fact NP-hard to determine $a'(G)$ even when G is a 2-degenerate subcubic graph. A graph is 2-degenerate if one can reduce the given graph to a graph on 2 vertices by iteratively removing vertices of degree at most 2. The class of graphs we have studied here (outerplanar graphs) are a non-trivial subclass of 2-degenerate graphs.

In this work we prove that $a'(G) \leq \Delta + 1$ for outerplanar graphs. This adds further evidence that the conjecture stated in [2] is perhaps true. The other known classes of graphs for which the conjecture has been verified include random regular graphs ([2], [11]), graphs having large girth [2], subcubic graphs [12] and grid-like graphs [10].

Our proof is constructive and we also describe an $O(n \log \Delta)$ time algorithm in section 3. For bounded values of Δ, the complexity is linear in n. Formally, we prove the following result.

Theorem 1 (Outerplanar). *Let $G = (V, E)$ be an outerplanar graph of maximum degree Δ. Then, $a'(G) \leq \Delta + 1$. Also, an acyclic $(\Delta + 1)$- edge colouring can be obtained in $O(n \log \Delta)$ time where n denotes the number of vertices in G.*

1.1   Definitions and Notation

Throughout the paper, we focus only on edge colourings and we often omit the word "edge" and simply say colourings. A connected graph G is said to be *2-connected* if the removal of a vertex does not produce a disconnected graph. If G is connected, then a *block* of G is a maximal 2-connected subgraph of G. For any vertex $v \in V(G)$, we use d_v to denote the degree of v. With respect to an acyclic colouring $\mathcal{C}$ of a graph H, let $\mathcal{C}_v$ denote the set of colours used by edges incident on any specific vertex $v \in V(H)$. We use the standard notation $[k]$ to denote the set $\{1, \ldots, k\}$.

Definition 1. *A graph is* outerplanar *if there exist a planar embedding of G in which all the vertices lie on the unbounded face.*

2 Colouring

We first state certain facts relating to outerplanar graphs and acyclic edge colourings. These standard results can be found in [14].

Fact 1. *Any subgraph of an outerplanar graph is also outerplanar.*

Fact 2. *Every outerplanar graph has a vertex of degree at most 2.*

Fact 3. *If G is an outerplanar graph on 4 or more vertices, then G has two non-adjacent vertices of degree at most 2.*

Fact 4. *Any two-connected outerplanar graph has a unique hamiltonian cycle.*

We will also be using the following easy to verify observation.

Fact 5. *If $a'(G) = \eta$ and if G has $H_1, \ldots, H_k$ as its blocks (maximal 2-connected subgraphs), then an acyclic η-edge colouring of G can be obtained from any collection of acyclic η-edge colourings of $H_1, \ldots, H_k$ after suitably permuting the colours of edges incident at cut-vertices.*

Proof. Note that every cycle in a graph should lie within a block. Hence, it is essentially sufficient to first obtain acyclic edge colourings of all blocks. Now, to extend this to an acyclic colouring of G, one only needs to rename (permute) the colours assigned to edges incident at cut-vertices. Now, using the block-cutpoint tree (defined in Section 3) to guide the renaming, one can extend the colouring of blocks to a colouring of G.   □

The following lemma gives some insight into the structural properties of outerplanar graphs.

Lemma 1. *Every two-connected outerplanar graph G has a vertex of degree two adjacent to a vertex of degree at most four.*

Proof. Since G is two-connected and outerplanar, there is always a vertex of degree 2 in G (using Fact 2).

Noting that no degree 2 vertex in G can be part of any chord edge of the unique hamiltonian cycle of G (Fact 4), it follows that every vertex in G can have at most 2 neighbours of degree exactly 2.

Supposing that each vertex of degree 2 has no neighbour having degree at most 4, consider the outerplanar subgraph H obtained by deleting all vertices of degree 2 from G (Fact 1). The degree of any vertex in H is at least 3. This contradicts Fact 2. Thus our claim holds.   □

A stronger version of the above is shown in [6] (Lemma 2 in this reference) which implies Lemma 1 but the proof is more complicated.

Lemma 2 (See [6]). *Let G be an outerplanar graph with minimum degree at least 2. Then G satisfies one of the following two properties :*

(a) There exists a vertex of degree 2 having a neighbour of degree ≤ 3.
(b) There exist two vertices of degree 2 having a common neighbour of degree 4.

We will frequently use the following powerful extension lemma in the proof.

Lemma 3 (Extension Lemma). *Let G be any graph with maximum degree Δ and u be a degree 2 vertex in G with neighbours v and w. Let C be any $[k]$-acyclic edge colouring of $G \setminus u$, for some $k > \Delta$. If $|C_v \cup C_w| < k$, C can be extended to get an acyclic edge colouring of G using k colours.*

Proof. Colour (u, v) using any $c \in [k] \setminus (C_v \cup C_w)$. For the edge (u, w), use some arbitrary colour $c' \in [k] \setminus (C_w \cup \{c\})$. Note that $|C_w| \leq \Delta - 1$ and hence c' can always be found. Since $c \notin C_w$, the colouring of the edges (u, v) and (u, w) cannot introduce any (c, c')-coloured cycle.   □

We now prove our main result.

Proof (Theorem 1). We use induction on the number of vertices. If G is a graph on ≤ 3 vertices, then $a'(G) \leq \Delta + 1$ clearly. Assume the statement is true for each outerplanar graph on fewer than $|V(G)|$ vertices. Using Fact 5, we can assume without loss of generality, that G is 2-connected, and hence has a unique hamiltonian cycle.

By Lemma 2, *either* (i) G has a vertex u of degree 2 having a neighbour with degree at most 3 *or* (ii) G has two vertices u and x each of degree 2 and have a common neighbour of degree 4. In each case, u is the vertex which will be removed for applying inductive hypothesis. For the rest of the proof, we assume that v and w denote the two neighbours of u with degree of v being at most 4.

By inductive hypothesis, $G \setminus u$ can be acyclically edge coloured using colours from $[\Delta(G \setminus u) + 1] \subseteq [\Delta + 1]$. Let C_v and C_w be the respective sets of colours used on edges incident at v and w in the acyclic colouring of $G \setminus u$. We have the following cases.

Case 1 ($\Delta(G \setminus u) = \Delta - 1$). Using inductive hypothesis, $G \setminus u$ can be acyclically edge coloured using colours from $[\Delta]$. Since $\Delta + 1 \notin C_v \cup C_w$, we can apply the Extension Lemma and obtain an acyclic $[\Delta + 1]$-colouring of G.

Hence, for the rest of the proof, we may assume that $\Delta(G \setminus u) = \Delta$.

Case 2 ($(v, w) \notin E$). In this case, we consider the graph $H = (G \setminus u) \cup (v, w)$. We have $\Delta(H) \leq \Delta$ and H is also 2-connected and outerplanar. By inductive hypothesis, we have an acyclic $(\Delta + 1)$-colouring of H from which we get an acyclic $(\Delta + 1)$-colouring of $G \setminus u$. In this colouring, the colour used on (v, w) is missing from $C_v \cup C_w$ and hence we can apply the Extension Lemma and obtain an acyclic colouring of G with $[\Delta + 1]$.

Henceforth, we assume that $\Delta(G \setminus u) = \Delta$ and also that $(v, w) \in E$.

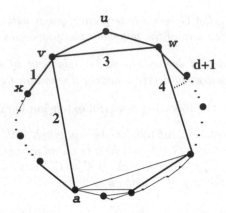

Fig. 1. Outerplanar graph with vertex u of degree 2

Case 3 ($d_v = 3$). Since $(v, w) \in E$, $\mathcal{C}_v \cap \mathcal{C}_w \neq \emptyset$. Then $|\mathcal{C}_v \cup \mathcal{C}_w| \leq |\mathcal{C}_v| + |\mathcal{C}_w| - 1 \leq 2 + (\Delta - 1) - 1 < \Delta + 1$. By applying the Extension Lemma, we get an acyclic colouring of G with $[\Delta + 1]$.

Case 4 ($d_v = 4$). By Lemma 2, there exists a vertex x such that u and x are both degree 2 vertices in G having v as a common neighbour. Since there can be no chord edge incident at a degree 2 vertex of a 2-connected outerplanar graph, it follows that u, v and x appear in that order in the unique hamiltonian cycle of G.

Using the Extension Lemma, we can assume without loss of generality that $\mathcal{C}_v \cup \mathcal{C}_w = [\Delta + 1]$. Also, $|\mathcal{C}_v \cap \mathcal{C}_w| \geq 1$ since $(v, w) \in E$. Since, $|\mathcal{C}_v| = 3$ and $|\mathcal{C}_w| \leq \Delta - 1$, it follows that $|\mathcal{C}_v \cap \mathcal{C}_w| = 1$. Without loss of generality, we assume that $\mathcal{C}_v = \{1, 2, 3\}$ and $\mathcal{C}_w = \{3, 4, \ldots, \Delta + 1\}$ and 3 is used on (v, w). See Fig. 1.

Without loss of generality, assume that the hamiltonian cycle edge (v, x) is coloured with 1. Since x has degree 2 in G, for some $c \in \mathcal{C}_w \setminus \{3\}$, there is no c-coloured edge incident at x and hence there is no $(1, c)$-coloured path joining v and w in $(G \setminus u) \setminus (v, w)$. Thus, we can safely colour (u, v) with c and (u, w) with 1 to get an acyclic edge colouring of G.

This completes the proof of the bound stated in Theorem 1. The description and analysis of the algorithm are presented in the next section.   □

3   Algorithmic Aspects

The proof of $a'(G) \leq \Delta + 1$ given above for outerplanar graphs is constructive. In this section, we show how to implement the various steps involved efficiently, leading to an $O(n \log \Delta)$ time algorithm. Here, n denotes the number of vertices. We also use m to denote the number of edges. The procedure is described in the pseudocode BlockColOP(B) given below. In order to keep the discussion simple and brief, we assume that the input B is a 2-connected outerplanar graph (also known as a block). This does not result in any loss of generality for the following reasons.

First, it is easy to see that blocks (maximal 2-connected subgraphs) of an outerplanar graph $G = (V, E)$ are also outerplanar. Moreover, since $|E| \leq 2|V|$ for outerplanar graphs, the blocks and cut-vertices of G can be computed in $O(n)$ time using standard search techniques like DFS. Once this is done, we compute the block-cutpoint graph $BC(G)$ of G in $O(n)$ time. This is a bipartite graph $H = (A, B, F)$ where A is the set of articulation vertices of G and B is the set of blocks of G. For each $a \in A$, $b \in B$, we join them by an edge if and only if $a \in V(b)$. It is easy to verify that $BC(G)$ is a forest in general and is a tree if G is connected.

We now invoke BlockColOP(B) for each block B and obtain an acyclic colouring of B in $O(m_B \log \Delta_B)$ time where m_B denotes the number of edges in B and Δ_B denotes the maximum degree of B. Since $m = \sum_B m_B$, this takes a total of $O(m \log \Delta) = O(n \log \Delta)$ time.

Now, since articulation vertices are shared by more than one block, we need to permute the colourings of edges incident at articulation vertices so as to remove potential conflicts among edges incident at articulation vertices. For this purpose, we first root the tree $BC(G)$ at an articulation vertex a of G and order the articulation vertices and blocks of G based on the preorder-traversal order of $BC(G)$.

For each articulation vertex a of G considered in this order, let B_0 be the parent of a and $B_1, \ldots, B_k$ be its children. When we come to process a, we distribute the remaining colours of $[\Delta + 1]$ *not* used on the edges from B_0 which are incident at a, to the remaining edges (from other blocks $B_1, \ldots B_k$) incident at a. For each block B considered in this order, we permute colour classes of edges in B so as to match the colours used on edges incident at a (the parent of B) with those distributed by a. This takes care of conflicts at each articulation vertex a of G. It is easy to see that this can be achieved in $O(n)$ time with suitable data structures.

Hence it suffices to show that 2-connected outerplanar graphs can be acyclically edge coloured in $O(n \log \Delta)$ time. Now onwards, we assume that G is a 2-connected outerplanar graph.

We assume the adjacency list representation for storing G. The two occurrences of each edge (i, j) (one in $Adj[i]$ and the other in $Adj[j]$) are linked to each other. The set of colours used so far on edges incident at a vertex u are stored in a height balanced binary search tree (BST) $Col(u)$ ordered by the colour values. In addition, we assume two queues $Q3$ and $Q4$ where $Q3$ is a queue on those vertices of degree 2 having a neighbour of degree ≤ 3. $Q4$ is a queue on those vertices of degree 2 having a neighbour v of degree 4 such that v has a neighbour of degree 2. All degrees are with respect to the graph being considered in the current recursive invocation. All data structures mentioned before are assumed to be globally available in each recursive call.

Algorithm 1. BlockColOP(B)

1: **if** B is a single edge (u, v) **then**
2:     colour (u, v) with 1 and **RETURN**.
3: **end if**
4: Find a vertex u having exactly two neighbours v and w in B such that *either* (i) degree of v in B is at most 3 *or* (ii) degree of v in B is exactly 4 and v has a neighbour x having degree 2 in B.
5: **if** $\Delta(B \setminus u) < \Delta(B)$ or if $(v, w) \notin E(B)$ **then**
6:     Obtain an acyclic $(\Delta(B') + 1)$-colouring of $B' = (B \setminus u) \cup \{(v, w)\}$ by invoking BlockColOP(B').
7:     From this, obtain an acyclic $(\Delta(B') + 1)$-colouring of $B \setminus u$.
8:     Applying the Extension Lemma, obtain an acylic $(\Delta(B) + 1)$-colouring of B and **RETURN**.
9: **end if**
10: **if** the degree of v in B is exactly 3 **then**
11:     Obtain an acylic $(\Delta(B \setminus u) + 1)$-colouring of $B \setminus u$ by invoking BlockColOP($B \setminus u$).
12:     Applying the Extension Lemma, obtain an acylic $(\Delta(B) + 1)$-colouring of B and **RETURN**.
13: **end if**
14: **if** the degree of v in B is exactly 4 **then**
15:     Obtain an acylic $(\Delta(B \setminus u) + 1)$-colouring of $B \setminus u$ by invoking BlockColOP($B \setminus u$).
16:     Colour (u, w) with the colour used for (v, x) and colour (u, v) with a colour $c \in C_w \setminus C_x$ to obtain an acylic $(\Delta(B) + 1)$-colouring of B and **RETURN**.
17: **end if**

3.1   Correctness and Complexity

Since BlockColOP is essentially the proof of Theorem 1 stated as an algorithm, the correctness follows immediately. So we focus on the complexity of the algorithm.

By adding the edge (v, w) to $B \setminus u$ whenever required, we ensure that the input graph to each recursive call is always 2-connected. Also, since each recursive call works on a graph with one vertex less than its parent call, there are at most n recursive calls.

One can build $Q3$ and $Q4$ initially once in the first invocation of BlockColOP in $O(n)$ time by scanning the adjacency lists. After this, for each recursive call, we only need to update $Q3$ and $Q4$ and do not need to compute them from scratch. It is easy to check that this update can be done in $O(1)$ time. Hence, total time required in all recursive calls for Step 4 is $O(n)$.

After u has been found in Step 4, checking each of the **if** conditions in Steps 5, 10 and 14 can be done in $O(1)$ time per recursive call. Step 7 involves removing the colour of the edge (v, w) from each of $Col(v)$ and $Col(w)$ *if* (v, w) is not part of $E(B)$ and has been explicitly added to B' to make it 2-connected. This can be done in $O(\log \Delta)$ time per recursive call.

We now need to estimate the time required for an application of the Extension Lemma. Recall from its proof that we need to find a colour $c \notin (Col(v) \cup Col(w))$ and also a colour $c' \notin (Col(w) \cup \{c\})$. For $j = 1, 2, \ldots$, we keep finding the j-th *smallest* colour which is *not* in $Col(w)$ until we find one which is not also in $Col(v)$. Since there exists such a colour and since $|Col(v)| \leq 3$, we don't need to go beyond $j = 4$. For each j, the j-th smallest colour which is absent from $Col(w)$ can be found in $O(\log \Delta)$ time by maintaining the size of each subtree at its root in the BST associated with $Col(w)$. Similarly, one can find c' also. Thus, the total time required for all applications of Extension Lemma is $O(n \log \Delta)$ since there are at most n recursive calls. Step 16 is similar to applying to Extension Lemma and this also requires the same time on the whole.

Thus, the overall time required by BlockColOP(B) is $O(n \log \Delta)$ in the worst case. Hence, an arbitrary outerplanar graph can be acyclically $(\Delta + 1)$-edge coloured in $O(n \log \Delta)$ time.

4   Conclusions

As mentioned in Section 1, it is NP-hard to determine $a'(G)$ even for 2-degenerate graphs. The class of graphs we have studied here (outerplanar graphs) are a non-trivial subclass of 2-degenerate graphs. We have also obtained tight estimates on $a'(G)$ for a few other subclasses of 2-degenerate graphs and we are pursuing further theoretical and algorithmic work in this direction. An interesting algorithmic question is to design (if it is possible) a linear, that is $O(n)$, time algorithm for $(\Delta + 1)$-acyclic edge colouring of outerplanar graphs.

It seems possible to handle graphs of bounded tree-width using a similar approach. We are currently working on it. We are also trying to extend these results to planar graphs.

References

1. N. Alon, C.J.H. McDiarmid, and B. Reed. Acyclic coloring of graphs. *Random Structures and Algorithms*, 2:277–288, 1991.
2. N. Alon, B. Sudakov, and A. Zaks. Acyclic edge colorings of graphs. *Journal of Graph Theory*, 37:157–167, 2001.
3. N. Alon and A. Zaks. Algorithmic aspects of acyclic edge colorings. *Algorithmica*, 32:611–614, 2002.
4. D Amar, A Raspaud, and O. Togni. All to all wavelength routing in all-optical compounded networks. *Discrete Mathematics 235*, pages 353–363, 2001.
5. B. Grünbaum. Acyclic colorings of planar graphs. *Israel J Math*, 14:390–408, 1973.
6. A. Hackmann and A. Kemnitz. List edge colorings of outerplanargraphs. *Ars Combinatoria*, 60:181–185, 2001.
7. A.V. Kostochka, E. Sopena, and X. Zhu. Acyclic and oriented chromatic numbers of graphs. *J. Graph Theory 24(4)*, pages 331–340, 1997.

8. M. Molloy and B. Reed. Further algorithmic aspects of lovaz local lemma. *30th Annual ACM Symposium on Theorey of Computing*, pages 524–529, 1998.
9. Rahul Muthu, N Narayanan, and C R Subramanian. Improved bounds on acyclic edge colouring. *Electronic Notes in Discrete Mathematics*, 19:171–177, 2005.
10. Rahul Muthu, N Narayanan, and C R Subramanian. Optimal acyclic edge colouring of grid like graphs. *Lecture Notes in Computer Science*, 4112:360–367, 2006.
11. J. Nešetřil and N. C. Wormald. The acyclic edge chromatic number of a random d-regular graph is $d+1$. *Journal of Graph Theory*, 49(1):69–74, 2005.
12. San Skulrattanakulchai. Acyclic colorings of subcubic graphs. *Information Processing Letters*, 92:161–167, 2004.
13. C R Subramanian. Analysis of a heuristic for acyclic edge colouring. *Information Processing Letters*, 99:227–229, 2006.
14. Douglas B West. *Introduction to Graph Theory*. Prentice Hall India, 2001.

Smallest Bipartite Bridge-Connectivity Augmentation (Extended Abstract)

Pei-Chi Huang[1], Hsin-Wen Wei[1], Wan-Chen Lu[1,*], Wei-Kuan Shih[1], and Tsan-sheng Hsu[2,*]

[1] Department of Computer Science, National Tsing-Hua University, Hsinchu, Taiwan
{peggy,bertha,wanchen,wshih}@rtlab.cs.nthu.edu.tw
[2] Institute of Information Science, Academia Sinica, Taipei, Taiwan
tshsu@iis.sinica.edu.tw

Abstract. This paper addresses two augmentation problems related to bipartite graphs. The first, a fundamental graph-theoretical problem, is how to add a set of edges with the smallest possible cardinality so that the resulting graph is 2-edge-connected, i.e., bridge-connected, and still bipartite. The second problem, which arises naturally from research on the security of statistical data, is how to add edges so that the resulting graph is simple and dose not contain any bridges.

1 Introduction

A graph is said to be *k-edge-connected* if it remains connected after the removal of any set of edges whose cardinality is less than k. Finding the smallest set of edges, the addition of which makes an undirected graph k-edge-connected, is a fundamental problem with many important applications that has been studied extensively; readers may refer to [5,7,19] for a comprehensive survey. Studies of augmentation problems in bipartite graphs can be found in [9,11,12]. In this paper, we focus on augmenting bipartite graphs. A graph is *componentwise 2-edge-connected* if each connected component is either 2-edge-connected, or it is an isolated vertex. Figure 1(a) shows an example of a bipartite graph. A smallest 2-edge-connectivity augmentation of (a) is shown in Figure 1(b), and a smallest componentwise 2-edge-connectivity augmentation of (a) is shown in Figure 1(c).

Note that there is a linear-time algorithm for the smallest bridge-connectivity augmentation problem on the general graph that does not have a bipartite constraint [4]. In [11], Jensen et al. proposed a polynomial time algorithm that solves the smallest bridge-connectivity augmentation problem on a graph that has partition constraints, such as bipartite graph, in $O(n(m + n \log n) \log n)$ time, where m is the number of distinct edges in the input graph. We are unaware of any previous results for the smallest componentwise bridge-connectivity augmentation problem.

---

* Supported in part by National Science Council (Taiwan) Grants NSC 94-2213-E-001-014 and NSC 95-2221-E-001-004.

M.-Y. Kao and X.-Y. Li (Eds.): AAIM 2007, LNCS 4508, pp. 153–166, 2007.

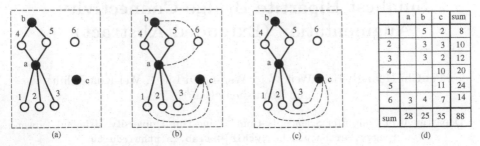

Fig. 1. (a) A bipartite graph. (b) A smallest 2-edge-connectivity augmentation of (a). (c) A smallest componentwise 2-edge-connectivity augmentation of (a). (d) A 2-dimensional cross-tabulated table with some suppressed cells.

Motivation

The related componentwise 2-edge-connectivity augmentation problem arises naturally from research on the security of statistical data [1,2,3,15]. To protect sensitive information in a cross-tabulated table, it is a common practice to suppress some of the cells in the table, so that the resulting table does not leak important or confidential information. This protection problem can be reduced to an augmentation problem in bipartite graphs [6,8,12,13,14,16,17,18].

Figure 1(a) and Figure 1(d) illustrate the relationship between our augmentation problem and the table protection problem. Figure 1(d) is a 2-dimensional cross-tabulated table with some suppressed cells. In the bipartite *suppressed graph* constructed from the table, the vertices correspond to the columns and rows, and the edges correspond to the suppressed cells, as shown in Figure 1(a). It has been proven [6] that the value of a suppressed cell can be revealed to an adversary if and only if it is a bridge in the constructed suppressed graph. Therefore, since there are three bridges in our suppressed graph, an adversary could infer the values of the three corresponding cells. For instance, let $C_{i,j}$ be the cell at the intersection of row i and column j, let $S_{*,j}$ be the sum of the cells in column j, and let $S_{i,*}$ be the sum of the cells in row i. Then, the value of $C_{1,a}$ must be 1 because it is equal to $S_{1,*} - C_{1,b} - C_{1,c}$. The value of $C_{5,b}$ is arbitrary. After suppressing three more cells, namely, $C_{1,c}$, $C_{2,c}$, and $C_{3,c}$, the values of the suppressed cells cannot be inferred. This corresponds to the smallest componentwise 2-edge-connectivity augmentation shown in Figure 1(c).

Our approach and results

We first solve the problem of a smallest 2-edge-connectivity augmentation of bipartite graphs, and then extend the proposed algorithms to deal with the componentwise 2-edge-connectivity case. To solve the first problem, we transform the input graph G into a well-known data structure called a *bridge-block forest*. A *block* of a graph G is a maximal 2-edge-connected subgraph (or component) of G. We assume B and W are the two bipartite sets of vertices in G. A block

Algorithm 1. Finding a smallest 2-edge-connectivity augmentation of a bipartite graph G

```
 1: procedure FS2Aug(G)
 2:     Let T = BB(G);
 3:     E = ∅;
 4:     repeat
 5:         switch (T)
 6:         Case 1: T is a tree
 7:             Case 1.1: T is an easy tree
 8:                 Case 1.1.1: T is an ETC tree
                        E′= ETCT(T);   {∗ Algorithm 2 ∗}
 9:                 Case 1.1.2: T is an anti-ETC tree with more than 4 leaves
                        E′= AETC(T);   {∗ Algorithm 3 ∗}
10:                 Case 1.1.3: T is an anti-ETC tree with at most 4 leaves
                        Use the solution shown in Figure 2 to find E′;
11:             Case 1.2: T is a general tree
12:                 Case 1.2.1: T has no hybrid leaves
                        E′= BGTWAug(T);   {∗ Algorithm 4 ∗}
13:                 Case 1.2.2: T has hybrid leaves
                        E′= HTAug(T);   {∗ Algorithm 6 ∗}
14:         Case 2: T is a forest
15:             Case 2.1: T contains no isolated vertices
16:                 Case 2.1.1: T is a light forest with |T_B| = |T_W|
                        E′= FTConversion(T);   {∗ Algorithm 7 ∗}
17:                 Case 2.1.2: T is a light forest with |T_B| > |T_W|
                        E′= BGTW_FTConversion(T);   {∗ Algorithm 8 ∗}
18:                 Case 2.1.3: T is a forest with hybrid leaves
                        E′= H_FTConversion(T);   {∗ Algorithm 9 ∗}
19:             Case 2.2: T contains a set of isolated vertices S
20:                 Case 2.2.1: T − S contains at least 2 white and 2 black vertices
                        Use the method in §4.2.1 to find E′;
21:                 Case 2.2.2: T − S contains either 1 white or 1 black vertex
                        Use the method in §4.2.2 to find E′;
22:                 Case 2.2.3: T − S is null
                        E′= ISOF(T);   {∗ Algorithm 10 ∗}
23:         Let E = E ∪ E′;
24:         Let T = BB(T ∪ E′);
25:     until Case 1 is executed
26:     return E;
27: end procedure
```

---

that only contains vertices in B (respectively, W) is called a *black* (respectively, *white*) block, while a block that contains both vertices in B and W is called a *hybrid* block. A vertex in the bridge-block forest is *white* if its corresponding block is white. Black and hybrid vertices in the bridge-block forest are defined similarly. Hereafter, we focus on a bridge-block forest, rather than a graph.

Let an *easy tree* be a tree with an equal number of black and white leaves and no hybrid leaves. Our main algorithm first solves the problem on an easy tree, and then solves it on a general tree. Finally, we solve the case where the input graph is a forest. In addition, the edge set added to the bridge-block forest by our algorithms can be transformed into the corresponding edge set added to the input graph G. The algorithms run in sequential liner time and $O(\log n)$ parallel time on an EREW PRAM using a linear number of processors. A high-level description of the algorithm for the 2-edge-connectivity case is given in Algorithm 1. The main result of this paper is stated in Theorem 1 and will be proved in the remaining sections. Due to space limitation, we omit some details which can be found in [10].

Theorem 1. *Algorithm 1 runs in sequential linear time and $O(\log n)$ parallel time on an EREW PRAM using a linear number of processors.*

2   Preliminaries

In this paper, all graphs are undirected, and have neither self-loops nor multiple edges. Let a graph $G = (V, E)$, where $|V| = n$ and $|E| = m$. Then, for a vertex set V', let $G - V'$ be G without the vertices and their adjacent edges in V'. Note that, for an edge set E', $G - E'$ denotes G without the edges in E', and $G \cup E'$ denotes G with the edges in E' added to it. An edge whose endpoints are a vertex u and a vertex v is denoted as (u, v). A *bipartite* graph is defined as a graph in which the set of vertices can be partitioned into two disjoint sets such that no edge connects vertices in the same set.

Two vertices of a graph are *2-edge-connected* if they are in the same connected component and remain so after the removal of any single edge. A set of vertices is *2-edge-connected* if each pair of its vertices is 2-edge-connected; similarly, a graph is *2-edge-connected* if its set of vertices is 2-edge-connected. A *bridge* is an edge of a graph G, the removal of which would increase the number of connected components of G by one. Given a graph G with at least three vertices, a smallest *2-edge-connectivity augmentation* of G, denoted by aug2e(G), is a set of edges with the minimum cardinality whose addition makes G 2-edge-connected. A graph is *componentwise 2-edge-connected* if it does not have a bridge. A smallest componentwise 2-edge-connectivity augmentation of G is denoted by augc2e(G). A *block* in a graph is an induced subgraph of a maximal 2-edge-connected subset of vertices. If a block consists of all the nodes in a connected component of G, it is called an *isolated block*. A *singular* connected component is one formed by an isolated vertex, and a *singular block* is one with exactly one vertex. The *bridge-block graph* of an undirected graph G, denoted by BB(G), is defined as follows. Each block is represented by a vertex of BB(G). When all the blocks in G are represented by vertices, BB(G) becomes a forest. Each bridge in G corresponds to an edge in BB(G) and vice versa. In this paper, let G be the input graph and we use T and BB(G), interchangeably to denote the bridge-block forest for an input graph G.

3   Case 1: When BB(G) Is a Tree

Assume BB(G) contains ℓ leaves that can be divided into the following three categories: B is a set of black leaves, W is a set of white leaves, and H is a set of hybrid leaves. Without loss of generality, we assume that $|B| \geq |W|$. Furthermore, we say that BB(G) is *B-dominated* if $|B| > |W| + |H|$.

Algorithm 2. ETC tree connection

1: **procedure** ETCT(T)   {* where T is an ETC tree with ℓ leaves *}
2:     Find i^* such that v_{i^*} and $v_{i^*+\ell/2}$ are in different colors;
3:     Let $V_{in} = \{v_{i^*+1}, v_{i^*+2}, \ldots, v_{(i^*+\ell/2)-1}\}$, and $V_{out} = \{v_1, v_2, \ldots, v_{i^*-1}\} \cup \{v_{(i^*+\ell/2)+1}, v_{(i^*+\ell/2)+2}, \ldots, v_\ell\}$;
4:     Number the black (respectively, white) leaves in V_{in} starting from 1 as $b_1, b_2, \ldots$ (respectively, $w_1, w_2, \ldots$);
5:     Number the black (respectively, white) leaves in V_{out} starting from 1 as $b'_1, b'_2, \ldots$ (respectively, $w'_1, w'_2, \ldots$);
6:     Let $E' = \{(b_i, w'_i) \mid \forall i\} \cup \{(b'_i, w_i) \mid \forall i\}$;
7:     **return** $E' \cup \{(v_{i^*}, v_{i^*+\ell/2})\}$;
8: **end procedure**

3.1 Lower Bound on aug2e(BB(G))

Let $\text{LOW}_{t2e}(\text{BB}(G)) = \max\{\lceil(|B| + |W| + |H|)/2\rceil, |B|\}$ when BB(G) is a tree.

Lemma 1. $|\text{aug2e}(\text{BB}(G))| \geq \text{LOW}_{t2e}(\text{BB}(G))$.

Corollary 1. *If* BB(G) *is* B-*dominated, then* $\text{LOW}_{t2e}(\text{BB}(G)) = |B|$.

3.2 Case 1.1: When BB(G) Is an Easy Tree

Recall that an easy bridge-block tree T for a bipartite graph is one with an equal number of white and black leaves and no hybrid leaves. We number the leaves of T via a depth-first ordering from 1 to ℓ, i.e., the number of leaves in T, and denote them by $v_1, v_2, \ldots, v_\ell$. Note that, since ℓ is even, $\text{LOW}_{t2e}(\text{BB}(G)) = \ell/2$. By Lemma 1, $|\text{aug2e}(T)| \geq \ell/2$. Our algorithm, described below, always adds $\ell/2$ edges. Thus, after adding edges, if we can prove the resulting graph is 2-edge-connected, the solution found is a smallest 2-edge-connectivity augmentation of T.

If T is an easy tree and there exists i such that v_i and $v_{i+\ell/2}$ are two different-colored leaves, we say that the tree is an *easy-to-connect* or *ETC* tree. An easy tree that is non-ETC is called an *anti-ETC* tree. Note that both ETC and anti-ETC trees are easy trees. Our algorithm considers three cases: (1) an ETC tree, (2) an anti-ETC tree with more than four leaves, and (3) an anti-ETC tree with at most four leaves.

Lemma 2. *Let T be the input tree and $T_{new} = T \cup E_{added}$, where E_{added} is a set of added edges. Then, each added edge $e \in E_{added}$ is not a bridge in T_{new}.*

Case 1.1.1: When BB(G) Is an ETC tree. Our algorithm for finding aug2e(BB(G)) when BB(G) is an ETC tree is shown in Algorithm 2.

Lemma 3. *For a subtree T' of T, let e_a be the antenna edge of T' and $T_{new} = T \cup E_{added}$, where E_{added} is a set of edges added to T. If there exists an edge $e = (v_a, v_b) \in E_{added}$, such that $v_a \in T'$ and $v_b \notin T'$ or vice versa, then e_a is not a bridge in T_{new}.*

Lemma 4. *Let E_{added} be the set of edges derived by Algorithm 2 and let $T_{new} = T \cup E_{added}$. Then T_{new} does not contain any bridge; that is, $E_{added} = \text{aug2e}(T)$.*

Algorithm 3. Anti-ETC tree connection

1: **procedure** AETC(T)   {* where T is an anti-ETC tree with ℓ leaves and $\ell > 4$ *}
2:       Find leaves v_a, v_{a+1}, $v_{a+\ell/2}$ and $v_{(a+1)+\ell/2}$ such that v_a and $v_{a+\ell/2}$ are black, and
             v_{a+1} and $v_{(a+1)+\ell/2}$ are white;
3:       Let $E_1 = \{(v_a, v_{(a+1)+\ell/2}), (v_{a+1}, v_{a+\ell/2})\}$;
4:       Let $V_{in} = \{v_{a+2}, v_{a+3}, \ldots, v_{(a+\ell/2)-1}\}$;
5:       Let $V_{out} = \{v_1, v_2, \ldots, v_{a-1}\} \cup \{v_{(a+1)+\ell/2+1}, v_{(a+1)+\ell/2+2}, \ldots, v_\ell\}$;
6:       Number the black (respectively, white) leaves in V_{in} starting from 1 as $b_1, b_2, \ldots$
             (respectively, $w_1, w_2, \ldots$);
7:       Number the black (respectively, white) leaves in V_{out} starting from 1 as $b'_1, b'_2, \ldots$
             (respectively, $w'_1, w'_2, \ldots$);
8:       Let $E' = \{(b_i, w'_i) \mid \forall i\} \cup \{(b'_i, w_i) \mid \forall i\}$;
9:       **return** $E' \cup E_1$;
10: **end procedure**

Case 1.1.2: When BB(G) is an anti-ETC tree with more than four leaves. In this case, we can find two consecutive leaves, denoted, respectively, by v_a and v_{a+1} ($a < \ell/2$), such that v_a and v_{a+1} are different colors. Without loss of generality, we assume that v_a is a black leaf; therefore, v_{a+1} is white. Furthermore, we can find $v_{a+\ell/2}$, which must be black, and $v_{(a+1)+\ell/2}$, which must be white. The steps of the proposed algorithm for this case are given in Algorithm 3.

Lemma 5. *Let E_{added} be the set of edges derived by Algorithm 3. Then, $T_{new} = T \cup E_{added}$ contains no bridges.*

Case 1.1.3: When BB(G) Is an anti-ETC tree with at most four leaves. Note that an easy tree has an even number of leaves; therefore, a tree can have either two leaves or four leaves in this case. Clearly an easy tree with two leaves must be an ETC tree. Hence, we only need to consider an anti-ETC tree with exactly four leaves. Depending on the tree structure, we have the solution for each case of an anti-ETC tree with exactly four leaves, as shown in Figure 2.

3.3   Case 1.2: When BB(G) Is a General Tree

Case 1.2.1: When BB(G) has no hybrid leaves. Note that, if BB(G) has no hybrid leaves and $|B| > |W|$, then we apply Algorithm 4.

Lemma 6. *Algorithm 4 is correct.*

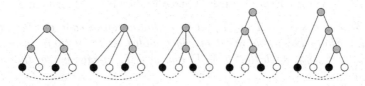

Fig. 2. All the possible cases of an anti-ETC tree with exactly four leaves

---

Algorithm 4. When the input has no hybrid leaves and $|B| > |W|$.

```
 1: procedure BGTWAUG(T)
 2:     Let b_i and w_i be, respectively, the ith black and white leaf;
 3:     Let V = b_1, b_2, ..., b_{|B|}, w_1, w_2, ..., w_{|W|} and V' = b_{|W|+1}, b_{|W|+2}, ..., b_{|B|};
 4:     Let T' = T - V';
 5:     if T' is an ETC tree then
 6:         E_1=ETCT(T');   {* Algorithm 2 *}
 7:     else if T' is an anti-ETC tree with more than 4 leaves then
 8:         Let E_1=AETC(T');   {* Algorithm 3 *}
 9:     else if T' is an anti-ETC tree with at most 4 leaves then
10:         Use the solution illustrated in Figure 2 to find E_1;
11:     end if
12:     if there is only one white vertex in T i.e., there is no white leaf in T then
13:         Let u be the white vertex in T and E_2 = {(b_i, u) | 1 ≤ i ≤ |B|};
14:     else
15:         Let u_1, u_2 be two white vertices in T;
16:         Let E_2 ={(b_i, u_j) | |W + 1| ≤ i ≤ |B|, j ∈ {1, 2} , where u_j is not the neighbor of
                b_i};   {* add edges between a white vertex and the remaining black leaves *}
17:     end if
18:     return E_1 ∪ E_2;
19: end procedure
```

Algorithm 5. H assignment

```
 1: procedure HASSIGN(T)   {* where T is a tree with hybrid leaves *}
 2:     if |B| > ⌈(|B| + |W| + |H|)/2⌉ then
 3:         All hybrid leaves are recolored white;
 4:     else
 5:         Arbitrarily select |B| − |W| hybrid leaves to be recolored white;
 6:         The remaining ⌊(|H| − |B| + |W|)/2⌋ hybrid leaves are recolored white;
 7:         The rest are recolored black;
 8:     end if
 9:     Let T' be the resulting tree;
10:     return T';
11: end procedure
```

Case 1.2.2: When BB(G) has hybrid leaves. Note that if an endpoint of an added edge is a hybrid leaf, the other endpoint of the edge can be either black or white. To handle this case, we first transform a tree with hybrid leaves into a tree without hybrid leaves using an algorithm called HASSIGN, described in Algorithm 5. Then, we apply Algorithm 4 to the recolored tree derived by Algorithm 5. The steps followed in this case are described in Algorithm 6.

Lemma 7. *Algorithm 6 is correct and optimal. That is,* $\text{aug2e}(T) = \text{aug2e}(T')$, *where* $T = \text{BB}(G)$ *and* T' *is the recolored tree returned by Algorithm 5.*

4   Case 2: When BB(G) Is a Forest

In this section, we present a number of algorithms that convert a forest into a tree. After this transformation, we can apply the algorithms presented in Section 3 to add edges such that no bridges exist in the final graph.

Algorithm 6. When T has hybrid leaves

1: **procedure** HTAUG(T)
2:     $T' =$ HASSIGN(T);   {* Algorithm 5 *}
3:     $E' =$ BGTWAUG(T');   {* Algorithm 4 *}
4:     **return** E';
5: **end procedure**

Recall that a leaf in a forest is a degree-1 vertex; and B, W, and H are, respectively, the sets of black, white, and hybrid leaves in BB(G). Without loss of generality, we assume that $|B| \geq |W|$. Let B', W', and H' be, respectively, the sets of isolated black, white, and hybrid vertices in BB(G). We now present a simple lower bound for $|\text{aug}2e(\text{BB}(G))|$.

Let $\text{LOW}_{f2e}(\text{BB}(G)) = \max\{2|B'|+|B|, 2|W'|+|W|, \lceil(2|B'|+2|W'|+2|H'|+ |B|+|H|+|W|)/2\rceil\} = p + \max\{|B|+|B'|-|W'|-|H'|, |W|+|W'|-|B'|- |H'|, \lceil(|B|+|H|+|W|)/2\rceil\}$, where p is the number of isolated vertices in BB(G). Note that if BB(G) is a tree, $\text{LOW}_{t2e}(\text{BB}(G)) = \text{LOW}_{f2e}(\text{BB}(G))$.

Lemma 8. $|\text{aug}2e(\text{BB}(G))| \geq \text{LOW}_{f2e}(\text{BB}(G))$.

4.1 Case 2.1: When BB(G) Contains No Isolated Vertices

In this subsection, we assume that BB(G) dose not have any isolated vertices. First, we consider the case where BB(G) dose not contain any hybrid leaves. This is called a *light forest*; otherwise, it is called a *general forest*. In a light forest, the trees can be classified into three different types: (1) $T_B = \{ T \mid T$, a tree with only black leaves in $T\}$; (2) $T_W = \{ T \mid T$, a tree with only white leaves in $T\}$; and (3) $T_{BW} = \{ T \mid T$, a tree with at least one black and one white leaf in $T\}$.

Without loss of generality, we assume that $|T_B| \geq |T_W|$. Next, we propose algorithms for two cases: (1) $|T_B| = |T_W|$, and (2) $|T_B| > |T_W|$. For the remainder of this section, let $F = \text{BB}(G)$.

Case 2.1.1 and Case 2.1.2: When BB(G) is a light forest and $|T_B| \geq |T_W|$. The steps for Case 2.1.1 and case 2.1.2 are shown in Algorithm 7 and Algorithm 8, respectively.

Theorem 2. $F \cup E_{added}$ *is a tree in which* E_{added} *is the set of edges derived by Algorithm 7 or 8. Furthermore,* $|\text{LOW}_{f2e}(\text{BB}(G))| = |\text{LOW}_{f2e}(\text{BB}(\text{BB}(G) \cup E_{added}))| + |E_{added}|$.

Case 2.1.3: When BB(G) has hybrid leaves. If one endpoint of an added edge is a hybrid leaf, the other endpoint of that edge can be either black or white. For a general forest, we first transform a forest with hybrid leaves into a forest without hybrid leaves using Algorithm 5. Then, we apply Algorithm 8 to convert a forest into a tree. The steps followed in this case are shown in Algorithm 9.

Lemma 9. *Algorithm 9 finds* $\text{aug}2e(\text{BB}(G))$ *when* BB(G) *is a forest containing no isolated vertices.*

Algorithm 7. Forest-Tree Conversion

1: **procedure** FTCONVERSION(F)   {* F is a light forest with $|T_B| = |T_W| = k$ and $|T_{BW}| = z$ *}
2:    Number each tree in T_B as $1, 3, \ldots, 2k - 1$;
3:    Number each tree in T_W as $2, 4, \ldots 2k$;
4:    Number each tree in T_{BW} as $2k + 1, 2k + 2, \ldots 2k + z$;
5:    Give two labels to each tree as follows:
6:    **for** the tree i from 1 to $2k + z$ **do**
7:        **if** the tree $\in T_B$ **then**
8:            Assign the labels $2i - 1$ and $2i + 1$ to two leaves chosen arbitrarily;
9:        **else if** the tree $\in T_W$ **then**
10:            Assign the labels $2i - 2$ and $2i$ to two leaves chosen arbitrarily;
11:        **else if** the tree $\in T_{BW}$ **then**
12:            Assign the labels $2i - 1$ and $2i$ to two different colored leaves. Here, $2i - 1$ is
                assigned to the black leaf and $2i$ is assigned to the white leaf;
13:        **end if**
14:    **end for**
15:    $E_1 = \{(v_{2j}, v_{2j+1}) \mid \text{for all labeled leaves } v_{2j} \text{ and } v_{2j+1}, 1 \leq j \leq 2k + z - 1\}$;
16:    **return** E_1;
17: **end procedure**

4.2   Case 2.2: When $BB(G)$ Contains Isolated Vertices

Recall that each isolated black (respectively, white) block is an isolated black (respectively, white) vertex in G. Let b'_i (respectively, w'_i) be the ith isolated black (respectively, white) vertex in G, and let $h'_{1,i}$, $h'_{2,i}$ be arbitrary black and white vertices, respectively, in the ith isolated hybrid block of G.

Let G' be the graph obtained by removing the vertices and edges from the isolated blocks of G. There are three cases, which we describe below.

Case 2.2.1: G' contains at least two white and two black vertices. Without loss of generality, we assume that $|B'| \geq |W'|$, which yields five sub-cases: (1) Case 2.2.1.1: $|W'| > 0$; (2) Case 2.2.1.2: $|W'| = 0$, $|B'| > 0$ and $|H'| > 0$; (3) Case 2.2.1.3: $|W'| = 0$, $|B'| = 0$ and $|H'| > 0$; (4) Case 2.2.1.4: $|W'| = 0$, $|B'| > 0$, $|H| + |W| > 0$, and $|H'| = 0$; (5) Case 2.2.1.5: $|W'| = 0$, $|B'| > 0$, $|H| + |W| = 0$, and $|H'| = 0$.

Case 2.2.2: G' contains either a white or a black vertex. Without loss of generality, we assume that G' contains exactly one white vertex, which yields two sub-cases: (1) Case 2.2.2.1: there is no white vertex in $G - G'$; (2) Case 2.2.2.2: there is a white vertex in $G - G'$.

Case 2.2.3: G' is null.

Case 2.2.1: G' has at least two white and two black vertices. Let E' be the set of added edges to be decided in each sub-case; $\hat{T} = BB(BB(G) \cup E')$; $\hat{B}'$, $\hat{W}'$, and $\hat{H}'$ be the respective sets of isolated black, white, and hybrid blocks in $\hat{T}$; and $\hat{B}$, $\hat{W}$, and $\hat{H}$ be the respective sets of black, white, and hybrid leaf-blocks in $\hat{T}$.

Case 2.2.1.1: $|W'| > 0$. Since we assume that $|B'| \geq |W'|$, $|B'| > 0$, let $E' = \{(b'_i, w'_i) \mid 1 \leq i \leq |W'|\}$. Then, $|\hat{B}| = |B| + |W'|$, $|\hat{W}| = |W| + |W'|$,

---

Algorithm 8. $|T_B| > |T_W|$ Forest-Tree Conversion

1: **procedure** BGTW_FTCONVERSION(F)   {* where F is a light forest with $|T_B| = k + x$,
       ($x \geq 1$), $|T_W| = k$, and $|T_{BW}| = z$ *}
2:     **if** $T_{BW} = T_W = \phi$ **then**
3:         Pick a leaf from each tree in T_B and number them as $b_1, b_2, \ldots, b_{k+x}$;
4:         Let $E_1 = \{(b_i, u) \mid 1 \leq i \leq k + x$ and let u be a white vertex of T_B with the
              number b_r, where $i \neq r$ and $1 \leq r \leq k + x\}$;
5:     **else**
6:         Find a subset T'_B of T_B, such that $|T'_B| = k$;
7:         Let $\overline{T'_B} = T_B - T'_B$ and $|\overline{T'_B}| = x$;
8:         Let $F' = T'_B \cup T_W \cup T_{BW}$;
9:         $E_1 =$ FTCONVERSION(F');   {* Algorithm 7 *}
10:        Pick a leaf from each tree in $\overline{T'_B}$ and number them as $b_1, b_2, \ldots, b_x$;
11:        Number the remaining white leaves of BB($F' \cup E_1$) as $w_1, w_2, \ldots w_y$;   {* as-
             suming there are y remaining white leaves *};
12:        **if** $x \leq y$ **then**
13:           Let $E_2 = \{(b_i, w_i) \mid 1 \leq i \leq x\}$;
14:        **else**
15:           Let $E_2 = \{(b_i, w_i) \mid 1 \leq i \leq y\} \cup \{(b_i, u) \mid y < i \leq x, u$ is an arbitrary white
              leaf.$\}$;
16:        **end if**
17:     **end if**
18:     **return** $E_1 \cup E_2$;
19: **end procedure**

Algorithm 9. When the input is a forest that has hybrid leaves

---

1: **procedure** H_FTCONVERSION(F)   {* where F is a forest with hybrid leaves *}
2:     $T' =$ HASSIGN(F);   {* Algorithm 5 *}
3:     $E' =$BGTW_FTCONVERSION(T')   {* Algorithm 8 *}
4:     **return** E';
5: **end procedure**

$\hat{H} = H$, $|\hat{W}'| = 0$, $|\hat{B}'| = |B'| - |W'|$, and $\hat{H}' = H'$. Thus, LOW$_{f2e}(\hat{T}) \geq$ LOW$_{f2e}(\text{BB}(G)) - |W'|$. We have reduced Case 2.2.1.1 to Case 2.2.1.2, Case 2.2.1.3, Case 2.2.1.4, Case 2.2.1.5 or Case 2.1.

Case 2.2.1.2: $|W'| = 0$, $|B'| > 0$ and $|H'| > 0$. Let $k = \min\{|B'|, |H'|\}$ and $E' = \{(b'_i, h'_{2,i}) \mid 1 \leq i \leq k\}$. Then, $|\hat{B}| = |B| + k$, $\hat{W} = W$, $|\hat{H}| = |H| + k$, $|\hat{W}'| = 0$, $|\hat{B}'| = |B'| - k$, and $|\hat{H}'| = |H'| - k$. Thus, LOW$_{f2e}(\hat{T}) \geq$ LOW$_{f2e}(\text{BB}(G)) - k$. We have reduced Case 2.2.1.2 to Case 2.2.1.3, Case 2.2.1.4, Case 2.2.1.5 or Case 2.1.

Case 2.2.1.3: $|W'| = 0$, $|B'| = 0$, and $|H'| > 0$. Let w, h, and b_i be, respectively, arbitrary white, hybrid, and ith black leaves in BB(G') if they exist. Let $k = \min\{|B|, |H'|\}$ and $E_1 = \{(b_i, h'_{2,i}) \mid 1 \leq i \leq k\}$. If $|H'| > k$, then let $|H''| = |H'| - k$ and $E_2 = \{(h'_{1,i}, h'_{2,i+1})) \mid 1 \leq i < \lfloor |H''|/2 \rfloor\}$. Furthermore, when $|H''|$ is odd, let $E_3 = \{(h'_{1,|H'|}, w)\}$ if w exists; otherwise, $E_3 = \{(h'_{1,|H'|}, h)\}$. Then, $|\hat{B}| = |B| - k$, $|\hat{H}| = |H| + k + |H''|$, $|\hat{H}'| = 0$, and $E' = E_1 \cup E_2 \cup E_3$. Thus, LOW$_{f2e}(\hat{T}) \geq$ LOW$_{f2e}(\text{BB}(G)) - |E'|$. We have reduced Case 2.2.1.3 to Case 2.1.

Case 2.2.1.4: $|W'| = 0$, $|B'| > 0$, $|H| + |W| > 0$, and $|H'| = 0$. Let $k = \min\{|B'|, |H| + |W|\}$; w_i be a white vertex in the ith leaf of $H \cup W$, and $E' = \{(b_i, w_i) \mid 1 \le i \le k\}$. Then, $|\hat{B}| = |B| + |k|$, $|\hat{W}| + |\hat{H}| = |H| + |W| - k$, $|\hat{W}'| = 0$, $|\hat{B}'| = |B'| - k$, and $|\hat{H}'| = 0$. Thus, $\text{LOW}_{f2e}(\hat{T}) \ge \text{LOW}_{f2e}(\text{BB}(G)) - k$, so we have reduced Case 2.2.1.4 to either Case 2.2.1.5 or Case 2.1.

Case 2.2.1.5: $|W'| = 0$, $|B'| > 0$, $|H| + |W| = 0$, and $|H'| = 0$. Now, we only have black leaves and isolated black vertices. Let w be a white vertex in G', and $E' = \{(b'_i, w) \mid 1 \le i \le |B'|\}$. Note that, in this case, $2|B'| + |B| > \lceil (2|B'| + 2|W'| + 2|H'| + |B| + |H| + |W|)/2 \rceil$. Therefore, $\hat{B} = B \cup B'$, $\hat{W} = \emptyset$, $\hat{H} = \emptyset$, $\hat{W}' = \emptyset$, $\hat{B}' = \emptyset$, and $\hat{H}' = \emptyset$, such that $\text{LOW}_{f2e}(\hat{T}) \ge \text{LOW}_{f2e}(\text{BB}(G)) - |E'|$. We have reduced Case 2.2.1.5 to Case 2.1.

Case 2.2.2: G' contains either one white or one black vertex. Without loss of generality, we assume that G' consists of exactly one white vertex w. Hence, $|H| = 0$ and G' is a star with center w; $\text{BB}(G)$ is also a star. There are two sub-cases: (1) $G - G'$ contains a white vertex, and (2) $G - G'$ does not contain a white vertex.

Case 2.2.2.1: there is no white vertex in $G - G'$. All isolated vertices in G are black, $|W'| = 0$ and $|H'| = 0$, such that $|\text{LOW}_{f2e}(\text{BB}(G'))| = |B|$. Let $E' = \{(b'_i, w) \mid \forall i\}$.

Lemma 10. *For Case 2.2.2.1, $\text{aug2e}(\text{BB}(G)) = \text{aug2e}(\text{BB}(G')) \cup E'$, and $\text{BB}(G) \cup \text{aug2e}(\text{BB}(G))$ is a multi-graph.*

Case 2.2.2.2: there is one white vertex in $G - G'$. Let b be a black leaf in $\text{BB}(G')$. Since $\text{BB}(G')$ is a star with a white center, b must exist. Let w' be a white vertex in an isolated block (i.e., in $G - G'$), and let $G'' = \text{BB}(G') \cup \{(w', b)\}$. Note that the number of isolated blocks in $\text{BB}(G) \cup \{(w', b)\}$ is one less than in $\text{BB}(G)$, and the number of black leaves in $\text{BB}(G'')$ is one less than in $\text{BB}(G')$. However, there is one more white leaf in $\text{BB}(G'')$ than in $\text{BB}(G')$. Thus, we have transformed Case 2.2.2.2 into Case 2.2.1.

Lemma 11. *For Case 2.2.2.2, $|\text{LOW}_{f2e}(\text{BB}(\text{BB}(G) \cup \{(w', b)\}))| = |\text{LOW}_{f2e}(\text{BB}(G))| - 1$.*

Case 2.2.3: G' is null. Let q_B, q_W, and q_H be the numbers of isolated black, white, and hybrid blocks, respectively. Without loss of generality, we assume that $q_B \ge q_W$. Our algorithm is shown in Algorithm 10.

Theorem 3. $|\text{LOW}_{f2e}(\text{BB}(G))| = |\text{LOW}_{f2e}(\text{BB}(\text{BB}(G) \cup E_{added}))| + |E_{added}|$, *where E_{added} is the set of added edges returned by Algorithm 10.*

5   Componentwise 2-Edge-Connectivity Augmentation

In this section, we present Algorithm 11, i.e., C2AUG, which solves the componentwise 2-edge-connectivity augmentation problem. If the leaves in a graph are not all black, i.e., there is a white or a hybrid leaf in the graph, we can use the minimum number of added edges to make all tree edges non-bridge edges.

Algorithm 10. When G' is null, i.e., $\mathrm{BB}(G)$ consists of isolated vertices

---

1: **procedure** ISOF(F) {∗ where F is a forest that consists of isolated vertices S ∗}
2:     **if** $q_B = 0$ **then**   {∗ q_H must be at least 2; ∗}
3:         Let $E_1 = \{(h'_{1,2i-1}, h'_{2,2i}) \mid 1 \le i \le \lfloor q_H/2 \rfloor\}$;
4:         **if** q_H is odd number **then**
5:             $E' = E_1 \cup (h'_{1,q_H-1}, h'_{2,q_H})$;
6:         **end if**
7:     **else**
8:         **if** $q_B > q_W + q_H$ **then**
9:             Let $E_1 = \{(b'_i, w'_i) \mid 1 \le i \le q_W\} \cup \{(b'_{i+q_W}, h'_{2,i}) \mid 1 \le i \le q_H\}$;
10:            $E' = E_1 \cup \{(b'_{i+q_W+q_H}, w) \mid 1 \le i \le q_B - q_W - q_H$ and w is a white vertex in $S\}$;
11:        **else if** $q_B = q_W + q_H$ **then**
12:            Let $E_1 = \{(b'_i, w'_i) \mid 1 \le i \le q_W\} \cup \{(b'_{i+q_W}, h'_{2,i}) \mid 1 \le i \le q_H\}$;
13:            $E' = E_1$;
14:        **else**
15:            Let $E_1 = \{(b'_i, w'_i) \mid 1 \le i \le q_W\} \cup \{(b'_{i+q_W}, h'_{2,i}) \mid 1 \le i \le q_B - q_W\}$;
16:            Let $E_2 = \{(h'_{1,2i-1+q_W-q_B}, h'_{2,2i+q_W-q_B}) \mid 1 \le i \le \lfloor(q_H + q_W - q_B)/2\rfloor\}$;
17:            **if** $(q_H + q_W - q_B)$ is odd **then**
18:                Let $E_2 = E_2 \cup (h'_{1,q_H-1}, h'_{2,q_H})$;
19:            **end if**
20:            $E' = E_1 \cup E_2$;
21:        **end if**
22:    **end if**
23:    **return** E';
24: **end procedure**

Algorithm 11. Componentwise 2-edge-connectivity augmentation

1: **procedure** C2AuG(G)
2:     Let $T = \mathrm{BB}(G)$;
3:     Let S be the set of isolated vertices in T;
4:     **if** there are leaves in T **then**
5:         **if** the leaves are not all black **then**
6:             $T' = T - S$;
7:         **else**   {∗ Without loss of generality, assume that all leaves are black. ∗}
8:             **if** there is an isolated vertex v in S whose corresponding block contains a white vertex **then**
9:                 Let $T' = T - S \cup \{v\}$;
10:            **else**
11:                $T' = T - S$;
12:            **end if**
13:        **end if**
14:        $E = \mathrm{FS2AuG}(T')$;   {∗ Algorithm 1 ∗}
15:    **else**
16:        $E = \emptyset$;
17:    **end if**
18:    **return** E;
19: **end procedure**

Theorem 4. *Algorithm 11 is correct and optimal. Furthermore, it runs in sequential linear time and $O(\log n)$ parallel time on an EREW PRAM using a linear number of processors.*

6   Concluding Remarks

We have considered two augmentation problems related to bipartite graphs. The first is a fundamental graph-theoretical problem. The second focuses on how to suppress the smallest amount of sensitive information in a cross-tabulated table, so that the resulting table does not leak important or confidential information. The latter is a fundamental issue concerning the security of statistical data. In both cases, after adding edges, the resulting graph is simpler than the input graph and does not contain any bridges. It can be either a simple graph or, if necessary, a multi-graph. The proposed approach determines whether or not such an augmentation is feasible. The algorithms can be trivially parallelized to run in optimal $O(\log n)$ time using a linear number of EREW processors. Details are omitted and can be found in [10].

References

1. N. R. Adam and J. C. Wortmann. Security-control methods for statistical database: A comparative study. *ACM Computing Surveys*, 21:515–556, 1989.
2. L. H. Cox. Suppression methodology and statistical disclosure control. *Journal of the American Statistical Association*, 75:377–385, 1980.
3. D. E. Denning and J. Schlörer. Inference controls for statistical databases. *IEEE Computer*, 16:69–82, July 1983.
4. K. P. Eswaran and R. E. Tarjan. Augmentation problems. *SIAM Journal on Computing*, 5:653–665, 1976.
5. A. Frank. Connectivity augmentation problems in network design. In J. R. Birge and K. G. Murty, editors, *Mathematical Programming: State of the Art 1994*, pages 34–63. The University of Michigan, 1994.
6. D. Gusfield. A graph theoretic approach to statistical data security. *SIAM Journal on Computing*, 17:552–571, 1988.
7. T.-s. Hsu. *Graph Augmentation and Related Problems: Theory and Practice*. PhD thesis, University of Texas at Austin, 1993.
8. T.-s. Hsu and M. Y. Kao. Security problems for statistical databases with general cell suppressions. In *Proceedings of the 9th International Conference on Scientific and Statistical Database Management*, pages 155–164, 1997.
9. T.-s. Hsu and M. Y. Kao. Optimal augmentation for bipartite componentwise biconnectivity in linear time. *SIAM Journal on Discrete Mathematics*, 19(2): 345–362, 2005.
10. P.-C. Huang, H.-W. Wei, W.-C. Lu, W.-K. Shih, and T.-s. Hsu. Smallest bipartite bridge-connectivity augmentation. Technical Report TR-IIS-06-016, Institute of Information Science, Academia Sinica, Nankang, Taipei, Taiwan, 2006.
11. J. B. Jensen, H. N. Gabow, T. Jordán, and Z. Szigeti. Edge-connectivity augmentation with partition constraints. *SIAM Journal on Discrete Mathematics*, 12:160–207, 1999.

12. M. Y. Kao. Linear-time optimal augmentation for componentwise bipartite-completeness of graphs. *Information Processing Letters*, pages 59–63, 1995.
13. M. Y. Kao. Data security equals graph connectivity. *SIAM Journal on Discrete Mathematics*, 9:87–100, 1996.
14. M. Y. Kao. Total protection of analytic-invariant information in cross-tabulated tables. *SIAM Journal on Computing*, 26:231–242, 1997.
15. J. P. Kelly, B. L. Golden, and A. A. Assad. Cell suppression: Disclosure protection for sensitive tabular data. *Networks*, 22:397–417, 1992.
16. F. M. Malvestuto and M. Moscarini. Censoring statistical tables to protect sensitive information: Easy and hard problems. In *Proceedings of the 8th International Conference on Scientific and Statistical Database management*, pages 12–21, 1996.
17. F. M. Malvestuto and M. Moscarini. Suppressing marginal totals from a two-dimensional table to protect sensitive information. *Statistics and Computing*, 7:101–114, 1997.
18. F. M. Malvestuto, M. Moscarini, and M. Rafanelli. Suppressing marginal cells to protect sensitive information in a two-dimensional statistical table. In *Proceedings of the 10th ACM SIGACT-SIGMOD-SIGACT Symposium on Principles of Database Systems*, pages 252–258, 1991.
19. H. Nagamochi. Recent development of graph connectivity augmentation algorithms. *IEICE Transactions on Information and System*, E83-D:372–383, 2000.

Approximation Algorithms for the Graph Orientation Minimizing the Maximum Weighted Outdegree*

Yuichi Asahiro[1] , Jesper Jansson[2,**] , Eiji Miyano[3] , Hirotaka Ono[2],
and Kouhei Zenmyo[3]

[1] Department of Social Information Systems,
Kyushu Sangyo University, Fukuoka 813-8503, Japan
asahiro@is.kyusan-u.ac.jp
[2] Department of Computer Science and Communication Engineering,
Kyushu University, Fukuoka 812-8581, Japan
{jj@tcslab.,ono@}csce.kyushu-u.ac.jp
[3] Department of Systems Innovation and Informatics,
Kyushu Institute of Technology, Fukuoka 820-8502, Japan
{miyano@,kouhei@theory.}ces.kyutech.ac.jp

Abstract. Given an undirected graph $G = (V, E)$ and a weight function $w : E \rightarrow \mathbb{Z}^+$, we consider the problem of orienting all edges in E so that the maximum weighted outdegree among all vertices is minimized. In this paper (1) we prove that the problem is strongly NP-hard if all edge weights belong to the set $\{1, k\}$, where k is any integer greater than or equal to 2, and that there exists no pseudo-polynomial time approximation algorithm for this problem whose approximation ratio is smaller than $(1 + 1/k)$ unless P=NP; (2) we present a polynomial time algorithm that approximates the general version of the problem within a factor of $(2 - 1/k)$, where k is the maximum weight of an edge in G; (3) we show how to approximate the special case in which all edge weights belong to $\{1, k\}$ within a factor of 3/2 for $k = 2$ (note that this matches the inapproximability bound above), and $(2 - 2/(k + 1))$ for any $k \geq 3$, respectively, in polynomial time.

1 Introduction

1.1 Problems and Summary of Results

Let $G = (V, E, w)$ be a simple, undirected and weighted graph, where V, E and w denote the set of nodes, the set of edges and a positive integral weight function $w : E \rightarrow \mathbb{Z}^+$, respectively. Throughout the paper, let $|V| = n$ and $|E| = m$ for the graph. An *orientation* Λ *of the graph* G is an assignment of a

---

* This work is partially supported by Grant-in-Aid for Scientific Research on Priority Areas No. 16092223, and by Grant-in-Aid for Young Scientists (B) No. 17700022, No. 18700014 and No. 18700015.
** Supported by JSPS (Japan Society for the Promotion of Science).

M.-Y. Kao and X.-Y. Li (Eds.): AAIM 2007, LNCS 4508, pp. 167–177, 2007.

direction to each edge $\{u, v\} \in E$, i.e., $\Lambda(\{u, v\})$ is either (u, v) or (v, u). The *weighted outdegree* of u is $d_\Lambda^+(u)$, where $d_\Lambda^+(u)$ denotes $\sum_{\substack{\{u,v\} \in E \\ \Lambda(\{u,v\})=(u,v)}} w(\{u, v\})$.

We consider the problem of finding an orientation such that the maximum weighted outdegree is minimum. This basic problem has several applications. For example, such orientations can be used to construct efficient dynamic data structures for graphs that support fast vertex adjacency queries under a series of edge insertions and edge deletions [3]. Also, it can be considered a variation of *art gallery problems* (e.g., [4,11]) and *unrelated parallel machine scheduling* (e.g., [10]). Especially, the polynomial time (in)approximability of the latter problem has been intensively studied, as discussed in the next subsection.

Previous studies show that our problem can be solved in polynomial time if all the edge weights are identical [1,9,15], while it is NP-hard in general [1]. Also, a $(2 - 1/\lceil L(G) \rceil)$-approximation algorithm with $O(m^2)$ running time was presented in [1], where $L(G) = \max_{H \subseteq G} \{\sum_{\{u,v\} \in E(H)} w(\{u, v\})/|V(H)|\}$.

In this paper, we consider the problem from the viewpoint of polynomial time approximability and inapproximability. Our results are summarized as follows:

- We present a $(2 - 1/k)$-approximation algorithm with running time $O(m^{3/2} \cdot \log m \cdot \log k \cdot \log \Delta^* + m^2)$, where k, m and Δ^* denote the maximum weight of the edges, the number of the edges and the optimal value, respectively.
- For special cases in which the weight of each edge is either 1 or k, a refined algorithm achieves a better approximation factor, $2 - 2/(k + 2)$, also with running time $O(m^{3/2} \cdot \log m \cdot \log k \cdot \log \Delta^* + m^2)$.
- We prove that there is no polynomial time approximation algorithm whose factor is smaller than $3/2$, unless P=NP. (More precisely, in case where weights of all the edges are either 1 or a positive integer $k \geq 2$, no pseudo-polynomial time algorithm achieves an approximation ratio smaller than $1 + 1/k$.) That is, for $k = 2$, the above algorithm is best possible with respect to the approximation ratio.

Note that the new $2 - 1/k$-approximation ratio in this paper and the previous $2 - 1/\lceil L(G) \rceil$ one in [1] are incomparable; sometimes the former is better than the latter, and vice versa. For example, we have an instance for which the latter algorithm outputs 5/3-factor solution, while the former achieves approximation ratio 1.5 (see Figure 6 in [1]). Due to space limitations, the formal proofs have been omitted in this paper. Please refer to the full paper for a complete version.

1.2   Related Work

Graph orientation itself is a quite basic, natural and important problem in graph theory and combinatorial optimization (see Chapter 61 of [13]). However, most of the studies consider the problems of finding an orientation with lower outdegree satisfying some special graph properties, such as high connectivity, small diameter, no-cycle and so on [2,5,8], and very few studies consider just the minimization of the maximum outdegree (or indegree) [1,15].

As mentioned in the previous subsection, another aspect of the minimization of the maximum outdegree is scheduling. For an undirected graph, let us consider the vertices as the machines and the edges as the jobs. Then our orientation problem can be regarded as a special case of the job assignment problem, in which the minimization of the maximum outdegree means to minimize the finishing time of all the jobs [12]. From the viewpoint of scheduling, our problem has some restriction, that is, 1) each job must be assigned to exactly one of pre-determined two machines, and 2) the processing time of each job does not depend on the machines. Therefore, our problem is a special case of *scheduling on unrelated parallel machines* ($R||C_{max}$ in the now-standard notation), given a set J of jobs, a set M of machines, and the time $p_{ij} \in \mathbb{Z}^+$ taken to process job $j \in J$ on machine $i \in M$, its goal is to find a job scheduling so as to minimize the makespan, i.e., the maximum processing time of any machine. In [10], Lenstra, et al. gave a polynomial time 2-approximation algorithm that is based on the LP-formulation for the general $R||C_{max}$ and its 3/2 inapproximability result (see also [14].)

Note that the 3/2 inapproximability result of Lenstra, et al. *cannot* be directly applied to the restricted assignment variant in which every job can be processed on a *constant number* of machines. In our problem, each job associated with an edge can be assigned only to one of the *two* machines associated with the two nodes of the edge, which means that their proof is not applicable to our case. Also note that their proof of inapproximability uses the assumption that the processing time of each job may vary depending on which machine it is processed on. Thus, our result provides a stronger inapproximability bound to the problem.

2 Preliminaries

2.1 Definitions

Let $G = (V, E, w)$ be a simple, undirected, weighted graph, where V, E, and w denote a set of vertices, a set of edges, and an integral weight function, $w : E \rightarrow \mathbb{Z}^+$, respectively. Let $w_{\max}$ and W be the maximum weight of edges and the total weight of edges, respectively. We denote the undirected edge whose endpoints are u and v where $u < v$ in lexicographic order by $e_{u,v}$, or simply $\{u, v\}$, and denote the directed edge (or *arc*) from u toward v, by (u, v). An *orientation* Λ of the undirected graph G is an assignment of direction to each edge $\{u, v\} \in E$, i.e., (u, v) or (v, u). A *directed path* P *of length* l from a vertex v_0 to a vertex v_l in a directed graph $G = (V, A, w)$ is a set $\{(v_{i-1}, v_i) \mid (v_{i-1}, v_i) \in A, i = 1, 2, \ldots, l$ and $v_i \neq v_j$ for any i and $j\}$ of arcs, which is also denoted by a sequence $\langle v_0, v_1, \ldots, v_l \rangle$ for simplicity. For the path P, the path of its reverse order is denoted by $\overline{P}$, i.e., $\overline{P} = \langle v_l, v_{l-1}, \ldots, v_0 \rangle$. Especially, a directed path P satisfying $v_l = v_0$ is called an *l-directed cycle*.

Let $d_\Lambda^+(v)$ and $d_\Lambda^-(v)$ under an orientation Λ denote the total weight of outgoing arcs and that of incoming arcs of a vertex v in the weighted directed graph $G(V, A, w)$, which we call the *weighted outdegree* and the *weighted indegree* of v, respectively. Throughout the paper, we use the words "outdegree" and "indegree" to represent these weighted degrees. Then the *cost* of an orientation Λ for a graph G is defined to be $\Delta_\Lambda(G) = \max_{v \in V}\{d_\Lambda^+(v)\}$. For an undirected graph

$G = (V, E)$ and a node $u \in V$, we define $\Gamma(u) = \{v \mid \{u, v\} \in E\}$, the set of neighbors of u. Given an orientation Λ of G, we define $\Gamma_\Lambda(u) = \{v \mid \{u, v\} \in E \text{ and } \Lambda(\{u, v\}) = (u, v)\}$, the set of neighbors of u on G under Λ.

Every orientation has the following trivial lower bound caused by the maximum weight of edges:

Proposition 1. ([1]) For a graph G and any orientation Λ, $\Delta_\Lambda(G) \geq w_{\max}$.   □

2.2 Problem and Basic Operations

The problem that we consider in this paper is the minimization of the maximum outdegree of a given undirected weighted simple graph. To specify the class of weight function of the graph, we formally define our problem as follows.

Problem: S-MINIMUM MAXIMUM OUTDEGREE (S-MMO)
Input: An undirected graph $G = (V, E)$ and a weight function $w : E \to S$, where S is a set of weights.
Output: An orientation Λ that minimizes $\max\{d_\Lambda^+(u) \mid u \in V\}$.

Namely, if we have no restriction about the weight function (just it should be a positive integral function), our problem is $\mathbb{Z}^+$-MMO. In this paper, we mainly consider the problem for the case of $S = \{1, 2, \ldots, k\}$. We also consider a special case in which the range of w is restricted to $S = \{1, k\}$ with $k \geq 2$.

Let OPT denote an optimal orientation. We say a graph orientation algorithm is a σ-approximation algorithm if $ALG(G)/OPT(G) \leq \sigma$ holds for any undirected graph G, where $ALG(G)$ is the objective value of a solution obtained by the algorithm for G, and $OPT(G)$ is that of an optimal solution. In the following we use $OPT(G)$ or Δ^* to denote the optimal value.

Here we introduce three basic operations; REVERSE, UP-TO-ROOTS and SOLVE-1-MMO.

- REVERSE does the following: *Given an orientation Λ of graph G and a directed path $P = \langle u_0, u_2, \ldots, u_l \rangle$ in G under Λ, update Λ by replacing P with $\overline{P}$, i.e., let $\Lambda(e_{u_i, u_{i+1}}) = (u_{i+1}, u_i)$ for $i = 0, \ldots, l - 1$.* Note that the outdegree for each vertex remains the same after the operation if P is a directed cycle and $w(e_{u_i, u_{i+1}})$'s are all identical. We call this operation REVERSECYCLE if $u_0 = u_l$.

- UP-TO-ROOTS determines an orientation Λ for a given simple forest G, in the following manner: *First fix an arbitrary root for each connected component of G (it is a tree). Then for every edge e, orient $\Lambda(e)$ towards the root of the tree containing e.* Note that for a forest with weighted edges UP-TO-ROOTS operation returns an optimal solution, whose value is $w_{\max}$ [1].

- SOLVE-1-MMO outputs an optimal orientation Λ for a given undirected graph G with identical weights. It is shown in [1] that the running time of SOLVE-1-MMO is $O(m^{3/2} \cdot \log(\Delta^*/k))$ for $\{k\}$-MMO, in which the log factor comes from the binary search.

3   Approximation Algorithms

In this section, we present three pseudo-polynomial time approximation algorithms for the S-MMO problem. The first and the second algorithms (in Sections 3.1 and 3.2) work for S-MMO with $S = \{1, 2, \ldots, k\}$, both of which are based on the replication of weighted edges, and their approximation ratios are 2 and $2 - 1/k$, respectively. The third algorithm (in Section 3.3) for $\{1, k\}$-MMO is a refined version of the second one, and its approximation ratio is $2 - 2/(k+1)$ for $k \geq 3$. In Section 3.4, we show how to improve the running times of the three approximation algorithms to polynomial time.

3.1   Majority Voting Algorithm

We first present a basic 2-approximation algorithm, named MAJORITY. Although MAJORITY can be considered a variation of Lenstra-Shmoys-Tardos algorithm [10] (LST, for short), which is based on the LP-rounding and has approximation factor 2, MAJORITY is combinatorial and provides basic ideas for the algorithms presented later. Also it is much faster than LST, by Corollary 1.

The idea of the algorithm is as follows: We replace each edge $e = \{u, v\}$ in G with $w(e)$ edges of weight 1 between u and v, and then we obtain an undirected multi-graph G' with $W = \sum_{e \in E} w(e)$ edges. We find an optimal MMO orientation Λ' for G', and then we decide an orientation of each weighted edge on G according to Λ' by the majority voting manner; in Λ', for each $e_{u,v} \in E$, some of replicated edges of $e_{u,v}$ are oriented from u to v and the others from v to u. Let us denote the number of edges from u to v (resp., from v to u) in Λ' by $f_{u \to v}$ (resp., $f_{v \to u}$). Since we assume the original graph is simple, $f_{u \to v} + f_{v \to u} = w(e_{u,v})$ holds. By using these, we decide the orientation Λ of the original G by the following manner: For $e_{u,v} \in E$,

$$\Lambda(e_{u,v}) := \begin{cases} (u, v) & \text{if } f_{u \to v} \geq f_{v \to u}, \\ (v, u) & \text{otherwise.} \end{cases} \qquad (1)$$

In the case of a tie the direction is determined according to a lexicographic order. We call this algorithm MAJORITY.

Algorithm MAJORITY

1. For graph G, construct G' by replacing each edge e with $w(e)$ edges.
2. Find an optimal orientation Λ' of G' by using SOLVE-1-MMO.
3. Decide the orientation Λ of G according to (1).
4. Return Λ.

Theorem 1. For $S = \{1, \ldots, k\}$, Algorithm MAJORITY approximates S-MMO within a factor of 2 and runs in $O(W^{3/2} \cdot \log \Delta^*)$ time.

Proof. Since Steps 1, 2 and 3 take $O(W)$, $O(W^{3/2} \log \Delta^*)$ and $O(W)$ time, respectively, the running time of MAJORITY is $O(W^{3/2} \log \Delta^*)$, in total. The approximation factor 2 is immediately obtained by the result of [10].   □

3.2   Cycle Canceling Algorithm

Here, we describe a new algorithm named CYCLE-CANCELING, which improves
MAJORITY; the approximation ratio is $2 - 1/k$.

Algorithm CYCLE-CANCELING

1. For graph G, construct G' by replacing each edge e with $w(e)$ edges.
2. Find an optimal orientation Λ' of G' by using SOLVE-1-MMO.
3. Decide the (partial) orientation Λ of G according to (2) and obtain,
   $G_{\Lambda'} = (V, F_{\Lambda'})$ as described later.
4. If $G_{\Lambda'}$ has an l-directed cycle with $l \geq 3$, apply REVERSECYCLE
   and go to 3.
5. For undecided edges of Λ, apply UP-TO-ROOTS.
6. Return Λ.

In the first and second steps of the algorithm, do as MAJORITY; construct
G' (replicate each edge) and then find an optimal orientation Λ'. After that we
decide the orientation of the original problem by

$$\Lambda(e_{u,v}) := \begin{cases} (u,v) & \text{if } f_{v \to u} = 0, \\ (v,u) & \text{if } f_{u \to v} = 0, \\ - & \text{otherwise,} \end{cases} \tag{2}$$

where $-$ means "not decided yet." Note that the direction of the edges decided
by this operation is essentially same as the one of Λ'; the cost of the orientation
does not change.

Here, we introduce a new operation, *cycle cancellation*, which updates the
orientation to more desirable orientation without changing the outdegrees of all
the nodes. To this end, we construct another undirected graph $G_{\Lambda'} = (V, F_{\Lambda'})$,
where $F_{\Lambda'} = \{e_{u,v} \in E \mid f_{u \to v} \neq 0 \text{ and } f_{v \to u} \neq 0 \text{ in } \Lambda'\}$. From $G_{\Lambda'}$, we find
an l-cycle with $l \geq 3$, say $C = \langle v_1, v_2, \ldots, v_l, v_1 (\equiv v_{l+1}) \rangle$, if exists. (From here,
when we mention l-cycles with $l \geq 3$, we just use "cycles" for simplicity, because
we do not consider 2-cycles in this paper.) Let $c = \min\{f_{v_i \to v_{i+1}} \mid i = 1, \ldots, l\}$,
which is a positive integer, by the definition of $F_{\Lambda'}$. We then go back to G'
and Λ' and apply REVERSECYCLE with size c to C; since there exist c cycles
of $\langle v_1, v_2, \ldots, v_l, v_1 (\equiv v_{l+1}) \rangle$ on G' under Λ', we can reverse the direction of
the edges along the c cycles. Note that the outdegree (or the indegree) of each
node in the resulting directed graph is equal to the one under Λ'; it is still an
optimal orientation in G' and can be updated as Λ'. For this new Λ', we apply
the equation (2), then go back to the beginning of this paragraph. Since at least
one edge on the cycle C satisfies $f_{v_i \to v_{i+1}} = 0$ by the REVERSECYCLE, the new
$F_{\Lambda'}$ is strictly smaller than the old $F_{\Lambda'}$; this step ends in at most $m - 2$ iterations.

After the several (or possibly no) iterations of the above procedure, $G_{\Lambda'}$ becomes a forest, and set $\mathcal{F} := G_{\Lambda'}$. Note that all the edges of $\mathcal{F}$ are not decided yet
by (2). The cycle cancellation itself implies that there always exists an optimal
solution Λ' for the relaxed problem such that Λ' has no cycles in $\mathcal{F}$. Then, we

have the nice tree structure, for which we can apply UP-TO-ROOTS operation that decides the orientation of all the remaining edges.

Theorem 2. For $S = \{1, \ldots, k\}$, Algorithm CYCLE-CANCELING approximates S-MMO within a factor of $(2 - \frac{1}{k})$ and runs in $O(W^{3/2} \log \Delta^* + m^2)$ time.

Proof. We first consider the running time. Steps 1 and 2 require the same time complexity as MAJORITY, i.e., $O(W^{3/2} \log \Delta^*)$ time. Each iteration of Steps 3 takes $O(m)$ time, and also each iteration of Steps 4 takes $O(m)$ time by the depth first search, and these steps can be iterated at most $m - 2$ times. Step 5 takes $O(m)$ time. In total, the running time is $O(W^{3/2} \log \Delta^* + m^2)$.

Next, we analyze the approximation factor. Let u^* be any critical node in G with respect to Λ, i.e., a node with maximum weighted outdegree under Λ. We now prove that $d_\Lambda^+(u^*) \leq (2 - \frac{1}{k}) \cdot OPT(G)$. First of all, note that $OPT(G) \geq k$ by Proposition 1 and also that $OPT(G) \geq OPT(G') = d_{\Lambda'}^+(x^*) \geq d_{\Lambda'}^+(u^*)$, where x^* is any critical node with respect to Λ'. Let $\mathcal{F}^*$ be the forest of rooted trees produced by UP-TO-ROOTS in Step 5. There are two possible cases to consider after the iterations of Steps 3 and 4:

1. u^* is a root in $\mathcal{F}^*$: [1] In this case, we immediately have $d_\Lambda^+(u^*) \leq d_{\Lambda'}^+(u^*)$ because zero or more of u^*'s outgoing edges in Λ' are reversed to obtain Λ, but none of its incoming edges in Λ' is reversed in Step 5. Then, recall that $d_{\Lambda'}^+(u^*) \leq OPT(G)$ by the above.
2. u^* is not a root in $\mathcal{F}^*$: In this case, let p denote the parent of u^* and $\mathcal{C}$ the set of children of u^* in $\mathcal{F}^*$, respectively. Clearly, we have

$$d_\Lambda^+(u^*) = d_{\Lambda'}^+(u^*) + f_{p \to u^*} - \sum_{v \in \mathcal{C}} f_{u^* \to v} \leq d_{\Lambda'}^+(u^*) + f_{p \to u^*},$$

which yields

$$\frac{d_\Lambda^+(u^*)}{OPT(G)} \leq \frac{d_{\Lambda'}^+(u^*) + f_{p \to u^*}}{OPT(G)} \leq \frac{d_{\Lambda'}^+(u^*)}{d_{\Lambda'}^+(u^*)} + \frac{f_{p \to u^*}}{k} \leq 1 + \frac{k-1}{k} = 2 - \frac{1}{k},$$

where the last inequality holds since $f_{p \to u^*} + f_{u^* \to p} \leq k$ and $f_{u^* \to p} \geq 1$.

In both cases, $d_\Lambda^+(u^*)$ is within the desired bound. The theorem follows. $\square$

Note that the analysis of Theorem 2 is tight; we can construct a worst-case example of CYCLE-CANCELING for $\{1, 3\}$-MMO (see the full-length version of our paper).

Remark: According to Theorem 2, the approximation factor of Algorithm CYCLE-CANCELING for $k = 2$ is $3/2$. This is actually the best possible in polynomial time for $k = 2$ (unless P=NP), as we shall see in Section 4.

---

[1] This case also handles the possibility that u^* is disconnected from all other vertices in $G_{\Lambda'}$.

3.3 Refined Cycle Canceling Algorithm

We now consider the special case of S-MMO in which $S = \{1, k\}$ for $k \geq 3$, and show that it can be approximated more efficiently than by Theorem 2. The key idea is to show that if all edge weights in G are either 1 or k, a slight modification to Algorithm CYCLE-CANCELING allows us to compute a stronger lower bound on an optimal solution which then yields an improved approximation factor.

As mentioned in the previous section, the cycle cancellation itself provides an optimal solution for the relaxed problem with a tree property. Here, we focus on Step 5 of the algorithm CYCLE-CANCELING, in which the naive application of UP-TO-ROOTS with arbitrary roots gives a worst-case example; this causes the approximation ratio to be $2 - 1/k$. Its reason is that some nodes having large outdegree under the orientation Λ' are not suitable for being root; if such a node is set to be a root, its outdegree will distribute to its neighbors, so that the neighbors have large outdegree under Λ compared to that under Λ'. To avoid such a bad situation, we introduce a simple procedure.

In the algorithm, do the same operations as CYCLE-CANCELING until Step 4, and obtain a forest $\mathcal{F}$. If there exists a leaf node u in $\mathcal{F}$ such that $f_{u \to v} \geq f_{v \to u}$ holds for its neighbor v, we fix the orientation of $e_{u,v}$ as (u, v) and remove $e_{u,v}$ from $\mathcal{F}$ (i.e., $\Lambda(e_{u,v}) := (u, v)$ and $\mathcal{F} = (V, F)$ with $F := F \setminus \{e_{u,v}\}$). We repeat this operation until no leaf node u satisfies $f_{u \to v} \geq f_{v \to u}$ where v is the neighbor node of u. Then we apply UP-TO-ROOTS.

Algorithm REFINED CYCLE-CANCELING

1-4. (Same as CYCLE-CANCELING).
4'.   While there exists a leaf u connecting to v such that $f_{u \to v} \geq f_{v \to u}$ in $\mathcal{F} = (V, F)$, let $\Lambda(e_{u,v}) := (u, v)$ and remove $e_{u,v}$ from F.
5.   For undecided edges of Λ, apply UP-TO-ROOTS to $\mathcal{F}$.
6.   Return Λ.

Theorem 3. For any $S = \{1, k\}$ where $k \geq 3$, Algorithm REFINED CYCLE-CANCELING approximates S-MMO within a factor of $(2 - \frac{2}{k+1})$ and runs in $O(W^{3/2} \log \Delta^* + m^2)$ time.

Proof. It is easy to see that adding Step 4' to Algorithm CYCLE-CANCELING in Section 3.2 does not increase the asymptotic running time. Therefore, the running time is $O(W^{3/2} \log \Delta^* + m^2)$.

To analyze the approximation factor of REFINED CYCLE-CANCELING, we proceed similarly as in the proof of Theorem 2. Let u^* be any critical node in G with respect to Λ, and let $\mathcal{F}^*$ be the forest of rooted trees produced by UP-TO-ROOTS in Step 5. Recall that $OPT(G) \geq k$ and $OPT(G) \geq OPT(G') \geq d_{\Lambda'}^+(u^*)$. There are two main cases:

1. u^* is a node which satisfies the condition in Step 4': Then, since $f_{p \to u^*} \leq \frac{k}{2}$ for the parent p of u^*,

$$\frac{d_\Lambda^+(u^*)}{OPT(G)} \leq \frac{d_{\Lambda'}^+(u^*) + f_{p \to u^*}}{OPT(G)} \leq \frac{d_{\Lambda'}^+(u^*)}{d_{\Lambda'}^+(u^*)} + \frac{f_{p \to u^*}}{k} \leq 1 + \frac{k/2}{k} = \frac{3}{2}.$$

2. u^* is a node which did not satisfy the condition in Step 4':
   (a) If u^* is a root in $\mathcal{F}^*$ then $d_\Lambda^+(u^*) \leq d_{\Lambda'}^+(u^*) \leq OPT(G)$ and we are done as before.
   (b) If not, consider the tree T in $\mathcal{F}^*$ that contains u^*. Let p be the parent of u^* in T and let $\langle u_1, u_2, \ldots, u_\ell \rangle$ be the path between any two leaves u_1 and u_ℓ in the undirected version of T. Since u_1 and u_ℓ satisfy $f_{u_1 \to u_2} < f_{u_2 \to u_1}$ and $f_{u_\ell \to u_{\ell-1}} < f_{u_{\ell-1} \to u_\ell}$, there must exist an intermediate node u_i such that $f_{u_{i-1} \to u_i} < f_{u_i \to u_{i-1}}$ and $f_{u_i \to u_{i+1}} \geq f_{u_{i+1} \to u_i}$. Next, because all edges in T have weight k, we know that $f_{v \to w} + f_{w \to v} = k$ for every edge $\{v, w\}$ in T, which means that $f_{u_i \to u_{i-1}} > k/2$ and $f_{u_i \to u_{i+1}} \geq k/2$. Thus, the outdegree of u_i is at least $f_{u_i \to u_{i-1}} + f_{u_i \to u_{i+1}} > k$, i.e., $OPT(G') \geq k + 1$. Plugging in this stronger lower bound gives us

$$\frac{d_\Lambda^+(u^*)}{OPT(G)} \leq \frac{d_{\Lambda'}^+(u^*) + f_{p \to u^*}}{OPT(G)} \leq \frac{d_{\Lambda'}^+(u^*)}{d_{\Lambda'}^+(u^*)} + \frac{f_{p \to u^*}}{k+1} \leq 1 + \frac{k-1}{k+1} = 2 - \frac{2}{k+1}.$$

Since $2 - \frac{2}{k+1} \geq 3/2$ for $k \geq 3$, the approximation is $2 - \frac{2}{k+1}$ for $k \geq 3$ in total. Note that the approximation ratio of REFINED CYCLE-CANCELING for $k = 2$ is $3/2$ (same as CYCLE-CANCELING) because Step 5 is not executed. □

The analysis of Theorem 3 is also tight; we can construct a worst-case example of REFINED CYCLE-CANCELING for $\{1, 3\}$-MMO.

3.4 Polynomial Time Computation of 1-MMO of G'

In this subsection, we show the technique of making Algorithms MAJORITY, CYCLE-CANCELING and REFINED CYCLE-CANCELING into polynomial time algorithms. Recall that in these algorithms, we have to solve 1-MMO for G', which is generated from G by replacing each edge e with $w(e)$ edges of weight 1, as a sub-procedure. Hence, as described in Section 3.1, the algorithm requires $O(W^{3/2} \cdot \log \Delta^*)$ time only to obtain an optimal solution of 1-MMO. However, the information that algorithms MAJORITY, CYCLE-CANCELING and REFINED CYCLE-CANCELING need is not the orientation itself but the values $f_{u \to v}$ and $f_{v \to u}$, which can be computed in polynomial time.

The idea is as follows: Instead of explicitly constructing G' and applying SOLVE 1-MMO, we solve a relaxed version of the problem by using a maximum network flow technique. The relaxed version means that for each edge, its orientation may be fractional. For example, edge $e = \{u, v\}$ with weight 2 may be oriented as (u, v) with weight 1.5 and (v, u) with weight 0.5. Although the relaxed optimal solution can contain fractional flows in some edges, the integral maximum flow problem is known to have an optimal solution of integral flows (flow integrality) and some standard algorithms find such solutions indeed (for example, [7] presents $O(m \min\{m^{1/2}, n^{2/3}\} \log(n^2/m) \log U)$-time algorithm, where U is the maximum capacity size). Thus, the solution can be regarded as an optimal solution of 1-MMO for G'. Although we omit the detail, the problem can be solved by computing $O(\log \Delta^*)$ times the maximum flow for a network of $m + n$ vertices and $3m$ arcs with the maximum capacity k, which leads the following.

Theorem 4. We can compute the $f_{u\to v}$ and $f_{v\to u}$ values of all the edges for 1-MMO of G' in $O(m^{3/2} \cdot \log m \cdot \log k \cdot \log \Delta^*)$ time.   □

Corollary 1. (a) The running time of Algorithm MAJORITY can be improved to $O(m^{3/2} \cdot \log m \cdot \log k \cdot \log \Delta^*)$ time, and also (b) the running time of Algorithms CYCLE-CANCELING and REFINED CYCLE-CANCELING can be improved to $O(m^{3/2} \cdot \log m \cdot \log k \cdot \log \Delta^* + m^2)$ time.   □

4   Inapproximability Results

It is shown that S-MMO is weakly NP-hard [1], but no result about the inapproximability is shown. In this section, we provide a proof of the strong NP-hardness of S-MMO, which also gives inapproximability results. More precisely, we give a reduction from a variation of 3-SAT problem, At-most-3-SAT(2L), to $\{1, k\}$-MMO. At-most-3-SAT(2L) is a restriction of 3-SAT where each clause includes at most three literals and each literal (not variable) appears at most twice in a formula. It can be easily proved that At-most-3-SAT(2L) is NP-hard by using problem [LO1] on p. 259 of [6].

Given a formula ϕ of At-most-3-SAT(2L) with n variables $\{v_1, \ldots, v_n\}$ and m clauses $\{c_1, \ldots, c_m\}$, we construct a graph G_ϕ including gadgets that mimic (a) literals, (b) clauses and (c) a special gadget. (a) Each literal gadget consists of two nodes labeled by v_i and $\overline{v_i}$ and one edge $\{v_i, \overline{v_i}\}$ between them, corresponding to variable v_i of ϕ. The weight of $\{v_i, \overline{v_i}\}$ is k. (b) Each clause gadget is one node labeled by c_j, corresponding to clause c_j of ϕ. The clause gadget c_j is connected to at most three nodes in the literal gadgets that have the same labels as the literals in the clause c_j, by edges of weight 1. For example, if $c_1 = x \vee \overline{y}$ is appeared in ϕ, then node c_1 is connected to nodes x and $\overline{y}$. (See Figure 1.) (c) The special gadget is a cycle of k nodes and k edges where each edge of the cycle

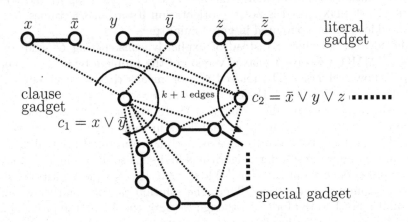

Fig. 1. Reduction from At-Most-3-SAT

has weight $k.^2$ If a clause consists of one (two or three, resp.,) variable(s), then it is connected to k (arbitrary $k - 1$ or $k - 2$, resp.,) nodes in the special gadget by edges of weight 1. Hence, the degree of every clause node is exactly $k + 1$.

We can prove the following:

Lemma 1. For the above construction of G_ϕ, the followings hold: (i) If ϕ is satisfiable, $OPT(G_\phi) \leq k$. (ii) If ϕ is not satisfiable, $OPT(G_\phi) \geq k + 1$.   □

From Lemma 1, we immediately obtain the following theorem.

Theorem 5. $\{1, k\}$-MMO is strongly NP-hard. Consequently, $\mathbb{Z}^+$-MMO is also strongly NP-hard.   □

Also the (in)satisfiability gap of Lemma 1 yields the following theorem.

Theorem 6. $\{1, k\}$-MMO (resp., $\mathbb{Z}^+$-MMO) has no pseudo-polynomial time algorithm whose approximation ratio is smaller than $1 + 1/k$ (resp., $3/2$), unless P=NP.   □

References

1. Y. Asahiro, E. Miyano, H. Ono, and K. Zenmyo, Graph orientation algorithms to minimize the maximum outdegree, *Proceedings of Computing: the Twelfth Australasian Theory Symposium (CATS 2006)*, pp. 11–20, 2006.
2. T. Biedl, T. Chan, Y. Ganjali, M. T. Hajiaghayi and D. R. Wood, Balanced vertex-orderings of graphs, *Discrete Applied Mathematics*, 48 (1), pp. 27–48, 2005.
3. G. S. Brodal and R. Fagerberg, Dynamic Representations of Sparse Graphs, *Proc. WADS1999, LNCS 1663*, pp. 342–351, 1999.
4. V. Chvátal, A combinatorial theorem in plane geometry, *J. Combinatorial Theory, series B*, 18, pp. 39–41, 1975.
5. F. V. Fomin, M. Matamala and I. Rapaport, Complexity of approximating the oriented diameter of chordal graphs, *J. Graph Theory*, 45 (4), pp. 255–269, 2004.
6. M. Garey and D. Johnson, *Computers and Intractability: A Guide to the Theory of NP-Completeness*, W. H. Freeman and Co., New York, 1979.
7. A. V. Goldberg and S. Rao, Beyond the flow decomposition barrier, *J. ACM*, 45(5)), pp. 783-797, 1998.
8. J. Kára, J. Kratochvíl, and D. R. Wood, On the complexity of the balanced vertex ordering problem, *Proc. COCOON2005, LNCS 3595*, pp. 849–858, 2005.
9. L. Kowalik, Approximation Scheme for Lowest Outdegree Orientation and Graph Density Measures, *Proc. ISAAC2006, LNCS 4288*, pp. 557–566, 2006.
10. J. K. Lenstra, D. B. Shmoys and Tardos, Approximation algorithms for scheduling unrelated parallel machines, *Mathematical Programming*, 46 (3), 259–271, 1990.
11. J. O'Rourke, *Art Gallery Theorems and Algorithms*, Oxford University Press, 1987.
12. M. Pinedo, *Scheduling: Theory, Algorithms, and Systems*, Prentice Hall, 2nd Ed., 2002.
13. A. Schrijver, *Combinatorial Optimization*, Springer, 2003.
14. P., Schuurman and G. J. Woeginger, Polynomial time approximation algorithms for machine scheduling: Ten open problems, *J. Scheduling*, 2, pp. 203–213, 1999.
15. V. Venkateswaran, Minimizing maximum indegree, *Discrete Applied Mathematics*, 143 (1-3), pp. 374–378, 2004.

---

[2] In case of $k = 2$, we prepare a cycle of 3 nodes as an exception to keep the simple property of the graph.

An Efficient Algorithm for the Evacuation Problem in a Certain Class of a Network with Uniform Path-Lengths

Naoyuki Kamiyama, Naoki Katoh, and Atsushi Takizawa

Department of Architecture and Architectural Engineering, Kyoto University
Kyotodaigaku-Katsura, Nishikyo-ku, Kyoto, 615-8540, Japan
{is.kamiyama,naoki,kukure}@archi.kyoto-u.ac.jp

Abstract. In this paper, we consider the evacuation problem for a network which consists of a directed graph with capacities and transit times on its arcs. This problem can be solved by the algorithm of Hoppe and Tardos [1] in polynomial time. However their running time is high-order polynomial, and hence is not practical in general. Thus it is necessary to devise a faster algorithm for a tractable and practically useful subclass of this problem. In this paper, we consider a dynamic network with a single sink s such that (i) for each vertex v the sum of transit times of arcs on any path from v to s takes the same value, and (ii) for each vertex v the minimum v-s cut is determined by the arcs incident to s whose tails are reachable from v. We propose an efficient algorithm for this network problem. This class of networks is a generalization of the grid network studied in the paper [2].

1 Introduction

The problem for finding the most effective plan to evacuate people to safe place has been modelled as an *evacuation problem* by using *dynamic network flow*. In the evacuation problem, we are given a directed graph $D = (V, A)$ which consists of a vertex set V of n vertices with supply $b(v)$ on every vertex v and an arc set A of m arcs with capacity $c(e)$ and transit time $\tau(e)$ on every arc e and a single sink $s \in V$. If we consider urban evacuation, vertices model buildings, rooms, exits and so on, and arcs model pathways or roads. For an arc e, capacity $c(e)$ represents the number of people which can traverse e per unit time, and transit time $\tau(e)$ represents the time required to traverse e. For any vertex v, supply $b(v)$ represents the number of people which exist at v. The evacuation problem asks to find the minimum time required to send all the supplies to a sink.

Given time horizon T, the decision problem of whether we can send all supplies to a sink within time horizon T can be transformed to the maximum-flow problem defined on the *time-expanded network* introduced by Ford and Fulkerson [3]. However the time-expanded network consists of $O(T)$ copies of original vertices and arcs and hence does not lead to the efficient algorithm.

The first polynomial time algorithm for the evacuation problem was proposed by Hoppe and Tardos [1]. However it requires to use the submodular function

M.-Y. Kao and X.-Y. Li (Eds.): AAIM 2007, LNCS 4508, pp. 178–190, 2007.

minimization as a subroutine. Hence the running time is high-order polynomial, and the algorithm is not practical in general. Therefore it is necessary to devise a faster algorithm for a tractable and practically useful subclass of this problem.

As a special case, Mamada et al. [4] gave $O(n \log^2 n)$ time algorithm for the tree network. Hall et al. [5] studied the case called *uniform path-lengths* where there exists a single sink s and for any vertex v the sum of transit times of arcs on any path from v to s takes the same value. They showed that in this case the time-expanded network can be condensed to the so-called *condensed time-expanded network* whose size is polynomial in the input size. Kamiyama et al. [2] have shown an $O(n \log n)$ time algorithm for a $\sqrt{n} \times \sqrt{n}$ grid network with uniform arc capacity. In this paper, we will generalize the class of networks for which the ideas developed in [2] can be applied, i.e., we consider a dynamic network with a single sink s such that (i) for each vertex v the sum of transit times of arcs on any path from v to s takes the same value, and (ii) for each vertex v the minimum v-s cut is determined by the arcs incident to s whose tails are reachable from v. The algorithm of [2] reduced the evacuation problem to the *min-max resource allocation problem* [6], but in this paper we reduce the evacuation problem to the *parametric flow problem* defined on a static network[1]. Although it is known [5] that the evacuation problem in the case of uniform path-lengths can be reduced to the parametric flow problem in which the capacity of a subset of arcs is a linear function of time horizon T, we prove that in our case the evacuation problem can be reduced to the special case of the parametric flow problem studied by [7] which can be solved more efficiently than the general parametric flow problem considered in [5]. Thus in the case where the input dynamic network satisfies (i) and (ii), our algorithm is faster than using the condensed time-expanded network. In particular, our algorithm becomes much faster when the in-degree of a sink is small or considered to be a constant which is often the case with road networks.

2   Preliminaries

Let $\mathbb{R}_+$ and $\mathbb{Z}_+$ denote the set of nonnegative reals and nonnegative integers, respectively. We will not distinguish between a singleton $\{x\}$ and its element x. For any finite set X, we define $|X|$ as the number of elements that belong to X.

Directed graph. We denote by $D = (V, A)$ a directed graph which consists of a vertex set V and an arc set A. A vertex u is said to be *reachable* to a vertex v when there is a path from u to v. We denote by $e = uv$ an arc e whose tail is u and head is v. For any $X, Y \subseteq V$, we define $\delta(X, Y) = \{e = xy \colon x \in X, y \in Y\}$, and we write $\delta^+(X)$ and $\delta^-(X)$ instead of $\delta(X, V - X)$ and $\delta(V - X, X)$, respectively. For any $u, v \in V$, we denote by $\lambda_D(u, v)$ the local arc connectivity from u to v in D. For any $W \subseteq V$, let $D[W]$ denote the directed subgraph of D induced by W. Throughout this paper, we assume that D is acyclic.

---

[1] In order to distinguish classical network from dynamic network, we call classical network *static network*.

Dynamic network. We denote by $\mathcal{N} = (D = (V, A), c, \tau, b, s)$ a dynamic network $\mathcal{N}$ which consists of the directed graph $D = (V, A)$, a capacity function $c\colon A \to \mathbb{R}_+$ which represents the upper bound for the rate of flow that enters an arc per unit time, a transit time function $\tau\colon A \to \mathbb{Z}_+$ which represents the time required to traverse an arc, a supply function $b\colon V \to \mathbb{R}_+$ which represents the supply of a vertex, and a single sink $s \in V$. In order to avoid complicated argument, we assume $\tau(e) > 0$ for any $e \in A$. In this paper, we use the following notations: (i) $c(W_1, W_2) = \sum_{e \in \delta(W_1, W_2)} c(e)$ for any $W_1, W_2 \subseteq V$, and (ii) $c(W) = c(W, V - W)$ and $b(W) = \sum_{v \in W} b(v)$ for any $W \subseteq V$. Since we consider evacuation to s, we assume that s has no leaving arcs and no supply, and any vertex is reachable to s. We define a *length* of a path p in D as $\sum_{e \in p} \tau(e)$. We define a *dynamic network flow* $f\colon A \times \mathbb{Z}_+ \to \mathbb{R}_+$ in $\mathcal{N}$ as follows. For any $e \in A$ and $\theta \in \mathbb{Z}_+$, we denote by $f(e, \theta)$ the flow rate entering e at the time step θ which arrives at the head of e at the time step $\theta + \tau(e)$. We call f a *feasible dynamic network flow* in $\mathcal{N}$ if it satisfies the following three conditions, i.e., capacity constraint **CC**, flow conservation **FC**, and demand constraint **DC** [4].

CC: For any $e \in A$ and $\theta \in \mathbb{Z}_+$, $0 \le f(e, \theta) \le c(e)$.

FC: For any $v \in V$ and $\Theta \in \mathbb{Z}_+$,

$$\sum_{e \in \delta^+(v)} \sum_{\theta = 0}^{\Theta} f(e, \theta) - \sum_{e \in \delta^-(v)} \sum_{\theta = 0}^{\Theta - \tau(e)} f(e, \theta) \le b(v).$$

DC: There exists $\Theta \in \mathbb{Z}_+$ such that

$$\sum_{e \in \delta^-(s)} \sum_{\theta = 0}^{\Theta - \tau(e)} f(e, \theta) = \sum_{v \in V} b(v). \tag{1}$$

For a feasible dynamic network flow f, let $\Theta(f)$ denote the minimum time step Θ satisfying (1). The *evacuation problem* asks to find the minimum value of $\Theta(f)$ among all feasible dynamic network flows f. Given a dynamic network $\mathcal{N}$, the evacuation problem $\mathrm{EP}(\mathcal{N})$ is formally defined as follows:

$$\mathrm{EP}(\mathcal{N})\colon \text{ minimize}\left\{\Theta(f)\colon f \text{ is a feasible dynamic network flow in } \mathcal{N}\right\}.$$

We define $\Theta(\mathcal{N})$ as the optimal value of $\mathrm{EP}(\mathcal{N})$. Given time horizon T, we define the *decision version of EP($\mathcal{N}$) with time horizon T* as the problem which determines whether there exists a feasible dynamic network flow f with $\Theta(f) \le T$ in $\mathcal{N}$. Throughout this paper, n and m denote $|V|$ and $|A|$, respectively.

Static network. We denote by $\mathcal{N}' = (D' = (V', A'), c', b', s')$ a *static network* $\mathcal{N}'$ which consists of the directed graph $D' = (V', A')$, a capacity function $c'\colon A' \to \mathbb{R}_+$ a supply function $b'\colon V' \to \mathbb{R}_+$, and a single sink $s' \in V'$. We call $f\colon A' \to \mathbb{R}_+$ a *feasible static network flow* in $\mathcal{N}'$ if it satisfies the following two conditions, i.e., capacity constraint **CC** and flow conservation **FC**.

CC: For any $e \in A'$, $0 \le f(e) \le c'(e)$.
FC: For any $v \in V' - s'$, $\sum_{e \in \delta^+(v)} f(e) - \sum_{e \in \delta^-(v)} f(e) = b'(v)$.
If there exists a feasible flow in $\mathcal{N}'$, $\mathcal{N}'$ is called *feasible*

Time-expanded network. To solve the decision version of EP($\mathcal{N}$) with time horizon T, Ford and Fulkerson [3] introduced the *time-expanded network* which is a static network such that for any $v \in V$ and $i = 0, 1, \ldots, T$, there is a vertex v_i, and for any $e = uv \in A$, $i = 0, 1, \ldots, T - \tau(e)$, there is an arc $e_i = u_i v_{i+\tau(e)}$ whose capacity is $c(e)$, and for any $v \in V$ and $i = 0, 1, \ldots, T - 1$, there is a *holdover arc* $v_i v_{i+1}$ with infinite capacity. For any $v \in V$ the supply of v_0 is set to $b(v)$ and the supplies of all the other vertices v_i for $i = 1, \ldots, T$ are set to zero. Let s_T be a sink in the time-expanded network (Fig. 1). Though we

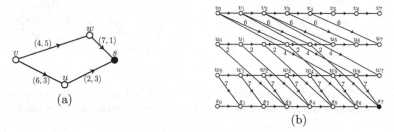

(a)

(b)

Fig. 1. (a) Dynamic network $\mathcal{N}$. (The pair of numbers attached to the arc indicates the capacity and the transit time.) (b) Time-expanded network with $T = 7$. (The number attached to the arc indicates the capacity.)

can decide whether the time-expanded network is feasible or not by solving the maximum-flow problem, the running time is pseudo-polynomial because the size of the time-expanded network is pseudo-polynomial in the input size.

2.1 Dynamic Networks with Uniform Path-Lengths

From here, we assume that any dynamic network satisfies uniform path-length condition. First we review the result due to Hall et al. [5]. They proved that EP($\mathcal{N}$) can be reduced to the *parametric flow problem* defined on the *condensed time-expanded network* whose size is polynomial in the input size.

We introduce necessary notations for $\mathcal{N} = (D = (V, A), c, \tau, b, s)$. For $v \in V$, we define l_v as the length of a path from v to s. Let us arrange the distinct values in $\{l_v : v \in V\}$ as $L_1 < \cdots < L_k$ where $L_1 = 0$ and k is the number of the distinct path-lengths in $\mathcal{N}$. Without loss of generality we assume that for any i with $2 \leq i \leq k$ $b(v) > 0$ for at least one vertex $v \in V$ with $l_v = L_i$. Let $L_{k+1} = T + 1$. We say a vertex v is at level i when $l_v = L_i$, which is denoted by $lev(v) = i$. We partition interval $[0, T]$ into $I_1, I_2, \ldots, I_k$ such that $I_i = [L_i, L_{i+1} - 1]$ holds for $i = 1, \ldots, k$. Moreover, let $P_s = \{v \in V : e = vs \in A\}$ and $R_v = \{w \in P_s : w$ is reachable from v in $D\}$ for $v \in V$. For example, for $\mathcal{N}$ in Fig. 1(a) with $T = 7$, we obtain $(l_s, l_w, l_u, l_v) = (0, 1, 3, 6)$. Thus, we have $k = 4$ and $I_1 = \{0\}$, $I_2 = \{1, 2\}$, $I_3 = \{3, 4, 5\}$, $I_4 = \{6, 7\}$.

The *condensed time-expanded network* $\mathcal{N}^c = (D^c = (V^c, A^c), c^c, b^c, s^c)$ for $\mathcal{N}$ with time horizon T is defined as follows. V^c is defined as $\{v_i : v \in V, i = lev(v), \ldots, k\}$. A^c consists of two types, i.e., $A^c = A_1^c \cup A_2^c$. $A_1^c = \{e_i = u_i v_i : e =$

$uv \in A, i = lev(u), \ldots, k\}$ and $A_2^c = \{v_i v_{i+1} \colon v \in V, i = lev(v), \ldots, k-1\}$. Arc $e_i \in A_1^c$ has the capacity $|I_i| c(e)$ where $|I_i|$ denotes the number of elements in I_i. An arc in A_2^c is a holdover arc whose capacity is infinity. For $v \in V$ the supply of $v_{lev(v)}$ is set to $b(v)$ and the supplies of all the other vertices v_i for $i = lev(v) + 1, \ldots, k$ are set to zero. $s^c = s_k$ holds (Fig. 2(a)).

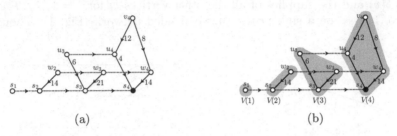

(a)   (b)

Fig. 2. (a) $\mathcal{N}^c$ for $\mathcal{N}$ in Fig. 1(a) with $T = 7$. (b) $V(i)$ for $\mathcal{N}^c$. (The number attached to the arc indicates the capacity, and holdover arcs are illustrated by dotted lines.)

For $i = 1, \ldots, k$, let $V(i) = \{v_i \in V^c \colon v \in V\}$ and $A(i) = \{e_i \in A_1^c \colon e \in A\}$. Notice that $V(i)$ for $i = 1, \ldots, k$ partitions V^c. It is easy to see that $A(i)$ is the arc set of $D^c[V(i)]$, i.e., the subgraph of D^c induced by $V(i)$ (Fig. 2(b)). From the definition of $\mathcal{N}^c$, we have the following fact.

Fact 1. *For any $i, j = 1, \ldots, k$ with $j - i \neq 1$, there is no arc connecting from $V(i)$ to $V(j)$. For any $i = 1, \ldots, k - 1$, $\delta(V(i), V(i+1)) = \{v_i v_{i+1} \colon v_i \in V(i)\}$ holds.*

From Fact 1, we can see that (i) $\mathcal{N}^c$ consists of k components such that for any $i = 1, \ldots, k$, the *i-th component* is a directed graph $D^c[V(i)]$ such that capacity of $e_i \in A(i)$ is $|I_i| c(e)$ (Fig. 2(b)), and (ii) consecutive components are connected by holdover arcs. Let $V_{\leq i} = \{v \in V \colon lev(v) \leq i\}$ for $i = 1, \ldots, k$.

Lemma 1. *(i) For any $i = 1, \ldots, k$, $D^c[V(i)]$ is isomorphic to $D[V_{\leq i}]$. (ii) For any $i = 1, \ldots, k$ and $u, v \in V_{\leq i}$, $\lambda_{D^c[V(i)]}(u_i, v_i) = \lambda_D(u, v)$.*

Proof. (i) follows from the definition of $D^c[V(i)]$. (ii) follows from $\lambda_{D[V_{\leq i}]}(u, v) = \lambda_D(u, v)$ for $i = 1, \ldots, k$ and $u, v \in V_{\leq i}$ and from (i). □

Hall et al. showed that a feasible dynamic flow f with $\Theta(f) \leq T$ exists in $\mathcal{N}$ if and only if $\mathcal{N}^c$ is feasible for time horizon T. Thus EP($\mathcal{N}$) can be solved by computing the minimum T such that $\mathcal{N}^c$ is feasible. By regarding T as the parameter we can reduce EP($\mathcal{N}$) to the *parametric flow problem* defined as follows.

Parametric flow problem. Given a static network $\mathcal{N}'=(D'=(V', A'), c', b', s')$ such that the capacity of $e \in A'$ is represented by $a_e + g_e \xi$ where a_e is a real constant, g_e is a nonnegative constant, and ξ is a nonnegative parameter, the parametric flow problem asks to find the minimum value of ξ such that $\mathcal{N}'$ is feasible. This problem can be solved in $O(|A'|^2|V'|\log(|V'|^2/|A'|))$ time by using the algorithm of [8].

Notice that from $L_{k+1} = T + 1$ $c^c(e_k) = |I_k|c(e) = (T - L_k + 1)c(e)$. The following theorem follows from $|V^c| = O(kn)$ and $|A^c| = O(km)$.

Lemma 2 ([5]). $EP(\mathcal{N})$ can be solved in $O(k^3 m^2 n \log(kn^2/m))$ time.

3   Evacuation Problem for a Fully Connected Network

A dynamic network $\mathcal{N} = (D = (V, A), c, \tau, b, s)$ is called *fully connected* if for each vertex $v \in V - s$ the minimum v-s cut is determined by the arcs incident to s whose tails are reachable from v. That is, the value of the minimum v-s cut is equal to $\sum_{e \in \delta(R_v, s)} c(e)$. In the subsequent discussion, we concentrate on the unit capacity case, i.e., the capacity of every arc is equal to one. In this case, $\mathcal{N}$ is fully connected if and only if $\lambda_D(v, s) = |\delta(R_v, s)|$ holds for any $v \in V - s$. The general capacity case can be treated similarly, which we will consider at the end of this section. In this section, we prove that $EP(\mathcal{N})$ for a fully connected network can be solved efficiently. This is a generalization of the result of [2]. We will prove that the problem can be reduced to to the *restricted parametric flow problem* defined in Section 3.2.

For the subsequent discussion, we will define *contraction* in a static network $\mathcal{N}' = (D' = (V', A'), c', b', s')$ and show the sufficient condition such that we can contract some vertex set in $\mathcal{N}^c$. The contraction of $X \subseteq V' - s'$ in $\mathcal{N}'$ is defined as the operation which consists of shrinking the vertices in X into a single vertex, eliminating loops, and combining multiple arcs by adding their capacities. For $X \subseteq V' - s'$, we call X *contractible* when $\mathcal{N}'_{/X}$ is feasible if and only if $\mathcal{N}'$ is feasible. We then give the sufficient condition such that X is contractible in $\mathcal{N}'$ (the proof is omitted).

Lemma 3. *For $X \subseteq V' - s'$, if there exists $Y \subseteq V' - s'$ with $X \subseteq Y$ such that $c'(Z) \geq c'(Y \cup Z)$ holds for any $Z \subseteq V' - s'$ with $X \cap Z \neq \emptyset$ and $X \not\subseteq Z$, X is contractible.*

3.1   Contraction in the Condensed Time-Expanded Network

For $\mathcal{N} = (D = (V, A), c, \tau, b, s)$ and $Q \subseteq P_s$, let $\mathcal{P}_Q = \{v \in V : R_v \subseteq Q\}$ and $\mathcal{P}_Q^* = \{v \in V : R_v = Q, \lambda_D(v, s) = |\delta(Q, s)|\}$ (Fig. 3(a)). If $\mathcal{N}$ is fully connected, $V - s = \bigcup_{Q \subseteq P_s} \mathcal{P}_Q^*$ holds. For any $W \subseteq V$ and $i = 1, \ldots, k$, let $W(i) = \{v_i \in V^c : v \in W\}$. The following theorem will be used in the subsequent discussion.

Theorem 1. *For any $Q \subseteq P_s$ and $i = 1, 2, \ldots, k$, $\mathcal{P}_Q^*(i)$ is contractible in $\mathcal{N}^c$.*

The following lemma will be used in the proof of Theorem 1 (the proof is omitted).

Lemma 4. $\delta^+(\bigcup_{i \leq j \leq k} \mathcal{P}_Q(j)) = \bigcup_{i \leq j \leq k} \delta(Q(j), s_j)$ *holds.*

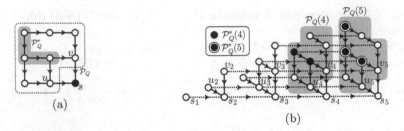

Fig. 3. (a) $\mathcal{P}_Q$ and $\mathcal{P}_Q^*$ with $Q = \{u, v\}$. (b) $\mathcal{P}_Q(i)$ and $\mathcal{P}_Q^*(i)$ with $Q = \{u, v\}$ and $i = 4, 5$. (The transit time in Fig 3(a) takes the same value.)

Proof. (**Theorem 1**) Let us fix $Q \subseteq P_s$ and i. We will use Lemma 3 to prove that $\mathcal{P}_Q^*(i)$ is contractible by setting $X = \mathcal{P}_Q^*(i)$ and $Y = \bigcup_{i \leq j \leq k} \mathcal{P}_Q(j)$. Thus, it is sufficient to prove $c^c(Z) \geq c^c(Y \cup Z)$ for any $Z \subseteq V^c - s^c$ with $X \cap Z \neq \emptyset$ and $X \nsubseteq Z$. In order to prove $c^c(Z) \geq c^c(Y \cup Z)$, it is sufficient to prove $c^c(Y) \leq c^c(Y \cap Z)$ since c^c is a submodular function. Recalling that every arc capacity is assumed to be one, $c^c(e_j) = |I_j|$ holds. Thus from Lemma 4, we have

$$c^c(Y) = \sum_{j=i}^{k} |\delta(Q(j), s_j)||I_j|. \tag{2}$$

Now we evaluate $c^c(Y \cap Z)$. From $X \cap Z \neq \emptyset$, let $v_i^* \in X \cap Z$. Since the capacity of holdover arc is infinity, we can assume $v_j^* \in Z$ holds for any $j = i+1, \dots, k$ since otherwise $c^c(Y \cap Z) = +\infty$ and the theorem clearly holds. We have

$$c^c(Y \cap Z) \geq \sum_{j=i}^{k} \sum_{e \in \delta(\mathcal{P}_Q(j) \cap Z, V(j) - (\mathcal{P}_Q(j) \cap Z))} c^c(e) \tag{3}$$

(the details are omitted). Since $\delta(\mathcal{P}_Q(j) \cap Z, V(j) - (\mathcal{P}_Q(j) \cap Z))$ is the set of arcs outgoing from $\mathcal{P}_Q(j) \cap Z$ in the j-th component, the following inequality holds for every j with $j = i, i+1, \dots, k$

$$\sum_{e \in \delta(\mathcal{P}_Q(j) \cap Z, V(j) - (\mathcal{P}_Q(j) \cap Z))} c^c(e) \geq \lambda_{D^c[V(j)]}(v_j^*, s_j)|I_j| \text{ (from } v_j^* \in \mathcal{P}_Q(j) \cap Z)$$

$$= \lambda_D(v^*, s)|I_j| \text{ (from Lemma 1(ii))} = |\delta(Q, s)||I_j| \text{ (from } v^* \in \mathcal{P}_Q^*). \tag{4}$$

Since $|\delta(Q(j), s_j)| \leq |\delta(Q, s)|$ holds, we have from (2), (3) and (4)

$$c^c(Y) = \sum_{j=i}^{k} |\delta(Q(j), s_j)||I_j| \leq \sum_{j=i}^{k} |\delta(Q, s)||I_j| \leq c^c(Y \cap Z). \qquad \square$$

3.2   The Restricted Parametric Flow Problem

In this problem, we are given a static network with multiple sinks $\mathcal{N}'' = (D'' = (V'', A''), c'', b'', S'')$ such that (i) S'' is a set of sinks, (ii) the capacity $c''(e)$ for an arc e incident to a sink is a linear function $a_e + g_e \xi$ where a_e is a constant, g_e is a nonnegative constant and ξ is a nonnegative parameter. The problem asks to find the minimum value of ξ such that $\mathcal{N}''$ is feasible where we define $f \colon A'' \to \mathbb{R}_+$ a feasible flow in $\mathcal{N}''$ when it satisfies **CC** and **FC** for any $v \in V'' - S''$. This problem can be transformed into a parametric maximum-flow problem studied by [7] by introducing a super source vertex q and arcs from q to every vertex v with $b''(v) > 0$ such that the capacity of qv is set to $b''(v)$. It is then easy to see that $\mathcal{N}''$ is feasible for a fixed ξ if and only if the maximum-flow value from q to S'' in the transformed problem is $\sum_{v \in V''} b''(v)$. Regarding ξ as a parameter, the maximum-flow value is a linear function in ξ.

Lemma 5 ([7]). *The maximum-flow value from q to S'' in the transformed network is a non-decreasing piecewise linear concave function $\kappa(\xi)$, and the largest breakpoint of $\kappa(\xi)$ can be found in the same time complexity as that of a single computation of the maximum-flow, i.e., $O(|A''||V''|\log(|V''|^2/|A''|))$.*

Lemma 6. *We can determine whether there exists ξ such that $\mathcal{N}''$ is feasible, and if there exists such ξ, the minimum such value can be found in $O(|A''||V''| \log(|V''|^2/|A''|))$.*

Proof. From the above discussion, $\mathcal{N}''$ is feasible when there exists ξ such that maximum-flow value in the transformed problem is equal to $\sum_{v \in V''} b''(v)$. On the other hand, the maximum-flow value in the transformed problem can not exceed $\sum_{v \in V''} b''(v)$. Thus when ξ is larger than the largest breakpoint the slope of $\kappa(\xi)$ is zero and $\kappa(\xi)$ is less than or equal to $\sum_{v \in V''} b''(v)$. Checking whether there exists ξ such that $\mathcal{N}''$ is feasible reduces to computing the largest breakpoint of $\kappa(\xi)$. Moreover, if there exists ξ such that $\mathcal{N}''$ is feasible, the minimum value of ξ such that $\mathcal{N}''$ is feasible is equal to the largest breakpoint of $\kappa(\xi)$. Thus, the lemma follows from Lemma 5. $\qquad\square$

As was defined in Section 2.1, in the condensed time-expanded network, the capacity of all arcs in the k-th component $D^c[V(k)]$ contains the parameter T, i.e., linear function of T. In Fig. 2(a), regarding T as the parameter, we have $c^c(u_4s_4) = 2(T-5)$, $c^c(w_4s_4) = 7(T-5)$, $c^c(v_4u_4) = 6(T-5)$, and $c^c(v_4w_4) = 4(T-5)$. Thus, the arcs which are not incident to a sink (i.e., v_4u_4 and v_4w_4) have the parametric capacity. Therefore, we can not reduce EP($\mathcal{N}$) for a general dynamic network with uniform path-lengths to the restricted parametric flow problem.

3.3   Reduction to the Restricted Parametric Flow Problem

Our reduction is constructed by the following lemmas. First, given a vertex set $\hat{V}$, a supply function $\hat{b}: \hat{V} \to \mathbb{R}_+$, a path-length function $\hat{l}: \hat{V} \to \mathbb{R}_+$, and a sink $\hat{s} \in \hat{V}$, let $\mathcal{N}(\hat{V}, \hat{b}, \hat{l}, \hat{s})$ be a set of dynamic networks $\hat{\mathcal{N}} = (\hat{D} = (\hat{V}, \hat{A}), \hat{c}, \hat{\tau}, \hat{b}, \hat{s})$

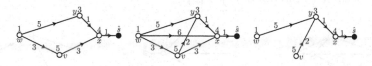

Fig. 4. Example of dynamic networks in $\mathcal{N}(\hat{V}, \hat{b}, \hat{l}, \hat{s})$. (The numbers attached to the arc and the vertex indicate the capacity and the supply, respectively.)

which satisfies (i) $|\delta^-(\hat{s})| = 1$, (ii) for any $v \in \hat{V}$ the length from v to $\hat{s}$ is equal to $\hat{l}_v$, and (iii) $\hat{c}(e) = 1$ for any $e \in \hat{A}$. Since we are only given path-length function but not the arc set or transit time of arcs, there may exist many possible networks which satisfy the given path-length function. For example, given $\hat{V} = \{\hat{s}, x, y, v, w\}$, $(\hat{b}(\hat{s}), \hat{b}(x), \hat{b}(y), \hat{b}(v), \hat{b}(w)) = (0, 4, 3, 5, 1)$, and $(\hat{l}_{\hat{s}}, \hat{l}_x, \hat{l}_y, \hat{l}_v, \hat{l}_w) = (0, 1, 2, 4, 7)$, all dynamic networks in Fig. 4 belong to $\mathcal{N}(\hat{V}, \hat{b}, \hat{l}, \hat{s})$.

Lemma 7. *For any $\hat{N} \in \mathcal{N}(\hat{V}, \hat{b}, \hat{l}, \hat{s})$, $\Theta(\hat{N})$ takes the same value regardless of the underlying network topology of $\hat{N}$.*

Proof. For any $\hat{N} = (\hat{D} = (\hat{V}, \hat{A}), \hat{c}, \hat{\tau}, \hat{b}, \hat{s}) \in \mathcal{N}(\hat{V}, \hat{b}, \hat{l}, \hat{s})$, $P_{\hat{s}}$ consists of a single element from $|\delta^-(\hat{s})| = 1$. Thus, $\hat{N}$ is fully connected because any $v \in \hat{V}$ is reachable to $\hat{s}$ by using the path of length $\hat{l}_v$. Since $\hat{l}_v$ does not depend on the choice of $\hat{N}$, the number of distinct values in $\{\hat{l}_v : v \in \hat{V}\}$ does not depend on the choice of $\hat{N}$. Let $\hat{k}$ denote this number. Let $\hat{N}^c$ be the condensed time-expanded network for $\hat{N}$. Since $\hat{N}$ is fully connected, $\hat{V}(i) - \hat{s}_i$ is contractible in $\hat{N}^c$ for any $i = 1, \ldots, \hat{k}$ from Theorem 1. Let $\hat{N}^* = (\hat{D}^* = (\hat{V}^*, \hat{A}^*), \hat{c}^*, \hat{b}^*, \hat{s}^*)$ be the one obtained by contracting $\hat{V}(i) - \hat{s}_i$ into a single vertex p_i for every $i = 1, \ldots, \hat{k}$ in $\hat{N}^c$. It is easy to see that arcs whose capacity is not infinity in $\hat{N}^*$ are $p_i\hat{s}_i$ with $i = 1, \ldots, \hat{k}$ and the capacity of $p_i\hat{s}_i$ is equal to $|\hat{I}_i|$ since the capacity of any arc is assumed to be one where $\hat{I}_i$ is defined for $\hat{N}$ in a manner similar to I_i for $\mathcal{N}$. It is easy to see that $|\hat{I}_i|$ does not depend on the choice of $\hat{N}$ from the definition of $\hat{I}_i$. Since $\hat{V}(i)$ does not depend the choice of $\hat{N}$, the supply of p_i does not depend on the choice of $\hat{N}$. From the above discussion, regardless of the choice of $\hat{N} \in \mathcal{N}(\hat{V}, \hat{b}, \hat{l}, \hat{s})$, $\hat{N}^*$ is the same. This completes the proof.   □

Fig. 5. $\hat{N}^*$ for dynamic network in Fig. 4. (The numbers attached to the vertex and the arc indicate the supply and the capacity, respectively.)

Form the proof of Lemma 7, we can see that for any $\hat{N} \in \mathcal{N}(\hat{V}, \hat{b}, \hat{l}, \hat{s})$ $\Theta(\hat{N})$ depends only on the sum of the supplies of vertices $v \in \hat{V}$ such that $lev(v)$ takes the same value, but not the supply of each vertex.

For $\mathcal{N} = (D = (V, A), c, \tau, b, s)$, let $\delta(P_s, s) = \{e^1, e^2, \ldots, e^d\}$, and $V^j = \{v \in V : v$ is reachable to the tail of $e^j\} \cup \{s\}$.

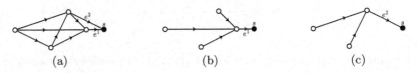

Fig. 6. (a) $D = (V, A)$, (b) $D^1 = (V^1, A^1)$, (c) $D^2 = (V^2, A^2)$.

Lemma 8 ([9]). *Given a dynamic network $\mathcal{N} = (D = (V, A), c, \tau, b, s)$, there exist d arc-disjoint s-rooted trees $D^j = (V^j, A^j)$ for $j = 1, \ldots, d$ such that D^j spans V^j and $A^j \subseteq A$ if and only if $\lambda_D(v, s) = |\delta(R_v, s)|$ holds for any $v \neq s$.*

Fig. 6(b) and (c) illustrate D^1 and D^2 of the directed graph D in Fig. 6(a).

Now let us fix $\{b^j : j = 1, \ldots, d\}$ such that (i) For any $v \in V$, $\sum_{j=1}^{d} b^j(v) = b(v)$ holds, and (ii) for any $v \in V$ and $j = 1, \ldots, d$ with $v \notin V^j$, $b^j(v) = 0$ holds. Intuitively speaking, $b^j(v)$ represents the assignment of the supply of v which reaches s through $D^j = (V^j, A^j)$. For a fully connected network $\mathcal{N} = (D = (V, A), c, \tau, b, s)$, let $\mathcal{N}^j = (D^j = (V^j, A^j), c^j, \tau^j, b^j, s)$ where c^j and τ^j respectively denote c and τ whose domain is restricted to A^j. Notice that from Lemma 7 $\Theta(\mathcal{N}^j)$ does not depend on the choice of A^j if b^j is fixed. Let f^j_{opt} be an optimal dynamic network flow in $\mathcal{N}^j$. Recalling that since $A^{j_1} \cap A^{j_2} = \emptyset$ holds with $j_1 \neq j_2$, the dynamic flow obtained by combining f^j_{opt} for all $j = 1, \ldots, d$ is feasible in $\mathcal{N}$.

Lemma 9. *Given a fully connected network $\mathcal{N} = (D = (V, A), c, \tau, b, s)$, under the constraint such that for each $v \in V$ the amount of $b(v)$ which reaches s through e^j is $b^j(v)$, $\Theta(\mathcal{N})$ is equal to $\max\{\Theta(\mathcal{N}^j) : j = 1, \ldots, d\}$.*

The proof of this lemma is almost the same as Theorem 3 in [2], and hence is omitted. From Lemma 9, we only need to determine b^j for $j = 1, \ldots, d$ to obtain $\Theta(\mathcal{N})$.

Lemma 10. *We can reduce $EP(\mathcal{N})$ for a fully connected network $\mathcal{N}$ to the restricted parametric flow problem.*

We will prove the lemma as follows.

For a fully connected network $\mathcal{N} = (D = (V, A), c, \tau, b, s)$, let $\mathcal{R}(\mathcal{N}) = (D_R = (V_R, A_R), c_R, b_R, S_R)$ be the static network with multiple sinks to which $EP(\mathcal{N})$ is reduced. First we consider $\mathcal{R}(\mathcal{N})$ in the case of $|\delta^-(s)| = 1$. In this case, $\mathcal{R}(\mathcal{N})$ is the same as $\hat{\mathcal{N}}^*$ defined in the proof of Lemma 7. Notice that the parameter T is contained only in the capacity of the arc which is incident to a sink $\hat{s}_{\hat{k}}$ by the definition of $\hat{I}_i$ (e.g. see Fig. 5). It is clear that in order to compute $\Theta(\mathcal{N})$ we need to compute T^* which is the minimum value of T such that $\mathcal{R}(\mathcal{N})$ is feasible, i.e., the solution of the restricted parametric flow problem defined on $\mathcal{R}(\mathcal{N})$. Notice that $\Theta(\mathcal{N}) = \lceil T^* \rceil$ holds.

From the above discussion, we can construct $\mathcal{R}(\mathcal{N})$ for the case of $|\delta^-(s)| > 1$ in three steps as follows. $\mathcal{R}(\mathcal{N})$ is constructed so that the minimum value of $\max\{\Theta(\mathcal{N}^j) : j = 1, \ldots, d\}$ among all b^j with $j = 1, \ldots, d$ is equal to $\lceil T^* \rceil$ where T^* is the same as defined above and we can compute an optimal allocation of the supplies b^j with $j = 1, \ldots, d$ which attains T^*, i.e. $\Theta(\mathcal{N})$. Let $V(i, Q) = \{v \in V : len(v) - i, R_v = Q\}$.

(i) We first construct *gadget* G^j separately for each $j = 1, \ldots, d$ which is the same as $\mathcal{R}(\mathcal{N}^j)$ with no supply (Fig. 7(a), (b), and (c)). Notice that the parameter T is common in all gadgets. (ii) For every nonempty $V(i, Q)$, we add vertices u_i^Q in V_R. The supply of u_i^Q (denoted by $b_R(u_i^Q)$) is defined as the sum of supplies in $V(i, Q)$. (iii) We add the arc from u_i^Q to the gadget G^j in A_R if $V^j \cap V(i, Q) \neq \emptyset$. Notice that the allocation of the supply of u_i^Q to the gadget G^j means that we allocate the supplies of $V(i, Q)$ to $\mathcal{N}^j$. We determine to which vertex in G^j u_i^Q is connected as follows. For any $j = 1, \ldots, d$, we arrange the distinct values $\{l_v : v \in V^j\}$ as $L_1^j < \cdots < L_{k^j}^j$. We connect u_i^Q to $p_{i'}$ in G^j with $L_{i'}^j = L_i$. Notice that from the way of construction of $\mathcal{R}(\mathcal{N})$ the parameter T is contained only in the capacity of the arc which is incident to s_{k^j} in each gadget G^j. Therefore, all arcs in A_R whose capacity contains the parameter T are incident to sinks S_R. Lemma 10 then follows from the way of construction of $\mathcal{R}(\mathcal{N})$. For example, in step(ii) $u_4^{\{x,y\}}$ in Fig. 7(d) is added to allocate the supply of v in Fig. 7(a). In step(iii), for $\mathcal{N}^1$ and $\mathcal{N}^2$ in Fig. 7(b), $k^1 = 5$ and $k^2 = 4$ hold, and $u_4^{\{x,y\}}$ in Fig. 7(d) is connected to p_4 in G^1 and p_3 in G^2. In Fig. 7(c) and (d), only $p_5 s_5$ in G^1 and $p_4 s_4$ in G^2 contain the parameter T.

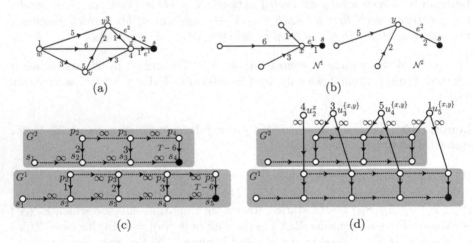

Fig. 7. (a) Dynamic network $\mathcal{N}$. (b) $\mathcal{N}^1$ and $\mathcal{N}^2$ for $\mathcal{N}$ with no supply. (c) Gadgets G^1 and G^2. (d) Vertices and arcs introduced to allocate supplies.

As was seen in Section 3.2, the restricted parametric flow problem defined on $\mathcal{R}(\mathcal{N})$ can be transformed into the parametric maximum-flow problem studied by [7] by adding the super source vertex q as well as arcs from q to all u_i^Q's in V_R such that the capacity of $q u_i^Q$ is set to $b_R(u_i^Q)$. Since in this parametric maximum-flow problem the capacities of all cuts except $\delta(q, V_R)$ diverge to ∞ from the way of construction of $\mathcal{R}(\mathcal{N})$ as T goes to ∞, the maximum flow value of the parametric maximum-flow problem is bounded by $\sum_{v \in V_R} b_R(v)$, i.e., there always exists T such that $\mathcal{R}(\mathcal{N})$ is feasible. Since we assume $b(v) > 0$ for at least one vertex v with $lev(v) = k$, $\Theta(\mathcal{N}) \geq L_k$ holds. Therefore, we need to consider only the case of $T \geq L_k$ in the parametric maximum-flow problem.

3.4   Time Complexity

Let η be the number of distinct combinations of the path-length from v to s and R_v, i.e., $\eta = |\{(l_v, R_v): v \in V\}|$. Notice that η is equal to the number of u_i^Q defined above and $\eta = O(n)$ holds.

Theorem 2. *The evacuation problem EP($\mathcal{N}$) for a fully connected network $\mathcal{N}$ can be solved in $O(|P_s|m + n \log n + d(dk + \eta)(k + \eta) \log n)$ time.*

Proof. The term $O(|P_s|m + n \log n)$ is the time required to construct $\mathcal{R}(\mathcal{N})$ (the details are omitted). The third term represents the time required to solve the restricted parametric flow problem. Let us evaluate the size of $\mathcal{R}(\mathcal{N})$. A single gadget has $O(k)$ vertices and $O(k)$ arcs. Since there exist d gadgets, the union of all gadgets has $O(dk)$ vertices and $O(dk)$ arcs. The number of vertices which is added to allocate the supplies is equal to η. The number of the arcs added to these vertices is clearly $O(d\eta)$. From the above discussion, we have $|V_R| = O(dk + \eta)$ and $|A_R| = O(dk + d\eta)$. From Lemma 6, this completes the proof.   □

Let us analyze the running time given in the above theorem in terms of m and n. Notice that the number of the arcs added to allocate the supplies is bounded by $O(m)$. This is because this number is at most $\sum_{v \in V - s} |\delta(R_v, s)|$ since $u_i^{R_v}$ is connected to at most $|\delta(R_v, s)|$ gadgets for $v \in V - s$ with $lev(v) = i$. Moreover, we have $\sum_{v \in V - s} |\delta(R_v, s)| \leq \sum_{v \in V - s} |\delta^+(v)| = m$ since the out-degree of v is no less than $|\delta(R_v, s)|$ from the fact that $\mathcal{N}$ is fully connected and the capacity of any arc is one. Next we prove that the union of all gadgets has $O(m)$ vertices and $O(m)$ arcs. Since $\mathcal{N}^j$ has $|V^j|$ vertices, the gadget G^j has $O(|V^j|)$ vertices and $O(|V^j|)$ arcs from the way of construction of G^j. Thus the number of vertices and arcs in the union of all gadgets are $O(\sum_{j=1}^{d} |V^j|)$, respectively. Since V^j is the union of a sink s and the set of vertices which are reachable to the tail of e^j, $\sum_{j=1}^{d} |V^j| = \sum_{v \in V - s} |R_v| + d$ holds (the term d represents the number of the copies of a sink). From $\sum_{v \in V - s} |R_v| \leq \sum_{v \in V - s} |\delta^+(v)| = m$ and $O(d) = m$, the number of vertices and arcs in the union of all gadgets are $O(m)$, respectively. Thus we have $|V_R| = O(m)$ and $|A_R| - (m)$ from $\eta = O(n)$, and the following corollary follows from Lemma 6.

Corollary 1. *The evacuation problem EP($\mathcal{N}$) for a fully connected network $\mathcal{N}$ can be solved in $O(m^2 \log n)$ time.*

If we simply apply the algorithm of [5], the time complexity is $O(k^3 m^2 n \log (kn^2/m))$. Our algorithm much improves the result of [5] in this case. In many practical cases, the in-degree of a sink can be considered as a constant. In this case, if we can regard d as a constant, the time complexity of our algorithm is $O(dm + d^2 n^2 \log n)$.

Integral capacity case. For this case, we can apply our algorithm by splitting arcs into ones whose capacity is one. In this case, we have $\mathcal{R}(\mathcal{N})$ which has $O(kn)$ vertices and $O(n^2)$ arcs by combining all gadgets corresponding to parallel arcs, and hence our algorithm can solve EP($\mathcal{N}$) in $O(kn^3 \log n)$ time. In the general capacity case, we can extend our algorithm similarly.

4   Conclusion and Remarks

In this paper, we generalize the class of networks to which the algorithm of [2] can be applied. Though the details are omitted, our algorithm can solve $EP(\mathcal{N})$ for a d-dimensional grid network with uniform capacity in $O(d^2 n + n \log n + d^3 3^{2d} n^{2/d} \log n)$ time. In particular, in the case of $d = 2$, $EP(\mathcal{N})$ can be solved in $O(n \log n)$ time. This time complexity matches the result of [2]. In the case where there exists a vertex v with $\lambda_D(v, s) < |\delta(R_v, s)|$ (called *deficient vertex*) in a 2-dimensional grid network with uniform capacity, this problem can be solved in $O(\sigma^3 n^{3/2} \log n)$ time by contracting the condensed time-expanded network according to Theorem 1 where σ is the number of deficient vertices.

Acknowledgement. This research is supported by the project *New Horizons in Computing*, Grant-in-Aid for Scientific Research on Priority Areas, MEXT Japan.

References

1. Hoppe, B., Tardos, É.: The quickest transshipment problem. Mathematics of Operations Research **25**(1) (2000) 36–62
2. Kamiyama, N., Katoh, N., Takizawa, A.: An efficient algorithm for evacuation problems in dynamic network flows with uniform arc capacity. In: Proc. AAIM2006. Volume 4041 of LNCS., Springer (2006) 231–242
3. Ford, L., Fulkerson, D.: Flows in Networks. Princeton University Press, Princeton, NJ (1962)
4. Mamada, S., Uno, T., Makino, K., Fujishige, S.: An $O(n \log^2 n)$ algorithm for the optimal sink location problem in dynamic tree networks. Discrete Applied Mathematics **154**(16) (2006) 2387–2401
5. Hall, A., Hippler, S., Skutella, M.: Multicommodity flows over time: Efficient algorithms and complexity. In: Proc. ICALP2003. Volume 2719 of LNCS., Springer (2003) 397–409
6. Ibaraki, T., Katoh, N.: Resource allocation problems under submodular constraints. In: Resource Allocation Problems : Algorithmic Approaches. MIT Press, Cambridge, MA (1988) 144–176
7. Gallo, G., Grigoriadis, M.D., Tarjan, R.E.: A fast parametric maximum flow algorithm and applications. SIAM J. Comput. **18**(1) (1989) 30–55
8. Radzik, T.: Parametric flows, weighted means of cuts, and fractional combinatorial optimization. In Pardalos, P., ed.: Complexity in Numerical Optimization. World Scientific, River Edge, NJ (1993) 351–386
9. Kamiyama, N., Katoh, N., Takizawa, A.: Generalization of theorem of Edmonds. In: Proc. KyotoCGGT2007. (2007) submitted.

Online OVSF Code Assignment with Resource Augmentation

Francis Y. L. Chin[1,*], Yong Zhang[1], and Hong Zhu[2,**]

[1] Department of Computer Science, The University of Hong Kong, Hong Kong
{chin,yzhang}@cs.hku.hk
[2] Institute of Theoretical Computing, East China Normal University, China
hzhu@sei.ecnu.edu.cn

Abstract. Orthogonal Variable Spreading Factor (OVSF) code assignment is a fundamental problem in Wideband Code-Division Multiple-Access (W-CDMA) systems, which play an important role in third generation mobile communications. In the OVSF problem, codes must be assigned to incoming code requests, with different data rate requirements, in such a way that they are mutually orthogonal with respect to an OVSF code tree. An OVSF code tree is a complete binary tree in which each node represents a code associated with the combined bandwidths of its two children. To be mutually orthogonal, each leaf-to-root path must contain at most one assigned code. In this paper, we focus on the online version of the OVSF code assignment problem, in the often-studied context of the single cell as well as in the more general context of the whole multi-cell cellular network (for which there are no known results). With the help of 1/8 and 11/8 extra bandwidth resources, we are able to give a 5-competitive algorithm in the single cell and the multi-cell context respectively, which means that the competitive ratio is a constant and not a function of the height of the OVSF tree and thereby improving upon past results.

1 Introduction

Wideband Code-Division Multiple-Access (W-CDMA) technology is one of the main technologies widely-developed in recent years for the implementation of third-generation (3G) cellular systems. We consider the well-studied problem of Orthogonal Variable Spreading Factor (OVSF) code assignment in W-CDMA systems [5,6,9,10,12].

OVSF is an implementation of CDMA wherein, before each signal is transmitted, the spectrum is spread according to a unique code, which is derived from an OVSF code tree. An OVSF code tree is a complete binary tree. Users have requests for different data rates, and we accommodate these different requests by assigning codes at different levels of the OVSF code tree, with the root being at

---

* This research was supported in part by Hong Kong RGC Grant HKU-7113/07E.
** This research was supported in part by National Natural Science Fund (grant no. 60496321).

M.-Y. Kao and X.-Y. Li (Eds.): AAIM 2007, LNCS 4508, pp. 191–200, 2007.
© Springer-Verlag Berlin Heidelberg 2007

the highest level and representing the entire bandwidth of the wireless system. The code at any node other than the root denotes the bandwidth half that of its parent in the tree. In any *legal assignment* in the code tree, no two assigned codes lie on a single path from the root to a leaf, i.e., any two assigned codes are *mutually orthogonal*. The subset of nodes in the code tree, which forms a legal assignment, is called a *code assignment* (CA). A node x is said to be *free* if there are no assigned nodes in every root-to-leaf path containing x, and thus making x an assigned node would still result in a legal assignment. For convenience, we use the words "code" and "node" interchangeably. Fig. 1 is an example of an OVSF code tree with the code assignment represented by the darkened nodes marked as c, d, e, g and i.

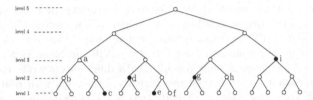

Fig. 1. An example of OVSF code tree, solid circles are the assigned codes

To illustrate the essence of the OVSF code assignment problem, consider the configuration shown in Fig. 1. Let $Req(x)$ denote the request to which code x is assigned, and let *reassign x to y* denote the reassignment of $Req(x)$ to code y and the freeing of code x (i.e. making x a free code). Suppose a level-2 code request arrives followed by a level-3 code request. If we assigned code b to the first request, we would have to make two code reassignments before we can assign code a to the second request, e.g. reassign b to h (and thereby freeing b) and reassign c to f (freeing c and consequently a). If, on the other hand, h were assigned to the first request, only one reassignment would be needed to satisfy the second request, i.e. reassign c to f (freeing c and consequently a), and then assign code a to the second request.

Since each reassignment requires processing overhead and may affect the quality of communications, a natural goal would be to minimize the number of reassignments. Note that this problem will not be too difficult and can be solved optimally by a greedy strategy if codes were never released. However, when codes can be released, the code tree can be fragmented so that many reassignments might be needed if a good assignment algorithm were not used.

In general, the algorithm for OVSF code assignment is expected to handle a sequence $\sigma = (C_1, C_2, \ldots, C_k, \ldots)$ of code operations over time, each operation C_k being either to *request* a code at a particular level or to *release* an assigned code. Note that, if the total bandwidth of any set of mutually orthogonal free codes is less than the bandwidth required by a code request, the new code request cannot be satisfied. In this paper, without loss of generality, we assume that all requests in σ can be satisfied (since we can easily check whether a new code request can be satisfied), and after each request is satisfied, a legal assignment results.

The OVSF code assignment problem is hard, and the approach has often been to produce heuristics, whose performance is measured by the approximation (or competitive) ratio, which compares the cost of the algorithm to the cost of optimal off-line scheme where cost is the total number of assignments or reassignments done by the algorithm.

The problem has been studied extensively in recent years, and we have the off-line and online versions of this problem.

- **Off-line CA Problem**

  Given a sequence σ of code operations, find a sequence of code assignments such that the total number of reassignments is minimum, assuming the initial code tree is empty. This problem was proved to be NP-hard by Marco Tomamichel [11], who also gave an exponential-time algorithm to solve it.

- **Online CA Problem**

  The operations in the sequence σ arrive through time. At any time $t > 0$, we only know about the operations until t and have no information about any future operation $C_{t'}$ with $t' > t$. The problem is to find a sequence of code assignments such that the total number of reassignments is minimum. Erlebach et al [5] gave an $O(h)$-competitive algorithm for this problem, where h is the height of the OVSF code tree. They also proved that the lower bound on the competitive ratio of this problem is 1.5. With resource augmentation [7], which means the online algorithm is allowed to use more bandwidth than the optimal scheme, a 4-competitive algorithm with a double-bandwidth code tree was given in [5] .

In this paper, we focus on the Online CA Problem. In Section 2, for the single-cell context, we give a new constant-competitive algorithm, using less extra bandwidth (less resource augmentation) to improve upon previous results. As far as we know, the Online CA Problem for multi-cell cellular networks has not been studied. In Section 3, we apply the techniques used in Section 2 in the context of the whole multi-cell cellular network to give a new constant-competitive algorithm.

2   Code Assignment in a Single Cell

Our online algorithm, Online-CA-Cell, makes use of extra resource in the form of an additional, albeit smaller, OVSF tree. Therefore, we talk about code assignments in the *main* OVSF tree and in a separate *extra* OVSF tree.

There are two properties that Online-CA-Cell seeks to maintain: (a) for any set of mutually orthogonal free codes in the main tree, there is at most one free code at each level; and (b) at each level of the extra tree, there is at most one assigned code. Fig. 2(a) shows an example of a main tree that have assigned codes, which are *sorted* and *compacted*. "Sorted" means that assigned codes are in non-decreasing order in terms of level (from left to right); and "compacted" means that, at each level, there is at most one free code. Fig. 2(b) gives an

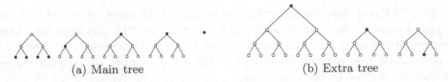

(a) Main tree   (b) Extra tree

Fig. 2. Structure with assigned codes (shown as solid circles)

example of an extra tree. Online-CA-Cell is comprised of the method for handling each code request and the method for handling each code release.

When a level-i code request arrives, Online-CA-Cell first tries to satisfy the request in the main tree. Note that, we need only to consider assignment or reassignment at level i or higher in the main tree, since the total bandwidth of any set of mutually orthogonal free codes at lower levels is not large enough to satisfy a request for level i (because there is at most one free code at every level). If there is no free code at level i in the main tree, Online-CA-Cell will then try to satisfy the request by assigning a code from the extra tree, and if the extra tree already contains an assigned code at level i, Online-CA-Cell will reassign the leftmost assigned code at the lowest level $j > i$ in the main tree so as to free its offsprings in the main tree in order to accommodate the new request and the request of the assigned level-i code in the extra tree. The pseudo-code for code request is as follows:

Code-Request(R,i)——*allocate a free code to satisfy the request R of level i*
if the main tree has a free code at level i **then**
    Assign that free code to R.
else if the extra tree contains no assigned node at level i **then**
    Assign the rightmost free code at level i in the extra tree to R.
else
    Let w be the leftmost assigned code at lowest level $j > i$ in the main tree.
    Apply **Code-Request**($Req(w),j$) so as to free up w and its offspring codes.
    Assign the leftmost free code of level i in the main tree to R.
    For those assigned codes in the extra tree from level i to level $j - 1$
        (if they exist) reassign them to the main tree.
end if

When a level-i code release of code x arrives, Online-CA-Cell might have to reassign the rightmost assigned code y at level i to x in order to maintain the compactness of the tree. The freeing up of y, however, could mean that there would be more than one free node at level i. So, part of the algorithm is to fix this up. The pseudo-code for code release is as follows:

Code-Release(x,i)——*release code x at level i*
Let y be the rightmost assigned code at level i (in either main or extra tree).
if $x = y$ **then** Free x
else Reassign y to x, freeing up y.

end if
if there are two free codes at level i of the main tree, they must be children of z **then** Apply **Code-Release($z,i+1$)**
end if

Let the amortized cost of each code request be 3 credits and each code release be 2 credits because a code assignment/reassignment costs 1 credit, while freeing a code costs 0. Next, we will define the potential function f (Fig. 3). We shall show that the amortized cost of a code request or a code release can pay for the assignment/reassignment costs and the change of potentials.

The intuition behind the definition of potential function f is the following. The seven different cases shown in Fig. 3 exhaust all possible configurations of the assigned codes at level i. Configurations C_2, C_5 and C_7, which have 0, 1 and more than one assigned codes in the main tree, respectively, have an assigned code in the extra tree and have a potential value of 2 credits to compensate for the reassignment cost of bringing the assigned code in the extra tree back to the main tree if needed. Their corresponding configurations without assigned codes in the extra tree (i.e. C_1, C_4 and C_6) do not carry any potential credits. The remaining configuration C_3, which is the only configuration having an assigned code and a free code in the main tree (thus no assigned code in the extra tree), is associated with 1 potential credit to compensate for the cost of the reassignment of the rightmost code upon any code release at this level or lower levels.

Lemma 1. *Assume each code request is associated with 3 credits. The number of credits at each level of the main and extra tree, as defined by the potential function f as given in Fig. 3, will be maintained after each code request as described in Online-CA-Cell.*

Proof. According to Code-Request(R,i), when a request R arrives at level i:

if the main tree has a free code at level i: This configuration may be C_1 or C_3, whose potential value is 0 or 1. Note that C_1 can have one or no free codes. After the assignment, the configuration is changed to C_4 or C_6 and the remaining number of credits is either 2 or 3 (*code request cost + configuration potential − assignment cost*), which is larger than the potential values of C_4 and C_6.

else if the extra tree contains no assigned code at level i: This configuration may be C_1, C_4 or C_6, whose potential value is 0. After the assignment, the configuration is changed to C_2, C_5 or C_7 whose potential value is 2. Thus, the amortized cost of the code request can cover exactly the assignment cost and change of potential.

else (There is an assigned code in the extra tree.) The configuration may be C_2, C_5 or C_7 with potential value 2. After the assignment and Code-Request(Req $(w),j$), the configuration is changed to C_6 with potential value 0. Since we have to do two assignment/reassignments in this level, 3 credits will be left behind to cover the reassignment costs at higher levels, i.e. Code-Request($Req(w),j$). Note that, for those levels between i and $j-1$ whose

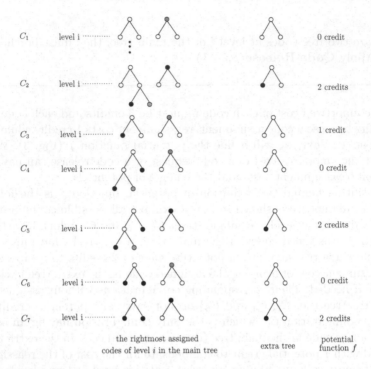

Fig. 3. Potential function of each level of the code tree, solid circle denotes an assigned code, empty circle denotes a free code, gray circle may or may not be an assigned code

assigned codes are reassigned to the main tree, the reassignment costs can be covered by the change of potential.   □

Lemma 2. *Assume each code release is associated with 2 credits. The number of credits at each level of the main and extra tree as defined by the potential function f as given in Fig. 3 will be maintained after each code release as described in Online-CA-Cell.*

Proof. According to Code-Release(x, i), when a code x of level i is released:

Reassign y to x where y is the rightmost assigned code at level i: This costs one credit if $y \neq x$ exists.

After the assignment, the configuration cannot be C_2, C_5 nor C_7.

If there is at most one free code at level i of the main tree: The configuration must be C_1, C_4, C_3 or C_6 derived respectively from C_4, C_5, C_6 or C_7. We can easily check that the 2 credits associated with the code release operation can cover one reassignment cost and the increase of potential value of C_3 (no increase of potential value for C_1, C_4 and C_6).

If there are two free codes at level i of the main tree: After Code-Release($z, i + 1$), the final configuration may be C_1, C_4 or C_6 with potential value 0, respectively derived from the initial configuration C_4, (C_3 or C_5), or

(C_3 or C_7). Since the code release operation has 2 credits and freeing a code costs nothing, the remaining credits in level i is at least 2 (from the change of potential), which can cover the reassignments at level $i + 1$, i.e. Code-Release$(z, i + 1)$.   □

Theorem 1. *Given a code tree T and extra tree T', which contains half the bandwidth of T, Algorithm Online-CA-Cell is 5-competitive.*

Proof. Since the extra tree can contain at most one code at all levels from 1 to $h - 2$, the total bandwidth of the extra tree is at most half of T. Note that an extra code at level $h - 1$ is not possible because that would imply the total bandwidth of all assigned codes is larger than the bandwidth of T.

Suppose there are m_1 code requests and m_2 code releases in the sequence. From Lemmas 1 and 2, we can see that, at any step, the credit of each level is at least 0. Thus, the total cost of Online-CA-Cell is at most $3m_1 + 2m_2 \leq 5m_1$ since the number of releases is at most the number of requests. The optimal cost is at least m_1, and so, the competitive ratio of Online-CA-Cell is at most 5.   □

With the observation that the algorithm is k-competitive without any extra resource if each code request/release involves only a fixed number k of levels, we have a new 5-competitive algorithm with a extra code tree of $1/8$ full bandwidth. We partition the main code tree into a *lower* part and an *upper* part. The lower part contains all the assigned codes from level 1 to $h - 4$ and its configuration is similar to the main tree and extra tree as described before. The extra tree contains codes from levels 1 to $h - 4$, and thus, the total bandwidth of the extra tree is $1/8$ of the code tree. The upper part contains the assigned codes from level $h - 3$ to $h - 1$ which are *sorted* and *compacted*. Between the lower part and upper part, there is no free code of level $h - 3$. The potential of configurations at level i ($1 \leq i \leq h - 4$) is the same as the potential function defined in Fig. 3, and each code request and code release is associated with 3 and 2 credits respectively. The algorithm for code request/release is the same as before for levels 1 to $h - 4$ and keeps the assigned codes sorted and compacted for levels $h - 3$ to $h - 1$.

When a code request (or release) arrives at level i ($1 \leq i \leq h - 4$), if the algorithm does not affect upper level $h - 3$, we can say the competitive ratio in this case is at most 5; otherwise, the algorithm in the lower part will give 3 (2 for release) credits to the upper part, which is enough to cover the cost of reassignments in the three levels from $h - 3$ to $h - 1$ (in the case of release, the two levels $h - 3$ and $h - 2$ since reassignment at level $h - 1$ is not possible). So, we can also say that the competitive ratio in this case is at most 5.

When a code request or release happens at level i ($h - 3 \leq i \leq h$), since the process does not affect the lower part, the competitive ratio in this case is at most 3. Thus, we have the following result.

Theorem 2. *Given a code tree T and extra tree T' which contains $1/8$ the bandwidth of T, we can have a 5-competitive algorithm.*

3   Code Assignment in Cellular Networks

The geographic area of a mobile communications network is usually divided into many small cellular regions and call requests may be made at any cell. The code assignment in a single cell can be considered as a special case of code assignment in multi-cell cellular networks. In this section, we study the online OVSF code assignment problem in multi-cell cellular networks.

In cellular networks, each cell contains a base station, which communicates with other base stations via a high-speed wired network. Communications between any two users (even within the same cell) must be established through base stations. When a call request arrives, the nearest base station must assign a code, whose bandwidth matches with the bandwidth of the call request and which is orthogonal to all the assigned codes in the OVSF code tree of its own cell and its neighboring cells in order to avoid interference.

Communications in cellular networks have been widely studied [1,2,3,4,8], but mainly focus on Frequency Division Multiplexing (FDM) networks, which are the backbone for 1G/2G wireless communications. This is the first study on OVSF code assignment in cellular network. We will give a constant-competitive algorithm for OVSF code assignment in cellular networks with resource augmentation [6].

The cellular network can be 3-colored with $\{R, G, B\}$ so that each cell has one color and neighboring cells have different colors. Each cell with a different color can use a different code tree for handling code requests and releases initiated at that cell. In order to achieve a constant-competitive ratio, a total of six code trees with full bandwidth will be needed using the resource augmentation approach given in [5], whereas 3.375 code trees with full bandwidth would be sufficient using the approach in Section 2. However, we will introduce here an algorithm with a competitive ratio of 5 that uses only $19/8 = 2.375$ code trees of full bandwidth. Fig. 4 depicts three code trees T_1, T_2 and T_3, with the heights of T_1

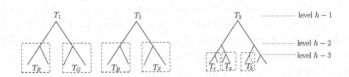

Fig. 4. Three code tree used by Online-CA-Network

and T_2 being h and the height of T_3 being $h - 2$. The off-line optimal algorithm uses only one code tree with height h. Let the left and right subtree of T_1 be T_R and T_G, and those of T_2 be T_B and T_S, and let the three equal-bandwidth subtrees with height $h - 4$ of T_3 be T_r, T_g and T_b.

The cell with color X will use T_X as its main tree and T_x as its extra tree. T_S will be shared by all cells. We will show in Theorem 3 that conflicts from different cells can never occur.

Online algorithm **Online-CA-Network** can be described as follows:

- For cells colored X ($X \in \{R, G, B\}$), perform Online-CA-Cell on T_X and T_x as the main and extra trees for all call operations in the cell colored X. In case the bandwidth of T_X is not enough, use T_S and T_x for those calls that cannot be accommodated by T_X.
- When a code request asks for the bandwidth of the whole code tree, assign the root of T_1, since there will be no other assigned code in this cell and its neighboring cells.

Theorem 3. *Online-CA-Network is a 5-competitive algorithm for the online OVSF code assignment problem with augmentation ratio of 19/8.*

Proof. As all neighboring cells use different code trees, there will be no interference as long as all the assigned codes in each cell are mutually orthogonal. Since Online-CA-Cell is applied to handle code operations in each cell, orthogonal assigned codes are ensured if Online-CA-Cell works properly with T_X and T_S, which is trivially true if T_S is not used by any other cells. Suppose T_S contains codes at level l assigned to call requests from a cell X. The total bandwidth of these codes must be greater than the total bandwidth of the free codes in T_X (the main tree is "sorted" and "compacted"). Since the bandwidth of T_X is $\mathcal{B}/2$, assuming that the total bandwidth of the code tree is $\mathcal{B}$, the total bandwidth of the assigned codes in the cell X is greater than $\mathcal{B}/2$. Since the off-line optimal algorithm uses only one code tree, the total bandwidth of neighboring cells is no more than $\mathcal{B}$. So the total bandwidth in each cell neighboring with X must be strictly less than $\mathcal{B}/2$. Therefore, we can ensure that T_S is used by at most one cell at any time, which means that the assigned codes in the main trees in cellular network do not affect each other.

As for each extra tree in T_3, the bandwidth is $1/8$ of T_1 (T_X and T_S), i.e., $\mathcal{B}/8$. From Section 2, we know this is sufficient to handle all the extra codes, and the three subtrees of T_3 in the whole cellular network do not affect each other.

Since the analysis and code operations are similar to that in Online-CA-Cell, the competitive ratio of Online-CA-Network in the cellular network should be the same as the competitive ratio of Online-CA-Cell, which is 5-competitive.   □

The following result applies in the special case where all code requests are at the same level.

Theorem 4. *There exists a 2-competitive algorithm with augmentation ratio of 2 that solves the code assignment problem when all requests are at the same level.*

Proof. We use two code trees T_1 and T_2. Similar to Fig. 4, let the two subtrees of T_1 be T_R and T_G, and let those of T_2 be T_B and T_S. Since all the requested bandwidth are the same, the extra trees as given in T_3 may not be needed.

Suppose there are m_1 code requests and m_2 code releases in a given sequence of code operations. Since each code request has 1 credit to pay for the assignment, and each code release has 1 credit to pay for the reassignment of the rightmost assigned code, the cost of our scheme is at most $m_1 + m_2 \le 2m_1$ since the number of releases is no more than the number of requests. Note that the cost of the optimal off-line algorithm is at least m_1, and so we can say that our scheme is 2-competitive. □

Acknowledgements. The authors thank Dr. Bethany M. Y. Chan for her efforts in making this paper more readable.

References

1. W. T. Chan, F. Y. L. Chin, D. Ye, Y. Zhang and H. Zhu. Frequency Allocation Problem for Linear Cellular Networks. In Proc. of the 17th Annual International Symposium on Algorithms and Computation (ISAAC 2006), LNCS 4288, 61-70.
2. W. T. Chan, F. Y. L. Chin, D. Ye, Y. Zhang and H. Zhu. Greedy Online Frequency Allocation in Cellular Networks. *Information Processing Letters*, 102(2007), 55-61.
3. W. T. Chan, F. Y. L. Chin, D. Ye and Y. Zhang. Online Frequency Allocation in Cellular Networks. To appear in Proc. of the 19th ACM Symposium on Parallelism in Algorithms and Architectures (SPAA '07).
4. I. Caragiannis, C. Kaklamanis, and E. Papaioannou. Efficient on-line frequency allocation and call control in cellular networks. *Theory Comput. Syst.*, 35(5):521–543, 2002. A preliminary version of the paper appeared in SPAA 2000.
5. T. Erlebach, R. Jacob, M. Mihalak, M. Nunkesser, G. Szabo and P. Widmayer. An algorithmic view on OVSF code assignment. In proc. of 21th Symposium on Theoretical Aspects of Computer Science (STACS 2004), LNCS 2996, pp. 270-281.
6. Xiang-Yang Li and Peng-Jun Wan. Theoretically Good Distributed CDMA/OVSF Code Assignment for Wireless Ad Hoc Networks. In Proc. the 11th Annual International Conference of Computing and Combinatorics (COCOON05), pp. 126-135.
7. B. Kalyanasundaram and K. Pruhs. Speed is as powerful as clairvoyance. In Proceedings of the 36th IEEE Symposium on Foundations of Computer Science, pages 214-221, 1995.
8. C. McDiarmid and B. A. Reed. Channel assignment and weighted coloring. *Networks*, 36(2):114-117, 2000.
9. T. Minn and K. Y. Siu. Dynamic assignment of orthogonal variable-spreadingfactor codes in W-CDMA. IEEE Journal on Selected Areas in Communications, 18(8):1429-1440, 2000.
10. Angelos N. Rouskas, Dimitrios N. Skoutas. OVSF codes assignemnt and reassignment at the forward link OFW-CDMA 3G systems. In Proc. of the 13 th IEEE International Symposium on Personal, Indoor and Mobile Radio Communications, 2002.
11. Marco Tomamichel. Algorithmische Aspekte von OVSF Code Assignment mit Schwerpunkt auf Offline Code Assignment. menuscript.
12. Peng-Jun Wan, Xiang-Yang Li and Ophir Frieder. OVSF-CDMA Code Assignment in Wireless Ad Hoc Networks. In Proc. of DIAL M-POMC 2004 joint workshop on Foundations of mobile computing, pp. 92-101, 2004.

Optimal Joint Rate and Power Allocation in CDMA Networks

R.J. Boucherie[1], A.I. Endrayanto[1], and A. F. Gabor[2]

[1] Faculty of Electrical Engineering, Mathematics and Computer Science,
University of Twente, 7500 AE Enschede, The Netherlands
[2] Faculty of Mathematics and Computer Science and EURANDOM,
Eindhoven University of Technology, 5600 MB, Eindhoven, The Netherlands

Abstract. In this paper we propose a polynomial time algorithm for the optimal rate and power allocation problem in a two cell CDMA network. We assume continuous rates and limited powers for the base stations.

1 Introduction

The inherit capacity restrictions due to scarce resources are fundamental problems in the operation of a wireless CDMA system. At the operational level (time scale of minutes), load fluctuations occur due to randomness in call generation and call lengths. At this time scale, load balancing is carried out via power and rate assignment as well as a reconfiguration of calls over cells. Power and rate assignment requires an underlying policy or network optimality criterion.

Common optimality criteria for CDMA network optimization are equal rate to all calls, or maximum total network data rate. Equal rates to all calls seems fair from a call perspective, but is rather inefficient in networks sustaining a normal load, mainly due to calls far away from base transmitter stations (BTS) causing a large amount of interference, and therefore a substantial reduction in network capacity. An important question in achieving maximum data rate is the assignment of data rates to individual calls. This assignment is clearly closely related to power assignment. This paper addresses, in an analytical setting, the joint power and rate assignment in two cells in a CDMA network.

Literature. The joint rate and power assignment problem for CDMA systems has received considerable attention over the last decades. Due to the complexity of the problems, several restrictions have been made, in order to obtain mathematically tractable models.

The most common simplifications are considering a cell in isolation, thus neglecting the interference effects, or assuming some extra properties of rates/powers, like unlimited rates or powers. For the simplified model of a *single cell* in isolation, down link power assignment schemes for maximizing the throughput or minimizing the total power in the cell are proposed in [8,5,12]. Resource assignment in a *multi cell* environment is more complex than in one cell, due to the interferences caused by users in adjacent cells. It has been studied in the framework of cell-breathing for fixed data rates, see e.g. the pioneering work of

M.-Y. Kao and X.-Y. Li (Eds.): AAIM 2007, LNCS 4508, pp. 201–210, 2007.
© Springer-Verlag Berlin Heidelberg 2007

[7], [11] that consider the uplink, that in the early days of CDMA was considered to be the bottle-neck. For the *down link*, joint rate and cell assignment is studied in [9] via a dynamic pricing algorithm under the assumption that each base station maximizes its total system utility, without considering the status of the other cells. In [2] an distributed algorithm for assigning base station transmitter (BTSs) powers such that the *common* rate of the users is maximized is described. In [10], Perron-Frobenius theory is used to design an approximate algorithm for a model with multiple rates, which permits the use of techniques from convex optimization. In [4], the authors propose a polynomial time approximation scheme for the joint rate and power assignment problem under the assumption that the rates allocated are *discrete* and the power of the base stations is *unlimited*.

Contribution and Outline. This paper proposes a fast and *exact* joint rate and power allocation algorithm in the *down link* of a telecommunication network formed by two cells, where the base stations transmit at *limited* powers. Thus, we incorporate in our model two important aspects of a CDMA network, namely interference and limited powers. We assume that the rates are *continuous* and may be chosen from a given interval. This assumption seems realistic, since in a CDMA system data rates may be rapidly modified in accordance with channel conditions, resulting in an average rate that lies in an interval.

Section 2 provides the model and describes the resulting optimization problem. Due to the impact of the interference between users in different cells, this problem is much more difficult then the rate/power optimization problem in one cell, and it is more difficult then the problem with unlimited powers. In Section 3 we show that despite its non-convexity, the optimal solutions can be very well characterized. We prove that the optimal rate allocations are monotonic in a function of the path loss . Based on this property, we show that in the optimal rate allocation, only 3 rates are given to users. In Section 4 we propose a polynomial time algorithm in the number of users that solves optimally the joint rate and power allocation problem. Our algorithm can be generalized to solve the optimal rate/power allocation problem in small networks, thus providing a first step into the direction of fast algorithms for resource allocation in a large network. We conclude with some remarks and open problems in Section 5.

2   Model

We consider a system with mobile users served by 2 base transmitter stations (BTSs), X and Y. Denote by U_X, respectively U_Y, the set of mobiles served by BTS X, respectively BTS Y. Let $l_{i,X}$ denote the path loss from BTS X to mobile i, let N_i be the thermal noise at the location of mobile i, and let ϵ_i denote the energy per bit to interference ratio requirement for mobile i. Let P_{iX} denote the transmission power of BTS X to mobile i, and P_X the maximum down link transmission power of BTS X. The power received by mobile i from BTS X is $P_{iX}^{rec} = P_{iX} l_{i,X}$. We assume that mobiles are served by a single BTS, which is a natural assumption for moving mobiles. A configuration of mobiles is feasible when for each mobile i served by BTS X, say, the energy per bit to interference ratio exceeds the threshold ϵ_i. If a configuration is feasible, then under perfect power control the energy per bit to interference ratio $\left(\frac{E_b}{I_0}\right)_i$ equals this threshold. Thus, assuming perfect power control, feasibility for a configuration in which mobile i is served by BTS X is characterized by,

$$\left(\frac{E_b}{I_0}\right)_i := \frac{W}{r_i} \frac{P_{iX} l_{i,X}}{\alpha l_{i,X}(\sum\limits_{j \in U_X} P_{jX} - P_{iX}) + l_{i,Y} \sum\limits_{j \in U_Y} P_{jY} + N_i} = \epsilon_i, \tag{1}$$

where U_X is the set of mobiles served by BTS X, W is the system chip rate, α is the down link orthogonality factor, and r_i is the data rate for mobile i.

Data rates can be assigned from the continuous interval $[r_{\min}, R_{\max}]$, with $r_{\min} > 0$. The optimization problem is to determine an assignment of rates and powers to mobiles that maximizes the total rate.

For each fixed number of mobile calls placed in the coverage area, the rate assignment problem can be formulated as the following optimization problem:

$$\max \sum_{i \in U} r_i$$

$$\text{s.t.} \left(\frac{E_b}{I_0}\right)_i = \epsilon_i, \quad i \in U,$$

$$P(n) \quad \sum_{i \in U_X} P_{iX} \leq P_X,$$

$$\sum_{i \in U_Y} P_{iY} \leq P_Y,$$

$$r_i \in [r_{\min}, R_{\max}], \quad i \in U,$$

$$P_{iX} \geq 0, \quad \forall i \in U_X \cup U_Y.$$

3 Characterization of an Optimal Rate Assignment

For clarity of presentation, we assume that all users have the same threshold $\epsilon_i = \epsilon$. Denote $V(r_i) = \frac{\epsilon r_i}{W + \alpha \epsilon r_i}$, and let

$$l_i = \begin{cases} \frac{l_{i,Y}}{l_{i,X}}, & \text{for } i \in U_X, \\ \frac{l_{i,X}}{l_{i,Y}}, & \text{for } i \in U_Y. \end{cases}$$

According to Lemma 1.1. in [4], $P(n)$ can be rewritten as:

$$P(n) : R(n) = \max \sum_{i \in U} r_i$$

$$\text{s.t.} \left(1 - \alpha \sum_{i \in U_X} V(r_i)\right) x - \sum_{i \in U_X} V(r_i) l_i y - \sum_{i \in U_X} V(r_i) l_{i,X}^{-1} N_i = 0, \quad (2)$$

$$- \sum_{i \in U_Y} V(r_i) l_i x + \left(1 - \alpha \sum_{i \in U_Y} V(r_i)\right) y - \sum_{i \in U_Y} V(r_i) l_{i,Y}^{-1} N^i = 0, \quad (3)$$

$$P_X - x \geq 0, \quad (4)$$

$$P_Y - y \geq 0, \quad (5)$$

$$x \geq 0, \quad (6)$$

$$y \geq 0, \quad (7)$$

$$R_{\max} - r_i \geq 0, \text{ for } i \in U_X \cup U_Y, \quad (8)$$

$$r_i - r_{\min} \geq 0, \text{ for } i \in U_X \cup U_Y. \quad (9)$$

Notice that this is neither a linear programming nor a convex programming problem. We assume that the rate assignment problem above has at least one feasible solution, or, in other words, that there exist powers P_X, P_Y, such that assigning minimum rate to all users is feasible.

For later reference, we also provide the Lagrangian. Let $\lambda \in \mathbb{R}^6$, μ, $\nu \in \mathbb{R}^{|U|}$ be the Lagrangian multipliers corresponding to equations (2)-(9). Denote by $r = (r_i)_{i \in U_X \cup U_Y}$ the vector of the rates allocated to users. The Lagrangian corresponding to $P(n)$ is

$$L(x, y, r, \lambda, \mu, \nu) = \sum_{i \in U} r_i$$

$$+ \lambda_1 ((1 - \alpha \sum_{i \in U_X} V(r_i)) x - \sum_{i \in U_X} V(r_i) l_i y - \sum_{i \in U_X} V(r_i) l_{i,X}^{-1} N_i))$$

$$+ \lambda_2 \left(- \sum_{i \in U_Y} V(r_i) l_i x + (1 - \alpha \sum_{i \in U_Y} V(r_i)) y - \sum_{i \in U_Y} V(r) l_{i,Y}^{-1} N_0 \right)$$

$$+ \lambda_3 (P_X - x) + \lambda_4 (P_Y - y) + \lambda_5 x + \lambda_6 y$$

$$+ \sum_{i \in U} \mu_i (R_{\max} - r_i) + \sum_{i \in U} \nu_i (r_i - r_{\min}).$$

Next we will characterize the optimal rate assignment. We start with a monotonicity property of the rates.

Theorem 1. *If $P(n)$ is feasible, and (x^*, y^*, r^*) is an optimal solution, then for any two calls i and j, say, in cell X,*

$$y^* l_i + l_{i,X}^{-1} N_i < y^* l_j + l_{j,X}^{-1} N_j \Rightarrow r_i^* \geq r_j^*. \quad (10)$$

A similar statement holds for cell Y.

Proof. Suppose there exist two calls $i, j \in U_X$ such that $l_i y^* + l_{i,X}^{-1} N_i < l_j y^* + l_{j,X}^{-1} N_j$ and $r_i^* < r_j^*$.

Define the following rate vector $\hat{r} \in \mathbb{R}^{|U_X| + |U_Y|}$:

$$\hat{r_k} = \begin{cases} r_k^*, & \text{for } k \in U_X \cup U_Y \setminus \{i, j\} \\ r_j^*, & \text{for } k = i, \\ r_i^*, & \text{for } k = j, \end{cases}$$

i.e., with rate assignment to calls i and j interchanged. As the total rate is unchanged, the throughput of the rate assignments r and $\hat{r}$ is the same. Let

$$\hat{x} = \frac{\sum_{i \in U_X} V(\hat{r_i})(l_i y^* + l_{i,X}^{-1} N_i)}{1 - \alpha \sum_{i \in U_X} V(\hat{r_i})}. \tag{11}$$

It can be easily seen that $\hat{x} < x^*$.

Note that $(\hat{x}, y^*, \hat{r})$ is not a feasible solution of $P(n)$, since it does not satisfy constraints (2) and (3). However, we can obtain a feasible solution by increasing the rates $\hat{r}$ for users in $U_X \setminus \{j\}$, until power x^* is reached in (11) or all rates in $U_X \setminus \{j\}$ are maximal. Denote by $(\tilde{r})_{U_X}$ the rate assignment obtained in this way. Suppose that $(\tilde{r_k})_{k \in U_X \setminus \{j\}} = (R_{\max})_{U_X \setminus \{j\}}$. By decreasing y^* such that (3) is satisfied, while the rates for users in U_Y remain the same, we obtain a power/rate allocation with a higher throughput then r^*. If x^* was reached in (11), then $(x^*, y^*, (\tilde{r_k})_{k \in U_X}, (r_k^*)_{k \in U_Y})$ is a feasible solution of $P(n)$ which gives a higher throughput then $(x^*, y^*, ((r_k^*)_{k \in U_X}, (r_k^*)_{k \in U_Y})$. This contradicts the fact that (x^*, y^*, r^*) is an optimal solution. Hence, it must be that $r_i^* \geq r_j^*$.

Denote by $h_1(x, y, r)$, ..., $h_6(x, y, r)$ the functions in the left hand side of constraints (2)-(7) and by $g_1(x, y, r), ..., g_{2|U_X| + 2|U_Y|}(x, y, r)$ the functions in the left hand side of constraints (8)-(9).

We will first review some optimization terminology (see [3]). If an inequality constraint of $P(n)$ is satisfied with equality in a feasible vector $(x, y, r) \in \mathbb{R}^{|U_X| + |U_Y| + 2}$ of $P(\mathbf{n})$, the constraint is *active* in (x, y, r). Denote by $A(x, y, r)$ the set of *active* inequalities in the point (x, y, r). A feasible vector (x, y, r) is *regular* if the gradients $\nabla h_1(x, y, r)$, $\nabla h_2(x, y, r)$ and $\nabla h_i(x, y, r)$, $\nabla g_j(x, y, r)$ for $i \in A(x, y, r) \bigcap \{3, 4, 5, 6\}$, $j \in A(x, y, r)$ are linearly independent.

Notice that $\nabla h_1(x, y, r)$, $\nabla h_2(x, y, r)$ are linearly independent for any feasible (x, y, r), so that all points with $A(x, y, r) = \emptyset$ are regular. Further, note that since $r_{\min} > 0$, $x \neq 0$ and $y \neq 0$ in the optimal solution. Moreover, since the objective function is linear, each optimum must be a global optimum.

We will start by characterizing the rate assignment for regular points. In the proofs that follow, we will make use of the Karush-Kuhn-Tucker (KKT) necessary conditions for a regular point to be an optimal solution (see [3]). They state that for a regular point (x^*, y^*, r^*) that is an optimum of $P(n)$ there exists an unique multiplier vector $(\lambda^*, \mu^*, \nu^*)$ such that:

(K1) $\nabla_{(x^*, y^*, r^*)} L(x^*, y^*, r_i^*, \lambda^*, \mu^*, \nu^*) = 0$, where L denotes the Lagrangian function corresponding to $P(\mathbf{n})$.

(K2) $\lambda_k^* \geq 0$, for $k \in \{3,4,5,6\}$, $\mu^* \geq 0$ and $\nu^* \geq 0$,

(K3) The Lagrangian multipliers corresponding to non active constraints are equal to 0.

Theorem 2. *If $P(n)$ is feasible and (x^*, y^*, r^*) a regular optimal solution, then*

a) $x^* = P_X$ or $y^* = P_Y$ or $r_i^* = R_{\max}$, for each call $i \in U_X \cup U_Y$.

b) If the rates of two calls $i, j \in U_X$ satisfy $r_{\min} < r_i < R_{\max}$ and $r_{\min} < r_j < R_{\max}$, then $l_i y^* + l_{i,X}^{-1} N_i = l_j y^* + l_{j,X}^{-1} N_j$ and $r_i = r_j$. A similar statement holds for cell Y.

Proof. a) Note that since the minimum rate can be ensured to all accepted users, constraints (2) and (3) imply that $x^* > 0$ and $y^* > 0$. Thus, based on condition (K3), we conclude that $\lambda_5^* = \lambda_6^* = 0$. Suppose that $x^* < P_X$, $y < P_Y$ and $r_{\min} \leq r_i < R_{\max}$ for a call $i \in U_X$, say.

From (K3), follows that $\lambda_3^* = \lambda_4^* = 0$ and that $\mu_i^* = 0$.

Moreover, (K1) imply that $\frac{\partial L}{\partial x}(x^*, y^*, r^*, \lambda^*, \mu^*, \nu^*) = 0, \frac{\partial L}{\partial y}(x^*, y^*, r_i^*, \lambda^*, \mu^*, \nu^*) = 0$, and $\frac{\partial L}{\partial r_i}(x^*, y^*, r^*, \lambda^*, \mu^*, \nu^*) = 0$. Hence,

$$
\begin{cases}
\lambda_1^* \left(1 - \alpha \sum_{i \in U_X} V(r_i^*)\right) - \lambda_2^* \sum_{i \in U_Y} V(r_i^*) l_i & = 0 \\
-\lambda_1^* \sum_{i \in U_X} V(r_i^*) l_i + \lambda_2^* \left(1 - \alpha \sum_{i \in U_Y} V(r_i^*)\right) & = 0. \\
1 + \nu_i^* - \mu_i^* - \lambda_1^* V'(r_{i*})(\alpha x^* + l_i y^* + l_{i,X}^{-1} N_i) = 0.
\end{cases}
\tag{12}
$$

Observe that the first two equations in λ_1^*, λ_2^* are linearly independent (recall constraints (2)-(3) and the assumption that a minimal rate assignment is feasible), so the only solution is $\lambda_1^* = \lambda_2^* = 0$.

Further, since $\mu_i^* = 0$, from the third equation in (12) follows that $\nu_i = -1$, which contradicts condition (K2), that $\nu_i^* \geq 0$.

Hence, in an optimal solution, either the rates of all users are maximal, or the power in one of the cells is maximal.

b) Suppose that there exist two different values $l_i y^* + l_{i,X}^{-1} N_i$, $l_j y^* + l_{j,X}^{-1} N_j$, respectively, for which the corresponding rates are $r_{\min} < r_i^* < R_{\max}$ and $r_{\min} < r_j^* < R_{\max}$. Without loss of generality, we assume that $l_i y^* + l_{i,X}^{-1} N_i < l_j y^* + l_{j,X}^{-1} N_j$. From Theorem 1, it follows that $r_i^* \geq r_j^*$.

Since $r_{\min} < r_i^* < R_{\max}$ and $r_{\min} < r_j^* < R_{\max}$, condition (K3) imply that

$$\mu_i = \mu_j = \nu_i = \nu_j = 0.$$

Hence, (12) implies that

$$\frac{V'(r_i^*)}{V'(r_j^*)} = \frac{\alpha x^* + l_j y^* + l_{j,X}^{-1} N_j}{\alpha x^* + l_i y^* + l_{i,X}^{-1} N_i}.$$

Our assumption $l_j y^* + l_{j,X}^{-1} N_j > l_i y^* + l_{i,X}^{-1} N_i$ implies that $V'(r_i^*) > V'(r_j^*)$. Since the function V' is decreasing, it follows that $r_i^* < r_j^*$, which contradicts Theorem

1. We conclude that if the rates of two users $i, j \in U_X$ satisfy $r_{\min} < r_i^* < R_{\max}$ and $r_{\min} < r_j^* < R_{\max}$, then $l_i y^* + l_{i,X}^{-1} N_i = l_j y^* + l_{j,X}^{-1} N_j$. Clearly, it then follows that $r_i^* = r_j^*$.

Corollary 1. *Let (x^*, y^*, r^*) be regular and an optimal solution of problem $P(n)$. Suppose that calls in cell X, respectively in cell Y are ordered in increasing order of their $l_i y^* + l_{i,X}^{-1} N_i$, respectively $l_j x^* + l_{j,Y}^{-1} N_j$ values. Then, there exists a positive number $A(y^*)$, such that for each $i \in U_X$ with $l_i y^* + l_{i,X}^{-1} N_i < A(y^*)$, $r_i^* = R_{\max}$ and for each $i \in U_X$ with $l_i y^* + l_{i,X}^{-1} N_i > A(y^*)$, $r_i^* = r_{\min}$. Moreover, there exists a positive number $B(x^*)$, such that for each $j \in U_Y$ with $l_j x^* + l_{j,Y}^{-1} N_j < B(x^*)$, $r_j^* = R_{\max}$ and for each $j \in U_Y$ with $l_j x^* + l_{j,Y}^{-1} N_j > B(x^*)$, $r_j^* = r_{\min}$.*

For a non regular point, the following theorem gives a complete characterization of the optimal power and rate assignment.

Theorem 3. *For each non regular point (x, y, r), the following conditions are satisfied:*
a) $x = P_X$ or $y = P_Y$
b) If $x = P_X$ and $y \neq P_Y$, then $r_i \in \{r_{\min}, R_{\max}\}$, for each $i \in U_X$.
c) If $y = P_Y$ and $x \neq P_X$, then $r_i \in \{r_{\min}, R_{\max}\}$, for each $i \in U_Y$.

Proof. Let (x, y, r) be a non regular point, feasible for $P(\mathbf{n})$. Consider the matrix M formed by the $\nabla h_1(x, y, r)$, $\nabla h_2(x, y, r)$ and $\nabla h_i(x, y, r)$, $\nabla g_j(x, y, r)$ for $i \in A(x, y, r) \bigcap \{3, 4, 5, 6\}$, $j \in A(x, y, r)$. Let K be the number of active inequality constraints. Notice that for a non-regular point it must be that $K > 0$, since $\nabla h_1(x, y, r)$, $\nabla h_2(x, y, r)$ are linearly independent. Clearly, $2 \leq rank(M) \leq K + 2$.

a) Suppose that $x \neq P_X$ and that $y \neq P_Y$. In other words, the active inequality constraints correspond to the constraints on rates. Then, matrix M has the following form:

$$M = \begin{pmatrix} 1 - \alpha \sum_{i \in U_X} V(r_i) & -\sum_{i \in U_X} V(r_i) l_i & A & \mathbf{0} \\ -\sum_{i \in U_Y} V(r_i) l_i & 1 - \alpha \sum_{i \in U_Y} V(r_i) & \mathbf{0} & B \\ \mathbf{0} & \mathbf{0} & C & \mathbf{0} \\ \mathbf{0} & \mathbf{0} & \mathbf{0} & D \end{pmatrix},$$

where the vectors $A \in \mathbb{R}^{|U_X|}$, $B \in \mathbb{R}^{|U_Y|}$ are defined as follows:

$$A = [-V'(r_i)(\alpha x + l_i y + l_{i,X}^{-1} N_i)]_{i \in U_X}, \qquad B = [-V'(r_i)(\alpha y + l_i x + l_{i,Y}^{-1} N_i)]_{i \in U_Y},$$

and the matrices $C \in \mathbb{R}^{|\{i \in U_X : g_i \in A(x,y,r)\}|} \times \mathbb{R}^{|\{i \in U_X\}|}$, $D \in \mathbb{R}^{|\{i \in U_Y : g_i \in A(x,y,r)\}|} \times \mathbb{R}^{|\{i \in U_Y\}|}$ are obtained from the diagonal square matrices with diagonal

$$diag(\overline{C}) = [I_{\{r_i = r_{\min}\}}) - I_{\{r_i = R_{\max}\}}]_{\{i \in U_X\}}, diag(\overline{D}) = [I_{\{r_i = r_{\min}\}}) - I_{\{r_i = R_{\max}\}}]_{\{i \in U_Y\}},$$

by deleting all rows for which the diagonal elements equals zero, where $I_{\{a\}} = 1$ if a is true, and 0 otherwise.

Clearly, $rank(C) + rank(D) = K$. Since constraints $\nabla h_1(x, y, r)$, $\nabla h_2(x, y, r)$ are linearly independent, it follows that $rank(M) = K + 2$, which contradicts the fact that (x, y, r) is non regular. Hence, in a non regular point, the power assigned to one of the cells has to be maximal.

b) Suppose that $x = P_X$ and $y \neq P_Y$ and that there exist $i \in U_X$ such that $r_{\min} < r_i < R_{\max}$. It can be proved that the rank of the matrix M is again $rank(M) = K + 2$, which contradicts the fact that (x, y, r) is non regular.

c) The proof is similar to b).

4 Algorithm for Optimal Rate and Power Assignment

Based on Theorems 1-3 and Corollary 1, we now propose on algorithm for finding the optimal solution of $P(n)$. The algorithm relies on a reduction of the optimization problem $P(n)$ to a series of optimization problems in $\mathbb{R}$. Notice that the algorithm considers the regular and non regular points.

If maximum rate to all users is feasible, then the optimal solution has been found. To check whether the maximum rate is feasible, one only has to check if the corresponding powers calculated from (2)-(3) satisfy $0 \leq x \leq P_X$ and $0 \leq y \leq P_Y$. If this is not the case, then the algorithm calculates the rate allocation achieving maximum throughput for the case when the power in cell X is maximal, respectively the power in cell Y is maximal. The algorithm will choose among these 2 allocations the one with higher throughput. Note that if the rates are known, from (2), (3) and (1) the powers of each user can be derived.

Next we will consider the case when in cell X the base station transmits at maximum power, i.e., $x^* = P_X$. The case $y^* = P_Y$ can be treated similarly. The algorithm provides a reduction of the optimization problem $P(n)$ that is based on a search procedure to find the values $B(x^*)$ and $A(y^*)$ introduced in Corollary 1 to obtain the set of mobiles at which the rate drops from $R_{\max}$ to $r_{\min}$ in both cells. As the set of mobiles for maximum power at cell X also depends on the power assignment in cell Y, these sets cannot be determined independently.

Order the locations in cell Y in increasing order of $l_j P_X + l_{j,Y}^{-1} N_j$.

According to Theorem 2 b) all users j in cell Y with rate $r_j \in (r_{\min}, R_{\max})$ are characterized by the same value of $l_j P_X + l_{j,Y}^{-1} N_j$ and have the same rate r_X. Let $B(P_X)$ be this value and $U_Y(B(P_X)) = \{j \in U_Y : l_j P_X + l_{j,Y}^{-1} N_j = B(P_X)\}$. From Theorem 1 and Theorem 2 follows that for each $j \in U_Y$ with $l_j P_X + l_{j,Y}^{-1} N_j < B(P_X)$, $r_j^* = R_{\max}$ and for each $j \in U_Y$ with $l_j P_X + l_{j,Y}^{-1} N_j > B(P_X)$, $r_j^* = r_{\min}$. Suppose that s users in $U_Y(B(P_X))$ have rate $R_{\max}$, v users have rate $r_{\min}$ and the rest have rate r_Y. The rate r_Y is unknown at this stage of the algorithm. The power assigned to cell Y, as a function of r_Y, can be determined from constraint (3), and is given by

$$y^*(r_Y) = \frac{\sum_{j \in U_Y \setminus U_Y(B(P_X))} V(r_j)(l_j P_X + l_{j,X}^{-1} N_j) + (sV(R_{\max}) + vV(r_{\min}) + tV(r_Y))B(P_X)}{1 - \alpha(\sum_{j \in U_Y \setminus U_Y(B(P_X))} V(r_j) + sV(R_{\max}) + vV(r_{\min})) + tV(r_Y)}.$$

Similarly, for a specific $y^*(r_Y)$, Theorem 2 b), implies that all the users i in cell X with $r_i \in (r_{\min}, R_{\max})$ are characterized by the same value of $l_i y^*(r_Y) + l_{i,Y}^{-1} N_0^i$, say $A(y^*(r_Y))$. Denote by $U_X(A,B) = \{i \in U_X : l_i y^*(r_Y) + l_{i,X}^{-1} N_i = A(y^*(r_Y))\}$. Then all $i \in U_X$ with $l_i y^*(r_Y) + l_{i,X}^{-1} N_i < A(y^*(r_Y))$, have rate $R_{\max}$ and all $i \in U_X$ with $l_i y^*(r_Y) + l_{i,X}^{-1} N_i > A(y^*(r_Y))$ have rate $r_{\min}$. Suppose that u users in $U_X(A,B)$ have rate $R_{\max}$, z users have a rate $r_X \in (r_{\min}, R_{\max})$ and the rest have rate $r_{\min}$. Then the rate r_X can be expressed from (2) as follows:

$$r_X(r_Y) = \frac{W}{\epsilon} \frac{P_X - \sum_{i \in U_X^z(A,B)} V(r_i)(\alpha P_X + l_i y^*(r_Y) + l_{i,X}^{-1} N_i)}{(z-1)\alpha P_X + zA(y^*(r_Y)) + \alpha \sum_{i \in U_X^z(A,B)} V(r_i)(\alpha P_X + l_i y^*(r_Y) + l_{i,X}^{-1} N_i)},$$

where $U_X^z(A,B)$ denotes the set of users in U_X with rate $r \in (r_{\min}, R_{\max})$.

Note that if $B(P_X), s, v, u, z$ were known, r_Y would be the only unknown. This suggests that by enumerating all the possible values of $B(P_X), s, v, u, z$, the problem could be reduced to an optimization problem in one variable, r_Y. The optimization problem is not easy to formulate due to the fact that the value of r_Y, more precisely $y^*(r_Y)$, is a decision variable in the assignment of $R_{\max}$ and $r_{\min}$ to users in U_X (see Corollary 3). However, it can be easily seen that only some values of $y^*(r_Y)$ induce a different rate allocation in cell X. Let

$$L = \{\frac{l_{j_1,X}^{-1} N_{j_1} - l_{j_2,X}^{-1} N_{j_2}}{l_{j_2} - l_{j_1}}, j_1, j_2 \in U_X\} \bigcap R^+.$$

Suppose that $L \neq \emptyset$. For all $y^*(r_Y) \in [L_i, L_{i+1})$ the ordering of mobiles in cell X, as determined by their value of $l_i y^*(r_Y) + l_{i,X}^{-1} N_i$ is the same, but for different intervals $[L_j, L_{j+1})$ this ordering may be different. Note that $V(r)$ is strictly increasing, so that $y^*(r_Y)$ is strictly increasing. As a consequence, each unknown $y^*(r_Y) \in [L_i, L_{i+1})$ yields a unique r_Y.

Hence, for $y^*(r_Y) \in [L_i, L_{i+1})$, $P(n)$ can be reduced to the following optimization problem in $\mathbb{R}$:

$$\begin{aligned} max \quad & zr_X(r_Y) + tr_Y \\ s.t. \quad & y^*(r_Y) \leq P_Y \\ & y^*(r_Y) \in [L_i, L_{i+1}] \\ & r_X(r_Y) \in [r_{\min}, R_{\max}] \\ & r_Y \in [r_{\min}, R_{\max}]. \end{aligned} \tag{13}$$

Thus, the original rate optimization problem can be reduced to $O(|U_X|^2)$ optimization problems in $\mathbb{R}$, one for each interval $[L_i, L_{i+1})$.

If $L = \emptyset$, then the order of the users in U_X does not depend on $y^*(r_Y)$ and we obtain a similar optimization problem to (13), without the second constraint.

Note that the optimization problems (13) are constraint optimization problems in one variable, which can be easily solved.

5   Conclusions

In this paper we have proposed an exact algorithm for the joint rate and power allocation problem in two cells of a CDMA network. We have analyzed several properties of the optimal solutions, based on which, we have proposed a polynomial time algorithm for solving the problem. Our results can be extended to non-decreasing utility functions at the cost of a rather involved notation. Moreover, the algorithm can be extended to iteratively solve the rate/power allocation problem in a small number of cells.

References

1. J.B. Andersen, T.S. Rappaport and S. Yoshida, Propagation measurements and models for wireless communications channels, *IEEE Commun. Mag.*, vol. 33, pp. 42-49, 1995
2. F. Berggren, Distributed power control for throughput balancing in CDMA systems, in *Proceedings of IEEE PIMRC*, vol. 1, pp. 24-28, 2001.
3. D. Bertsekas, Nonlinear programming, Athena scientific, 2003
4. R.J. Boucherie, A. Bumb, A.I. Endrayanto, G.J. Woeginger, A combinatorial approximation algorithm for CDMA downlink rate allocation, in: Tellecomunications Planning : Innovations in pricing, network design and management, S. Raghavan, G. Anandalingam eds., Springer, pp. 175-193, 2006.
5. X. Duan, Z. Niu, J. Zheng, Downlink Transmit Power Minimization in Power-Controlled Multimedia CDMA Systems, in *Proceedings of IEEE 13th Int. Symposium Personal, Indoor and Mobile Radio Communication*, 2002.
6. A.I. Endrayanto, J.L van den Berg, R.J Boucherie, An analytical model for CDMA downlink rate optimization taking into account uplink coverage restrictions, *Performance Evaluation*, Vol 59. pp. 225-246, 2005.
7. S. V. Hanly. An algorithm of combined cell-site selection and power control to maximize cellular spread spectrum capacity, *IEEE Journal on Selected Areas in Communications*, Vol. 13 no. 7 pp. 1332–1340, 1995.
8. J.-W. Lee, R.R. Mazumdar and N.B. Shroff, Downlink power allocation for multiclass wireless systems. in *IEEE/ACM Trans. Netw.*, vol. 13, pp.854-867, 2005.
9. J.-W. Lee, R.R. Mazumdar and N.B. Shroff, Joint resource allocation and base station assignment for the downlink in CDMA networks, *IEEE/ACM Trans. Netw.*, vol. 14, no. 1, pp.1-14, 2006.
10. D. O 'Neill, D. Julian and D. Boyd, Seeking Foschini's Genie: Optimal Rates and Powers in Wireless Networks, to appear in *IEEE Transactions on Vehicular Technology*.
11. R. D. Yates, A Framework for Uplink Power Control in Cellular Radio Systems, *IEEE Journal of Selected Areas in Communications*, Vol. 13, no. 7, pp. 1341-1347, 1995.
12. Z. Yin and J.Xie, Joint power and rate allocation for the downlink in wireless CDMA networks, The 14th IEEE International Symposium on Personal, Indoor and Mobile Communication Proceedings, pp.326-330, 2003

Suppressing Maximum Burst Size Throughout the Path with Non-work Conserving Schedulers

Hongkyu Jeong[1], Kyoung Y. Bae[2], and Jinoo Joung[2]

[1] Samsung Advanced Institute of Technology, Kiheung, Korea
paul.jeong@samsung.com
[2] Sangmyung University, Seoul, Korea

Abstract. Because of the scalability problem, the aggregation of flows and the queueing/scheduling based on those flow-aggregates is unavoidable in Quality of Service (QoS) architectures for large scale networks. We investigate the effect of flow aggregation on the end-to-end delay bounds. It has observed that with traditional work-conserving schedulers, the maximum burst size of each *flow* increases linearly as it traverses the network. The increased maximum burst size does not affect the delay bound of a flow in cases where the schedulers are flow-based. In cases where deaggregation and aggregation take places in the middle of the network, however, the increased maximum burst size affects severely in terms of delay bound. This is in fact the case for the most of real network deployments since at the edge of a subnetwork the flows have to be deaggregated and then handed over to another subnetwork. We suggest a simple alternative to the existing work-conserving scheduler, the Smoothing-DRR (S-DRR) server, which is based on the Deficit Round Robin (DRR) server. S-DRR has a non-work conserving characteristic. S-DRR is proved to suppress the maximum burst size of the aggregated flow to a constant throughout the path, so that the delay bound is only linearly proportional to the hop counts.

1 Introduction

QoS characteristics of the network with Integrated Services (IntServ) [1] architecture have been well studied and understood by numerous researches in the past decade. Providing the allocated bandwidths, or service rates, or simply rates of an output link to multiple sharing flows plays a key role in this approach. Among a myriad of scheduling algorithms, we focus on the deficit round robin (DRR) [2], because the sorted priority scheduling algorithms, including Packetized Generalized Processor Sharing (PGPS), suffer from the complexity, which is $O(\log N)$ at best while N is the number of active flows in a scheduler [3]. The DRR, with many other rate-providing servers, is proved to be a Latency-Rate server [4], or simply $\mathcal{LR}$ server. All the work-conserving servers that guarantee rates exhibit this property and can therefore be modeled as $\mathcal{LR}$ servers. It was shown that the maximum end-to-end delay experienced by a packet in a network of $\mathcal{LR}$ servers can be calculated from only the latencies of the individual servers

M.-Y. Kao and X.-Y. Li (Eds.): AAIM 2007, LNCS 4508, pp. 211–220, 2007.
© Springer-Verlag Berlin Heidelberg 2007

on the path of the flow, and the traffic parameters of the flow that generated the packet. More specifically for a leaky-bucket constrained flow,

$$D_i \leq \frac{\sigma_i - L_i}{\rho_i} + \sum_{j=1}^{N} \Theta_i^{S_j}, \tag{1}$$

where D_i is the delay of flow i within a network, σ_i and ρ_i are the well known leaky bucket parameters, the maximum burst size and the average rate, respectively, L_i is the maximum packet length and $\Theta_i^{S_j}$ is the latency of flow i at the server S_j.

There is a significant volume of researches for networks with flow aggregation. End-to-end delay bounds with using *fair aggregator* was investigated [5]. It was shown that under condition that the scheduler is *fair*, the delay of an aggregated flow is bounded. A fair scheduler, however, should be able to refrain itself from transmitting packets at full link capacity whenever one or more flows are not active, i.e. do not have packets to transmit at the moment. This mandates a non-work conserving type of scheduler behavior to bound the delay. Using Guaranteed Rate ($\mathcal{GR}$) servers [6] as *fair aggregator* was also investigated [7], and the maximum end-to-end delay was obtained. It was concluded that the aggregated scheduling provides even better delay performance than per-flow scheduling. Contrary to the work with $\mathcal{GR}$ servers [7], we still find the traditional per-flow scheduling performs better in general. This is because that the aggregated scheduling does not protect the flow under observation from other flows within the aggregate. If we have sufficiently large amount of burst from other flows through aggregation and deaggregation, then the aggregated scheduling performs quite poorly. This is obvious when we consider DiffServ as an extreme QoS architecture where aggregation and deaggregation occur in *every* node. Finally we compare the end-to-end delays in each networks we analyzed.

2   Previous Works

We describe $\mathcal{LR}$ servers and its properties. The concept and the primary methodology for the analysis of $\mathcal{LR}$ servers are suggested by Stiliadis [8]. A *server* is a commonly used terminology which in convention means the combination of a scheduler and a transmitter that reside in a output port controller of a switch or a router. We assume a packet switch (router) where a set of flows share a common output link. We denote with ρ_i the bandwidth, or the rate allocated to flow i. We assume that the switches (routers) are store-and-forward devices. Let $A_i(\tau, t)$ denote the arrivals from flow i during the interval $(\tau, t]$ and $W_i(\tau, t)$ the amount of service received by flow i during the same interval. In a system based on the fluid-flow model both $A_i(\tau, t)$ and $W_i(\tau, t)$ are continuous functions of t. In the packet-by-packet model, however, we assume that $A_i(\tau, t)$ increases only when the last bit of a packet is received by the server; likewise $W_i(\tau, t)$ is increased only when the last bit of the packet in service leaves the server. We further denote that a flow i is *backlogged* when one or more packets of i are waiting for service. In other words, if $A_i(0,t) - W_i(0,t)$ is larger than zero then the flow i is backlogged at t. Therefore a *backlogged period* of flow i is any period of time during which packets belong to flow i are continuously queued in the server. Here we continue with some formal definitions on more time intervals.

Definition 1. *A server busy period is a maximal interval of time during which the server is never idle.*

During a server busy period the server is always transmitting packets.

Definition 2. *A flow i busy period is a maximal interval of time $(\tau_1, \tau_2]$ such that for any time $t \in (\tau_1, \tau_2]$, packets of flow i arrive with rate greater than or equal to ρ_i or, $A_i(\tau_1, t) \geq \rho_i(t - \tau_1)$.*

Now we are ready for the definition and the primary characteristics of $\mathcal{LR}$ servers.

Definition 3. *A server S belongs in the class $\mathcal{LR}$ if and only if for all times t after time τ that the jth busy period started and until the packets that arrived during this period are serviced $W_{i,j}^S(\tau, t) \geq \max\left(0, \rho_i(t - \tau - \Theta_i^S)\right)$. Θ_i^S is the minimum non-negative number that satisfies the above inequality.*

Lemma 1. *If S is an $\mathcal{LR}$ server, and flow i is leaky bucket constrained with parameters (σ_i, ρ_i), then the followings hold.*

1. *If $Q_i^S(t)$ is the backlog of flow i at time t, $Q_i^S(t) \leq \sigma_i + \rho_i\Theta_i^S$.*
2. *If D_i^S is the delay of any packet in flow i in server S, $D_i^S \leq \frac{\sigma_i}{\rho_i} + \Theta_i^S$.*
3. *The output traffic of flow i from S conforms to the leaky bucket model with parameters $(\sigma_i + \rho_i\Theta_i^S, \rho_i)$.*

Proof. See the proof of theorem 3.1 of [8].

Using lemma 1, in [9] it was shown that a FIFO server is an $\mathcal{LR}$ server for the individual flows in an aggregate, as the following.

Lemma 2. *Under a condition that all the input flows are leaky-bucket constrained, during a flow i busy period a FIFO server can provide service to flow i as the following: $W_i^S(\tau, t) \geq \max\left(0, \rho_i(t - \tau - \frac{\sigma - \sigma_i}{r} - \frac{L_i}{r})\right)$, where τ is a starting time of flow i busy period.*

Proof. See the proof of lemma 5 of [9].

Next we describe the detailed behavior of DRR, the server we will use in the numerical analysis section, as a representative $\mathcal{LR}$ server. A DRR scheduler maintains a deficit counter per each flow, thus per each queue. A flow i is assigned with a quantum value ϕ_i, which represents a relative amount of service a flow will receive. A round is defined as a time interval during which all the active flows receive service opportunities, one per each flow. We will call this service opportunity a *turn* of a flow. At the start of the flow i's turn, the deficit value δ_i is incremented as much as the quantum value of the flow, ϕ_i. The size of the head packet of the flow i then is compared with the δ_i. If δ_i is larger or equal to the head packet size, then the head packet gets service and leave the queue. Whenever a packet is served, δ is decremented as much as the size of the served packet. The second head packet of the queue, which now becomes the head packet is then compared with the δ_i again. This process continues until the δ_i becomes smaller than the head packet. When this happens the next flow enters a turn and the packets within this flow will be served. Using this policy, the DRR can achieve $O(1)$ complexity, given that the ϕ_i is set to be greater than or equal to the maximum packet size of the flow i, for all i [10]. This is because otherwise a flow may not receive a service at all during a turn, and the amount of calculation required for serving a packet in a flow increases consequently.

In an accompanying research [4], DRR is proved to be an $\mathcal{LR}$ server. The latency of the DRR server is given as

$$\Theta_i^{\text{DRR}} = \frac{3F - 2\phi_i}{r}, \tag{2}$$

where F is defined as the frame size, which is the sum of all ϕ_i over i, and r is the output link capacity. Note that F does not represent the actual number of bytes served during a specific round, but the average number of bytes served in a round.

A simple yet efficient improvement to the DRR, the DRR with Instant Service (DRR-IS), was later proposed. DRR-IS reduces the latency by about 30% without any additional complexity [11]. By providing immediate access to the server for a newly backlogged flow or queue, the DRR-IS reduces the latency to

$$\Theta_i^{\text{DRR-IS}} = \frac{2F + \phi_{\max} - \phi_i}{r}, \tag{3}$$

where $\phi_{\max}$ is the maximum quantum size at the server.

3   Delay Bounds in Networks of $\mathcal{LR}$ Servers with Flow Aggregations

A rationale for providing an amount of reserved service rate to an individual flow in IntServ architecture is to protect the flow from other data traffic from unpredictable sources that request best-effort service to the network, or malicious users that purposefully violate the input constraints. All the $\mathcal{LR}$ servers successfully achieve this mission, at the cost of the complexity of per flow scheduling and queueing. If we have a confidence in some of flows, however, that they never violate the promised leaky bucket parameters, or the network itself can shape the incoming traffic to conform to these parameters, then those trusted flows can be aggregated into a fatter flow while still be guaranteed for QoS, therefore we can greatly reduce the scheduling and queueing complexity in a server.

We consider a series of switches each with $\mathcal{LR}$ servers, a part of which is depicted in figure 1. The $(n-1)$th server from the network entrance, $\mathcal{S}(n-1)$, generates output traffic $I_{out}^{\mathcal{S}(n-1)}$, or equivalently an aggregated flow of several

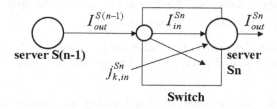

Fig. 1. Flow aggregation and deaggregation in a switch

elemental flows including i, which is the flow under observation. The next switch also has many output ports therefore many servers, including Sn to which i is destined. Among the flows within $I_{out}^{S(n-1)}$, some of flows are switched to this server. Let us denote by I_{in}^{Sn} such a set of flows. Note that $I_{in}^{Sn} \subset I_{out}^{S(n-1)}$. In Sn, I_{in}^{Sn} is considered as a single flow, and queued into a single queue and served accordingly. There are other elemental flows or aggregated flows from the same or other input ports, which share the server Sn with I_{in}^{Sn} and then aggregated with it, thus are consisted in I_{out}^{Sn}. Note that for the aggregation Sn does nothing more than a normal $\mathcal{LR}$ server does. Also note that there may be other background flows that share the servers but are not aggregated into flow I_{out}^{Sn}. The amount of service given to the aggregated flow I_{in}^{Sn} is bounded as follows, because it is served by an $\mathcal{LR}$ server: $W_{I,in}^{Sn} \geq \max\left(0, \rho_{I,in}^{Sn}(t - T_0 - \Theta_{I,in}^{Sn})\right)$, where T_0 is a starting time of a flow I_{in}^{Sn} busy period, $\Theta_{I,in}^{Sn}$ is the latency of the aggregated flow I_{in}^{Sn} at Sn, and $\sigma_{I,in}^{Sn} = \sum_{k \in I_{in}^{Sn}} \sigma_k^{Sn}$, $\rho_{I,in}^{Sn} = \sum_{k \in I_{in}^{Sn}} \rho_k$, where σ_k^{Sn} is the maximum burst size of a flow k within I_{in}^{Sn}. The delay of a packet within flow I_{in}^{Sn} is also bounded by the lemma 1 as the following: $D_{I,in}^{Sn} \leq \sigma_{I,in}^{Sn}/\rho_{I,in}^{Sn} + \Theta_{I,in}^{Sn}$. We are interested in the output traffic characteristics of the flow i within I_{out}^{Sn}. Let $W_i^{Sn}(\tau, t)$ be the service given to the packets that belong to flow i, at the server Sn during a time interval $[\tau, t)$. We argue the following.

Lemma 3. *Under a condition that all the input flows within I_{in}^{Sn} are leaky-bucket constrained, during a flow i busy period an $\mathcal{LR}$ server Sn can provide service to flow i, within aggregated flow I_{in}^{Sn}, as the following. $W_i^{Sn}(T_0, t) \geq \max\left(0, \rho_i(t - T_0 - (\sigma_{I,in}^{Sn} - \sigma_i^{Sn})/\rho_{I,in}^{Sn} - \Theta_{I,in}^{Sn})\right)$, where T_0 is a starting time of flow i busy period.*

Proof. The proof takes the same steps with the proof of lemma 2. Since an aggregated flow is queued and scheduled as if they were in a FIFO server with some latency, lemma 2 can be considered as a special case of this lemma with the latency L_i/r and the allocated rate r. We omit the details. $\square$

The following theorem is a direct consequence from the definition of $\mathcal{LR}$ servers and lemma 3.

Theorem 1. *An $\mathcal{LR}$ server with an aggregated flow, under condition that all the input flows within the aggregated flow are leaky bucket constrained, is an $\mathcal{LR}$*

server for individual flows with latency given as the following. $\Theta_i^S = (\sigma_{I,in}^S - \sigma_i^S)/\rho_{I,in}^S + \Theta_{I,in}^S$, *where* I_{in}^S *is the aggregated flow.*

From lemma 1 and theorem 1, the following corollary can be claimed, so that the maximum burst size at each server through the network can be calculated recursively.

Corollary 1. *The output traffic of flow* i *within an aggregated flow* I_{out}^S *from an* $\mathcal{LR}$ *server* S *conforms to leaky bucket model with parameters* $(\sigma_i^S + \rho_i\Theta_i^S, \rho_i)$, *where* σ_i^S *is the maximum burst size of flow* i *into the server* S.

The maximum end-to-end delay of a network of $\mathcal{LR}$ servers with aggregated flows can be obtained by the following sets of equations:

$$D_i \leq \frac{\sigma_i - L_i}{\rho_i} + \sum_{n=1}^{N} \Theta_i^{Sn},$$

$$\Theta_i^{Sn} = \frac{\sigma_{I,in}^{Sn} - \sigma_i^{Sn}}{\rho_{I,in}^{Sn}} + \Theta_{I,in}^{Sn}, \quad \sigma_{I,in}^{Sn} = \sum_{k \in I_{in}^{Sn}} \sigma_k^{Sn},$$

$$\sigma_i^{Sn+1} = \sigma_i^{Sn} + \rho_i\Theta_i^{Sn}, \quad \text{for } n \geq 1, \text{ and } \sigma_i^{S1} = \sigma_i. \tag{4}$$

4   Flow Aggregation with Non-work Conserving DRR Servers

The problem of aggregation and deaggregation is that the burst size linearly increases as hops are passed by, so that the delay bound in a node increases linearly and therefore the end-to-end delay increases quadratically. DRR with a special modification, however, can smooths its output flow so that the burst size remains constant. Let us observe the following.

Lemma 4. *Assume that flow* i *is continuously backlogged during* $(\alpha, \beta]$. *Let* k *be the number of DRR turns given to* i *during the interval* $(\alpha, \beta]$. *The service given by a DRR server to* i *during this period,* $W_i(\alpha, \beta)$, *is bounded as* $k\phi_i - \delta_i^k \leq W_i(\alpha, \beta) \leq k\phi_i + \delta_i^0$, *where* δ_i^k *is the deficit value of* i, *at the end of the* kth *round, counting from the first round within* $(\alpha, \beta]$.

Proof. See the proof of lemma 2 of [2]. $\qquad\qquad\square$

Lemma 5. *For any interval* $(\alpha, \beta]$, *within a backlogged period, the service given by a DRR server to flow* i *is bounded as* $W_i(\alpha, \beta) \leq \rho_i \frac{F}{f}(\beta - \alpha) + 2\phi_i$, *where* f *is the sum of quantum values of flows that are continuously backlogged during this interval.*

Proof. Let t_k indicate the time that the kth round finishes in the DRR server, counting from time α. Further let $t_0 = \alpha$. For each interval of time $(t_{k-1}, t_k]$, $t_k - t_{k-1} \geq \frac{1}{r}\sum_{j \in B}\left(\phi_j + \delta_j^{k-1} - \delta_j^k\right)$, where B is defined to be the set of

flows that are continuously backlogged during this interval. By summing over k,
$t_k - t_0 \geq \frac{f}{r}k + \frac{1}{r}\sum_{j \in B}\left(\delta_j^0 - \delta_j^k\right) \geq \frac{f}{r}(k-1)$, since $\sum_{j \in B}\delta_j^0 \geq 0$ and $\sum_{j \in B}\delta_j^k \leq f$. Equivalently $k \leq \frac{r}{f}(t_k - t_0) + 1$. From lemma 4, for the interval $[t_0, t_k)$, $W_i(t_0, t_k) \leq k\phi_i + \delta_i^0 \frac{r}{f}(t_k - t_0)\phi_i + \phi_i + \delta_i^0 \leq \rho_i \frac{F}{f}(t_k - t_0) + 2\phi_i$, since $\rho_i/r = \phi_i/F$. We conclude that, for an arbitrary time β, between t_k and t_{k+1}, the lemma holds because $W_i(\alpha, \beta) = W_i(\alpha, t_k) \leq \rho_i \frac{F}{f}(t_k - \alpha) + 2\phi_i \leq \rho_i \frac{F}{f}(\beta - \alpha) + 2\phi_i$.   □

The service upper bound we calculated is in fact as tight as possible. It is sufficient to show an example that satisfies the above relation. Suppose a tiny packet of size $\Delta\phi_i$ is served within a round started with zero deficit and therefore the deficit value after the round is $\phi_i - \Delta\phi_i$. During the next round, two packets with sizes $\phi_i - \Delta\phi_i$ and ϕ_i may be served. If $\Delta\phi_i \ll \phi_i$ then we can argue $2\phi_i$ is immediately served. Afterwards, the service is offered with constant rate $\rho_i F/f$. Therefore the upper bound is tight, and this value is indeed the latency of the $\mathcal{LR}$ server.

Corollary 2. *During an interval $(\alpha, \beta]$, where all the flows in the server are continuously backlogged, the service upper bound is given as $W_i(\alpha, \beta) \leq \rho_i(\beta - \alpha) + 2\phi_i$.*

We suggest a variation of DRR scheduling algorithm that we call the Smoothing DRR, which is again an $\mathcal{LR}$ server. By using this scheduler, all the queues are virtually always backlogged, so that corollary 2 is applicable for network analysis. In this case the output flows from all the Smoothing DRR server $\mathcal{S}n$ conform to leaky bucket model with parameters $\left(\min(2\phi_i, \sigma_i^{\mathcal{S}n}), \rho_i\right)$. The basic idea behind this algorithm is that whenever the queue is empty, generate a *null packet* with the amount of quantum size plus deficit value, and then serve it. If a packet belongs to the queue arrives during the service of a null packet, then stop serving the null packet, decrement the deficit as much as the served amount from the null packet, and turn to the normal operation mode. The important part is that the transmitter ignores any null packet, so that no real transmission occurs for a null packet. The detailed algorithm is described in algorithm 1..

5   Numerical Results

We focus on a residential network environment, where the maximum number of hops and the number of flows are confined and predictable. Moreover in such networks the demand for real-time service is strong, especially for video and high quality audio applications. IEEE 802.1 Audio/Video Bridging Task Group [12] defines a bound for the end-to-end delay to be 2ms in a network of 7 hops for stringent audio and video applications [13]. We assume the 100Mbps Fast Ethernet links are used across the network.

If we are to transmit the MPEG-2 Transport Streams (TS) data whose lengths are fixed at 188 bytes with 12 bytes RTP fixed header, 4 bytes RTP video-specific header, 8 bytes UDP header, 20 bytes IP header and finally 26 bytes Ethernet header and trailer including preamble, then the maximum packet length in this

---

Algorithm 1. Smoothing DRR

Ensure: D is deficit value and Q is quantum size
Require: Upon a queue's turn starts
    $D \Leftarrow D + Q$
    if there is no packet queued **then**
3:    enqueue a null packet with size D.
    end if
    while packet(s) is queued **do**
6:    **if** D is greater than or equal to the head packet size **then**
        start serving the head packet
        if the null packet is being served and a new packet for the queue has arrived
        during the service **then**
9:        stop immediately the service of the null packet and drop the null packet
        $D \Leftarrow (D-$ the number of bits that have been served from the null packet$)$
        Go to 6
12:    **else**
        complete the packet's service
        $D \Leftarrow (D-$ the size of the packet just served$)$
15:    **end if**
    end if
    end while
18: Release the server from the queue.

---

case becomes 258 bytes. Considering the extended headers fields and Ethernet inter-frame gap, we set our maximum packet length at 300 bytes.

Consider a network of arbitrary topology whose maximum radius is seven hops. We refer a hop by a switch, therefore a server. Consider the longest path where seven DRR servers are in series. In this longest path, at each server there are nine flows with the average rate of 10Mbps, including the flow under observation, i. The maximum burst size of i at the entrance of the network is 300 bytes. Note that the other flows do not have any burst size constraints. The flows other than i at different servers may or may not be the same ones. In this scenario the latencies at all the servers are identical and is 0.528ms, as given in equation (2). The end-to-end delay with seven servers is again obtained from equation (1) and is 3.696ms.

Now consider a network with an aggregated flow which comprises flow 1 and 2. This aggregated flow traverses a network of LR servers in series. In every server there are seven others flows, demanding 10Mbps per each flow. Again the flows 1 and 2 are constrained with leaky buckets at the entrance of the network with the parameters (300 bytes, 10Mbps). We find the delay bound of this network, by lemma 1, to be 4.632ms. Next we consider the case where the aggregated flow in the previous scenario is deaggregated into flows 1 and 2 at the input port of switch 7 to different output ports thus different servers, and each is confronted with other eight flows there. This scenario is depicted in figure 2. The delay bound in this case is 8.040ms. We then repeat with the scenarios where four among nine flows are aggregated instead of just two. Finally we apply the Instant Service policy to the Smoothing DRR.

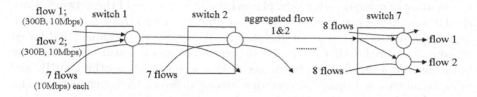

Fig. 2. Seven DRR servers in series. Flows 1 and 2 are aggregated at switch 1 and deaggregated at switch 7: Delay bound is 8.040ms.

Table 1. Summary of performance comparison

Scenario	1	2	3	4	5	6	7	8	9	10
Delay bound	4.200ms	3.024ms	4.632ms	4.116ms	8.040ms	7.140ms	4.680ms	6.060ms	3.744ms	5.604ms
Applicable to	Any case				Where leaky bucket conformance of some flows are certain, those trusted flows can be aggregated together.					

Scenario 1   IntServ with DRR, 9 flows at each server; The flow i is with leaky bucket parameters (300bytes, 10Mbps).
Scenario 2   IntServ with DRR-IS
Scenario 3   Aggregation of 2 flows with DRR, Flows 1&2 take the same path; Agg. @ SW1
Scenario 4   Aggregation of 4 flows with DRR, Flows 1, 2, 3 & 4 take the same path; Agg. @ SW1
Scenario 5   Aggregation & Deaggregation of 2 flows with DRR; Agg. @ SW1, Deagg. @ SW7
Scenario 6   Aggregation & Deaggregation of 4 flows with DRR
Scenario 7   Aggregation & Deaggregation of 2 flows with S-DRR
Scenario 8   Aggregation & Deaggregation of 4 flows with S-DRR
Scenario 9   Aggregation & Deaggregation of 2 flows with S-DRR-IS
Scenario 10 Aggregation & Deaggregation of 4 flows with S-DRR-IS

We compare a number of scenarios with various schemes of flow aggregations and examine their delay performances. Table 1 summarizes the results. IntServ successfully protects the flows from each other's burst size variations, therefore is considered to have the predictable and robust performance. When there is only an aggregation at the entrance of the network, the performance degradation is not significant. In the case with aggregation of four flows, it even reduces the delay bound. When a deaggregation takes place in the middle of the path, however, the delay bound is large due to the maximum burst size of the flow under observation that have increased to a significant level while traveling the path. The aggregation in the middle of a network will also lengthen the delay bound with the same reason. This degradation is somewhat mitigated by S-DRR.

6   Conclusion

We have analyzed the servers with aggregated flows and obtained an iterative method to calculate end to end delay bounds. The delay bounds in such networks depend on many parameters, including burst size of other flows within the aggregate and the number of aggregation/deaggregation. In networks with traditional work-conserving servers the delay bound depends especially on how far the aggregation or deaggregation take places from the entrance of the network, since the farther the aggregation point the larger the maximum burst size, which can be interpreted as the degree of uncertainty. We suggested a simple DRR-based scheduling algorithm that is without the characteristics of linearly incrementing burst size. We studied the performance of this server, the Smoothing DRR, and have found that it is quite satisfactory. among different QoS architectures and server choices.

References

1. Braden R., Clark D. and Shenker S.: Integrated Services in the Internet Architecture – an Overview. IETF Request for Comments, RFC-1633. (1994)
2. Shreedhar M. and Varghese G.: Efficient fair queueing using deficit round-robin. IEEE/ACM Trans. Networking, vol. 4, no. 3. (1996) 375–385
3. Golestani S.: A Self-clocked Fair Queueing Scheme for Broadband Applications. In Proc. IEEE INFOCOM. (1994)
4. Stiliadis D. and Varma A.: Latency-Rate servers – A general model for analysis of traffic scheduling algorithms. IEEE/ACM Trans. Networking, vol. 6, no. 5. (1998)
5. Cobb J. A.: Preserving quality of service guarantees in spite of flow aggregation. IEEE/ACM Transactions on Networking, vol. 10, no. 1. (2002) 43–53
6. Goyal P., Lam S. S., and Vin H. M.: Determining end-to-end delay bounds in heterogeneous networks. In Proc. Workshop on Network and Operating Systems upport for Digital Audio and Video (NOSSDAV95). (1995) 287–298
7. Sun W. and Shin K. G.: End-to-End Delay Bounds for Traffic Aggregates Under Guaranteed-Rate Scheduling Algorithms. IEEE/ACM Transactions on Networking. Vol. 13, No. 5. (2005)
8. Stiliadis D.: Traffic Scheduling in Packet-Switched Networks – Analysis, Design and Implementation. Ph.D. Dissertation. U.C. Santa Cruz. (1996)
9. Joung J., Choe B-S., Jeong H., and Ryu H.: Effect of Flow Aggregation on the Maximum End-to-End Delay. In Proc. International Conference on High Performance Computing and Communications, HPCC-06. Also in Lecture Notes in Computer Science, Vol. 4208. Springer-Verlag (2006) 426–435
10. Kanhere S. S. and Sethu H.: On the latency bound of deficit round robin. in Proceedings of the ICCCN, Miami. (2002)
11. Joung J., Shin D., Feng F., and Jeong H.: Instant Service Policy and Its Application to Deficit Round Robin. In Proc. AAIM'06. Also in Lecture Notes in Computer Science, Vol. 4042. Springer-Verlag (2006) 114–125
12. Audio/Video Bridging Task Group website, http://www.ieee802.org/1/pages/avbridges.html
13. Feng F. and Garner G. M.: Meeting Residential Ethernet Requirements – A Simulation Study. IEEE 802.1 Audio/Video Bridging Task Group. (2005)

How to Play the Majority Game with Liars

Steve Butler[1], Jia Mao[2,*], and Ron Graham[2,**]

[1] Dept. of Mathematics, University of California, San Diego,
La Jolla, CA 92093-0112
sbutler@math.ucsd.edu

[2] Dept. of Computer Science and Engineering, University of California, San Diego,
La Jolla, CA 92093-0404
jiamao@cs.ucsd.edu, graham@ucsd.edu

Abstract. The *Majority* game is a two player game with a questioner **Q** and an answerer **A**. The answerer holds n elements, each of which can be labeled as 0 or 1. The questioner can ask questions comparing whether two elements have the same or different label. The goal for the questioner is to ask as few questions as possible to be able to identify a single element which has a majority label, or in the case of a tie claim there is none. We denote the minimum number of questions **Q** needs to make, regardless of **A**'s answers, as q^*. In this paper we consider a variation of the Majority game where **A** is allowed to lie up to t times, i.e., **Q** needs to find an *error-tolerant* strategy. In this paper we will give upper and lower bounds for q^* for an adaptive game (where questions are processed one at a time), and will find q^* for an oblivious game (where questions are asked in one batch).

1 Introduction

The well-studied *Majority* game consists of two players: a questioner **Q** and an answerer **A**. Initially **A** holds a set of n elements, each of which will have a binary label (e.g., 0 or 1), and **Q** asks questions as to whether two elements have the same or different labelling. In the game, **Q**'s goal is to identify one element of the majority label (or in the case of a tie, claim that there is none), while **A**'s goal is to block **Q** from identifying such an element. If after no more than q questions **Q** can identify a majority element then **Q** wins, otherwise **A** wins. We say **Q** has a *winning strategy* of length q if **Q** can always win the game with at most q questions, regardless of what **A** answers. The goal is to design strategies for **Q** with minimal q, denoted by q^*.

1.1 History

The earliest variant of the *Majority* problem was originally proposed by Moore in the context of fault-tolerant system design in 1981 [11]. A number of different variants were subsequently proposed and analyzed. This problem resurfaced after

---

* Partially supported by an NSF graduate fellowship.
** Research partially supported by NSF Grant CCR-0310991.

M.-Y. Kao and X.-Y. Li (Eds.): AAIM 2007, LNCS 4508, pp. 221–230, 2007.

about twenty years in a military application where communication needs to be minimized to locate one sensor that has not been corrupted among a group of sensors [6].

There have been several variations of the majority game studied in past literature. Variations have included examining different k (the number of permissible labels); considering *adaptive* or *oblivious* versions of the game (in an adaptive game, **Q** learns the answer to each question before asking the next question, while in an oblivious game, **Q** asks all questions in one batch before getting any answers from **A**); and whether or not a majority label is known to exist or not. (See [1,3,6,7,9,14,17].)

In the adaptive case with 2 labels, Saks and Werman [14] were the first to prove a tight bound of the minimum length of a winning strategy for **Q** to be $n - \mu_2(n)$, where $\mu_2(n)$ is the number of 1s in n's binary expansion. Different proofs for the same result were subsequently given in [3] and [17]. When k is unknown, a tight bound of $\lceil 3n/2 \rceil - 2$ was given in [9]. The average case of the same setting was analyzed in [4]. Similar bounds were proven for randomized algorithms [10].

In the oblivious case, when k is unknown, the optimal winning strategy for **Q** is much harder to design or analyze. If the existence of a majority label is not known a priori, **Q** needs $\Omega(n^2)$ many questions [15]. However, if a majority label is known to exist, by using a special type of graph, called Ramanujan graph, there is a constructive strategy for **Q** that uses no more than $(1 + o(1))27n$ queries. The constant 27 can be further improved to 19.5 if only existence of such a strategy is desired [7].

1.2   Error Tolerance

In past literature, **A** is always a *malevolent but truthful* adversary in the sense that as long as their current answer is *consistent* with previously given answers, they will want to win the game. However, an *error-tolerant* feature is desired when the answers to the queries in the application may be faulty due to communication errors. In this paper we address this issue by putting the *Majority* game in a broader context of fault-tolerant communication, namely, searching games with errors. Generally, these games are more difficult to analyze, but have a much wider range of applications compared to perfect information two person games. One such famous game is the *Rényi-Ulam liar game* [13,16]. For a comprehensive overview of this topic, we refer the reader to a recent survey [12].

In this paper we consider bounded error tolerance for the *Majority* game, where **A** is allowed to *lie up to a fixed number t times*. More precisely, this means that after **Q** specifies an element as being in the majority or state there is none, **A** has the freedom to flip up to t answers of the previously asked queries and reveal a labelling that is consistent with this modified set of answers. Now that **A** can lie how much will this handicap **Q**? What is the new minimal q^* and what strategy should **Q** adopt to achieve this bound?

In this paper we will begin to answer some of these questions for the *Majority* game with binary labels. We will give upper and lower bounds for q^* by producing strategies for $\mathbf{Q}$ and $\mathbf{A}$ in various versions of the game. For the sake of exposition, some of the justification for these strategies will be omitted from this paper and will appear in a longer version of this paper. We summarize our results in the table below, where t is the number of lies and n is the number of objects.

Game	n is odd	n is even
Adaptive $t = 1$	$n \leq q^* \leq n+1$	$n+1 \leq q^* \leq n+2$
Adaptive $t > 1$	$\lceil \frac{t+3}{4}n - \frac{t+1}{4} \rceil \leq q^* \leq \left(\frac{t+1}{2} + o(1)\right)n$	$\frac{t+1}{2}n \leq q^* \leq \left(\frac{t+1}{2} + o(1)\right)n$
Oblivious $t \geq 1$	$q^* = \lceil (t + \frac{1}{2})n \rceil$	

The upper bound in the adaptive case for $t > 1$ holds only when $t = o(n^{1/2})$. A more precise statement of the upper bound is given by Theorem 4. For the case when t is large in comparison to n then a better upper bound might be given by the oblivious upper bound, this of course is dependent on the size of t.

Also note that there is a difference between the case when n is odd and n is even. This is due to the fact that when n is odd there *must* be a majority element which gives additional information $\mathbf{Q}$ can use in forming a strategy.

2 Adaptive Setting

We can adapt the known strategy for the game with no lies to give a simple upper bound for q^*. Recall that for the majority game of binary labels with no lies the optimal winning strategy takes $(n - \mu_2(n))$ queries [14].

Theorem 1. *In the adaptive Majority game on n elements with binary labels and at most t lies,*

$$q^* \leq (t + 1)(n - \mu_2(n)) + t$$

where $\mu_2(n)$ is the number of 1s in n's binary expansion.

Proof. Let $\mathbf{Q}$ ask the same queries as in the best strategy to play the *Majority* game with no lies, only that each query is repeated until $(t+1)$ answers agree (at which point we know the relationship between the two elements) before going to the next query. Because $\mathbf{A}$ is not allowed to lie more than t times, the total number of queries $\mathbf{Q}$ needs to ask is $(t + 1)(n - \mu_2(n))$ plus at most t.   □

This simple bound for q^* can be improved by $\mathbf{Q}$ using error detection and correction from the answers of $\mathbf{A}$.

2.1 Preliminaries

To keep track of the game as it progresses, we use an auxiliary (multi-)graph. The n objects are the vertices and edges correspond to comparisons between elements. (We will allow queries to be repeated and any multiple queries are represented by multi-edges.) As the game progresses **Q** will give **A** an edge and ask **A** to color it blue if the two elements have the same label and red it the two elements have different labels.

One of the most important tools used in the formation of a strategy is to use the auxiliary graph to detect lies. For instance, note that if **A** were truthful then every cycle would always have an even number of red edges. But when **A** can lie this no longer needs to be the case and we have the following observation.

Observation 1. *A cycle can have an odd number of red edges if and only if an odd number of edges in the cycles correspond to lies.*

This follows by noting that the number of red edges corresponds to the number of times along the cycle that we switch labels. Initially if there were no lies we would have to switch an even number of times (i.e., we must return to the same label we started with). Now for every lie we either increase or decrease by 1 the number of switches made.

From the observation any cycle which contains an odd number of red edges must contain a lie, we will refer to such cycles as invalid, otherwise we say that the cycle is valid. It is important to notice though that just because a cycle is valid it does not imply that there are no lies, only that there are an even number of them.

2.2   Majority Game with at Most $t = 1$ Lie

Theorem 1 gives an upper bound of $\left(2(n - \mu_2(n)) + 1\right)$ for **Q**'s adaptive strategy when **A** is allowed to lie once. This bound can be improved by using *validity checking* of the auxiliary graph. (For the case involving one lie we have that a cycle is valid if and only if it has no lies, in general this will not hold.)

Theorem 2. *In the adaptive Majority game on n elements with binary labels and at most 1 lie*

$$q^* \leq \begin{cases} n + 1 & \text{if } n \text{ is odd,} \\ n + 2 & \text{if } n \text{ is even.} \end{cases}$$

Proof. We give a two-stage strategy for **Q** satisfying the bounds.

Stage 1:   In the first stage **Q** starts by growing long blue paths by the following rule: connect the ends of two blue paths which have an equal number of vertices. This continues until either there are no two paths with the same number of vertices or we get a red edge.

In the first case **Q** takes the longest blue path and closes it up by a single question, if the cycle is blue then **Q** identifies any element on the cycle as a majority element (i.e., in such a case it is easy to see that the cycle contains more than half of the elements and all of them must have the same label), otherwise if the edge is red (i.e., the cycle is invalid so contains a lie) **Q** goes to the second stage.

In the second case there is a path with one single red edge in the middle. **Q** then asks a single question to close it up to form a cycle. If the new edge is red, then the coloring is valid and so the cycle has an equal number of each label, **Q** then removes this component from the graph and continues as before. If the new edge is blue (i.e., the cycle is invalid so contains a lie) **Q** goes to the second stage.

Stage 2: Starting stage 2 we already know that **A** has used their lie and so all subsequent answers must be true. We initially have one cycle with a single red edge (denoted by $\{u, v\}$) and possibly several blue paths. The first step in this stage is to connect one vertex from each blue path to u. The graph now consists of a cycle with a tree attached to u. At this point **Q** has asked exactly n questions.

Because the lie lies in the cycle, all edges involved in the tree reflect truthful responses. In particular, **Q** can determine how many of the vertices of the cycle must have the same label as u in order for u to be the majority element, denote this number by k. Starting at u and going in the opposite direction of v, **Q** counts out k vertices. Denote the kth vertex by w. (If the cycle contains fewer than k vertices then u is in the minority and it will be easy to identify a majority element in the tree. Similarly, if $k \leq 0$ then u has a majority label.)

Q now queries the edge $\{u, w\}$. If the edge is blue then all the vertices between u and w have the needed label and we can conclude that u is a majority element. On the other hand if the edge is red then there is a lie somewhere between u and w and so u cannot be a majority element. In the case when n is even we then only need to compare u with the vertex that precedes w to test if there is a tie. If the edge is blue then there is a tie, while if the edge is red then w is a majority element.

The result now follows by counting the number of queries used.   □

To find a lower bound for q^* we need to give a strategy for **A**. Since **A** can adopt the same strategy as in the game with no lies we have that $q^* \geq n - \mu_2(n)$. However, **A** can do better as shown in the next theorem.

Theorem 3. *In the adaptive Majority game on n elements with binary labels and at most 1 lie*

$$q^* \geq \begin{cases} n & \text{if } n \text{ is odd,} \\ n + 1 & \text{if } n \text{ is even.} \end{cases}$$

The case for n is odd will follow from Theorem 5 with $t = 1$. The proof for n is even will be found in the longer form of this paper.

2.3 Majority Game with at Most $t \geq 2$ Lies

We now consider the game where **A** is allowed to lie up to $t \geq 2$ times. We first start by establishing an upper bound.

Theorem 4. *In the adaptive Majority game on n elements with binary labels and at most $t \geq 2$ lies,*

$$q^* \leq \frac{t+1}{2}n + 6t^2 + 2t + 3\log n.$$

Proof. We give a sketch of the strategy here. More details and justification can be found in the longer form of this paper. The strategy for $\mathbf{Q}$ will be to use two rounds. The first round will be "oblivious" in that $\mathbf{Q}$ will always ask the same set of questions (this round will use $(t+1)n/2$ questions). In the second round $\mathbf{Q}$ then uses the answers from the first round to find and correct all lies.

We first consider the case $n > 2t$ with n even.

In the first stage $\mathbf{Q}$ forms an n-cycle with the n vertices and asks $\lfloor t/2 \rfloor$ questions on each edge of the cycle. $\mathbf{Q}$ then makes

$$t + 1 - 2\left\lfloor \frac{t}{2} \right\rfloor = \begin{cases} 1 & \text{if } t \text{ is even,} \\ 2 & \text{if } t \text{ is odd,} \end{cases}$$

queries between opposite vertices of the cycles (we will refer to these queries as spokes). An example is shown in Figure 1.

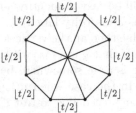

Fig. 1. "Oblivious" first round queries   Fig. 2. Looking for lies in the spokes

This strategy has the following useful property. If $\mathbf{A}$ has lied then there is either an invalid 2-cycle either in the spokes or along the exterior; or there is an invalid 4-cycle of the form $v_1 v_1' v_2' v_2$ (see Figure 2); or there is an invalid $n/2$ cycle of the form $v_1 v_1' v_2' \ldots v_1$ (i.e., lying along half of the outer cycle plus a spoke).

We can quickly find and remove all errors from invalid 2-cycles and 4-cycles. Namely, for each such cycle we make $2t$ queries on each of 3 edges and take the majority answer on each edge, if we have not yet found the lie from this then the remaining edge is a lie and we then can correct. Thus we would need at most $6t^2$ queries to correct these lies.

If after correcting these queries there is still an invalid $n/2$ cycle then it must be the case that there are two opposite intervals "saturated" with lies (i.e., all queries between $u_1 u_2$ and $u_1' u_2'$ in Figure 2 are lies). In particular, there can only be at most one lie left. We now lift up the $n/2$ cycle (which will have only one lie) and locate the lie by using a splitting technique. Namely we join two opposite pairs of vertices with three edges and take the majority answer and use this to split the cycle into two smaller cycles, one of which will be invalid (and which contains the lie of the $n/2$ cycle). We then continue this process of cutting in half each time until we have located the lie. In particular, this technique will locate the lie in $3 \log n$ steps.

$\mathbf{Q}$ is now finished because they can detect and correct lies given by $\mathbf{A}$ and relate all elements together. In particular, $\mathbf{Q}$ has used at most $(t+1)n/2 + 6t^2 + 3\log n$ queries to accomplish this.

For the case n odd $\mathbf{Q}$ sets aside a single element and runs the procedure and then at the end connects the element back into the graph by making at most $2t+1$ queries relating the odd element out with some arbitrary element.

Finally for the case $n \le 2t$ we can simply build a tree where we keep asking questions on each edge until we get $t+1$ responses which agree. In particular, we would need at most $2t^2 + t$ queries in such a case.

Putting it all together gives the desired result. $\square$

To establish the lower bound, we will make use of the following general observation.

Observation 2. *If the coloring is valid (i.e., no lies are detected), then* $\mathbf{Q}$ *will not be able to determine the correct relationship for an element which is involved in no more than* t *queries.*

This observation follows by noting that since the coloring is valid and $\mathbf{A}$ is allowed to change the color of up to t edges, then $\mathbf{A}$ can change all the queries involved with a vertex of low degree (i.e., no more than t) and still produce an admissible labelling.

Theorem 5. *In the adaptive majority game on* n *elements with binary labels and at most* $t \ge 1$ *lies,*

$$
q^* \ge \begin{cases} \dfrac{t+1}{2}n & \text{if } n \text{ is even,} \\[2mm] \left\lceil \dfrac{t+3}{4}n - \dfrac{t+1}{4} \right\rceil & \text{if } n \text{ is odd.} \end{cases}
$$

Proof. For the case when n is even, $\mathbf{A}$ can use the following strategy: Initially assign half of the elements with label 0 and the other half with label 1 and answer all of the questions truthfully. If $\mathbf{Q}$ makes fewer than $(\frac{t+1}{2})n$ queries, then by degree considerations there is a vertex with degree at most t. Based on the above observation, $\mathbf{Q}$ will be unable to determine the correct relationship of that vertex with the remaining elements and in particular will not be able to distinguish between a tie and the existence of a majority element.

The case for n is odd will be found in the longer form of this paper. $\square$

3   Oblivious Setting

In the oblivious setting, $\mathbf{Q}$ has to specify all the edges in the auxiliary graph G before $\mathbf{A}$ colors any of them. This implies that $\mathbf{Q}$ has to accomplish *detection* and *location* of lies simultaneously. We have another important observation.

Observation 3. *In the Majority game of binary labels with at most t lies, if an edge e is part of 2t cycles that pairwise edge-intersect only at e (though they might share many vertices in common), then a lie is located at e if and only if at least $(t+1)$ of these cycles are invalid.*

This observation follows by noting that if an edge corresponds to a truthful answer then there can be at most t of the $2t$ cycles intersecting at e which can be invalid. On the other hand if the edge corresponded to a lie then at most $t-1$ of the $2t$ cycles intersecting at e can be valid, or equivalently, at least $t+1$ invalid cycles.

Theorem 6. *In the oblivious Majority game on n elements with binary labels and at most $t \geq 1$ lies,*

$$q^* = \left\lceil (t+\frac{1}{2})n \right\rceil.$$

Proof. We first establish the upper bound. The observation above implies that if we can construct an auxiliary graph for **Q** such that for any particular edge we can find $2t$ cycles that are pairwise edge-joint only at that edge, we can locate and thus correct all possible lies with no more queries needed.

We handle the base cases first. For $n = 2$, we use $(2t+1)$ edges for the same query. For $n = 3$, the query graph is a triangle with one query asked t times and the other two queries each asked $(t+1)$ times.

For even $n \geq 4$, we construct a multigraph as shown in Figure 3 where all edges in the outer cycle are multi-edges (repeated t times) and single edges (or spokes) connect each pair of opposite vertices. The total number of edges is therefore $(t+\frac{1}{2})n$. For odd $n \geq 3$, first construct a graph as in the $n+1$ case and then contract a set of edges on the outer cycle, an example is shown in Figure 4. In this case it can be checked that there are $\lceil (t+\frac{1}{2})n \rceil$ edges in the graph.

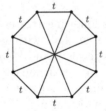
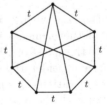

Fig. 3. Oblivious graph, n even   **Fig. 4.** Oblivious graph, n odd

For each spoke, we can find t cycles using one half of the outer cycle and another t using the other half. For each outer edge e, first we can find $(t-1)$ small cycles by joining it with the other $(t-1)$ multiedges with the same endpoints, then we use edges in the outer cycle to obtain another $(t-1)$ cycles. We need two more cycles and these are constructed using the spokes and the remaining unused edges of the outer cycle as shown in Figure 5. Because each edge lies in

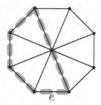

Fig. 5. The remaining two cycles for a side edge e

at least $2t$ cycles pairwise edge-joint only at that edge, all lies can be located and hence corrected. Establishing the upper bound.

For the lower bound, we again consider the degrees. If **Q** asks fewer than $\lceil (t+\frac{1}{2})n \rceil$ then since **Q**'s strategy is oblivious, **A** can examine the entire auxiliary graph and find a vertex v with degree at most $2t$. **A** can split the remaining vertices into two sets U and V as equally as possible (i.e., $||U| - |V|| \leq 1$) with label 0 and 1 respectively. For queries not involving v, **A** answers truthfully. For queries involving v, **A** answers half of the queries as if v is labelled 0 and the other half as if v is labelled 1. This is possible because **A** is allowed to lie up to t times. Now **Q** cannot distinguish which half are lies and hence cannot determine the label for v which is essential because the other vertices are in an (almost) exact balance. This establishes the lower bound and concludes the proof.   □

4 Conclusion and Remarks

Motivated by the practical need of an *error-tolerant* feature, we have concentrated on optimizing **Q**'s strategy in the presence of lies (or errors) for binary labels in the *Majority* game. We point out that when the number of lies is upper bounded by a constant t, **Q** can still win the game with a linear (in n) number of questions. Upper and lower bounds on the length of **Q**'s optimal strategy were derived in both the adaptive setting and the oblivious setting.

Consideration of fault-tolerance may also be useful for the many other variants of the *Majority* game, such as when the number of different labels is more than two. A natural generalization of the Majority game is the *Plurality* game where **Q** wants to identify one element of the *plurality* label (most frequently occurring), still using only pairwise equal/unequal label comparisons of elements. Much attention has been given to designing adaptive strategies (deterministic or randomized) for fixed or unknown k (see [2,5,8,10,15]). We remark here that the same reasoning of Theorem 1 applies to existing bounds for all variants of the *Majority* game (including the *Plurality* game) if the maximum number of lies allowed t is fixed. The new upper bounds will only be at most worse by a multiplicative constant $(t+1)$ and an additive constant t.

In this paper, we gave a complete picture for the *oblivious* setting in the *Majority* game with a constant bounded number of lies. In the *adaptive* setting, however, there are still various gaps between the upper and lower bounds obtained. Closing these gaps would be interesting to pursue. In the meantime, other types of error-tolerance may also be considered, such as bounded error fraction or random errors.

Acknowledgement

We thank Joel Spencer, Robert Ellis and anonymous referees on an earlier version of this paper for helpful discussions and suggestions.

References

1. M. Aigner, "Variants of the majority problem", *Applied Discrete Mathematics* **137** (2004), 3–25.
2. M. Aigner, G. De Marco, M. Montangero, "The plurality problem with three colors and more", *Theoretical Computer Science* **337** (2005), 319–330.
3. L. Alonso, E. Reingold, R. Schott, "Determining the majority", *Information Processing Letters* **47** (1993), 253–255.
4. L. Alonso, E. Reingold, R. Schott, "Average-case complexity of determining the majority", *SIAM J. Computing* **26** (1997), 1–14.
5. L. Alonso, E. Reingold, "Determining plurality", *manuscript*, 2006.
6. F. R. K. Chung, R. L. Graham, J. Mao, and A. C. Yao, "Finding favorites", *Electronic Colloquium on Computational Complexity (ECCC)* (2003) 078, 15pp.
7. F. R. K. Chung, R. L. Graham, J. Mao, and A. C. Yao, "Oblivious and adaptive strategies for the majority and plurality problems", *COCOON 2005*, 329–338; *Algorithmica*, to appear.
8. Z. Dvorak, V. Jelinek, D. Kral, J. Kyncl, and M. Saks, "Three optimal algorithms for balls of three colors", *STACS 2005*, Lecture Notes in Comp. Sci. **3404**, Springer, Berlin, 2005, 206–217.
9. M. J. Fischer and S. L. Salzberg, "Finding a majority among n votes", *J. Algorithms* **3** (1982), 375–379.
10. D. Kral, J. Sgall, and T. Tichỳ, "Randomized strategies for the plurality problem", *manuscript*, 2005.
11. J. Moore, "Proposed problem 81-5", *J. Algorithms* **2** (1981), 208–210.
12. A. Pelc, "Searching games with errors — fifty years of coping with liars", *Theoret. Comput. Sci.* **270** (2002), 71–109.
13. A. Rényi, "On a problem in information theory", *Magyar Tud. Akad. Mat. Kutató Int. Közl.* **6** (1961), 505–516.
14. M. Saks and M. Werman, "On computing majority by comparisons", *Combinatorica* **11** (1991), 383–387.
15. N. Srivastava, and A. D. Taylor, "Tight bounds on plurality", *Information Processing Letters* **96** (2005), 93–95.
16. S. M. Ulam, **Adventures of a mathematician**, Charles Scribner's Sons, New York, 1976, xi+317pp.
17. G. Wiener, "Search for a majority element", *J. Statistical Planning and Inference* **100** (2002), 313–318.

On Satisfiability Games and the Power of Congestion Games

Vittorio Bilò

Dipartimento di Matematica "Ennio De Giorgi", Università del Salento
Provinciale Lecce-Arnesano, P.O. Box 193, 73100 Lecce, Italy
`vittorio.bilo@unile.it`

Abstract. We introduce and study satisfiability games, a new class of games that can be seen as the non-cooperative version of classical maximum satisfiability problems. We give several results involving these games and mainly focus on their expressiveness. In particular, we show that there exists a strong correspondence between satisfiability games and congestion games. As one of the consequences of our results, we show that each game is isomorphic to a congestion game with player specific payoffs. Thus, each other game can be defined as a particular specialization of congestion games with player specific payoffs and this paper can be considered as a first effort in outlining a classification of non-cooperative games.

1 Introduction

The study of non-cooperative games is receiving more and more attention due to its tight relationship with that of unregulated networks, such as the Internet. Lots of results have been achieved in the last years about the hardness of computing pure and/or mixed Nash equilibria and their relative price of anarchy for several particular games, but, despite this effort, there is a certain lack of general results. A classification of non-cooperative games is thus being hotly encouraged. The main aspects that one may want to take into account when classifying games are essentially three: their expressiveness (i.e., their capability to model several non-cooperative scenarios), the complexity of computing their Nash equilibria, and the quality of such equilibria (i.e., price of anarchy and price of stability).

Studying the expressiveness of a class of games and, in particular, its relationships with that of another class, requires the use of the notion of equivalence among games. Comparing the computational complexity of computing their Nash equilibria and their quality, instead, requires the use of the weaker notion of reducibility among games, (see Subsection 1.1 for formal definitions).

In this paper we deal with the notion of expressiveness of games whose study, however, can give us also some insights into the other two aspects. To this aim, we introduce satisfiability games and show how they can be considered as the most general class of non-cooperative games. The interesting point is that, despite their power, satisfiability games have a very simple structure and thus can be likely used to prove general results for other classes of games.

M.-Y. Kao and X.-Y. Li (Eds.): AAIM 2007, LNCS 4508, pp. 231–240, 2007.
© Springer-Verlag Berlin Heidelberg 2007

We embed our study on satisfiability games into a more complex and interesting process of classification of non-cooperative games. In particular, we show a strong relationship tying together satisfiability games and congestion games as well as their generalization with player specific payoffs. It turns out that this latter class can be elected as the representative class of all games because of its generality and its well-known properties and characterization. Each of the other games studied in the literature and each of the other ones that may be introduced can be defined as a particular specialization of congestion games with player specific payoffs, thus allowing the definition of a hierarchy of games and making easier the achievement of general results.

1.1 Definitions

Given a set U, a sequence $a = (u_1, \ldots, u_k)$ of elements of U, an index $i \in \{1, \ldots, k\}$ and an element $u \in U$, we write $(a_{-i}, u) = (u_1, \ldots, u_{i-1}, u, u_{i+1}, \ldots, u_k)$ to denote the sequence obtained from a by replacing u_i with u.

Strategic games. A strategic game $\mathcal{G}$ is a triple $\mathcal{G} = (P, S_{i \in P}, \omega_{i \in P})$ where P is a set of n players, S_i is the set of strategies available to player i and $\omega_i : S_1 \times \ldots \times S_n \mapsto I\!R$ is his payoff function. We will assume throughout the paper that ω_i models a benefit for player i, thus each player aims to maximize his payoff. Of course, for the case in which the payoffs are costs for the players one can always be consistent with this definition by taking the negative of ω_i for each player i.

States and improving steps. The set $S = S_1 \times \ldots \times S_n$ is called the set of states or strategy profiles of $\mathcal{G}$. Consider a state $\sigma = (\sigma_1, \ldots, \sigma_n) \in S$ and a strategy $s \in S_i$. The action of changing his strategy from σ_i to s is called an improving step performed by player i if $\omega_i(\sigma_{-i}, s) > \omega_i(\sigma)$.

Pure Nash equilibria. A state σ is a pure Nash equilibrium if no player possesses an improving step, that is, $\forall i \in P$ and $\forall s \in S_i$ it holds $\omega_i(\sigma_{-i}, s) \leq \omega_i(\sigma)$.

Mixed strategies. A mixed strategy for player i is a probability distribution Y_i defined over the set S_i of his pure strategies. Given a mixed strategy Y_i for player i, the support of Y_i, denoted as $support(Y_i)$, is the set of strategies $s \in S_i$ for which $Y_i(s) > 0$. A mixed strategy $Y = (Y_1 \ldots, Y_n)$ is a sequence of mixed strategies Y_i for each player $i \in P$. The support of a mixed strategy Y is defined as the set of states $support(Y) = support(Y_1) \times \ldots \times support(Y_n)$. The payoff of player i yielded by a mixed strategy Y is defined as $\omega_i(Y) = \sum_{\sigma \in support(Y)} (\omega_i(\sigma) \cdot Prob_Y(\sigma))$, where $Prob_Y(\sigma) = \prod_{i=1}^{n} Y_i(\sigma_i)$.

Mixed Nash equilibria. A mixed strategy Y is a (mixed) Nash equilibrium if $\forall i \in P$ and for any probability distribution Z defined over S_i it holds $\omega_i(Y_{-i}, Z) \leq \omega_i(Y)$. A fully mixed Nash equilibrium Y is a mixed Nash equilibrium such that $support(Y_i) = S_i$ for each $i \in P$.

Price of anarchy and stability. One of the main concerns when dealing with non-cooperative systems is to bound their inefficiencies due to the lack of coordination among the players. More formally, given a function $\gamma : S \mapsto I\!R$, called the social function, let σ^* be a state optimizing γ. On the other hand, given a mixed strategy Y, the social value of Y is defined as $\gamma(Y) = \sum_{\sigma \in support(Y)}(\gamma(\sigma) \cdot Prob_Y(\sigma))$. The price of anarchy $\rho(\mathcal{G}, \gamma)$ of game $\mathcal{G}$ for the social function γ is defined as $\rho(\mathcal{G}, \gamma) = \inf_{Y \in NE(\mathcal{G})} \frac{\gamma(Y)}{\gamma(\sigma^*)}$, while the price of stability $\alpha(\mathcal{G}, \gamma)$ of game $\mathcal{G}$ for the social function γ is defined as $\alpha(\mathcal{G}, \gamma) = \sup_{Y \in NE(\mathcal{G})} \frac{\gamma(Y)}{\gamma(\sigma^*)}$, where $NE(\mathcal{G})$ denotes the set of Nash equilibria of $\mathcal{G}$.

Equivalence among games. Two games $\mathcal{G} = (P, S_{i \in P}, \omega_{i \in P})$ and $\mathcal{G}' = (P, S'_{i \in P}, \omega'_{i \in P})$ are equivalent if there exists a $1-1$ mapping $g_i : S_i \mapsto S'_i$ such that $\omega_i(\sigma) =_{def} \omega_i(\sigma_1, \ldots, \sigma_n) = \omega'_i(g_1(\sigma_1), \ldots, g_n(\sigma_n)) =_{def} \omega'_i(g(\sigma))$ for each player i.

Reduction among games. A reduction from game $\mathcal{G} = (P, S_{i \in P}, \omega_{i \in P})$ to game $\mathcal{G}' = (P', S'_{i \in P}, \omega'_{i \in P})$ is a function $g : S \mapsto S'$ such that a state σ is a Nash equilibrium for $\mathcal{G}$ if and only if the state $g(\sigma)$ is a Nash equilibrium for $\mathcal{G}'$. If g can be computed in polynomial time with respect to the dimensions of $\mathcal{G}$, the reduction is called polynomial time reduction between $\mathcal{G}$ and $\mathcal{G}'$. Clearly the equivalence between $\mathcal{G}$ and $\mathcal{G}'$ implies a reduction from $\mathcal{G}$ to $\mathcal{G}'$.

Congestion games. A congestion game is a 4-tuple $(P, E, S_{i \in P}, d_{e \in E})$, where P is the set of n players, E is a set of m resources, $S_i \subseteq 2^{|E|}$ for each $i \in P$ and $d_e : I\!N \mapsto I\!R$ for each $e \in E$. Strictly speaking each player in a congestion game can choose among different subsets of resources, each resource e has an associated delay function d_e returning the delay experienced by any player using e in terms of the number of players using it. Once defined $n_e(\sigma) = |\{i \in P : e \in \sigma_i\}|$ as the number of players using resource e in state σ, the payoff function of player i is defined as $\omega_i(\sigma) = \sum_{e \in \sigma_i} d_e(n_e(\sigma))$. Congestion games were introduced by Rosenthal [18] who proved that they always possess pure Nash equilibria by defining the potential function $\Phi_R(\sigma) = \sum_{e \in E} \sum_{i=1}^{n_e(\sigma)} d_e(i)$. A classical social function associated to congestion games is the sum of the delays experienced on the resources, that is, $\gamma(\sigma) = \sum_{e \in E} d_e(n_e(\sigma))$. The generalization in which the delay functions can differ among the players, i.e., $\omega_i(\sigma) = \sum_{e \in \sigma_i} d_e^i(n_e(\sigma))$, is called congestion games with player specific payoffs, while the restriction in which each player can choose only one resource is called singleton congestion games.

Potential games. These are essentially all games possessing a potential function $\Phi : S \mapsto I\!R$. The existence of a potential function implies not only that potential games always admit pure Nash equilibria, but the stronger property that they do converge to an equilibrium starting from any state in a finite number of improving steps (finite improvement path (FIP) property). There are three types of potential games:

1. *Exact potential* games, where $\forall \sigma \in S$ and $\forall s \in S_i$, it holds $\Phi(\sigma) - \Phi(\sigma_{-i}, s) = \omega_i(\sigma) - \omega_i(\sigma_{-i}, s)$.
2. *Weighted potential* games, where $\exists \beta = (\beta_1, \ldots, \beta_n)$ such that $\forall \sigma \in S$ and $\forall s \in S_i$, it holds $\Phi(\sigma) - \Phi(\sigma_{-i}, s) = \beta_i(\omega_i(\sigma) - \omega_i(\sigma_{-i}, s))$
3. *General potential* games, where $\forall \sigma \in S$ and $\forall s \in S_i$, it holds $\Phi(\sigma) - \Phi(\sigma_{-i}, s) < 0$ if $\omega(\sigma) - \omega(\sigma_{-i}, s) < 0$.

The class of potential games generalizes that of congestion games, since, as shown by Monderer and Shapley [15], congestion games are equivalent to exact potential games and vice versa.

Cut games. These games, also known as party affiliation games, have been first introduced in [7,13] and then further studied in [12,4]. There is an undirected graph $G = (V, E)$ with edge weights $d : E \mapsto I\!R$. We assume here, for simplicity of notation, that $d_{ij} = 0$ if $(i, j) \notin E$. Any vertex in V is a player whose strategies are in the set $\{1, -1\}$. Thus, a state of this game can be seen as a cut in G. The payoff function is $\omega_i(\sigma) = \sum_{j:\sigma_j \neq \sigma_i} d_{ij}$, that is the contribution of player i in the cut. The usual social function associated with this game is the total contribution of the cut, that is, $\gamma(\sigma) = \sum_{(i,j) \in E:\sigma_i \neq \sigma_j} d_{ij}$.

Satisfiability games. Consider a pair $(X_{i \in P}, \mathcal{C},)$, where each player i owns a set of ℓ_i variables X_i, with the property that $X_i \cap X_j = \emptyset$ when $i \neq j$ and $\bigcup_{i \in P} X_i = X =_{def} \{x_1, \ldots, x_\ell\}$, and $\mathcal{C} = \{C_1, \ldots, C_m\}$ is a set of m clauses which can be any boolean formula defined over the literals yielded by all the variables in X and their negations. Each clause C_j has a weight c_j and consists of m_j literals. The set of strategies for each player i is equal to $S_i \subseteq \{0, 1\}^{\ell_i}$. Thus, a state of this game $\sigma = (\sigma_1, \ldots, \sigma_n)$ can be seen as an assignment of values to the variables in X and, in particular, we denote by $\sigma_i(x) \in \{0, 1\}$ the value assigned to variable $x \in X_i$ by player i in state σ. We denote with $\mathcal{C}_k^-$ the set of clauses containing $\overline{x}_k$, with $\mathcal{C}_k^+$ the set of clauses containing x_k, and with $\mathcal{C}^i = \bigcup_{x_k \in X_i} (\mathcal{C}_k^- \cup \mathcal{C}_k^+)$ the set of clauses containing at least one of the literals induced by X_i. Let $C_j(\sigma)$ be the result of the evaluation of C_j under state σ, the payoff function of player i is $\omega_i(\sigma) = \sum_{C_j \in \mathcal{C}^i : C_j(\sigma)=1} c_j$. The social function is $\gamma(\sigma) = \sum_{C_j \in \mathcal{C} : C_j(\sigma)=1} c_j$. Throughout the paper we will deal with CSG and DSG, that is, games in which each clause is in disjunctive normal form and conjunctive normal form, respectively. Associating with each clause C_j an n-tuple of weights $(c_j^1, \ldots, c_j^n)$, one for each player, we obtain satisfiability games with player specific payoffs, while defining $S_i = \{0, 1\}^{\ell_i}$ instead of $S_i \subseteq \{0, 1\}^{\ell_i}$, we obtain unconstrained satisfiability games. When $\ell_i = 1$ for all $i \in P$, we have singleton satisfiability games. An interesting case is when S_i contains only the ℓ_i strategies in which player i is allowed to set to 1 one and only one of his variables. We call this special case *restricted* satisfiability games.

1.2   Related Works

Non-cooperative games and their Nash equilibria [16] have been studied since 1950 before the affirmation of the Internet gave them new life and favored the interest of computer science researchers. Congestion games were defined by Rosenthal [18] who introduced the idea of potential functions to show that these games always possess pure Nash equilibria. This idea was exploited some years later by Monderer and Shapley [15] who defined the class of potential games and proved that the class of congestion games and that of exact potential games coincide. Singleton congestion games with player specific payoffs were studied by Milchtaich [11] who proved that they no longer admit a potential function, but still possess pure Nash equilibria in the case in which the delay functions associated with each resource is decreasing for each player.

The FIP property of potential games clearly yields a naive algorithm for computing pure Nash equilibria, unfortunately its complexity can be exponential in the dimensions of the game. In fact, repeatedly performing improving steps can be seen as an application of the local search technique in several games, PLS-completeness of the computation of pure Nash equilibria for congestion games was proved in [7]. The problem of computing Nash equilibria is a very interesting algorithmic issue. The general case in which $n \geq 3$ has been proved to be PPAD-complete [5,6] by exploiting the first polynomial time reductions among games known in the literature and PPAD-completeness of the two player case was recently shown in [2].

The idea of comparing the optimal performances of an unregulated system with those achieved by a Nash equilibrium dates back to the works of Korilis and Lazar [8], La and Anantharam [10], and Shenker [19]. However, it has been with the paper by Koutsoupias and Papadimitriou [9] that the notion of price of anarchy, as a worst case measure, took its form. The notion of price of stability has been later introduced in [1]. Since then, several papers have studied the price of anarchy and/or stability of different non-cooperative games.

1.3   Our Results

We first address the study of satisfiability games and show that they are congestion games, thus possessing pure Nash equilibria and the FIP property. We also show that CSG with player specific payoffs are instances of congestion games with player specific payoffs. Moreover, we prove that the problem of computing a fully mixed Nash equilibrium either for singleton and restricted CSG and DSG with player specific payoffs can be solved in polynomial time for the case in which each clause has at most two literals.

We then study separately the two classes of games. For CSG we prove that each game has an equivalent restricted CSG with player specific payoffs. As a consequence, we have that congestion games with player specific payoffs can be seen as the most general class of games. A similar result has been obtained by Monderer [14] through the definition of multipotential games. In particular, this means that any extension to a congestion game with player specific payoffs gives rise to a congestion game with player specific payoffs. Moreover, we show that each congestion game has an equivalent CSG, thus strengthening the equivalence between CSG and congestion games.

With respect to DSG, we first show that each cut game has an equivalent singleton DSG in which all clauses have exactly two literals. This gives evidence that fully mixed Nash equilibria can be computed in polynomial time for cut games. Then, we provide the characterization of the price of anarchy of unconstrained DSG by showing that it is exactly $1/2$ for both pure and mixed Nash equilibria. This generalizes the results known for cut games since singleton DSG are clearly also unconstrained and the equivalence among the two games holds also for their social functions.

Finally, as an application of our results, we define an extension of congestion games that we call congestion games with free riders and prove that they are still congestion games.

1.4 Paper Organization

In the next section we present general results, that is, results holding for all satisfiability games, or holding for both CSG and DSG. In Sections 3 and 4 we address CSG and DSG respectively, while in Section 5 we present an application of our results given by the definition of congestion games with free riders. Finally, in the last section we provide a final discussion of our achievements and open problems.

2 General Results

We start with a preliminary result showing that satisfiability games are instances of congestion games, thus possessing pure Nash equilibria and the FIP property.

Theorem 1. *Each satisfiability game is a congestion game.*

As a corollary we obtain the following results.

Corollary 1. *The price of stability of satisfiability games is 1 for both pure and mixed Nash equilibria.*

Corollary 2. *The problem of computing a Nash equilibrium for satisfiability games is in PLS for both pure and mixed strategies.*

We now show how the computation of a fully mixed Nash equilibrium can be performed in polynomial time for either singleton and restricted CSG and DSG with player specific payoffs in which the number of literals occurring in each clause is at most two.

Theorem 2. *The problem of computing a fully mixed Nash equilibrium for either singleton and restricted CSG and DSG with player specific payoffs in which each clause has at most two literals is solvable in polynomial time.*

3 Conjunctive Satisfiability Games

In this section we focus on CSG by analyzing, in particular, their generality in the sense that they can be used to effectively represent all other non-cooperative games. This result is quite intuitive since CSG can be thought as an alternative way to represent non-cooperative games in standard form.

Theorem 3. *Each non-cooperative game admits an equivalent restricted CSG with player specific payoffs whose clauses have exactly n literals.*

Since we have shown in the previous section that satisfiability games with player specific payoffs are also congestion games with player specific payoffs, we get the following interesting result.

Corollary 3. *The following three classes of games coincide:*

1. *Non-cooperative games,*
2. *(Restricted) CSG with player specific payoffs,*
3. *Congestion games with player specific payoffs.*

Thus, both the class of congestion games with player specific payoffs and that of CSG with player specific payoffs can be considered as a candidate to represent the class of all possible non-cooperative games. We now show that congestion games and CSG are further on tied, since their equivalence persists even when we do not allow player specific payoffs.

Theorem 4. *Each congestion game admits an equivalent CSG.*

Thus, we obtain the following corollaries.

Corollary 4. *The class of CSG coincides with that of congestion games.*

Corollary 5. *The problem of computing a pure Nash equilibrium for CSG is PLS-complete.*

Corollary 6. *There exists a polynomial time reduction from congestion games in which the maximum number of players sharing the same resource is constant to CSG.*

Before concluding the section, we observe that even for singleton CSG there may exist Nash equilibria σ yielding an unbounded price of anarchy for the social function γ.

Proposition 1. *The price of anarchy of singleton congestion games can be unbounded for the social function γ.*

A similar result was proved in [3] for the social functions $\gamma_1(\sigma) = \sum_{i \in P} \omega_i(\sigma)$ and $\gamma_2(\sigma) = \max_{i \in P} \omega_i(\sigma)$.

4   Disjunctive Satisfiability Games

We first show that the class of DSG includes that of cut games.

Theorem 5. *Each cut game admits an equivalent singleton DSG.*

Since the reduction of cut games to DSG gives life to clauses with exactly two literals, it follows that the problem of computing fully mixed Nash equilibria for cut games is solvable in polynomial time.

It is known that the price of anarchy for the cut game with respect to the social function γ is $\frac{1}{2}$ for pure Nash equilibria, since a pure Nash equilibrium can be seen as a solution of the local search algorithm for the MAX-CUT problem. We generalize such a result to DSG by obtaining the same bound on the function γ with respect to either pure and mixed Nash equilibria.

Theorem 6. *For any unconstrained DSG $\mathcal{G}$ it holds $\rho(\mathcal{G}, \gamma) \geq \frac{1}{2}$.*

We have shown a lower bound on the price of anarchy of mixed Nash equilibria for satisfiability games with respect to the social function γ. This clearly implies a lower bound also for pure Nash equilibria. On the other side, an upper bound on the price of anarchy of pure Nash equilibria implies an upper bound also for mixed ones. Thus, in order to show that the analysis carried out in the previous theorem is tight, we present a family of instances of singleton DSG whose price of anarchy is at most $\frac{1}{2}$ for the social function γ with respect to pure Nash equilibria.

Example 1. Given the set of variables $X = \{x_1, \ldots, x_n\}$, construct $\mathcal{C}$ as follows. There are n singleton clauses $C_{1i} = x_i$, for $1 \leq i \leq n$, and n clauses involving all the n variables $C_{ni} = (x_1, \ldots, \overline{x}_i, \ldots, x_n)$, for $1 \leq i \leq n$, in which only the ith variable appears negated. All clauses have the same weight that we assume to be equal to 1. The set X_i is equal to $\{x_i\}$. The state σ^* in which $\sigma_i^* = 1$ for every $1 \leq i \leq n$, satisfies all clauses; thus $\gamma(\sigma^*) = 2n$. The state σ in which $\sigma_i = 0$ for every $1 \leq i \leq n$, satisfies only the n clauses involving all variables; thus $\gamma(\sigma) = n$. State σ can be easily verified to be a pure Nash equilibrium. In fact, for every player i it holds $\omega_i(\sigma) = n$. Now, if player i changes his strategy from 0 to 1, he looses the contribution of clause C_{ni} and gain that of clause C_{1i}. Since the payoff stays the same for every player, σ is a pure Nash equilibrium and $\frac{\gamma(\sigma)}{\gamma(\sigma^*)} = \frac{1}{2}$.

From the above example and the result of Theorem 6, we obtain the following characterization of the price of anarchy of unconstrained DSG.

Theorem 7. *For unconstrained DSG it holds $\rho(\mathcal{G}, \gamma) = \frac{1}{2}$ even for unweighted clauses for both pure and mixed Nash equilibria.*

5   Congestion Games with Free Riders

We have proved that each game is a congestion game with player specific payoffs. This task was achieved by showing the equivalence between the class of non-cooperative games and that of CSG with player specific payoffs and the equivalence between the latter class with that of congestion games with player specific payoffs. We can prove that each DSG with player specific payoffs is indeed a congestion game with player specific payoffs in an alternative way by introducing the definition of a generalized model for congestion games which we call congestion games with free riders.

In this kind of games, resources can have two types of users. The first class of users includes players using resource e as in the classic definition of congestion games and achieving a contribution equal to $d_e(n_e(\cdot))$. The second class of users includes players, that we call free riders, which, at a first instance, only declare that they are willing to use resource e, but become real users (thus contributing in the definition of $n_e(\cdot)$ and receiving the value $d_e(n_e(\cdot))$ in their payoffs) only if the number of players using e (that is without considering the contribution of free riders) reaches a particular threshold t_e.

More formally, a congestion game with free riders $(P, E, S_{i \in P}, d_{e \in E}, t_{e \in E})$ is defined as follows. $(P, E, S_{i \in P}, d_{e \in E})$ is a traditional congestion game. But now each strategy $s_i = (e'_{s_i}, e''_{s_i}) \in S_i$ contains two sets of resources: the ones (e'_{s_i}) properly used by player i and the ones (e''_{s_i}) for which i is a free rider. Finally, for each $e \in E$, t_e is the threshold of e. Given a state σ, we define $n'_e(\sigma) = |\{i \in P | e \in e'_{\sigma_i}\}|$ as the number of players using resource e without counting free riders and

$$n_e(\sigma) = \begin{cases} |\{i \in P | e \in \sigma_i\}| & \text{if } n'_e(\sigma) \geq t_e, \\ n'_e(\sigma) & \text{otherwise.} \end{cases}$$

The payoff obtained by player i is now defined as $\omega_i(\sigma) = \sum_{e \in e'_{\sigma_i}} d_e(n_e(\sigma)) + \sum_{e \in e''_{\sigma_i} : n'_e(\sigma) \geq t_e} d_e(n_e(\sigma))$. It is easy to see that DSG are special instances of congestion games with free riders. It suffices setting $E = \mathcal{C}$. For each player i the mapping function becomes $g_i(\sigma_i) = (\bigcup_{C_j \in \mathcal{C}^i : C_j(\sigma_i) = 1} e_j, \bigcup_{C_j \in \mathcal{C}^i : C_j(\sigma_i) = 0} e_j)$. For each $e \in E$, we set $t_e = 1$, $d_e(0) = 0$ and $d_e(k) = c_e$ for any $k \geq 1$.

We now show that the class of congestion games and that of congestion games with free riders coincide.

Theorem 8. *For each congestion game with free riders there is an equivalent congestion game and viceversa.*

Thus, we have that congestion games with free riders are congestion games. This means that congestion games with free riders and player specific payoffs are congestion games with player specific payoffs. Since DSG with player specific payoffs are congestion games with free riders with player specific payoffs, we have shown that they are also congestion games with player specific payoffs.

6   Conclusions

We have introduced and studied satisfiability games. A special attention has been devoted to their expressiveness and their equivalence with congestion games. Congestion games, in particular, is a well-known and studied class of games, with some nice properties making it particular attracting and indicated to be chosen as the representative class of all non-cooperative games. Thus, in an ideal process of classification of games, one can define new and interesting classes of games as a particular specialization of congestion games with player specific payoffs obtained by considering, for instance, monotonic delay functions, continuous delays functions, monotonic and continuous delay functions, linear delay functions, and so on. Other changes can be applied to the definition itself of congestion games. For example, one can define the payoff achieved by a player as the maximum (and no longer as the sum) of the delays experienced on the chosen resources. We have proved that this does not enrich the expressiveness of congestion games, but however, it can make them more malleable and allow simpler and useful reductions. Up to now we know that the class of congestion games can be partitioned in that of congestion games with player specific payoffs and that of proper congestion games. We know that this latter class of games is also characterized by the fact that it contains all and only the games admitting an exact potential. What about games admitting a weighted potential? Are there other well characterized classes of congestion games? For example, how about weighted (linear) congestion games?

References

1. E. Anshelevich, A. Dasgupta, E. Tardos, and T. Wexler. Near-Optimal Network Design with Selfish Agents. In *Proceedings of STOC 2003*. ACM Press, pp. 511-520, 2003.
2. X. Chen and X. Deng. Settling the Complexity of 2-Player Nash Equilibrium. In *Proceedings of FOCS 2006*. IEEE Computer Society, pp. 261.252, 2006.
3. G. Christodoulou, E. Koutsoupias, and A. Nanavati. Coordination Mechanisms. In *Proceedings of ICALP 2004*, Volume 3142 of LNCS. Springer, pp. 345-357, 2004.
4. G. Christodolou, V. S. Mirrokni, and A. Sidiropolous. Convergence and Approximation in Potential Games. In *Proceedings of STACS 2006*, Volume 3884 of LNCS. Springer, pp. 349-360, 2006.
5. K. Daskalakis, P. W. Goldberg, and C. H. Papadimitriou. The Complexity of Computing a Nash Equilibrium. In *Proceedings of STOC 2006*. ACM Press, pp. 71-78, 2006.
6. K. Daskalakis and C. H. Papadimitriou. Three-Player Games Are Hard *Electronic Colloquium on Computational Complexity (ECCC)*, (139), 2005.
7. A. Fabrikant, C. H. Papadimitriou, K. Talwar. The Complexity of Pure Nash Equilibria. In *Proceedings of STOC 2004*, ACM Press, pp. 604-612, 2004.
8. Y. Korilis and A. Lazar. On the Existence of Equilibria in Non-cooperative Optimal Flow Control. *Journal of the ACM*, 42(3):584,613, 1995.
9. E. Koutsoupias and C. H. Papadimitriou. Worst-case Equilibria. In *Proceedings of STACS 1999*. Volume 1653 of LNCS. Springer, pp. 404-413, 1999.
10. R. La and V. Anantharam. Optimal Routing Control: Game Theoretic Approach. In Proceedings of the 1997 CDC Conference, 1997.
11. I. Milchtaich. Congestion Games with Player Specific Payoff Functions. *Games and Economic Behavior*. 13:111-124, 1996.
12. V. S. Mirrokni. Approximation Algorithms for Distributed and Selfish Agents. Massachusetts Institute of Technology, June, 2005.
13. V. S. Mirrokni and A. Vetta. Convergence Issues in Competitive Games. In *Proceedings of APPROX 2004*. Volume 3122 of LNCS. Springer, 183-194, 2004.
14. D. Monderer. Multipotential Games. In *Proceedings of IJCAI 2007*, to appear.
15. D. Monderer and L. S. Shapley. Potential Games. *Games and Economic Behavior*. 14:124-143, 1996.
16. J. Nash. Equilibrium Points in *n*-person Games. In *Proceedings of the National Academy of Sciences*, volume 36, pp. 48-49, 1950.
17. J. Nash. Non-cooperative Games. *Annals of Mathematics*, 54(2):286-295, 1951.
18. R. W. Rosenthal. A Class of Games Possessing Pure-Strategy Nash Equilibria. *International Journal of Game Theory*, 2:65-67, 1973.
19. S. J. Shenker. Making Greed Work in Networks: a Game-Theoretic Analysis of Switch Service Disciplines. *IEEE/ACM Transactions on Networking*, 3(6):819-831, Dec. 1995.

The Complexity of Algorithms Computing Game Trees on Random Assignments

ChenGuang Liu* and Kazuyuki Tanaka

Mathematical Institute, Tohoku University, Sendai 980-8578, Japan
liu@mail.tains.tohoku.ac.jp, tanaka@math.tohoku.ac.jp

Abstract. The complexity of algorithms for computing game trees on random assignments has been given substantial attention in the literature. In this line, we investigate the complexity of algorithms that compute a special class of game trees T_2^k from a new perspective — eigen-distribution. This particular distribution is defined as the worst distribution on assignments to variables of T_2^k regarding a best algorithm. In this paper, we show the eigen-distribution on assignments for T_2^k in two separate cases, where the assignments to leaves are independently distributed (ID) and correlated distributed(CD). Then we use eigen-distribution to derive the tight bound of the complexity of algorithms for T_2^k.

1 Introduction

In the present work, we are interested in the full binary AND-OR tree T_2^k, where the subscript 2 means "binary", and k is the number of rounds (one level AND gate followed by one level OR gate). This sort of tree can be viewed as a 0-1 game between two players, the 0-player and the 1-player, played as follows: The 0-player starts the game by moving down through the left or right child of the root. In general, at each internal node, the player that corresponds to its label ($\wedge$ for 0-player, and $\vee$ for 1-player) picks a child of the node. The goal of each player is to reach a leaf that is labeled with its name, that is, "0" for 0-player and "1" for 1-player. It is easy to see that,

Fact 1. *Given a full binary AND-OR tree T_2^k, 0-player wins the corresponding 0-1 game if and only if T_2^k evaluates to 0.*

By *computing a tree* T_2^k, we mean evaluating the value of the root of T_2^k, finding whether the 0-player wins or loses in running the corresponding 0-1 game. At the beginning of computing, the leaves are assigned with $0's$ and $1's$, but are "covered" so that one cannot see how they are labeled. In computing, a basic step consists of querying the value of one of the leaves to find whether it is labeled 0 or 1, and the operations repeat until the value of the root can be determined. The cost/complexity associated with this computation is only the number of leaves queried by the algorithm; all the other computations are cost free.

---

* Corresponding author.

M.-Y. Kao and X.-Y. Li (Eds.): AAIM 2007, LNCS 4508, pp. 241–250, 2007.

Over the years a number of game tree algorithms have been invented. Among them, the Alpha-Beta pruning algorithm has been proven quite successful, and no other algorithm has achieved such a wide-spread use in practical applications as the Alpha-Beta pruning algorithm. A precise formulation of the Alpha-Beta pruning algorithm can be found in [3]. The relation between the Alpha-Beta pruning algorithm and the two other well-known game tree algorithms PVS, and SSS^* was investigated in [1]. For the purposes of our discussion, we restrict ourselves to the Alpha-Beta pruning algorithm as done in [4,5,6,8].

In the literature, the tree T_2^k on the random assignment has been extensively investigated [2,4,8,11]. A traditional analytic model for T_2^k evaluation is one in which the assignment to the leaves is independently and identically distributed (IID). Assuming each leaf receives a 0 with probability p, Pearl[4] showed that, when $k \approx \infty$, the value of the root of tree T_2^k is almost a sure 0 or a sure 1, depending on whether p is higher or lower than some fixed-point probability $p = \frac{\sqrt{5}-1}{2}$. Later, Pearl[5] also proved that, when $p = \frac{\sqrt{5}-1}{2}$, increasing k by one extra round would increase the computational complexity by a factor $\Re = \frac{\sqrt{5}+3}{2}$. At the same time, Tarsi[8] showed that this branching factor is optimal.

For T_2^k, the most well-known result in the global distribution showed that,

Theorem 1 (Saks and Wighderson[6]). *For tree T_2^k on n variables, the randomized complexity is given by*

$$R(T_2^k) = \Theta(n^{\log_2(\frac{1+\sqrt{33}}{4})}).$$

Moreover, Saks and Wigderson[6] conjectured that this is the largest possible gap between the deterministic complexity and the randomized complexity for any Boolean decision tree. This conjecture is still wide open at the moment.

Besides the randomness on the assignments investigated in the above works, Yao[11] also observed another kind of randomness inside the algorithm itself, and constructed a bridge between these two randomness. Yao showed that the well-known Minimax theorem by von Neumann[10] implies that the distributional complexity is a lower bound on the complexity of randomized algorithms that compute the same tree. Moreover, Yao stated that the case for Las Vegas complexity is universal. That is,

Theorem 2 (Yao's Principle[11]). *In computing any tree T (including T_2^k), we have*

$$R(T) = P(T).$$

where $P(T)$ (resp. $R(T)$) is the distributional (resp. randomized) complexity for tree T.

Yao's Principle is a useful and valid tool in randomized complexity analysis. Since its introduction, it has been extensively studied in the literature. In applying Yao's Principle, a key step is to compute the distributional complexity of tree. However, no effective computing method for distributional complexity has been reported at the moment. One of the main motivations for doing the present work is to prepare for developing the effective techniques to compute the distributional complexity.

In this paper, we investigate the complexity of algorithm for tree T_2^k on the random assignment from the eigen-distribution perspective. We are interested in two separate cases: 1) the assignments to leaves are independently distributed; and 2) the assignments to leaves are correlated.

2 Definitions and Notations

Let n be the number of leaves of tree T_2^k, i.e., $n = 2^{2k}$. Let A_D be a deterministic algorithm to compute T_2^k and $\omega = \omega_1\omega_2\cdots\omega_n$ an assignment to the leaves $\{l_1, l_2, \cdots, l_n\}$ of tree T_2^k. By $C(A_D, \omega)$, we denote the number of leaves queried by A_D computing T_2^k on ω. Let $\mathcal{W}$ is the set of assignments, and p_ω^d the probability of ω over $\mathcal{W}$ with respect to distribution d. The average complexity $C(A_D, d)$ of a deterministic algorithm A_D on assignments with distribution d is defined by

$$C(A_D, d) = \sum_{\omega \in \mathcal{W}} p_\omega^d C(A_D, \omega).$$

Let $\mathcal{D}$ be the set of distributions, and $\mathcal{A}_\mathcal{D}(T_2^k)$ the set of deterministic algorithms computing tree T_2^k. The **distributional complexity** $P(T_2^k)$ computing tree T_2^k is defined by

$$P(T_2^k) = \max_{d \in \mathcal{D}} \min_{A_D \in \mathcal{A}_\mathcal{D}(T_2^k)} C(A_D, d).$$

Among the set $\mathcal{D}$, the special distribution δ such that $P(T_2^k)$ is achieved, that is,

$$\min_{A_D \in \mathcal{A}_\mathcal{D}(T_2^k)} C(A_D, \delta) = P(T_2^k),$$

is called an **eigen-distribution** on assignments for tree T_2^k in this paper.

When the assignments to leaves are restricted in the IID case, by $\hat{P}(T_2^k)$ we denote the corresponding distributional complexity. It is clear that $\hat{P}(T_2^k) \leq P(T_2^k)$. In the IID case, since each leaf is assigned a 0 with the same probability, we may analyze the eigen-distribution by studying the probability of a single leaf receiving a 0. By ϱ, we denote the probability of a leaf receiving a 0 in the eigen-distribution on assignments for T_2^k in the IID case, and call it **distributional probability** throughout this paper.

A randomized algorithm, denoted by A_R, is a probability distribution over that family of deterministic algorithms. For an assignment ω and a randomized algorithm A_R that has probability p_{A_D} to proceed exactly as a deterministic algorithm A_D, the cost of A_R on ω, $C(A_R, \omega)$, is defined as the expected number of the queried leaves:

$$C(A_R, \omega) = \sum_{A_D \in \mathcal{A}_\mathcal{D}(T_2^k)} P_{A_D} C(A_D, \omega).$$

We denote by $\mathcal{A}_R(T_2^k)$ the family of randomized algorithms computing T_2^k. For tree T_2^k, the **randomized complexity** $R(T_2^k)$ is defined by:

$$R(T_2^k) = \min_{A_R \in \mathcal{A}_R(T_2^k)} \max_{\omega \in \mathcal{W}} C(A_R, \omega).$$

3   The Complexity in the ID Case

For a tree T_2^k, a natural distribution on assignments to choose is the independent distribution. Hence, we first investigate the eigen-distribution in the case where the assignments to leaves are independently distributed (ID).

The left and right sub-trees of T (a simple notation for T_2^k) are designated as T_L and T_R, respectively. Let $\mu^d(T)$ denote the minimum expected number of the queried variables over all deterministic algorithms computing T with respect to distribution d, and $\mu^d(T_L)$ (resp. $\mu^d(T_R)$) be the analogous quantity for the left sub-tree T_L (resp. the right sub-tree T_R). By $p_{(T,d)}$ we denote the probability of returning a 0 at the root of T in the distribution d, and $p^L_{(T,d)}$ (resp. $p^R_{(T,d)}$) be the analogous quantity for the left sub-tree T_L (resp. the right sub-tree T_R). Then, the $\mu^d(T)$ and $p_{(T,d)}$ for T on the assignments that are independently distributed can be computed by the following recursive equations:

$$\begin{cases} \mu^d(T) = \min\{\mu^d(T_L) + p^L_{(T,d)} \cdot \mu^d(T_R), \mu^d(T_R) + p^R_{(T,d)} \cdot \mu^d(T_L)\} \\ p_{(T,d)} = p^L_{(T,d)} \cdot p^R_{(T,d)} \end{cases}$$

if the root of T is labeled by $\vee$, and

$$\begin{cases} \mu^d(T) = \min\{\mu^d(T_L) + \mu^d(T_R) - p^L_{(T,d)} \cdot \mu^d(T_R), \\ \qquad\qquad \mu^d(T_L) + \mu^d(T_R) - p^R_{(T,d)} \cdot \mu^d(T_L)\} \\ p_{(T,d)} = 1 - [(1 - p^L_{(T,d)}) \cdot (1 - p^R_{(T,d)})] \end{cases}$$

if the root of T is labeled by $\wedge$, with the initial conditions

$$\begin{cases} \mu^d(T) = 1 \\ p_{(T,d)} \in [0, 1] \end{cases}$$

if T contains a single node.

From this inductive computation, it is not hard to see that:

Proposition 1. *For any tree T_2^k on assignments that are independently distributed, the $\max_d \mu^d(T_2^k)$ is achieved only if the assignments are also identically distributed.*

Therefore, in investigating the eigen-distribution in the ID case, the work can be restricted in the special IID case. In the following, we focus on the probability of assignment for a single leaf in replace of the distribution d on assignments to all leaves. By p_k, we denote the probability of returning a 0 at a node labeled $\wedge$ at the k-th round of tree T_2^k, and μ_k^p the minimum expected number of leaves evaluated to this over all deterministic algorithms with respect to the probability p that each leave is set to 0. Set $p \in [0, 1]$. Then, in the IID case, we obtain the recurrence

$$\begin{cases} p_k = -p_{k-1}^4 + 2p_{k-1}^2 \\ \mu_k^p = (-p_{k-1}^3 - p_{k-1}^2 + 2p_{k-1} + 2) \times \mu_{k-1}^p \end{cases}$$

with initial conditions

$$\begin{cases} p_0 = p \\ \mu_0^p = 1 \end{cases}$$

Clearly, by the definition of distributional complexity, the distributional complexity computing T_2^k in the IID case is given by $\hat{P}(T_2^k) = \max_{p \in [0,1]} \mu_k^p$. Moreover, it is not hard to see that, this recurrence and Fact 1 provide a simpler proof for the results of Pearl[4]. Here, we use this recurrence to prove that

Theorem 3. *In the IID case, for any tree T_2^k on n leaves, we have*

(1) The distributional probability $\varrho \in [\frac{\sqrt{7}-1}{3}, \frac{\sqrt{5}-1}{2}]$, and ϱ is an strictly increasing function on round $k \in [1, \infty)$;

(2) The distributional complexity $\hat{P}(T_2^k) = \Theta(n^{\log_2 \frac{1+\sqrt{5}}{2}})$.

Proof. We sketch the proof here. Set $n^{\lambda_k^p} = \mu_k^p$. Since $n = 2^{2k} = 4^k$ for tree T_2^k, we obtain $\lambda_k^p = \frac{\log_4 \mu_k^p}{k}$. Then, we have

$$\begin{cases} \lambda_1^p = \lambda_2^p = \lambda_3^p = \cdots = \lambda_\infty^p = \frac{1}{2} & \text{for } p = 0 \\ \log_4 \frac{34+14\sqrt{7}}{27} > \lambda_1^p > \lambda_2^p > \lambda_3^p > \cdots > \lambda_\infty^p > \frac{1}{2} & \text{for } p \in (0, \frac{\sqrt{7}-1}{3}) \\ \log_4 \frac{34+14\sqrt{7}}{27} = \lambda_1^p > \lambda_2^p > \lambda_3^p > \cdots > \lambda_\infty^p > \frac{1}{2} & \text{for } p = \frac{\sqrt{7}-1}{3} \\ \begin{cases} \log_2 \frac{1+\sqrt{5}}{2} < \lambda_1^p < \cdots < \lambda_{m-1}^p < \lambda_m^p < \log_4 \frac{34+14\sqrt{7}}{27}, \\ \log_2 \frac{1+\sqrt{5}}{2} \leq \lambda_n^p < \cdots < \lambda_{m+1}^p < \lambda_m^p < \log_4 \frac{34+14\sqrt{7}}{27}, \\ \log_2 \frac{1+\sqrt{5}}{2} > \lambda_{n+1}^p > \lambda_{n+2}^p > \lambda_{n+3}^p > \cdots > \lambda_\infty^p > \frac{1}{2}, \\ \text{for some } n > m > 1. \end{cases} & \text{for } p \in (\frac{\sqrt{7}-1}{3}, \frac{\sqrt{5}-1}{2}) \\ \lambda_1^p = \lambda_2^p = \lambda_3^p = \cdots = \lambda_\infty^p = \log_2 \frac{1+\sqrt{5}}{2} & \text{for } p = \frac{\sqrt{5}-1}{2} \\ \log_2 \frac{1+\sqrt{5}}{2} > \lambda_1^p > \lambda_2^p > \lambda_3^p > \cdots > \lambda_\infty^p > \frac{1}{2} & \text{for } p \in (\frac{\sqrt{5}-1}{2}, 1) \\ \lambda_1^p = \lambda_2^p = \lambda_3^p = \cdots = \lambda_\infty^p = \frac{1}{2} & \text{for } p = 1 \end{cases}$$

Let $\lambda_k = \max_{p \in [0,1]} \{\lambda_k^p\}$. To sum up, we have

$$\begin{cases} \lambda_k^P = \log_4 \frac{34+14\sqrt{7}}{27} & \text{for } p = \frac{\sqrt{7}-1}{3}, \ k = 1 \\ \lambda_k^P \in (\frac{1}{2}, \log_4 \frac{34+14\sqrt{7}}{27}) & \text{for } p = (\frac{\sqrt{7}-1}{3}, \frac{\sqrt{5}-1}{2}), \ k \geq 1 \\ \lambda_k^P = \log_2 \frac{1+\sqrt{5}}{2} & \text{for } p = \frac{\sqrt{5}-1}{2}, \ k \geq 1 \end{cases}$$

Since $\max_{p \in [0, \frac{\sqrt{7}-1}{3}]} \lambda_k^p = \lambda_k^{\frac{\sqrt{7}-1}{3}}$, then we obtain the distributional probability $\varrho \in [\frac{\sqrt{7}-1}{3}, \frac{\sqrt{5}-1}{2}]$ and the distributional complexity $\hat{P}(T_2^k) = \Theta(n^{\log_2 \frac{1+\sqrt{5}}{2}})$. $\square$

However, the result of $\hat{P}(T_2^k) = \Omega(n^{log_4 \frac{3+\sqrt{5}}{2}})$ proved in the above theorem is slightly less than $R(T_2^k) = \Theta(n^{log_2(1+\sqrt{33})/4})$ proposed by Saks and Wigderson[6]. This fact doesn't meet Yao's Principle. The reason lies in that the distribution in the IID case is not the worst case over the global distribution. In the next section, we will investigate the eigen-distribution in the case where the assignments to leaves are correlated. The worst case in the global distribution comes from it.

4   The Complexity in the CD Case

In this section, we first introduce a reverse assigning technique to form two particular sets of assignments, called 1-set and 0-set, where the assignments to leaves are highly correlated. By E^1-*distribution* (resp. E^0-*distribution*), we denote the distribution on assignments of 1-set (resp. 0-set) such that the complexity of any deterministic algorithm is equal. Then, we prove that E^1-distribution is the unique eigen-distribution for tree T_2^k.

By $\wedge^z$ (resp. $\vee^z$) we denote the result z of operation AND (resp. OR) on two Boolean variables x, y, where $x, y, z \in \{0, 1\}$. In generally, we have Table 1:

Table 1. The general operation on two variables

operation result	variable x,	variable y	operation result	variable x,	variable y
$\wedge^1$	1	1	$\vee^0$	0	0
$\wedge^0$	1	0	$\vee^1$	0	1
$\wedge^0$	0	1	$\vee^1$	1	0
$\wedge^0$	0	0	$\vee^1$	1	1

In the above table, we remove the fourth row, that is, we preclude the possibility of both inputs to a $\wedge$ being 0 and to a $\vee$ being 1. Then, we obtain the following modified version of Table 1:

Table 2. The modified operation on two variables

operation result	variable x,	variable y	operation result	variable x,	variable y
$\wedge^1$	1	1	$\vee^0$	0	0
$\wedge^0$	1	0	$\vee^1$	0	1
$\wedge^0$	0	1	$\vee^1$	1	0

With respect to the modified operations, we propose a reverse assigning technique (RAT) as follows:

Methodology 1 (Reverse assigning technique)
   1) Assign a 1 (resp. 0) to the root of tree T_2^k.
   2) From the root to the leaves, assign a 0 or 1 to each child of any internal node as follows until all of the leaves were assigned:

- *for the node labeled ∧ with value 1, assign 1s to all its children;*
- *for the node labeled ∨ with value 0, assign 0s to all its children;*
- *for the node labeled ∧ with value 0, assign at random a 0 to one of its children and a 1 to the other one;*
- *for the node labeled ∨ with value 1, assign at random a 1 to one of its children and a 0 to the other one.*

3) Form the 1-set (resp. 0-set) by collecting all possible assignments.

Following from this technique RAT, we can form a special class of sets of assignments with highly correlated properties. For tree T_2^1, the 1-set and 0-set of assignments are as follows. By the definitions of E^1-distribution and E^0-distribution, it is not hard to see that, in E^1-distribution and E^0-distribution, the probability of each class of assignments is 0.25 (see Fig. 1).

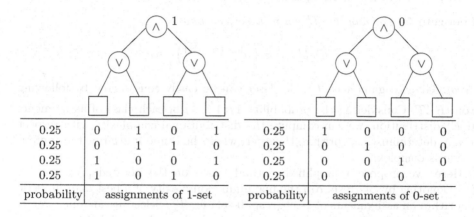

probability	assignments of 1-set				probability	assignments of 0-set			
0.25	0	1	0	1	0.25	1	0	0	0
0.25	0	1	1	0	0.25	0	1	0	0
0.25	1	0	0	1	0.25	0	0	1	0
0.25	1	0	1	0	0.25	0	0	0	1

Fig. 1. The 1-set and 0-set for tree T_2^1

By w_i^h we denote the i-th assignment of h-set in this figure. A little more consideration shows that,

Proposition 2. *For 1-set and 0-set formed by applying RAT on tree T_2^k, the following propositions hold:*

1) all assignments of these two set can be constructed by combining the assignment $w_1^1, w_2^1, w_3^1, w_4^1$ and $w_1^0, w_2^0, w_3^0, w_4^0$.

2) in E^1-distribution(or E^0-distribution), the probability of each assignment from 1-set (or 0-set) is equal to $1/(4^{\frac{4^k-1}{3}})$.

For the E^1-distribution on assignments of 1-set formed by following RAT, we have,

Theorem 4. *For any tree T_2^k, the E^1-distribution is the unique eigen-distribution.*

Proof. Following from the RAT, we can see that, for any deterministic algorithm that computes T_2^k, the worst assignments must be included in the assignments of 1-set or 0-set. Moreover, any assignment not included in them is not the worst assignment. For any tree T_2^k(one level AND gate followed by one level OR gate), the computation cost in E^1-distribution is larger than that in E^0-distribution. By the definition of distributional complexity, it is clear that the eigen-distribution can not be equal to E^0-distribution for any tree T_2^k. Therefore, for any tree T_2^k, the assignments such that the distributional complexity is achieved must be in 1-set.

For a tree T_2^k, supposing a distribution on assignments of 1-set that is different from the E^1-distribution be the eigen-distribution, in such distribution, we can easily find a deterministic algorithm such that the computational complexity is less than that in the E^1-distribution. Hence, this supposed distribution is not the worst one, which contradicts with the definition of eigen-distribution.   □

In such a special distribution, the distributional complexity is:

Theorem 5. *For any tree T_2^k on n leaves, we have*

$$P(T_2^k) = \Theta\left(n^{\log_2\left(\frac{1+\sqrt{33}}{4}\right)}\right).$$

Proof. Given a game tree T_2^k, the 1-set can be easily constructed by following from RAT. Considering the probability $1/(4^{\frac{4^k-1}{3}})$ of each class of assignments in E^1-distribution, we can compute the distributional complexity with respect to any deterministic algorithm. However, when the round k is large, the computation is complex.

Here, we propose a simplified method based on the iterated properties of T_2^k. Denoting by β_k (resp. α_k) the maximum complexity of a best deterministic algorithm on assignments of 1-set (resp. 0-set) for T_2^k. Select at random a deterministic algorithm, e.g., the algorithm that reads the nodes from left to right. To get a recurrence equation for β_k and α_k, we associate a 1 (resp. 0) occurring in the assignments of 1-set and 0-set for T_2^1 with the β_{k-1} (resp. α_{k-1}) for T_2^k. See Fig.2, where the leaves corresponding β_{k-1} and α_{k-1} with underline are queried by the algorithm. Considering the fact that the probability of each assignment in E^1-distribution and E^0-distribution for T_2^1 is 0.25, we have the recurrence

$$\begin{aligned}
\beta_k &= \tfrac{1}{4}\cdot[\alpha_{k-1}+\beta_{k-1}+\alpha_{k-1}+\beta_{k-1}] & \alpha_k &= \tfrac{1}{4}\cdot[\beta_{k-1}+\alpha_{k-1}+\alpha_{k-1}] \\
&\quad+\tfrac{1}{4}\cdot[\alpha_{k-1}+\beta_{k-1}+\beta_{k-1}] & &\quad+\tfrac{1}{4}\cdot[\alpha_{k-1}+\beta_{k-1}+\alpha_{k-1}+\alpha_{k-1}] \\
&\quad+\tfrac{1}{4}\cdot[\beta_{k-1}+\alpha_{k-1}+\beta_{k-1}] & &\quad+\tfrac{1}{4}\cdot[\alpha_{k-1}+\alpha_{k-1}] \\
&\quad+\tfrac{1}{4}\cdot[\beta_{k-1}+\beta_{k-1}] & &\quad+\tfrac{1}{4}\cdot[\alpha_{k-1}+\alpha_{k-1}] \\
&= \tfrac{1}{4}\cdot[4\cdot\alpha_{k-1}+8\cdot\beta_{k-1}] & &= \tfrac{1}{4}\cdot[9\cdot\alpha_{k-1}+2\cdot\beta_{k-1}] \\
&= \alpha_{k-1}+2\beta_{k-1} & &= \tfrac{9}{4}\alpha_{k-1}+\tfrac{1}{2}\beta_{k-1}
\end{aligned}$$

To sum up, we have the recurrence

$$\begin{cases} \alpha_k = \tfrac{9}{4}\alpha_{k-1}+\tfrac{1}{2}\beta_{k-1} \\ \beta_k = \alpha_{k-1}+2\beta_{k-1} \end{cases}$$

Clearly, the initial conditions $\alpha_0 = \beta_0 = 1$.

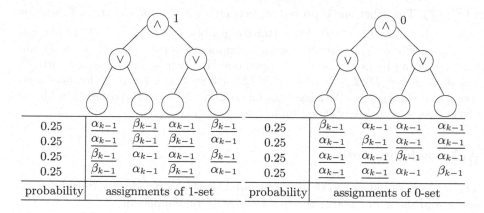

probability	assignments of 1-set				probability	assignments of 0-set			
0.25	α_{k-1}	β_{k-1}	α_{k-1}	β_{k-1}	0.25	β_{k-1}	α_{k-1}	α_{k-1}	α_{k-1}
0.25	α_{k-1}	β_{k-1}	β_{k-1}	α_{k-1}	0.25	α_{k-1}	β_{k-1}	α_{k-1}	α_{k-1}
0.25	β_{k-1}	α_{k-1}	α_{k-1}	β_{k-1}	0.25	α_{k-1}	α_{k-1}	β_{k-1}	α_{k-1}
0.25	β_{k-1}	α_{k-1}	β_{k-1}	α_{k-1}	0.25	α_{k-1}	α_{k-1}	α_{k-1}	β_{k-1}

Fig. 2. The 1-set and 0-set for tree T_2^k

Solving this recurrence, we obtain $\alpha_k = \beta_k = \Theta\left(n^{\log_2(\frac{1+\sqrt{33}}{4})}\right)$. By the definition of the distributional complexity, we have $P(T_2^k) = \Theta\left(n^{\log_2(\frac{1+\sqrt{33}}{4})}\right)$. □

This well-known result was first proved by Saks and Wigderson[6] in another way, but our proof based on eigen-distribution seems to be much simpler than their original.

In the CD cases, we have $P(T_2^k) = R(T_2^k)$, which meets the famous Yao's principle. This result shows that the E^1-distribution on assignments of 1-set formed by RAT is the worst distribution over all possible cases for tree T_2^k.

5   Conclusions

In this paper, we have investigated the eigen-distribution on random assignments to leaves for a tree T_2^k in the ID and CD cases. Furthermore, we used eigen-distribution deriving the tight bounds of the complexity of algorithms for T_2^k in the two cases, separately. The results show that the way based on eigen-distribution to analyze the complexity of algorithms for T_2^k is valid and effective. A new work along this paper is to study the eigen-distribution for general game trees, including not only the uniform game trees but also non-uniform game trees. This work will appear in the future literature.

Moreover, the present work is strictly limited to the Las Vegas case. In the Las Vegas case, Yao[11] showed that for any tree T_2^k, $R(T_2^k) = P(T_2^k)$. By $R^\epsilon(T_2^k)$ (resp. $P^\epsilon(T_2^k)$) we denote the randomized complexity (resp. distributional complexity) in the Monte Carlo case, where ϵ is the error probability at most. Corresponding to the duality theorem $R(T_2^k) = P(T_2^k)$ in the Las Vegas case, an obvious question whether there exists the analogous equation of $R^\epsilon(T_2^k) = P^\epsilon(T_2^k)$ is raised. In the literature, there are two results associated with this problem. The one is due to Yao[11]: for any tree T_2^k, and for $\epsilon \in [0, \frac{1}{2}]$, $R^\epsilon(T_2^k) \geq$

$\frac{1}{2}P^{2\epsilon}(T_2^k)$. The other one is proved by Vereshchagin [9]: for any tree T_2^k, and for any $\epsilon \in [0, 1]$, $R^{\epsilon}(T_2^k) \leq 2P^{\frac{\epsilon}{2}}(T_2^k)$. But the problem of $R^{\epsilon}(T_2^k) \overset{?}{=} P^{\epsilon}(T_2^k)$ is still wide open. If the eigen-distribution on assignments to leaves for T_2^k in the Monte Carlo case can be ascertained, this open question can be solved since Santha[7] has proved that $R^{\epsilon}(T_2^k) = (1 - \delta\epsilon)R(T_2^k)$, where δ is 1 (resp. 2) for one-sided (resp. two-sided) error. We suggest to investigate the eigen-distribution in Monte Carlo case as a future research topic.

Acknowledgments

The authors would like to thank Professor Takeshi Yamazaki for fruitful discussions, and the anonymous referees for many useful comments.

References

1. Bruin, A. de, Pijls, W., Plaat, A.: Solution Tress as a Basis for Game Tree Search. Technical Report EUR-CS-94-04, Department Computer Science, Erasmus University Rotterdam. (1994)
2. Karp, R., Zhang, Y.: Bounded Branching Process and AND/OR Tree Evaluation. Random Structures Algorithms. **7**(1995) 97-116
3. Knuth, D.E., Moore, R. W.: An analysis of alpha-beta pruning. Artificial Intelligence. **6** (1975) 293-326
4. Pearl, J.: Asymptotic Properties of Minimax Tress and Game-Searching Procedures. Artificial Intelligence. **14**(1980) 113-138
5. Pearl, J.: The Solution for the Branching Factor of the Alpha-Beta Pruning Algorithm and its Optimality. Communications of the ACM. **25**(1982) 559-564
6. Saks M., Wigderson A.: Probabilistic Boolean Decision Trees and the Complexity of Evaluating Game Trees. In Proceedings of 27th Annual IEEE Symposium on Foundations of Computer science(FOCS). (1986) 29-38
7. Santha M.: On the Monte Carlo Boolean Decision Tree Complexity of Read-once Formulae. In Proceedings oF 6th Annual Conference on Structrure in Complexity Theory. (1991) 180-187
8. Tarsi M.: Optimal Search on Some Game Trees. Journal of the ACM. **30**(1983) 389-396
9. Vereshchagin N.: Randomized Boolean Decision Tress: Several Remarks. Theoretical Computer Science. **207**(1998) 329-342
10. von Neumann J.: Zur Theorie der Gesellschaftsspiele. Mathematische Annalen. **100**(1928) 295-320
11. Yao, A. C.-C.: Probabilistic Computations: Towards a Unified Measure of Complexity. In Proceedings of 18th Annual IEEE Symposium on Foundations of Computer science(FOCS). (1977) 222-227

An Efficient, and Fast Convergent Algorithm for Barrier Options

Tian-Shyr Dai[1,*] and Yuh-Dauh Lyuu[2]

[1] Department of Information and Finance Management, National Chiao-Tung
University, 1001 Ta Hsueh Road, Hsinchu, Taiwan 300, ROC
d88006@csie.ntu.edu.tw
[2] Department of Computer Science
Information Engineering, National Taiwan University

Abstract. A barrier option is an option whose payoff depends on
whether the price path of the underlying asset ever reaches certain predetermined price levels called the barriers. A single- (double-) barrier
option is a barrier option with one (two, respectively) barrier(s). No
simple and exact closed-form pricing formula for double-barrier options
has been reported in the literature. This paper proposes a novel tree
model that can price both single- and double-barrier options efficiently
and accurately. This tree model achieves the high efficiency by combinatorial techniques and numerical accuracy by hitting the barriers exactly.
Numerical experiments are given to verify the superiority of our method.

Keywords: barrier option, combinatorics, option pricing, tree.

1 Introduction

A barrier option is an option whose payoff depends on whether the stock price
reaches a certain predetermined level (the barrier) before the maturity date.
A double- and the single-barrier options are barrier options with two and one
barriers, respectively. A knock-in barrier option comes into existence if the stock
price reaches the barrier(s) before the maturity date, while a knock-out one
ceases to exist if the stock price reaches the barrier(s) before the maturity.

When the payoff functions for the single barrier options follow some standard
forms, analytical pricing formulas are derived in [8]. The valuation of double-
barrier options is discussed in [5,11,6]. However, there are no simple, exact closed-
form formulas for these options. The pricing formula can be expressed as an
infinite series of cumulative normal distributions. Although truncation of this
series is necessary numerically, it can lead to large pricing errors [6].

A tree model is popular for pricing barrier options. This research is interesting
since it can price barrier options with nonstandard payoff functions, such as
power payoff functions. However, there may not be closed-form formulas for
such options. A tree divides the time span from now to the option's maturity

---

* He was supported by NSC grant 94-2213-E-033-024.

M.-Y. Kao and X.-Y. Li (Eds.): AAIM 2007, LNCS 4508, pp. 251–261, 2007.

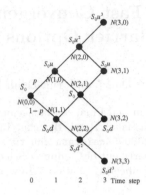

Fig. 1. The CRR Tree

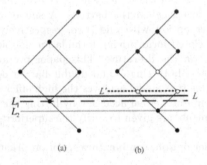

Fig. 2. Pricing a Single-Barrier Option by the CRR Tree Model. The barrier is denoted by L. (a) The effective barrier is L_1. (b) The effective barrier is L_2.

into n time steps and simulates the stock price discretely at each step. Take the 3-time-step CRR tree [2] illustrated in Fig. 1 as an example. Let the stock price at time step 0 be S_0. From an arbitrary node with price S', the CRR tree says that the stock price after one time step equals $S'u$ (the up move) with probability p and $S'd$ (the down move) with probability $1 - p$, where $ud = 1$. Hence the price resulting from j down moves and $i - j$ up moves from time step 0 equals $S_0 u^{i-j} d^j$ with probability $\binom{i}{j} p^{i-j} (1 - p)^j$. This node at time step i is denoted as $N(i, j)$ for simplicity. Some details will be introduced in Section 2.

Pricing barrier options on a CRR tree makes the computed prices oscillate significantly as a function of n [1]. This is because the barrier being assumed by the tree varies with n. Consider the pricing of a single-barrier option with barrier L in Fig. 2. In the 2-time-step tree of panel (a), the stock price can not hit the barrier L exactly. Instead, a near price on the CRR tree like L_1 becomes the barrier adopted. Call it the effective barrier. Similarly, in the 3-time-step tree of panel (b), the effective barrier is changed to L_2. The pricing results oscillate mainly because the effective barrier fluctuates with n. To solve this problem, Boyle and Lau suggest to pick a proper n for which the tree has a layer close to L [1]. This method reduces the errors dramatically. However, their method can not be easily adapted to handle two barriers since it is next to impossible to pick an n that will tailor to *both* barriers.Derman et al. price the single-barrier options by interpolation [3]. They first calculate the option value for each hallow node by moving the barrier outward to L_2. Then they calculate the value for each hallow node by moving the barrier inward to L'. We call L' and L_2 the inner and outer barriers to L, respectively. The option value for each hollow node is obtained by interpolating the two values mentioned above.

Ritchken alleviate the oscillation problem [10] by using "stretch parameter(s)" to tune the structure of his trinomial tree. Thus a layer of his tree can be made to coincide with each barrier. However, the branching probabilities of Ritchken's trinomial tree model are not guaranteed to be valid. More seriously, when a barrier is very close to the initial stock price, a large number of time steps is required and his algorithm becomes inefficient. Figlewski and Gao suggest the adaptive mesh model (AMM) to solve this "barrier-too-close" problem for pricing single-barrier options [4]. But no efforts have been carried out to extend the AMM to price double-barrier options.

All aforementioned approaches are not efficient enough since each node of the tree must be evaluated during backward induction to obtain the price. As the number of nodes of a tree is proportional to n^2, the backward induction also runs in $O(n^2)$ time. Lyuu provides an $O(n)$−time combinatorial algorithm for pricing single-barrier options on the CRR tree [7]. However, the pricing results oscillate significantly since the CRR tree is not guaranteed to hit the barrier.

This paper proposes a novel tree model, the tri-binomial tree (the TB tree hereafter), for pricing barrier options. The TB tree draws mainly from [7,10]. First, to alleviate the oscillation problem, the TB tree is guaranteed to have a layer that coincides with each barrier. Second, the TB tree is built on the CRR tree and the computation can be made linear in n by combinatorial techniques. We also prove that TB tree is constructible and valid. Numerical results show that our approach can achieve the same level of accuracy with much less computational time than other tree approaches mentioned above.

2   Preliminaries

We assume that the option initiates at time 0 and matures at time T. The exercise price for this option is denoted by X. Let S_t denote the stock price at time t, where $0 \leq t \leq T$. S_t follows the log-normal diffusion process:

$$S_{t+dt} = S_t \cdot \exp[(r - 0.5\sigma^2)\, dt + \sigma\, dW_t],   \tag{1}$$

where W_t is the standard Wiener process, r is the risk-free interest rate per annum, and σ denotes the volatility of the stock price.

A tree model divides the time span from time 0 to time T, into n time steps and simulates the price discretely at each time step. A tree converges to the stock price process mentioned in Eq. (1) if the first and the second moments of the stock price process are matched at each node of the tree. Consider the CRR

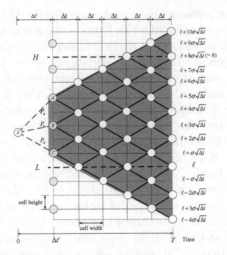

Fig. 3. The TB tree for Pricing Double-Barrier Options. L and H are denoted by thick dotted lines. The S_0-log-prices for the nodes at maturity are listed next to these nodes.

tree illustrated in Fig. 1. To match the first two moments of the stock price, u is set as $e^{\sigma\sqrt{T/n}}$. The probability p is set to $(e^{rT/n} - d)/(u - d)$.

The payoffs of the barrier options at time T are defined as follows. Define $S_{\inf} = \inf_{0 \le t \le T} S_t$. The payoff of a down-and-out single-barrier call option with barrier L is

$$\text{Payoff} = \begin{cases} 0, & \text{if } S_{\inf} \le L, \\ \max(S_T - X, 0), & \text{otherwise.} \end{cases}$$

Define $S_{\sup} = \sup_{0 \le t \le T} S_t$. The payoff of a knock-in double-barrier call option with barriers L and H is

$$\text{Payoff} = \begin{cases} \max(S_T - X, 0), & \text{if } S_{\sup} \ge H \text{ or } S_{\inf} \le L, \\ 0, & \text{otherwise.} \end{cases}$$

This paper focuses on the pricing aforementioned barrier options. The extension to other types of barrier options is straightforward.

3   Construction of the TB Tree

The TB tree is built on the CRR tree and is guaranteed to hit each barrier. We will first show how to construct an TB tree. The $O(n)$-time combinatorial algorithm for pricing double-barrier options will be introduced in section 4.

Consider the TB tree for a double-barrier option with barriers H and L in Fig. 3. The underlying is a CRR tree in shadow. The first two time steps of the CRR tree is truncated, and this CRR tree begins with three nodes: A, B, and C. These three nodes are connected by S at time 0. The TB tree has two following features: (1) It has two layers that coincide with H and L. (2) The branching probabilities P_u, P_m, and P_d are valid. To make the truncated CRR tree hit H and L, the length of a time step Δt should satisfy some specific constraint. We will explain how to pick a proper Δt later. The length of the the first time step of the TB tree $\Delta t'$ is the remaining amount of time to make the whole tree span T. A, B, and C are finally selected among the light gray nodes at time $\Delta t'$ to make P_u, P_m, and P_d valid and to make the TB tree hit H and L.

Now we determine Δt. Assume that the stock price at S is S_0. Define the S_0-log-price of price V' as $\ln(V'/S_0)$ for convenience. A S_0-log-price of z therefore implies a price of $S_0 e^z$. Note that the difference between the S_0-log-prices of two adjacent nodes like A and B is $2\sigma\sqrt{\Delta t}$, because the upward and the downward multiplication factors of the CRR tree are $e^{\sigma\sqrt{\Delta t}}$ and $e^{-\sigma\sqrt{\Delta t}}$, respectively. The S_0-log-prices of H and L are $h = \ln(H/S_0)$ and $\ell = \ln(L/S_0)$, respectively. To make sure the truncated CRR tree has two layers that coincide with H and L, $\frac{h-\ell}{2\sigma\sqrt{\Delta t}}$ must be some integer k. Assume that we try to construct an m-time-step tree. The length of each time step $\Delta\tau = T/m$, but $\frac{h-\ell}{2\sigma\sqrt{\Delta\tau}}$ may not be an integer. So we pick a Δt that is close to $\Delta\tau$ and that makes $\frac{h-\ell}{2\sigma\sqrt{\Delta t}}$ an integer. We pick $\Delta t = \left(\frac{h-\ell}{2\kappa\sigma}\right)^2$, where $\kappa = \left\lceil \frac{h-\ell}{2\sigma\sqrt{\Delta\tau}} \right\rceil$. Note that the number of time steps of TB tree is no longer equal to, but close to, m as we change the length of each time step. Let the truncated CRR tree has $\lfloor\frac{T}{\Delta t}\rfloor - 1$ time steps. The length of the first time step $\Delta t'$ is the remaining amount of time to make the whole tree span T: $\Delta t' = T - \left(\lfloor\frac{T}{\Delta t}\rfloor - 1\right)\Delta t$. It is easy to verify that $\Delta t \leq \Delta t' < 2\Delta t$.

Finally, we select A, B, and C that are connected by S. Note that three branches are required for S to match the first two moments of the logarithmic stock price process; a binomial branch from S does not have enough degrees of freedom. By Eq. (1), the mean μ and the variance Var of the S_0-log-prices of A, B, and C equal $(r - \sigma^2/2)\Delta t'$ and $\sigma^2\Delta t'$, respectively. Let the S_0-log-price of B be $\hat{u}$. To make the truncated CRR tree hit H and L, the following must be satisfied for some integer j:

$$\hat{u} = \begin{cases} \ell + 2j\sigma\sqrt{\Delta t}, & \text{if the truncated CRR tree has even number of time steps,} \\ \ell + (2j+1)\sigma\sqrt{\Delta t}, & \text{otherwise.} \end{cases}$$

Those nodes whose S_0-log-prices satisfy the above constraint at time $\Delta t'$ are colored in light gray in Fig. 3. We choose A, B, and C to make P_u, P_m, and P_d valid. Recall that the difference between two adjacent nodes' S_0-log-prices is $2\sigma\sqrt{\Delta t}$. There exists a unique light gray node whose S_0-log-price lies in the interval $[\mu - \sigma\sqrt{\Delta t}, \mu + \sigma\sqrt{\Delta t})$. We select this node as B. For example, the S_0-log-price of B is $\ell + 3\sigma\sqrt{\Delta t}$ in Fig. 3. The S_0-log-price of B is closest to μ among those of the light gray nodes. The S_0-log-prices of A and C are $\hat{u} + 2\sigma\sqrt{\Delta t}$ and $\hat{u} - 2\sigma\sqrt{\Delta t}$, respectively. Define $\beta \equiv \hat{u} - \mu$, $\alpha \equiv \beta + 2\sigma\sqrt{\Delta t}$, $\gamma \equiv \beta - 2\sigma\sqrt{\Delta t}$. Note that $\beta \in [-\sigma\sqrt{\Delta t}, \sigma\sqrt{\Delta t})$ and that $\alpha > \beta > \gamma$. The branching probabilities are derived by solving the following equalities

$$P_u \alpha + P_m \beta + P_d \gamma = 0, \tag{2}$$
$$P_u \alpha^2 + P_m \beta^2 + P_d \gamma^2 = \text{Var}, \tag{3}$$
$$P_u + P_m + P_d = 1. \tag{4}$$

Eqs. (2) and (3) match the first two moments of the logarithmic stock price, and Eq. (4) ensures that P_u, P_m, P_d as probabilities sum to one. A proof given in Appendix A shows that the inequalities $0 \le P_u, P_m, P_d \le 1$ are satisfied.

We now develop an efficient and accurate algorithm for pricing double-barrier options on the TB tree. First, an $O(n)$-time algorithm is developed to price a double-barrier option on the CRR tree in section 4. This algorithm can be used to evaluate the option values of A, B, and C efficiently since each of these nodes can be viewed as the root node of a CRR tree that begins at time $\Delta t'$. The final pricing result of our TB tree is $e^{-r\Delta t'} (P_u \times V_A + P_m \times V_B + P_d \times V_C)$, where V_X denotes the option value at X.

Pricing a single-barrier option with barrier L by the TB tree is much simpler! We simply construct the truncated CRR tree that has a layer coinciding with L. Note that we do not need to adjust the length of a time step since it needs to match only L (instead of L and H).

4 An $O(n)$-Time Combinatorial Algorithm on a CRR Tree

We first derive a useful combinatorial formula by the reflection principle and the inclusion-exclusion principle. This formula is used to build up the pricing algorithm. We focus on the knock-in double-barrier call options. The extension to other types of double-barrier options is straight forward.

A Combinatorial Formula
A combinatorial formula is derived with the help of the lattice in Fig. 4. This lattice reflects the structure of the CRR tree. The x- and y- coordinates denotes the time step of the tree and the stock price level, respectively. To fit the price movement on the CRR tree, each path can move from vertex (i, j) to vertex $(i+1, j+1)$ (the up move) or vertex $(i+1, j-1)$ (the down move). Now consider the following problem: How many price paths starting form A will reach either H or L before arriving at B? Without loss of generality, we assume that $a, b \ge 0$. A simplified problem is discussed first: How many paths moving form A to B will hit barrier H before one hit of barrier L? One such path may hit barrier H at J and barrier L at K. We can reflect the path $\widehat{AJ}$ (marked by solid curve) with respect to the H-axis to get $\widehat{A_1 J}$ (marked by dash curve). Each path from A to J maps to a unique path from A_1 to J. Thus the number of paths from A to J equals to the number of paths from A_1 to J. The reflection principle says that the number of paths staring from A and hitting H before reaching B equals to the number of paths moving from A_1 to B. The reflection principle can be applied more than once. The curve $\widehat{A_1 K}$ can be reflected with respect to the

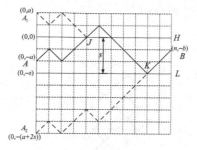

Fig. 4. Count the Number of Paths That Hit Barrier H and L by the Reflection Principle

L-axis to obtain $\widehat{A_2K}$. Again, the number of paths staring from A_1 and hitting L before reaching B equals to the number of paths moving from A_2 to B. Thus the number of paths moving from A to B and reaching H at least once before reaching L is equal to the number of paths moving from A_2 to B. Assume that x up moves and y down moves are required to move from A_2 to B. Thus we have $x + y = n$ and $x - y = a - b + 2s$. We get $x = \frac{n+a-b+2s}{2}$ by solving the above equations. So the answer to the aforementioned problem is

$$\binom{n}{\frac{n+a-b+2s}{2}} \quad \text{for even, non-negative } n + a - b. \tag{5}$$

Note that a path counted by Eq. (5) may hit L first before hitting H. The point is that among the hits, one hit of H must appear before one hit of L.

The problem of counting the number of paths that will hit either H or L before arriving at B is now within reach. First, a function f is constructed to map each path to a string. This string contains the information about the barrier hitting sequence. For example, $f(\widehat{AB}) = HHL$ since $\widehat{AB}$ hits H twice before hitting L.

Next, we define α_i as the set of paths whose f value *contains* $\overbrace{H^+L^+H^+\cdots}^{i}$. L^+ and H^+ denote a sequence of Ls and Hs, respectively. Obviously, $\widehat{AB}$ belongs to set α_1 and set α_2. Similarly, define β_i as the set of paths whose f value contains $\overbrace{L^+H^+L^+\cdots}^{i}$. Thus the path $\widehat{AB}$ belongs to set β_1. The number of elements in set α_i and β_i can be calculated by repeatedly using the reflection principle. The number of elements in each set is listed as follows:

$$|\alpha_i| = \begin{cases} \binom{n}{\frac{n+a+b+(i-1)s}{2}} & \text{for odd } i \\ \binom{n}{\frac{n+a-b+is}{2}} & \text{for even } i \end{cases} \qquad |\beta_i| = \begin{cases} \binom{n}{\frac{n-a-b+(i+1)s}{2}} & \text{for odd } i \\ \binom{n}{\frac{n-a+b+is}{2}} & \text{for even } i \end{cases}$$

$$\tag{6}$$

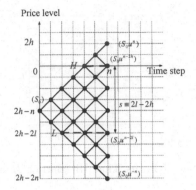

Fig. 5. Put a CRR Tree on a Lattice. The coordinate of the root node of the CRR tree is $(0, 2h - n)$. The values in parentheses denote the stock prices.

Note that each path that hits the barrier may belong to more than one set. For example, $\widehat{AB}$ belongs to α_1, α_2, and β_1. The inclusion-exclusion principle is then applied to calculate the number of paths that moves from A (with coordinate $(0, -a)$) to B $(n, -b)$ and that hit either H or L at least once as follows:

$$\mathbf{N}(a, b, s) = \sum_{i=1}^{\lceil \frac{n}{s} \rceil} (-1)^{i+1} (|\alpha_i| + |\beta_i|). \tag{7}$$

The Pricing Algorithm

The construction of the pricing algorithm can be divided into several different cases. For some degenerate cases like $S_0 \leq L$, $S_0 \geq H$, and $X \geq H$, the value of a knock-in double-barrier option can be proved to equal the value of a vanilla call option. These degenerate cases can be directly priced by any analytical or numerical method that price a vanilla call option. So we further focus on the case $L < S_0 < H$ and $X < H$ from now on. The pricing result is obtained by summing the value contributed by each node at the maturity of the CRR tree. Combinatorial techniques is applied to implement our pricing algorithm efficiently. We first put the CRR tree on a lattice as displayed in Fig. 5. Assume that the barrier H and L equal to $S_0 u^{n-h} d^h (= S_0 u^{n-2h})$ and $L = S_0 u^{n-l} d^l (= S_0 u^{n-2l})$, respectively. The exercise price X satisfies the following equality $S_0 u^{n-a} d^a \leq X < S_0 u^{n-a+1} d^{a-1}$ for some integer a.

Now we analyze the value contributed by a price path that reaches $N(n, j)$. The probability for this path is $p^{n-j}(1-p)^j$. The payoff at $N(n, j)$ is $(S_0 u^{n-j} d^j - X)^+$. Thus the value contributed by this path is $p(j) \equiv e^{-rT} p^{n-j}(1-p)^j (S_0 u^{n-j} d^j - X)^+$, if this price path hits the barrier. Note that the value contribution by $N(n, j)$ is $V(j) \equiv \binom{n}{j} p(j)$ if $N(n, j)$ is above H or below L (inclusive). This is because all the price paths that reach $N(n, j)$ must also hit the barrier.

The pricing algorithm can be divided into two following cases. For convenience, the term "terminal node" refers to the node of the tree at maturity date.

Case 1. $L < X < H$: The option value can be decomposed into two parts: (1) the value contributed by the terminal nodes between X and H, and (2) the value contributed by the terminal nodes above H (inclusive). The first part can be computed by accumulating the values contributed by the terminal nodes, says $N(n, j)$ $(a > j > h)$, between X and H. The number of paths that reach one of the barriers before reaching $N(n, j)$ is $\mathbf{N}(n - 2h, 2j - 2h, 2l - 2h)$ (see Eq. (7)). The value contributed by such a path is $p(j)$. Therefore, the value contributed by node $N(n, j)$ is $\mathbf{N}(n - 2h, 2j - 2h, 2l - 2h)p(j)$. The first part of the option value is $P_0 \equiv \sum_{j=h+1}^{a-1} \mathbf{N}(n - 2h, 2j - 2h, 2l - 2h))p(j)$. The second part of the option value is computed by accumulating the values contributed by the terminal nodes, says $N(n, i)$ $(0 \leq i \leq h)$, above the barrier H (inclusive). The value contributed by $N(n, i)$ is $V(i)$. Therefore, the second part of the option value is $P_1 = \sum_{i=0}^{h} V(i)$. The value of a knock-in double-barrier call option is $P_0 + P_1$.

Case 2. $X \leq L$: The option value is decomposed into three parts: (1) the value contributed by the terminal nodes between L and H, (2) the value contributed by the terminal nodes above H (inclusive), and (3) the value contributed by the terminal nodes between L (inclusive) and X. The first part of the option value is expressed as $P_0' \equiv \sum_{j=h+1}^{l-1} \mathbf{N}(n - 2h, 2j - 2h, 2l - 2h)p(j)$. The second part of the option value equals P_1. The third part is computed by accumulating the values contributed by the terminal nodes, says $N(n, k)$ $(l \leq k < a)$, between L (inclusive) and X to get $P_1' \equiv \sum_{k=l}^{a} V(k)$. The option value is $P_0' + P_1 + P_1'$.

5   Experimental Results

We first compare the performance among the TB tree, Lyuu's algorithm [7], and the Ritchken's trinomial tree [10] in Table 1. All programs are run on a Pentium-4 2.8GHz computer. Note that Lyuu's algorithm oscillates significantly. Both Ritchken's trinomial tree and the TB tree converge monotonically to the true value 5.9968. But the TB tree can achieve each level of accuracy with fewer computational time than the Ritchken's trinomial tree. Thus we conclude that the TB tree is superior to both Lyuu's and the Ritchken's approaches.

Table 1. Convergence Rate for Pricing a Down-and-Out Single-Barrier Option

Time (sec)	Ritchken		Lyuu		TB	
	n	Value	n	Value	n	Value
0.001	100	5.9997	1000	6.1002	500	5.9980
0.004	200	5.9986	4000	6.1998	2000	5.9972
0.016	400	5.9980	16000	6.0829	8000	5.9969

The initial stock price is 95, the exercise price is 100, the risk-free rate is 10% per annum, the volatility of the stock price is 25%, the time to maturity is 1 year, and the barrier is 90. Time denotes the computational time in seconds. n and Value denotes the number of time steps and the pricing result of each tree model.

Table 2. Convergence Rate Comparison between AMM and the TB tree

	AMM			TB	
Level	n	Inner Barrier	Outer Barrier	n	Value
2	671	0.000001	0.000015	655	0.000003
1	2686	0.000001	0.000005	2625	0.000003

The initial stock price and the exercise price is 100, the risk-free rate is 10%, the volatility of the stock price is 30%, the time to maturity is 1 year, and the two barriers are 99.5 and 120, respectively. The pricing results of the TB tree converge to 0.000003.

The TB tree can accurately solve the barrier-too-close problem for pricing double-barrier options while AMM can not as illustrated in Table 2. Level denotes the AMM level. The number of steps of AMM is determined by the AMM level. The Inner Barrier and the Outer Barrier denote the results computed by moving the upper barrier (120) to the inner barrier and the outer barrier, respectively. Note that AMM undervalues the option if the upper barrier moves down to the inner barrier and overvalues the option if the upper barrier moves up to the upper barrier. Each pricing result of the TB tree is properly selected so the number of steps of the TB tree is approximately equal to that of the AMM. Obviously, the TB tree provides more accurate results than the AMM.

6   Conclusion

The TB tree that can efficiently and accurately price barrier options is proposed in this paper. It is mainly composed of the CRR tree so the computation can be speeded up by combinatorial techniques. It produces accurate pricing results since it has a layer to coincide with each barrier. Numerical results show that the TB tree are superior to other existing approaches.

References

1. BOYLE P., AND LAU, S. "Bumping Against the Barrier with the Binomial Method," *Journal of Derivatives*, 1 (1994), pp. 6–14.
2. COX J., ROSS, S., AND RUBINSTEIN, M. "Option Pricing: A Simplified Approach." *Journal of Financial Economics*, 7 (1979), pp. 229–264.
3. DERMAN, E., KANI, I., ERGENER, D., AND BARDHAN, I. "Enhanced Numerical Methods for Options with Barriers," Working Paper, Goldman Sachs. (1995).
4. FIGLEWSKI, S., AND GAO, B. "The Adaptive Mesh Model: A New Approach to Efficient Option Pricing." *Journal of Financial Economics*, 53 (1999), pp. 313–351.
5. GEMAN H., AND YOR, M. "Pricing and Hedging Double-Barrier Options: A Probabilistic Approach." *Mathematical Finance,* 6 (1996), pp. 365–378.
6. LUO, L. "Various Types of Double Barrier Options." *Journal of Computational Finance*, 4 (2001), pp. 125–138.
7. LYUU Y.D. "Very Fast Algorithms for Barrier Option Pricing and the Ballot Problem." *Journal of Derivatives*, 5 (1998), pp. 68–79.

8. MERTON, R. "Theory of Rational Option Pricing." *Bell Journal of Economics and Management,* 4 (1973), pp. 141–183.

9. REINER, E. AND RUBINSTEIN, M. "Breaking Down the Barriers." *Risk,* 4 (1991), pp. 28–35.

10. RITCHKEN, P. "On Pricing Barrier Options." *Journal of Derivatives,* 3 (1995), pp. 19–28.

11. SIDENIUS, J. "Double Barrier Options: Valuation by Path Counting." *Journal of Computational Finance,* 1 (1998), pp. 63–79.

A   Validity of Risk-Neutral Probabilities

Define $\det = (\beta - \alpha)(\gamma - \alpha)(\gamma - \beta)$, $\det_u = (\beta\gamma + \text{Var})(\gamma - \beta)$, $\det_m = (\alpha\gamma + \text{Var})(\alpha - \gamma)$, and $\det_d = (\alpha\beta + \text{Var})(\beta - \alpha)$. Then Cramer's rule applied to Eq. (2)–(4) gives $P_u = \det_u/\det$, $P_m = \det_m/\det$, and $P_d = \det_d/\det$. Note that $\det < 0$ because $\alpha > \beta > \gamma$. To ensure that the branching probabilities are valid, it suffices to show that P_u, P_m, $P_d \geq 0$. As $\det < 0$, it is sufficient to show $\det_u, \det_m, \det_d \leq 0$ instead. Finally, as $\alpha > \beta > \gamma$, it suffices to show that $\beta\gamma + \text{Var} \geq 0$, $\alpha\gamma + \text{Var} \leq 0$, and $\alpha\beta + \text{Var} \geq 0$ under the premise $\beta \in [-\sigma\sqrt{\Delta t}, \sigma\sqrt{\Delta t})$. Indeed,

$$\beta\gamma + \text{Var} = \beta^2 - 2\beta\sigma\sqrt{\Delta t} + \sigma^2\Delta t' \geq \beta^2 - 2\beta\sigma\sqrt{\Delta t} + \sigma^2\Delta t = (\beta - \sigma\sqrt{\Delta t})^2 \geq 0,$$

$$\alpha\gamma + \text{Var} = \beta^2 - 4\sigma^2\Delta t + \sigma^2\Delta t' \leq \beta^2 - 4\sigma^2\Delta t + 2\sigma^2\Delta t = \beta^2 - 2\sigma^2\Delta t \leq 0,$$

$$\alpha\beta + \text{Var} = \beta^2 + 2\beta\sigma\sqrt{\Delta t} + \sigma^2\Delta t' \geq \beta^2 + 2\beta\sigma\sqrt{\Delta t} + \sigma^2\Delta t = (\beta + \sigma\sqrt{\Delta t})^2 \geq 0,$$

as desired.

An Ingenious, Piecewise Linear Interpolation Algorithm for Pricing Arithmetic Average Options

Tian-Shyr Dai[1], Jr-Yan Wang[2], and Hui-Shan Wei[3]

[1] Department of Information and Finance Management, National Chiao Tung University, Hsinchu, Taiwan
cameldai@mail.nctu.edu.tw
[2] Graduate School of Finance, National Taiwan University of Science and Technology, Taipei, Taiwan
jywang@mail.ntust.edu.tw
[3] Department of Finance, National Central University, Taoyuan County, Taiwan

Abstract. Pricing arithmetic average options continues to intrigue researchers in the field of financial engineering. Since there is no analytical solution for this problem until present, developing an efficient numerical algorithm becomes a promising alternative. One of the most famous numerical algorithms for pricing arithmetic average options is introduced by Hull and White [10]. In this paper, motivated by the common idea of reducing the nonlinearity error in the adaptive mesh model [7] and the adaptive quadrature numerical integration method [6], the logarithmically equally-spaced placement rule in the Hull and White's model is replaced by an adaptive placement method, in which the number of representative average prices is proportional to the degree of curvature of the option value as a function of the arithmetic average price. Numerical experiments verify the superior performance of our method in terms of reducing the interpolation error. In fact, it is straightforward to apply this method to any pricing algorithm with the techniques of augmented state variables and the piece-wise linear interpolation approximation.

Keywords: Arithmetic average options, logarithmically equally-spaced placement, adaptive placement.

1   Introduction

Asian options are path dependent securities whose payoff depends on the average of the underlying prices during the option life. They were originally issued in 1987 by Banker's Trust Tokyo on crude oil contracts, and hence the name "Asian" option. Asian options are commonly traded in a thinly traded market to prevent price manipulation. Besides, Asian options are less expensive than comparable vanilla options, because the volatility of the average value of an underling asset is lower than the volatility of the value of the underling asset. In practice, end-users of commodities, energies, or foreign currencies tend to be exposed to average

M.-Y. Kao and X.-Y. Li (Eds.): AAIM 2007, LNCS 4508, pp. 262–272, 2007.

prices over time, so Asian options are also attractive for them. This is because Asian options are often used as they more closely replicate the requirements of end-users exposed to price movements on the underlying asset.

To this date, more and more financial instruments include the average feature from Asian options, for example, structure notes issued by international banks, the contracts of convertible bonds in Taiwan, etc. If the underlying price process follows the geometric Brownian motion, the analytical pricing formula for geometric average options is feasible since the product of lognormally distributed prices remains to follow the lognormal distribution. Based upon this observation, Kemna and Vorst proposed an analytical solution for European geometric average options [11]. Unfortunately, it is still analytically intractable to price arithmetic average options due to the lack of proper mathematical representation for the sum of lognormal random variables. Thus many researches were devoted to deal with the distribution of the sum of lognormal random variables and derive approximated pricing formulae for arithmetic average options. Several works along this direction include the fast Fourier transformation in [2], the Edgeworth series expansion in [16], the reciprocal Gamma distribution in [13], the Laplace transform inversion in [9], etc.

The tree-based model is a possible alternative to price arithmetic average options. However, the naive pricing method based on the tree model which is able to derive the exact value of the arithmetic average options by recording all possible arithmetic average prices is simply intractable due to the exponential growth of the number of possible arithmetic average prices with respect to the number of time steps. In this paper, the exact option value stands for the option value derived from a tree-based model without any interpolation error.

To overcome the problem of the exponential growth of the number of possible arithmetic average prices, Dai and Lyuu develop a trinomial-tree pricing model for arithmetic average options that guarantees the convergence to the exact option value [5], in which the notion of integrality of stock prices is employed to reduce the time complexity of recording all possible arithmetic average prices to be sub-exponential. However, it is still intractable to price arithmetic average options via this model when the number of time steps is large.

On the other hand, in Hull and White's model [10], instead of keeping track of all possible arithmetic average prices, representative average prices are logarithmically equally-spaced placed between the maximum and minimum arithmetic average prices for each node, and the piece-wise linear interpolation is adopted to derive the corresponding option values for nonexistent average prices during the backward induction. Therefore, the interpolation error occurs and whether the interpolation error vanishes is uncertain for all but a single scenario in which the number of representative average prices for each node and the number of time steps for the tree model are well collocated, see [8].

Along with the line of [10], Neave and Turnbull [14] suggest using the conditional frequency distribution to adjust the number of representative average prices for each node. Cho and Lee [3] replace the uniform allocation of the number of representative average prices in the Hull and White's model with the

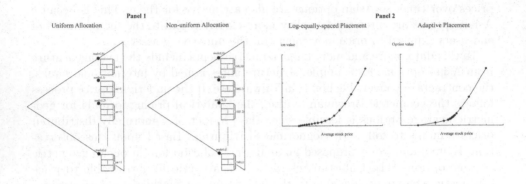

Fig. 1. The illustration of our adaptive placement method. Hull and White [10] adopted the combination of the uniform allocation and logarithmically equally-spaced placement rules in their pricing algorithm, i.e. m=100. Other modifications of the Hull and White's model focus on devising more efficient non-uniform allocation rules, i.e. $M(i, j)$ is different for each $node(i, j)$. However, the logarithmically equally-spaced placement rule is a common component in these models. In our adaptive placement method, the number of representative average prices is proportional to the degree of curvature of the option value function and an efficient non-uniform allocation of representative average prices is achieved automatically.

distribution of the number of possible geometric average prices. Klassen [12] proposes a revised version of the algorithm of [10], in which only a set of average prices at each node is considered, and the grid space for the logarithm of the arithmetic average prices is a pre-specified function of the time to maturity, the time steps, and the volatility of the stock price process. Although these methods of adjusting the allocation of the number of the representative average prices over the tree exhibit superior convergence rate to exact option values than the Hull and White's model, their major disadvantages are the absence of the economic meanings and the guarantee of the convergence of the interpolation error.

A different point of view is adopted in [1] and [4] to improve the convergence rate of the tree-based models for pricing arithmetic average options. Instead of recording the maximum and minimum arithmetic average prices, a more compact range is derived such that the interpolation error can be reduced effectively. Moreover, for European-style arithmetic average options, the optimal allocation of the representative average prices over the tree is derived in [4] to minimize the accumulated interpolation error of the option value.

Dedicated to devising the allocation of representative average prices over the tree to reduce the interpolation error, the suggested allocation rules of the above modifications are no longer uniformly distributed but are contingent on the probability reaching the underlying node, the time to maturity of the underlying node, the number of time steps in the tree model, and the volatility of the underlying process. The differences between uniform and non-uniform allocations are illustrated in Panel 1 of Fig. 1.

With the uniform allocation rule being replaced, the logarithmically equally-spaced placement rule proposed by Hull and White is still retained in the afore-mentioned works. Aiming at simultaneously guaranteeing the convergence of the interpolation error and improving the efficiency, we proposed a novel aspect to minimize the interpolation error by replacing the logarithmically equally-spaced placement rule with an adaptive placement method, in which more represen-tative average values are needed in the area around which the option value function of the arithmetic average price is with higher degree of curvature, and fewer representative average values are placed where the option value function is with lower degree of curvature. The ideas of the adaptive placement method and the logarithmically equally-spaced rules are illustrated in Panel 2 of Fig. 1. To achieve this goal, the adaptive placement method is actually designed to govern the linear interpolation error between each pair of adjacent representative aver-age prices under a limit criterion. Moreover, our method forms automatically an efficient non-uniform allocation of representative average prices over the tree.

2   Arithmetic Average Options

In this paper, the non-dividend-paying underlying stock price in the risk neutral world is assumed to follow the geometric Brownian motion: $dS_t/S_t = rdt + \sigma dZ$, where r is the risk free rate, σ is the volatility of the asset price, and Z is a Wiener process. Suppose that the stock price is sampled at the time points $0 = t_0 < t_1 < \cdots < t_n = T$ during the life of the arithmetic average options. If the corresponding stock prices are $S_{t_0}, S_{t_1}, \cdots, S_{t_n}$, the arithmetic average price from time 0 to t is $A(t) = (\sum_{i=0}^{l} S_{t_i})/(l+1)$, where $t_l \leq t < t_{l+1}$. In addition, the exercise value of the arithmetic average call considered in this paper at time t is $\max(A(t) - X, 0)$, where X is the strike price of the arithmetic average call. Furthermore, the stock price is assumed to be sampled periodically, which is often the case in the real world, and therefore $t_i = i\Delta t$ and $\Delta t = T/n$.

The Hull and White's Model
In the field of option pricing, the binomial-tree model divides the time horizon of an option into n discrete time steps and discretizes the stock prices at each time step. In Panel 1 of Fig. 2, it is shown that the stock price at time step 0 is S_0 (at $node(0,0)$), and the stock price can either move up to S_0u (at $node(1,0)$) or down to S_0d (at $node(1,1)$) at the first time step, where $u = \exp(\sigma\sqrt{\Delta t})$ is the magnitude of a upward movement for the stock price, and $d = \exp(-\sigma\sqrt{\Delta t})$ is the magnitude of a downward movement for the stock price. Similarly, each stock price can either move up or move down at subsequent time steps.

It is in theory possible to employ the binomial-tree model to calculate exact values of arithmetic average options by recording all possible average values reaching each node. Unfortunately, if the option life is divided into n periods, the number of all possible arithmetic average prices is 2^n, which implies that the computation complexity is intractable even for small values of n.

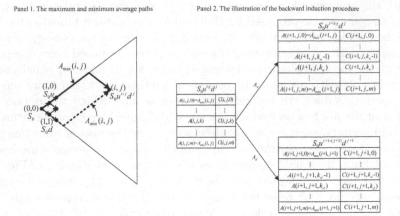

Fig. 2. The illustration of the Hull and White's model. In Panel 1, the $node(i,j)$ stands for the node at time point i with j cumulative down movements and the $S_0 u^{i-j} d^j$ is the corresponding stock price. $A_{max}(i,j)$ ($A_{min}(i,j)$) is the maximum (minimum) average stock price among all possible paths from $node(0,0)$ to $node(i,j)$. In Panel 2, for each possible average price $A(i,j,k)$, it is necessary to find the corresponding A_u and A_d and then to derive the option values C_u and C_d by the piece-wise linear interpolation. The continuation value for $A(i,j,k)$ is $C(i,j,k) = (p \cdot C_u + (1-p) \cdot C_d)e^{-r\Delta t}$.

One of the most famous tree-based models to price arithmetic average options efficiently is proposed in [10]. In their algorithm, to avoid tracking all possible arithmetic average prices of each node, only the maximum and the minimum arithmetic average prices of all traversed paths for each node are calculated, which is illustrated in Panel 1 of Fig. 2.

For $node(i,j)$ with the stock price $S_0 u^{i-j} d^j$ for $0 \le j \le i \le n$, the maximum arithmetic average price is contributed by a price path starting with $i-j$ consecutive up movements followed by j consecutive down movements, whose value is $A_{max}(i,j) = (S_0 \frac{1-u^{i-j+1}}{1-u} + S_0 u^{i-j} d \frac{1-d^j}{1-d})/(i+1)$. Likewise, the value of the corresponding minimum arithmetic average price can be calculated from a price path starting with j consecutive down movements followed by $i-j$ consecutive up movements: $A_{min}(i,j) = (S_0 \frac{1-d^{j+1}}{1-d} + S_0 d^j u \frac{1-u^{i-j}}{1-u})/(i+1)$. Once equipped with the knowledge about the maximum and minimum arithmetic average prices for each node, the logarithmic space between $A_{max}(i,j)$ and $A_{min}(i,j)$ is divided into m equal-length sub-intervals and $m+1$ representative average prices are obtained via $A(i,j,k) = \exp\left(\frac{m-k}{m}\ln(A_{max}(i,j)) + \frac{k}{m}\ln(A_{min}(i,j))\right)$.

After building the tree and the table of representative average prices for each node, we decide the payoff of each representative average price of the nodes at maturity first. Next, the option value is derived via the backward induction procedure. The backward induction procedure from $node(i+1,j)$ and $node(i+1,j+1)$ to $node(i,j)$ is illustrated in Panel 2 of Fig. 2.

For $A(i,j,k)$, the evolutions of the arithmetic average price at the next time point are $A_u = [(i+1)A(i,j,k)+S_0 u^{i+1-j} d^j]/(i+2)$, and $A_d = [(i+1)A(i,j,k)+$

$S_0 u^{i+1-(j+1)} d^{(j+1)}]/(i+2)$. Suppose that A_u is inside the range $[A(i+1,j,k_u)$, $A(i+1,j,k_u-1)]$. The option value C_u for the arithmetic average price A_u is approximated by the linear interpolation $C_u = w_u C(i+1,j,k_u) + (1-w_u)C(i+1,j,k_u-1)$, where $w_u = (A(i+1,j,k_u-1)-A_u)/(A(i+1,j,k_u-1)-A(i+1,j,k_u))$. Similarly, the option value of C_d for the arithmetic average price A_d is derived from $C_d = w_d C(i+1,j+1,k_d) + (1-w_d)C(i+1,j+1,k_d-1)$, where $w_d = (A(i+1,j+1,k_d-1)-A_d)/(A(i+1,j+1,k_d-1)-A(i+1,j+1,k_d))$, if A_d is inside the range $[A(i+1,j+1,k_d), A(i+1,j+1,k_d-1)]$. As a consequence, the continuation value for $A(i,j,k)$ is $C(i,j,k) = (p \cdot C_u + (1-p) \cdot C_d)e^{-r\Delta t}$.

Some Modifications for the Hull and White's Model

The interpolation error is inevitable in the Hull and White's model due to the limited number of representative average prices at each node and employing the piece-wise linear interpolation to find option values for nonexistent average prices. The brute-force method via increasing the number of representative average prices for each node is able to enhance the accuracy for the option values of course, but meanwhile it is accompanied with unacceptable computation time. In Section 4, in addition to the Hull and White's model, the performance of some modifications, including inserting the strike price into the average price table, applying the quadratic interpolation, and tightening the range for representative average prices,[1] will be compared to that of our adaptive placement method.

3 Our Models

The goal of our adaptive placement method is to intelligently reduce the interpolation error for pricing arithmetic average options in the tree-based model. Motivated by the common idea of dealing with the nonlinearity error in the Figlewski and Gao's adaptive mesh model [7] and the adaptive quadrature numerical integration method [6], our method differs from the Hull and White's method in the sense that more representative average prices are placed in the range where the option value function is with higher degree of curvature and fewer representative average prices are placed in the range where the option value function is with lower degree of curvature (see Fig. 1).

---

[1] For European fixed-strike-price arithmetic average calls, according to [1], for nodes at time point i, if some average price A is larger than the upper bound $(n+1)X/(i+1)$, because this path is sure to be in the money at maturity, the corresponding expected option value can be calculated directly via

$$e^{-r(n-i)\Delta t}[(i+1)A - (n+1)X + S_0 u^{i-j} d^j e^{r\Delta t}\frac{1 - e^{r(n-i)\Delta t}}{1 - e^{r\Delta t}}]/(n+1).$$

Therefore, the range $[A_{min}(i,j), A_{max}(i,j)]$ can be curtailed to $[\min(A_{min}(i,j), (n+1)X/(i+1)), \min(A_{\max}(i,j), (n+1)X/(i+1))]$, and whenever the average price is above the upper bound, the corresponding expected option value can be derived via the above equation without any interpolation error.

The details of our adaptive placement method are elaborated as follows. For any $A \in [A_1, A_2]$, where A_1 and A_2 stand for any pair of adjacent representative average prices in the table of average prices for some node, the option value for A can be derived from the linear interpolation $C = C_1 \cdot \frac{A - A_2}{A_1 - A_2} + C_2 \cdot \frac{A - A_1}{A_2 - A_1}$, where C_1 and C_2 are corresponding option values for A_1 and A_2. By the mean-value theorem, the error term of the linear interpolation caused from the nonlinearity of the option value function in $[A_1, A_2]$ can be expressed as

$$\frac{C''(\xi)}{2!} \cdot (A - A_1) \cdot (A - A_2), \text{ for some number } \xi \text{ between } A_1 \text{ and } A_2. \quad (1)$$

Our adaptive placement method is designed to examine whether the linear interpolation error between each pair of adjacent representative average prices in Eq. (1) is below some pre-specified limit. Once the error of the linear interpolation inside the range of $[A_1, A_2]$ is not negligible, i.e. $C''(\xi)$ is too large or the distance between A_1 and A_2 is too far, we divide $[A_1, A_2]$ into finer subsets by inserting an extra representative average price inbetween and then repeat the same procedure of examining the error of the linear interpolation for each subset. Once the value of the error term between any pair of adjacent representative average prices is smaller than the predefined threshold (termed the *second order error criterion* in our method), this examining-and-dividing process is stopped.

In practice, another constant termed the *precision criterion* is also defined to represent the threshold of negligible refinement for both average prices and option values in our method. The above examining-and-dividing process is also terminated when the difference between adjacent representative average prices or their corresponding option values is smaller than this minimum criterion. The purpose of introducing the *precision criterion* is to prevent possibly infinite dividing caused from the non-differentiable point. Within each examination of the linear interpolation error, we approximate $C''(\xi)$ in Eq. (1) by the second order numerical differentiation. For any pair of adjacent representative average prices A_1 and A_2, the midpoint $A = (A_1 + A_2)/2$ is employed together to approximate the error term of the linear interpolation for this range.

The steps to price arithmetic average options in this paper are described as follows. During the phase of building the stock price tree, only the maximum and minimum average prices for each node are recorded as representative average prices. Meanwhile, we also determine whether the strike price is needed to be inserted into the range between the maximum and the minimum average prices. As a consequence, there will be two or three representative average prices for each node after building the stock price tree. Since the number of possible arithmetic average prices for the nodes at the initial three time steps is not larger than three, it is not necessary to perform the above procedure and instead we record all possible average prices for these nodes.

After building the stock price tree, the tables of representative average prices for all nodes are mainly constructed during the phase of backward induction We take an example to illustrate the examining-and-dividing process of our adaptive placement method step by step. Suppose $S_0 = X = 50$, $n = 40$, $T = 1$, $r = 10\%$, $\sigma = 80\%$, and both the *second order error criterion* and the

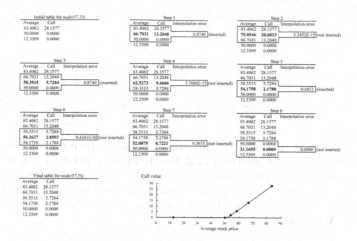

Fig. 3. The numerical example for the examining-and-dividing process. The values of parameters in this example are $S_0 = X = 50$, $n = 40$, $T = 1$, $r = 10\%$, $\sigma = 80\%$, and the *second order error criterion* and the *precision criterion* are both 0.5. These figures illustrate the examining-and-dividing process for $node(37, 25)$. Once the approximate linear interpolation error is larger than 0.5, the pair of the average price and the call value in boldface will be inserted into the table of representative average prices of $node(37, 25)$. The approximate linear interpolation error for any pair of adjacent representative prices in the final table is bounded by the *second order error criterion*.

precision criterion are 0.5. For $node(37, 25)$, the examining-and-dividing process is sketched in Fig. 3. Inside the frame of each step, there are three pairs of representative average prices and the corresponding call values, and we also report the linear interpolation error when these three pairs of representative average prices and option values are considered.

When the backward induction progresses to $node(37, 25)$, only 83.4062, 50, and 12.3309 are in the initial table of representative average prices, and their corresponding option values are 28.1577, 0, and 0 respectively. In step 1, the pairs of (83.4062, 28.1577), ((83.4062+50)/2=66.7031, 13.2048), and (50, 0) are considered to approximate the linear interpolation error for the range between 83.4062 and 50. Because the approximate linear interpolation error 0.8740 is larger than the *second order error criterion*, the pair of the average price 66.7031 and the corresponding call value 13.2048 should be inserted into the table of representative average prices. In step 2, the approximated linear interpolation error in the range between 83.4062 and 66.7031 is 3.2452E-15, which is smaller than the *second order error criterion* 0.5. Therefore, we do not insert the pair of the average price 75.0546(=(83.4062+66.7031)/2) and the corresponding option value 20.6813 since the linear interpolation works pretty well in the range between 83.4062 and 66.7031. Following the same reasoning, we can derive the final table of representative average prices and their corresponding option values through steps 3 to 8, in which the approximate linear interpolation error for any pair of adjacent representative average prices is below the *second order error criterion*.

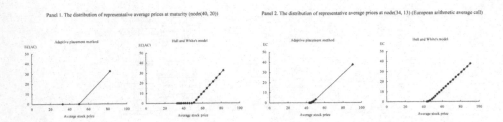

Fig. 4. Comparisons of the distributions of representative average prices of the adaptive placement method and the Hull and White's model. For the readability of this figure, the values of parameters are specified as: $S_0 = X = 50$, $n = 40$, $T = 1$, $r = 10\%$, $\sigma = 30\%$, the *second order error criterion* is 0.01, and the *precision criterion* is 0.001. In addition, the number of representative average prices in the Hull and White's model is 20.

In addition, the option value as the function of the arithmetic average price is plotted in Fig. 3. Since the performance of the piece-wise linear interpolation is poor around where the option value function is with high degree of curvature, our algorithm places more representative average prices in these areas to reduce the linear interpolation error. On the other hand, due to the satisfactory performance of the piece-wise linear interpolation for dealing with the option value function with low degree of curvature, our algorithm argues that less representative average prices placed in those areas will be sufficient.

4   Numerical Results

Comparisons with the Hull and White's Model

The differences between the logarithmically equally-spaced placed rule in the Hull and White model [10] and the adaptive placement rule in our model are shown in Fig. 4. In Panel 1 of Fig. 4, for some node at maturity, it is easily found that there is no linear interpolation error for both linear segments, and therefore it is not necessary to insert any representative average price. However, the Hull and White's model still employs $m + 1$ representative average prices for each node at maturity. For each linear segment, our method provides the interpolated results as accurate as those in the Hull and White's model, but near the kink, our method will outperform the Hull and White's model except the strike price happens to be one of representative average prices in their model.

In Panel 2 of Fig. 4, it is clear that the logarithmically equally-spaced placement in the Hull and White's model places too many representative average prices on the region with low degree of curvature, but only a few representative average prices are needed in our adaptive placement method to derive interpolated results with sufficient accuracy in this region. On the contrary, to deal with regions with high degree of curvature, the Hull and White's model generates unexpected large pricing errors due to large interpolation error.

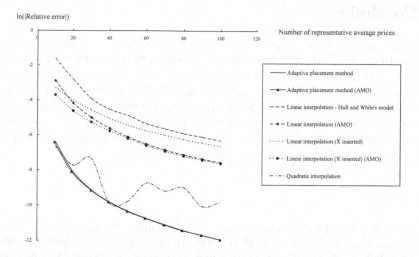

ln(|Relative error|)

Number of representative average prices

— Adaptive placement method

—▲— Adaptive placement method (AMO)

– – – Linear interpolation - Hull and White's model

– ◆ – Linear interpolation (AMO)

······ Linear interpolation (X inserted)

·· ◆ ·· Linear interpolation (X inserted) (AMO)

– · – · Quadratic interpolation

Fig. 5. The rates of convergence of seven different methods of pricing arithmetic average calls. Note that our adaptive placement method converges faster than other methods with respect to the number of representative average prices. Note also that except our adaptive placement method, the convergence rates of other methods are improved when the algorithm of AMO is applied.

Convergence Rates with the Number of Representative Average Prices

This section compares the rate of convergence with respect to the number of representative average prices for different methods. The values of parameters in our numerical example are as follows: $S_0 = X = 50$, $T = 1$, $r = 10\%$, $\sigma = 80\%$, $n = 40$, and different numbers of representative average prices are examined. In order to obtain a better understanding of the rates of convergence, an analysis on the relative error and the number of representative average prices is performed. Plots of ln(|relative error|) in relation to the number of representative average prices for the arithmetic average calls are in Fig. 5.

Obviously, the Hull and White's model converges poorly, but the relative error decreases significantly when the strike price X is inserted as a representative average price. This is because for nodes near maturity, the kink is near where the average price is equal to the strike price, and the piece-wise linear interpolation is inclined to overestimate the option value around the kink.

Note that the improvement is minor when combining the AMO algorithms [1] with our adaptive placement method. The idea of the AMO algorithm is to derive the option values of the arithmetic average prices higher than some threshold without incurring any interpolation error and meanwhile concentrate representative average prices on a smaller range to further reduce the interpolation error. However, for the region above the threshold, the interpolation error is in fact very small, so our adaptive placement method already places fewer representative average prices in the region above the threshold, and automatically concentrates on dealing with the region below the threshold. Thus the concept of the AMO algorithm is already nested in our adaptive placement method.

5   Conclusion

This paper proposes the adaptive placement method to price arithmetic average options. In our method, the representative average prices are placed effectively to reduce the interpolation error. Numerical results show that our adaptive placement method is superior to other methods in reducing interpolation error. Thus this method can be employed to price arithmetic average options efficiently and accurately. In fact, this novel technique can be applied to any other algorithms with augmented state variables and the piece-wise linear interpolation approximation like GARCH option pricing algorithm in [15].

References

1. Aingworth, D., Motwani, R., Oldham, J. D.: Accurate Approximations for Asian Options. Proc. 11th Annual ACM-SIAM Symposium on discrete Algorithms (2000)
2. Carverhill, A. P., Clewlow, L. J.: Valuing Average Rate (Asian) Options. Risk 3 (1990) 25-29
3. Cho, H. Y., Lee, H. Y.: A Lattice Model for Pricing Geometric and Arithmetic Average Options. J. Financial Engrg. 6 (1997) 179-191
4. Dai, T.-S., Huang, G.-S., Lyuu, Y.-D.: An Efficient Convergent Lattice Algorithm for European Asian Options. Appl. Math. and Comput. 169 (2005) 1458-1471
5. Dai, T.-S., Lyuu, Y.-D.: An Exact Subexponential-Time Lattice Algorithm for Asian Options. Proc. ACM-SIAM Symposium on Discrete Algorithms (SODA04) New Oreleans (2004) 710-717
6. Faires, J. D., Burden, R.: Numerical Methods. 3rd edn. Brooks/Cole Press. (2003)
7. Figlewski, S., Gao, B.: The Adaptive Mesh Model: A New Approach to Efficient Option Pricing. J. Financial Econ. 53 (1999) 313-351
8. Forsyth, P. A., Vetzal, K. R., Zvan, R.: Convergence of Numerical Methods for Valuing Path-Dependent Options Using Interpolation. Rev. Derivatives Res. 5 (2002) 273-314
9. Geman, H., Yor, M.: Bessel Processes, Asian Options, and Perpetuities. Math. Finance 3 (1993) 349-375
10. Hull, J. C., White, A.: Efficient Procedures for Valuing European and American Path-Dependent Options. J. Derivatives 1 (1993) 21-31
11. Kemna, A., Vorst, T.: A Pricing Method for Option Based on Average Asset Values. J. Banking and Finance 14 (1990) 113-129
12. Klassen, T. R.: Simple, Fast and Flexible Pricing of Asian Options. J. Comput. Finance 4 (2001) 89-124
13. Milevsky, M. A., Posner, S. E.: Asian Option, the Sum of Lognormals and the Reciprocal Gamma Distribution. J. Financial and Quant. Anal. 33 (1998) 409-422
14. Neave, E. H., Turnbull, S. M.: Quick Solutions for Arithmetic Average Options on a Recombining Random Walk. Actuarial Approach For Financial Risks 2 (1994) 717-740
15. Ritchken, P., Trevor, R.: Pricing Options under Generalized GARCH and Stochastic Volatility Processes. J. Finance. 54 (1999) 377-402
16. Turnbull, S. M., Wakeman, L. M. A Quick Algorithm for pricing European Average Option. J. Financial and Quant. Anal. 26 (1991) 377-389

Optimal Order Allocation with Discount Pricing

Boris Goldengorin[1], John Keane[2], Viktor Kuzmenko[3], and Michael Tso[4]

[1] Faculty of Economic Sciences, University of Groningen
P.O. Box 800, 9700 AV Groningen, The Netherlands
Department of Applied Mathematics
Khmelnitsky National University, Ukraine
[2] School of Computer Science, University of Manchester
Oxford Road, P.O. Box 88, Manchester M60 1QD, UK
[3] Glushkov Institute of Cybernetics
National Academy of Sciences of Ukraine
pr. Akademika Glushkova, 40, Kiev 03187, Ukraine
[4] School of Mathematics, University of Manchester,
Oxford Road, P.O.Box 88, Manchester M60 1QD, UK

Abstract. We consider optimal order allocation (or procurement) problems with discount pricing and fixed charges. Such a problem is faced for example by an internet trading agent who seeks to fulfil an order for specified amounts of several products from a pre-arranged list of suppliers, taking into account availability and price. We present a mixed integer programming (MILP) formulation assuming that suppliers impose a discount schedule with multiple price breaks including fixed charges. We show that a modified capacitated facility location problem (CFLP) model is appropriate for the general case under consideration and outline a Lagrangean relaxation approach improved by open and close penalties. Our experimental results show that problems arising in practice can be handled within seconds by either LINGO or XPress-MP software.

Keywords: Online pricing; Discount schedules; Location problems; Branch and bound.

1 Introduction

This study is motivated by the case of a trader in pharmaceuticals over the internet who has to respond in real time to an enquiry requesting a price quotation for specified amounts of set of products. The trader, who carries no stock, operates in a competitive market and aims to source the order at least cost from a pre-arranged set of suppliers who operate price schedules which have been negotiated with the trader in advance. Assuming that no supplier has uniformly the best price for all products, the trader has to consider allocating (partitioning) the order amongst the different suppliers, and the construction of an efficient algorithm to determine the optimal breakdown of an order into component suborders is the subject of this paper.

It will not in general be practicable or economic for the trader to order each product from the supplier quoting the best price due to fixed costs that limit in

M.-Y. Kao and X.-Y. Li (Eds.): AAIM 2007, LNCS 4508, pp. 273–284, 2007.

practice the number of suppliers to be used. Suppliers commonly impose a fixed charge to cover the costs of administration and delivery, which may be waived on orders exceeding a certain threshold of order value. They may also give quantity discounts depending on the order size, resulting in a discount schedule containing a sequence of price breaks. It is understood that the enquiry will be converted to a firm order only if the resulting price quotation is acceptable. Therefore the trader requires an efficient algorithm to determine an optimal procurement taking into account supplier/product list prices, availability and any applicable discounts.

We refer to this as the buyer's decision problem (BDP). The type of discount we consider in this paper is known as an "all units business volume discount" (Sadrian and Yoon, [1]) in contrast to the "total quantity discount" structures assumed in previous work by Goossens et al. [2]. We develop in Section 2 a basic MILP model akin to the uncapacitated facility location (UFL) model (see e.g. [3]) but with additional binary switches to incorporate a fixed discount structure. We illustrate by means of a numerical example the "buy more for less" feature of the model. In Section 3 we extend the model to discount schedules incorporating multiple breakpoints by the use of "pseudo-suppliers" and compare the model to a capacitated facility location problem (CFLP) (see e.g. [4]). In Section 4 we outline a Lagrangean heuristic and in Section 5 we provide explicit open and close penalties useful for fathoming nodes of a branch and bound (BnB) tree. Section 6 reports some computer experiments using LINGO and XPress-MP. Finally Section 7 contains some concluding remarks.

2   A Basic MILP Model

We provide in this section a mixed integer linear program (MILP) formulation for the buyer's decision problem when each supplier has a fixed charge which is discounted when the order value exceeds a single threshold.

Define the model parameters as follows. Let

I be the set of suppliers $I = \{1, .., m\}$
J be the set of products comprising the order $J = \{1, .., n\}$
D_j = demand (number of units) for item j $(j \in J)$
c_{ij} = unit cost for item j purchased from supplier i $(i \in I)$
f_i = fixed cost for use of (opening) supplier i
V_i = value threshold for discount on f_i
s_i = discount on f_i for orders above V_i $(s_i \leq f_i)$

and for $i \in I$ and $j \in J$, introduce the following decision variables :

x_{ij} number of units of product j to order from supplier i
$$y_i = \begin{cases} 1 \text{ if supplier } i \text{ used} \\ 0 \text{ otherwise} \end{cases}$$
$$z_i = \begin{cases} 1 \text{ if supplier } i \text{ discount applies} \\ 0 \text{ otherwise} \end{cases}$$

We then formulate the buyer's decision problem as follows:

$$\text{Minimize } \sum_{i \in I} \left\{ f_i y_i - s_i z_i + \sum_{j \in J} c_{ij} x_{ij} \right\}$$

$$
\begin{array}{lll}
\text{subject to} & \sum_{i \in I} x_{ij} = D_j & j \in J \quad \text{(a)} \\
& x_{ij} \le D_j y_i & i \in I, j \in J \text{ (b)} \quad \text{(BDP1)} \\
& \sum_{j \in J} c_{ij} x_{ij} \ge V_i z_i & i \in I \quad \text{(c)} \\
\\
& y_i, z_i \in \{0, 1\} & i \in I \quad \text{(d)} \\
& x_{ij} \in Z^+ & i \in I, j \in J \text{ (e)}
\end{array}
$$

REMARKS

1. The integrality of $\{x_{ij}\}$ cannot be assumed in any optimal solution to the relaxed problem with $x_{ij} \in \mathbb{R}^+$ as we show (see Note 1 of an example below).
2. Discounts may make it sometimes cheaper to supply more than the demand for any product. A formulation that allows an optimal solution to exceed demand for any product by some pre-specified margin ε is obtained by substituting for BDP1 (a),(b)

$$
\begin{array}{ll}
\sum_{i \in I} x_{ij} \ge D_j & j \in J \quad \text{(a1)} \\
x_{ij} \le (D_j + \varepsilon) y_i & i \in I, j \in J \text{ (b1)}
\end{array}
$$

(see Note 2 of example below).
3. A *minimum order value* imposed by a supplier i can be modelled by making f_i so large that this supplier will never be used without discount s_i being applied.
4. The "stockout" condition $x_{ij} = 0$ for some $i \in I, j \in J$ can be modelled by making c_{ij} large. This type of information may only be apparent after an order from supplier i is placed, when it could be incorporated in a post-optimality analysis.
5. Stock availability at supplier i may also be incorporated by upper bound constraints of the form $x_{ij} \le S_{ij}$.

Proposition 1. *The binary switches y_i, z_i in the model BDP1 are a correct encoding of the suppliers' fixed cost structure.*

Proof. We can prove the model is correct by considering the behaviour of the binary switch z_i for some $i \in I$ for the following cases.

CASE 1

If $y_i = 0$ then $x_{ij} = 0 \ \forall \ j$ from BDP1 (a). Thus $\sum_{j \in J} c_{ij} x_{ij} = 0$ and so $z_i = 0$ necessarily from BDP1 (c) since $V_i > 0$. We therefore never apply the discount when the supplier is closed.

CASE 2

If $y_i = 1$ and $\sum_{j \in J} c_{ij} x_{ij} < V_i$ then $z_i = 0$ from BDP1 (b) and no discount is applied (as we require).

CASE 3

If $y_i = 1$ and $\sum_{j \in J} c_{ij} x_{ij} \geq V_i$ then both $z_i = 0$ and $z_i = 1$ are feasible for BDP1 (c). However the minimization objective ensures that a discount will be applied where possible, ensuring that $z_i = 1$ (as we require).   ∎

Example 1. *An order specifies 7 and 3 units respectively of two products P1, P2 which may be sourced from 2 suppliers S1, S2. For "small" orders valued less than £50 supplier S1 imposes a carriage charge of £10 while supplier S2 charges just £5. Neither supplier charges carriage on orders over £50 in value, thus $s_i = f_i$ $(i = 1, 2)$ in this example. The unit costs and demands for each product are given in the following table, with the least cost supplier indicated for each product:*

Suppliers	Unit cost P1	P2	Surcharge f_i	Value threshold V_i
S1	7	6	10	50
S2	9	3	5	50
Demand	7	3		

If the demand is met by a single supplier, the costs are £67 and £72 from S1, S2 respectively. The overall minimum cost solution is achieved by sourcing 7 units of P1 and 1 unit of P2 from S1 and the remaining 2 units of P2 from S2 at a total cost of £66. This is seen to be optimal as follows:

Ignoring fixed costs, the minimum supply (transportation) policy is

	P1	P2	Value
S1	7	0	49
S2	0	3	9

Thus £58 is a lower bound on the total "variable" cost of the order. However fixed costs of £15 are payable with this policy giving a total cost of £73. (Note that fixed costs cannot be avoided completely in this example since to spend more than £50 with each supplier is clearly sub-optimal.) We then seek to avoid some fixed costs by redistributing some amount either of P1 to S2 or of P2 to S1 in order to cross some supplier's threshold of minimum value.

A fixed cost saving of £10 results from transferring one unit of P2 to S1 (for an increase of £3 in variable cost). The net saving is £7 and the resulting total cost is £66. We note that the net saving cannot exceed £5 by removing the fixed charge for S2. Therefore the minimum cost of supply is £66 and the optimal supply policy is

	P1	P2	Value
S1	7	1	55
S2	0	2	6

NOTES

1. *If the integer constraints on the supply from S1, S2 are relaxed, the optimal fractional solution is to source 7 units of P1 and $\frac{1}{6}$ unit of P2 from S1 and the remaining $2\frac{5}{6}$ units of P2 from S2 at a total cost of £$63\frac{1}{2}$.*

2. *If the demand for P2 is cancelled, the minimum cost to supply precisely 7 units of P1 is £59 (from S1). However, a less costly solution is to exceed the demand by supplying one unit of P2 from S1, resulting in a total cost of £55.*

3   Generalization of BDP1

We generalize the basic model to the case that supplier $s \in S$ specifies a discount schedule based on q *price bands* $I_k^s = [V_{k-1}^s, V_k^s), k \in K = \{1, ..., q\}$ with $0 = V_0^s < V_1^s ... < V_q^s$. For notational convenience we consider q to be the same for all suppliers, but in practice this need not be so. Let c_{1j}^s be the list price per unit of item j from supplier s and let x_{sj} be the corresponding amount ordered. The *value* of an *order* $x = (x_j)$ placed with supplier s is $V = \sum_{j \in J} c_{1j}^s x_j$. We will assume that the discounted price $\pi(x)$ that supplier s charges for the order has the linear form

$$\pi(x) = f_k^s + \alpha_k^s V(x)$$

if $V(x) \in I_k$ with $\alpha_1^s = 1$ and $\alpha_{k-1}^s \geq \alpha_k^s, f_{k-1}^s \geq f_k^s, (k = 2, ..., q)$ being the (generally monotonic) business value discount factors applied by supplier s.

Let x_{kj}^s denote the quantity of product j ordered from supplier s in some price band k. The *value* of the order placed with supplier s is $V^s = \sum_{j \in J} c_{1j}^s x_{kj}^s$ and hence the discounted *price* of the order is $f_k^s + \sum_{j \in J} c_{kj}^s x_{kj}^s$ where $c_{kj}^s = \alpha_k^s c_{1j}^s$ and k depends on the price band $I_k^s = [V_{k-1}^s, V_k^s)$ within which V^s falls.

To reflect the fact that k is uniquely determined by the order value V^s, we define the Boolean variables $\{y_k^s\}$ taking the values $y_k^s = 1$ if $V^s \in I_k^s$ and $y_k^s = 0$ otherwise. So as $\sum_{k \in K} y_k^s = 1$, we re-label the *supplier/price band* pair (s, k) as the ith *pseudo-supplier* and define the index set P^s appropriately so that $\sum_{i \in P^s} y_i = 1$ with $I = \cup P^s = \{1, ..., m\}$.

A corresponding re-indexing of $f_k^s = f_i, c_{kj}^s = c_{ij}, x_{kj}^s = x_{ij}$ and $I_k^s = [L_i, U_i)$ leads to the following more general formulation of the buyer's decision problem.

$$\min_{y_i, x_{ij}} \sum_{i \in I} \left\{ f_i y_i + \sum_{j \in J} c_{ij} x_{ij} \right\}$$

s.t.

$$
\begin{array}{lll}
L_i y_i \leq \sum_{j \in J} c_{ij}^0 x_{ij} \leq U_i y_i & \forall i \in I & \text{(a)} \\
\sum_{i \in P^s} y_i \leq 1 & \forall s \in S & \text{(b)} \\
\sum_{i \in I} x_{ij} \geq D_j & \forall j \in J & \text{(c)} \\
y_i \in \{0, 1\} & \forall i \in I & \text{(d)} \\
x_{ij} \in \mathbb{Z}^+ & \forall i \in I, j \in J & \text{(e)}
\end{array}
\qquad \text{(BDP2)}
$$

where

P^s is the set of pseudo-suppliers i corresponding to supplier s , and $c_{ij}^0 = c_{1j}^s$ is the unit list price of item j for pseudo-supplier i where $i \in P^s$.

The objective function in BDP2 is precisely that of a standard (either simple or capacitated) plant location model. Constraints BDP2(a) ensure that the total value of the suborder placed with pseudo-supplier i falls within pseudo-supplier i's price band. Note that in contrast to the usual CFLP model BDP2(a) contains both lower and upper bounds. Constraints BDP2(b) ensure that at most one price list per supplier can appear in an optimal solution. The demand constraints BDP2(c) are inequalities which allow for the possibility of exceeding demand which can sometimes, as shown by the example above, be advantageous to the buyer and reflect the feature "buy more-for-less" (see e.g. Goossens et al., [2]).

In practice stock limitations may restrict availability of some product j from some supplier s, creating an upper bound on x_{ij} for each pseudo-supplier $i \in P^s$. A supplier s may also for commercial reasons wish to restrict availability of some product j within a certain price band k. We therefore include in our formulation of BDP2 the upper bound constraints

$$0 \le x_{ij} \le S_{ij} \ i \in I, j \in J \tag{UB}$$

for given constants $\{S_{ij}\}$.

We observe that x_{ij} are formulated here as integer variables, but in other applications x_{ij} may be regarded as continuous, so that BDP2(e) may be replaced by $x_{ij} \in \mathbb{R}^+$. We also note that the formulation BDP2 contains some redundancy. It is clear that an optimal solution will automatically employ no more than one price band from any supplier. We can therefore technically remove all the upper bounds in BDP2(a) and in fact we may drop all the constraints BDP2(b) from the formulation. Retaining the additional constraints however allows tighter dual bounds in the solution approach by Lagrangean relaxation which we outline in the next section.

4 A Lagrangean Heuristic

Following [5], [6], [7] and [8], [9] in the context of location problems we define a Lagrangean relaxation of BDP2 by incorporating the demand constraints BDP2(c) in the form $D_j - \sum_{i \in I} x_{ij} \le 0 \ (j \in J)$ into the objective function with an associated vector of Lagrange multipliers $\lambda = (\lambda_1, \ldots, \lambda_n)$ where $\lambda_j \ge 0, j \in J$. The Lagrangean Dual Problem (LDP) corresponding to BDP2 can then be stated as

$$LDP: \quad \max_{\lambda \ge 0} F(\lambda)$$

where

$$F(\lambda) = \min_{y_i, x_{ij}} \left\{ \sum_{i=1}^{m} f_i y_i + \sum_{i=1}^{m} \sum_{j=1}^{n} (c_{ij} - \lambda_j) x_{ij} + \sum_{j=1}^{n} D_j \lambda_j \right\} \tag{LD}$$

subject to BDP2 and UB.

At any node of the BnB tree we consider a partition of I, the index set of $\{y_i\}$, into sets K_0, K_1, K_2 such that

$$y_i = 0, \quad i \in K_0, \qquad y_i = 1, \quad i \in K_1, \qquad y_i \in \{0,1\}, \quad i \in K_2$$

That is K_0, K_1 are the pseudo-suppliers fixed closed, open respectively; K_2 are the undetermined pseudo-suppliers. Letting $P_L = |K_1|$ and $P_U = |K_1 \cup K_2|$ we obtain explicit cardinality bounds

$$P_L \leq \sum_{i=1}^{m} y_i \leq P_U \tag{CB}$$

on the total number of actual suppliers used.

The solution to the subproblem of LDP for prescribed λ can be achieved by solving two knapsack problems. The first knapsack problem is defined for each non-closed pseudo-supplier $i \in K_1 \cup K_2$ and determines the contribution to the dual function (LD) from pseudo-supplier i if open :

$$\alpha_i(\lambda) = f_i + \min_{x_{ij}} \sum_{j=1}^{n} (c_{ij} - \lambda_j) x_{ij} \tag{KP1}$$

subject to BDP2(a),(e) and UB. We solve the continuous relaxation of this problem by a greedy heuristic, first ordering $\{x_{ij}\}_{j \in J}$ by non-decreasing value of the ratio $(c_{ij} - \lambda_j)/c_{ij}^0$ then setting the components of the solution x_{ij}^* in turn to their maximum value subject to BDP2(a) and UB.

The second knapsack problem is a minimization problem on the set of Boolean variables $\{y_i\}_{i \in I}$

$$F(\lambda) = \min_{y_i} \sum_{i \in K_2} \alpha_i(\lambda) y_i + \sum_{i \in K_1} \alpha_i(\lambda) + \sum_{j \in J} D_j \lambda_j \tag{KP2}$$

subject to BDP2(b),(d) and CB. The solution procedure is briefly outlined. For each supplier s, let $\beta_s = \min_{i \in P^s \cap K_2} \{\alpha_i(\lambda)\}$ and form the corresponding list of pseudo-suppliers $i_1, i_2, \ldots$ in non-decreasing order of β_s. Define the sequence of partial sums $\{S_t\}$ by

$$S_0 = \sum_{i \in K_1} \alpha_i(\lambda) + \sum_{j=1}^{n} D_j \lambda_j,$$

$$S_1 = S_0 + \alpha_{i_1}(\lambda),$$

$$\vdots$$

$$S_t = S_{t-1} + \alpha_{i_t}(\lambda).$$

The smallest value of S_{t^*} such that $P_L \leq t^* \leq P_U$ provides an optimal solution y^* to (KP2) and hence the solution to (LD) subject to the given constraints. Let Z_U be the value of the *incumbent* i.e. of the best feasible solution found so far. We decide the branch is fathomed if $S_{t^*} \geq Z_U$, otherwise we continue to develop this node.

5   Open and Close Penalties

Rules that seek to determine whether each y_i variable can be fixed at 0 or 1 in subsequent branchings from a node of a BnB tree were developed by Khumawala et al. [10] for the CFLP. Many authors including Beasley [7], Wu et al. [9], Klose et al. [11] have employed such rules in Lagrangean heuristics for location problems (see Sridaran [12] for a review). We provide explicit expressions in this section for the so-called open and close penalties appropriate to the model BDP2 (see Chan [13]).

Let M^* denote the set of pseudo-suppliers that have $y_i^* = 1$ in the solution S_{t^*} to the Lagrangean subproblem of LDP.

5.1   Close Penalties

For an open pseudo-supplier $i \in K_2 \cap M^*$, consider the change in $F(\lambda)$, say $\Delta F(\lambda)|_{y_i=0}$, as a result of setting $y_i = 0$. Let $\alpha_{\min}$ be the smallest value of $\alpha_l(\lambda)$ with $l \in K_2 \backslash M^*$ such that pseudo-supplier l can be feasibly opened in the solution to LDP. If closing i causes the lower limit on cardinality of M^* to be violated, l is forced to enter M^*. Otherwise forcing pseudo-supplier i closed allows l to be opened if S_{t^*} is reduced as a result. As a result of these considerations the increase in $F(\lambda)$ due to closing i is

$$\Delta F(\lambda)|_{y_i=0} = \begin{cases} -\alpha_i(\lambda) + \alpha_{\min}, & \text{if } |M^*| = P_L \\ -\alpha_i(\lambda) + \min\{0, \alpha_{\min}\}, & \text{if } |M^*| > P_L \end{cases} \tag{C}$$

If $F(\lambda) + \Delta F(\lambda)|_{y_i=0} \geq Z_U$ we discard the current subproblem with $y_i = 0$ added and fix $y_i = 1$ in all subsequent completions of this branch, i.e. we transfer i from K_2 to K_1.

5.2   Open Penalties

For each closed pseudo-supplier $i \in K_2 \backslash M^*$, we calculate the change $\Delta F(\lambda)|_{y_i=1}$ in $F(\lambda)$ as a result of setting $y_i = 1$. We need to distinguish between three cases.

CASE 1: $i \in P_s$ and supplier s is already represented in M^* by some pseudo-supplier l with $y_l^* = 1$ (i.e. P_s is already open in y^* at some other level of discount). This case results in a forced exchange of i with l.

CASE 2: Opening pseudo-supplier i violates the upper limit on cardinality of M^*, then some open pseudo-supplier $l \in K_2 \backslash M^*$, say, must be closed.

CASE 3: Opening pseudo-supplier i allows the *possibility* that some supplier $l \in K_2 \backslash M^*$ be closed.

Let $\alpha_{\max}$ be the maximum of $\alpha_l(\lambda)$ over $l \in K_2 \cap M^*$. The net change in $F(\lambda)$ can be summarized as:

$$\Delta F(\lambda)|_{y_i=1} = \begin{cases} \alpha_i(\lambda) - \alpha_l(\lambda), & \text{if } P_s \text{ is open at } l \\ \alpha_i(\lambda) - \alpha_{\max}, & \text{if } |M^*| = P_U \\ \alpha_i(\lambda) - \max\{0, \alpha_{\max}\}, & \text{otherwise} \end{cases} \tag{O}$$

If $f(\lambda) + \Delta F(\lambda)|_{y_i=1} \geq Z_U$ we fix $y_i = 0$ in all subsequent completions from this node and transfer i from K_2 to K_0.

6   Experiments with LINGO and Xpress-MP

We have evaluated the ability of standard commercial software to solve relatively small scale examples which however are typical of some real-world applications. We used the basic formulation BDP1 assuming a single price break for each supplier. Test data were generated using pseudo-random numbers to mimic real data (see [14] for further details). *Prices* were generated by superimposing

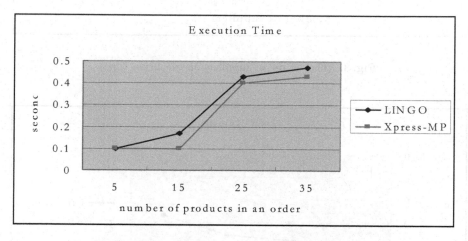

Fig. 1. Effect of order size on average execution time

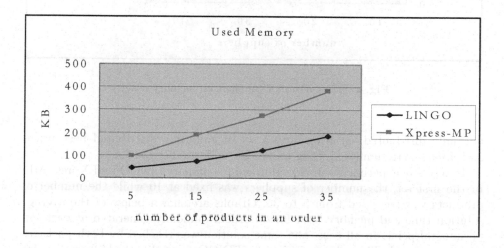

Fig. 2. Effect of order size on memory usage

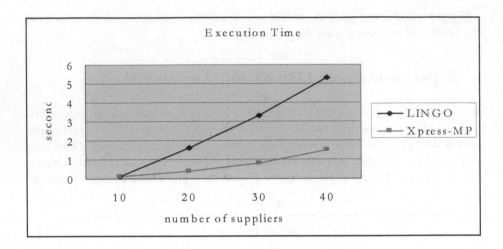

Fig. 3. Effect of no. of suppliers on execution time

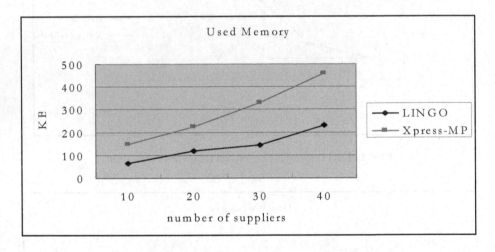

Fig. 4. Effect of no. of suppliers on memory usage

normally distributed variation onto a set of uniformly distributed base prices and *orders* were simulated using an integer uniform distribution.

Two sets of experiments were performed to compare LINGO and Xpress-MP. In the first set, the number of suppliers was fixed at 10 while the number of products was increased from 5 to 35. Graphs are shown below of the average solution time and memory used for 30 orders randomly generated at each experimental point. In all cases the same solution was reached by both packages within a second. The solution times for LINGO were slightly higher than for Xpress-MP, however Xpress-MP used more memory.

In the second set of experiments, the number of products was fixed at 10 and the number of suppliers varied between 10 and 40. Again 30 order instances were generated. and the average time taken to solve to optimality and the average memory usage were compared. Xpress-MP solved the 40 suppliers instances in an average time of less than 1 second, while LINGO took over 5 seconds on average. Xpress-MP's greater solution efficiency was perhaps related to its greater use of memory.

7   Summary and Future Research Directions

We have shown that an order allocation problem typically faced by trading intermediaries on the internet may be formulated as a mixed integer linear program (MILP) resembling a capacitated location problem, in which the capacities depend on pre-specified thresholds of order values. A Lagrangean heuristic is developed and explicit formulae for open and close penalties are provided. Our computational study using LINGO and XPress-MP show that many practically interesting instances are handled within seconds on a standard PC.

Example 1 (see Section 2) suggests that it may be possible to develop an effective rearrangement heuristic to determine a minimum cost order either exactly or approximately. We plan to investigate such a heuristic and perform computational experiments to examine whether such a heuristics can improve the performance of existing software. The algorithms developed and tested in this paper employ a depth first search (DFS) scheme. Such schemes economize on memory. However best first search (BFS) schemes are more useful if we wish to generate the minimum number of subproblems. The open and close penalties developed in this paper can easily be implemented for branch and bound (BnB) algorithms using BFS and it may be interesting to compare the efficiency of BnB algorithms for BDP developed using a BFS scheme.

Acknowledgements. We thank the three anonymous referees for their valuable comments and suggestions which have improved the presentation of our paper. The first and third authors gratefully acknowledge the financial support from the Manchester Institute of the Mathematical Sciences (MIMS), the School of Computer Science of the University of Manchester and the Research Institute (SOM) of the University of Groningen which have been invaluable in promoting this collaboration.

References

[1] Sadrian, A., Yoon, Y.: A procurement decision support system in business volume discounts environments. Operations Research **42** (1994) 14–23

[2] Goossens, D., Maas, A., Spieksma, F., van der Klundert, J.: Exact algorithms for procurement problems under a total quantity discount structure. Technical Report OR 0452, Faculty of Economics and Applied Economics, K.U. Leuven (2004)

[3] Krarup, J., Pruzan, P.M.: The simple plant location problem: Survey and synthesis. Eur. J. Oper. Res. **12** (1983) 36–81

[4] Cornuejols, G., Sridharan, R., Thizy, J.: A comparison of heuristics and relaxations for the capacitated plant location problem. Eur. J. Oper. Res. **50** (1991) 280–297

[5] Christofides, N., Beasley, J.: Extensions to a Lagrangean relaxation approach for the capacitated warehouse location problem. Eur. J. Oper. Res. **12** (1983) 19–28

[6] Beasley, J.: An algorithm for solving large capacitated warehouse location problems. Eur. J. Oper. Res. **33** (1988) 314–325

[7] Beasley, J.: Lagrangean heuristics for location problems. Eur. J. Oper. Res. **65** (1993) 383–399

[8] Correia, I., Captivo, M.: A Lagrangean heuristic for a modular capacitated location problem. Ann. Oper. Res. **122** (2003) 141–161

[9] Wu, L.Y., Zhang, X.S., Zhang, J.L.: Capacitated facility location problem with general setup cost. Comput. Oper. Res. **33** (2006) 1226–1241

[10] Akinc, U., Khumawala, B.M.: An efficient branch and bound algorithm for the capacitated warehouse location problem. Manage. Sci. **23** (1977) 585–594

[11] Klose, A., Drexl, A.: Facility location models for distribution system design. Eur. J. Oper. Res. **162** (2005) 4–29

[12] Sridharan, R.: The capacitated plant location problem. Eur. J. Oper. Res. **87** (1995) 203–213

[13] Chan, W.: An extension to the class of warehouse location problems via a discount scheme. Master's thesis, Department of Mathematics, UMIST (2001)

[14] Zhang, Y.: Optimization Modelling with LINGO and Xpress-MP. Master's thesis, Department of Computation, UMIST (2004)

Convex Hulls of Point-Sets and Non-uniform Hypergraphs

Hanno Lefmann

Fakultät für Informatik, TU Chemnitz, D-09107 Chemnitz, Germany
lefmann@informatik.tu-chemnitz.de

Abstract. For fixed integers $k \geq 3$ and hypergraphs $\mathcal{G}$ on N vertices, which contain edges of cardinalities at most k, and are uncrowded, i.e., do not contain cycles of lengths $2, 3$, or 4, and with average degree for the i-element edges bounded by $O(T^{i-1} \cdot (\ln T)^{(k-i)/(k-1)})$, $i = 3, \ldots, k$, for some number $T \geq 1$, we show that the independence number $\alpha(\mathcal{G})$ satisfies $\alpha(\mathcal{G}) = \Omega((N/T) \cdot (\ln T)^{1/(k-1)})$. Moreover, an independent set I of size $|I| = \Omega((N/T) \cdot (\ln T)^{1/(k-1)})$ can be found deterministically in polynomial time. This extends a result of Ajtai, Komlós, Pintz, Spencer and Szemerédi for uncrwoded uniform hypergraphs. We apply this result to a variant of Heilbronn's problem on the minimum area of the convex hull of small sets of points among n points in the unit square $[0,1]^2$.

1   Introduction

An independent set I in a graph or hypergraph $\mathcal{G} = (V, \mathcal{E})$ with vertex-set V and edge-set $\mathcal{E}$ is a subset of the vertex-set V, which does not contain any edges, i.e., $E \not\subseteq I$ for each edge $E \in \mathcal{E}$. The largest size of an independent set in $\mathcal{G}$ is the independence number $\alpha(\mathcal{G})$. For graphs $G = (V, E)$ with average degree $t := 2 \cdot |E|/|V| \geq 1$ Turán's theorem gives $\alpha(G) \geq |V|/(2 \cdot t)$. Turán's theorem for hypergraphs says, see [20]: If $\mathcal{G} = (V, \mathcal{E}_k)$ is a k-uniform hypergraph, i.e., all edges have cardinality k, with average degree $t^{k-1} := k \cdot |\mathcal{E}_k|/|V| \geq 1$, then $\alpha(\mathcal{G}) \geq ((k-1)/k) \cdot (|V|/t)$. An independent set $I \subseteq V$ in $\mathcal{G}$ achieving this lower bound can be found deterministically in time $O(|V| + |\mathcal{E}_k|)$. For uncrowded k-uniform hypergraphs $\mathcal{G} = (V, \mathcal{E}_k)$, i.e., $\mathcal{G}$ contains no cycles of length $2, 3$, or 4, Ajtai, Komlós, Pintz, Spencer and Szemerédi [1] improved this lower bound by a factor of $\Theta((\log t)^{1/(k-1)})$. Several applications of this result have been found, see [5]. Here we extend this result from [1] to non-uniform uncrowded hypergraphs:

Theorem 1. *Let $k \geq 3$ be a fixed integer. Let $\mathcal{G} = (V, \mathcal{E}_3 \cup \cdots \cup \mathcal{E}_k)$ be an uncrowded hypergraph on $|V| = N$ vertices, where $\mathcal{E}_i$ is the set of all i-element edges in $\mathcal{G}$, such that the average degrees $t_i^{i-1} := i \cdot |\mathcal{E}_i|/|V|$ for the i-element edges satisfy $t_i^{i-1} \leq c_i \cdot T^{i-1} \cdot (\ln T)^{(k-i)/(k-1)}$ for some number $T \geq 1$ with constants c_i, where $0 < c_i < 1/8 \cdot \binom{k-1}{i-1}/(10^{(3(k-i))/(k-1)} \cdot k^2)$, $i = 3, \ldots, k$.*

Then, for some constant $C_k > 0$ the independence number $\alpha(\mathcal{G})$ satisfies

$$\alpha(\mathcal{G}) \geq C_k \cdot (N/T) \cdot (\ln T)^{1/(k-1)}. \tag{1}$$

M.-Y. Kao and X.-Y. Li (Eds.): AAIM 2007, LNCS 4508, pp. 285–295, 2007.

An independent set $I \subseteq V$ with $|I| = \Omega((N/T) \cdot (\ln T)^{1/(k-1)})$ can be found deterministically in time $o(N \cdot T^{4k-4})$.

The corresponding result also holds for linear hypergraphs $\mathcal{G}$, which have the property that they do not contain cycles of length 2, i.e., each two distinct edges have at most one vertex in common, provided that $\mathcal{G}$ does not contain any 2-element edges. Theorem 1 is best possible up to a constant factor for a certain range $k < T < N$, as can be seen by considering random non-uniform hypergraphs $\mathcal{G} = (V, \mathcal{E}_3 \cup \cdots \cup \mathcal{E}_k)$ on $|V| = N$ vertices.

As an application we consider a variant of Heilbronn's problem for the convex hull of sets of points in the unit square $[0,1]^2$. The original problem of Heilbronn asks for a distribution of n points in $[0,1]^2$ such that the minimum area of a triangle determined by three of these n points achieves its largest value. For this problem, the points $1/n \cdot (i \bmod n, i^2 \bmod n)$, $i = 0, \ldots, n-1$, where n is a prime, give the lower $\Omega(1/n^2)$ on the minimum area of a triangle. This lower bound has been improved in [12] by a factor $\Omega(\log n)$, see [6] for a deterministic polynomial time algorithm. Upper bounds on the minimum area of a triangle among n points in $[0,1]^2$ were given by Roth [15,16,17,18] and Schmidt [19] and, the currently best upper bound $O(2^{c\sqrt{\log n}}/n^{8/7})$, $c > 0$ a constant, is due to Komlós, Pintz and Szemerédi [11].

Variants of Heilbronn's triangle problem in higher dimensions were investigated in [2,3,4,7,8,13]. A generalization of Heilbronn's triangle problem to k points, see Schmidt [19], asks, given an integer $k \geq 3$, for the supremum $\Delta_k(n)$ over all distributions of n points in $[0,1]^2$ of the minimum area of the convex hull determined by some k of n points. In [6] it has been shown that $\Delta_k(n) = \Omega(1/n^{(k-1)/(k-2)})$ for fixed $k \geq 3$, and any integers $n \geq k$; for $k = 4$ this was proved in [19]. This has been improved in [14] to $\Delta_k(n) = \Omega((\log n)^{1/(k-1)}//n^{(k-1)/(k-2)})$ for fixed $k \geq 3$. Currently, for $k \geq 4$ only the upper bound $\Delta_k(n) = O(1/n)$ is known.

Here we show for fixed integers $k \geq 3$, that one can achieve these lower bounds simultaneously for $j = 3, \ldots, k$ by a single configuration of n points in $[0,1]^2$.

Theorem 2. *Let $k \geq 3$ be a fixed integer. For integers $n \geq k$ there exists a configuration of n points in $[0,1]^2$, such that, simultaneously for $j = 3, \ldots, k$, the area of the convex hull of any j of the n points is $\Omega((\log n)^{1/(j-1)}/n^{(j-1)/(j-2)})$.*

By considering the standard $L \times L$-grid for a suitable integer $L \geq n$ one can also give a polynomial time algorithm which achieves the lower bounds from Theorem 2 on the areas of the convex hulls. (Details are omitted.)

2   Uncrowded and Linear Hypergraphs

Definition 1. *A hypergraph is a pair $\mathcal{G} = (V, \mathcal{E})$ with vertex-set V and edge-set $\mathcal{E}$, where $E \subseteq V$ for each edge $E \in \mathcal{E}$. For a hypergraph $\mathcal{G}$ the notation $\mathcal{G} = (V, \mathcal{E}_2 \cup \cdots \cup \mathcal{E}_k)$ indicates that $\mathcal{E}_i$ is the set of all i-element edges in $\mathcal{G}$, $i = 2, \ldots, k$. For a vertex $v \in V$ let $d_i(v)$ denote the number of i-element edges*

$E \in \mathcal{E}_i$ which contain v, i.e., $d_i(v)$ is the degree of v for the i-element edges in $\mathcal{G}$. The independence number $\alpha(\mathcal{G})$ of $\mathcal{G} = (V, \mathcal{E})$ is the largest size of a subset $I \subseteq V$ which contains no edges from $\mathcal{E}$. A j-cycle in a hypergraph $\mathcal{G} = (V, \mathcal{E})$ is a sequence $E_1, \ldots, E_j$ of distinct edges from $\mathcal{E}$, such that $E_i \cap E_{i+1} \neq \emptyset$, $i = 1, \ldots, j-1$, and $E_j \cap E_1 \neq \emptyset$, and a sequence $v_1, \ldots, v_j$ of distinct vertices with $v_{i+1} \in E_i \cap E_{i+1}$, $i = 1, \ldots, j-1$, and $v_1 \in E_1 \cap E_j$. An unordered pair $\{E, E'\}$ of distinct edges $E, E' \in \mathcal{E}$ with $|E \cap E'| \geq 2$ is a 2-cycle. A 2-cycle $\{E, E'\}$ in $\mathcal{G} = (V, \mathcal{E}_2 \cup \cdots \cup \mathcal{E}_k)$ with $E \in \mathcal{E}_i$ and $E' \in \mathcal{E}_j$ is called $(2; (g, i, j))$-cycle if and only if $|E \cap E'| = g$, $2 \leq g \leq i \leq j$ but $g < j$. The hypergraph $\mathcal{G}$ is called linear if it does not contain any 2-cycles, and $\mathcal{G}$ is called uncrowded if it does not contain any 2-, 3-, or 4-cycles.

For uncrowded k-uniform hypergraphs with average degree t^{k-1} the Turán bound on the independence number has been improved in [1] by a factor $\Theta((\log t)^{1/(k-1)})$, see [5] and [10] for a deterministic polynomial time algorithm.

Theorem 3. Let $k \geq 3$ be a fixed integer. Let $\mathcal{G} = (V, \mathcal{E}_k)$ be an uncrowded k-uniform hypergraph on $|V| = N$ vertices and with average degree $t^{k-1} := k \cdot |\mathcal{E}_k|/N$. Then, for some constant $C_k > 0$, the independence number $\alpha(\mathcal{G})$ satisfies $\alpha(\mathcal{G}) \geq C_k \cdot (N/t) \cdot (\log t)^{1/(k-1)}$.

To prove Theorem 3, in [1] the following central lemma has been used to construct iteratively a large independent set in a hypergraph, which we use in our arguments too; see [10] for a deterministic polynomial time algorithm.

Lemma 1. Let T and N be large positive integers. Let s be an integer with $0 \leq s \leq (\ln T)/10^2$. Let $w_s := (s+1)^{1/(k-1)} - s^{1/(k-1)}$ and $\varepsilon := 10^{-6}/\ln T$. Let $N/(2 \cdot e^s) \leq n \leq N/e^s$ and $T/(2 \cdot e^s) \leq t \leq T/e^s$.

Let $\mathcal{G} = (V, \mathcal{E}_2 \cup \cdots \cup \mathcal{E}_k)$ be an uncrowded hypergraph with $|V| = n$ vertices, where for each vertex $v \in V$ the degrees $d_i(v)$ for the i-element edges satisfy $d_i(v) \leq \binom{k-1}{i-1} \cdot s^{(k-i)/(k-1)} \cdot t^{i-1}$, $i = 2, \ldots, k$.

Then, one can find in time $O(n \cdot t^{4(k-1)})$ an independent set $I \subseteq V$ in $\mathcal{G}$, a subset $V^* \subset V$ with $V^* \cap I = \emptyset$, and a hypergraph $\mathcal{G}^* = (V^*, \mathcal{E}_2^* \cup \cdots \cup \mathcal{E}_k^*)$ such that

(i) $\alpha(\mathcal{G}) \geq |I| + \alpha(\mathcal{G}^*)$ and (ii) $|I| \geq 0.99 \cdot \frac{n \cdot w_s}{e \cdot t}$ and (iii) $|V^*| \geq \frac{n \cdot (1-\varepsilon)}{e}$

(iv) $d_i^*(v) \leq \binom{k-1}{i-1} \cdot (s+1)^{(k-i)/(k-1)} \cdot (t \cdot (1+\varepsilon)/e)^{i-1}$ for each vertex $v \in V^*$, where $d_i^*(v)$ denotes the degree of v for the i-element edges in $\mathcal{G}^*$, $i = 2, \ldots, k$.

Lemma 2. Let $k \geq 3$ be a fixed integer. Let $\mathcal{G} = (V, \mathcal{E}_2 \cup \cdots \cup \mathcal{E}_k)$ be a hypergraph with $|V| = N$ and $N \geq 65 \cdot (\ln k)^{1000/998}$, where the average degrees $t_i^{i-1} := i \cdot |\mathcal{E}_i|/N$ for the i-element edges in $\mathcal{E}_i$ fulfill $t_i^{i-1} \leq c_i \cdot T^{i-1} \cdot (\ln T)^{(k-i)/(k-1)}$ for some number $T > 1$ and for some constants $c_i > 0$ with $c_i < 1/8 \cdot \binom{k-1}{i-1}/(10^{(3(k-i))/(k-1)} \cdot k^2)$, $i = 2, \ldots, k$.

Then, for $s := 10^{-3} \cdot \ln T$, one can find in time $O(|V| + \sum_{i=2}^{k} |\mathcal{E}_i|)$ an induced subhypergraph $\mathcal{G}^* = (V^*, \mathcal{E}_2^* \cup \cdots \cup \mathcal{E}_k^*)$ on $|V^*| = n$ vertices with $\mathcal{E}_i^* := \mathcal{E}_i \cap [V^*]^i$,

$i = 2, \ldots, k$, such that $(3/4) \cdot N/e^s \le n \le N/e^s$ and for each vertex $v \in V^*$ the degrees $d_i^*(v)$ for the i-element edges in $\mathcal{G}^*$ satisfy

$$d_i^*(v) \le \binom{k-1}{i-1} \cdot s^{\frac{k-i}{k-1}} \cdot (T/e^s)^{i-1}. \tag{2}$$

Proof. We pick vertices with probability $p := 1/e^s$ uniformly at random and independently of each other from the vertex-set V in $\mathcal{G}$. Let V^* be the random set of chosen vertices of expected size $E[|V^*|] = p \cdot N$. With $s = 10^{-3} \cdot \ln T$ and $T = O(N)$, we have by Chernoff's inequality for $N \ge 65 \cdot (\ln k)^{1000/998}$:

$$\text{Prob } (E[|V^*|] - |V^*| > N/(8 \cdot e^s)) \le e^{-\frac{N^2/(64 \cdot e^{2s})}{N}} = e^{-N/(64 \cdot e^{2s})} < 1/k. \tag{3}$$

Let $\mathcal{E}_i^* := \mathcal{E}_i \cap [V^*]^i$, $i = 2, \ldots, k$, and let $\mathcal{G}^* = (V^*, \mathcal{E}_2^* \cup \cdots \cup \mathcal{E}_k^*)$ be the on V^* induced random subhypergraph of $\mathcal{G}$. For $i = 2, \ldots, k$, we have for the expected numbers $E[|\mathcal{E}_i^*|] = p^i \cdot |\mathcal{E}_i| = p^i \cdot N \cdot t_i^{i-1}/i \le p^i \cdot c_i \cdot T^{i-1} \cdot (\ln T)^{(k-i)/(k-1)} \cdot N/i$. By Markov's inequality it is Prob $(|\mathcal{E}_i^*| > k \cdot E[|\mathcal{E}_i^*|]) \le 1/k$, hence with (3) there exists a subhypergraph $\mathcal{G}^* = (V^*, \mathcal{E}_2^* \cup \cdots \cup \mathcal{E}_k^*)$ of $\mathcal{G}$ such that for $i = 2, \ldots, k$:

$$|V^*| \ge (7/8) \cdot N/e^s \quad \text{and} \quad |\mathcal{E}_i^*| \le k \cdot p^i \cdot c_i \cdot T^{i-1} \cdot (\ln T)^{(k-i)/(k-1)} \cdot N/i. \tag{4}$$

Let n_i be the number of vertices $v \in V^*$ with degree $d_i^*(v) \ge 8 \cdot e^s \cdot k^2 \cdot p^i \cdot c_i \cdot T^{i-1} \cdot (\ln T)^{(k-i)/(k-1)}$ for the i-element-edges in $\mathcal{G}^*$, $i = 2, \ldots, k$. By (4) we infer $n_i \le N/(8 \cdot k \cdot e^s) \le |V^*|/(7 \cdot k)$, thus $\sum_{i=2}^k n_i < |V^*|/7$. We delete these vertices from V^* and obtain a subset $V^{**} \subseteq V^*$ with $|V^{**}| \ge (6/7) \cdot |V^*|$. For the induced subhypergraph $\mathcal{G}^{**} = (V^{**}, \mathcal{E}_2^{**} \cup \cdots \cup \mathcal{E}_k^{**})$ of $\mathcal{G}^*$ with $\mathcal{E}_i^{**} := \mathcal{E}_i \cap [V^{**}]^i$, $i = 2, \ldots, k$, we infer with (4) for each vertex $v \in V^{**}$:

$$|V^{**}| \ge (3/4) \cdot N/e^s \text{ and } d_i^{**}(v) \le 8 \cdot k^2 \cdot c_i \cdot (T/e^s)^{i-1} \cdot (\ln T)^{(k-i)/(k-1)},$$

where $d_i^{**}(v)$ is the degree of v for the i-element edges in $\mathcal{G}^{**}$. For $s := 10^{-3} \cdot \ln T$ and $c_i < 1/8 \cdot \binom{k-1}{i-1}/(10^{(3(k-i))/(k-1)} \cdot k^2)$, $i = 2, \ldots, k$, we have

$$d_i^{**}(v) \le 8 \cdot k^2 \cdot c_i \cdot (T/e^s)^{i-1} \cdot (\ln T)^{\frac{k-i}{k-1}} \le \binom{k-1}{i-1} \cdot s^{\frac{k-i}{k-1}} \cdot (T/e^s)^{i-1},$$

which proves (2). By possibly deleting some more vertices and all incident edges we obtain $(3/4) \cdot N/e^s \le |V^{**}| \le N/e^s$. This probabilistic argument can be derandomized by using the method of conditional probabilities and yields a deterministic algorithm with running time $O(|V| + \sum_{i=2}^k |\mathcal{E}_i|)$. □

We prove Theorem 1 with an approach similar to that in [1]. The difference between their arguments and ours is, that we do not apply Lemma 1 step by step from the beginning, but use first Lemma 2 to jump to a suitable subhypergraph:

Proof. Apply Lemma 2 with $s := 10^{-3} \cdot \ln T$ to the hypergraph $\mathcal{G} = (V, \mathcal{E}_2 \cup \cdots \cup \mathcal{E}_k)$ on N vertices and obtain an induced subhypergraph $\mathcal{G}_{s-1} := (V_{s-1}, \mathcal{E}_{2;s-1} \cup \cdots \cup \mathcal{E}_{k;s-1})$ on n vertices with $\mathcal{E}_{i;s-1} := \mathcal{E}_i \cap [V_{s-1}]^i$, $i = 2, \ldots, k$, and with $(3/4) \cdot$

$N/e^s \leq n \leq N/e^s$, and for each vertex $v \in V_{s-1}$ its degree $d_{i;s-1}(v)$ in $\mathcal{G}_{s-1}$ for the i-element edges in $\mathcal{E}_{i;s-1}$ satisfies $d_{i;s-1}(v) \leq \binom{k-1}{i-1} \cdot s^{(k-i)/(k-1)} \cdot (T/e^s)^{i-1}$. Set $n_{s-1} := n$ and $t_{s-1} := T/e^s$. By iteratively applying Lemma 1 as in [1] with $\varepsilon := 10^{-6}/\ln T$ to the hypergraphs $\mathcal{G}_{r-1}$, we obtain for $r = s, \ldots, 10^{-2} \cdot \ln T$ independent sets $I_r \subseteq V_{r-1}$ and hypergraphs $\mathcal{G}_r = (V_r, \mathcal{E}_{2;r} \cup \cdots \cup \mathcal{E}_{k;r})$ with $|V_r| = n_r$, where $(3/4) \cdot N \cdot (1 - \varepsilon)^{r+1-s}/e^{r+1} \leq n_r \leq N/e^{r+1}$ with numbers $t_r \leq T \cdot (1 + \varepsilon)^{r+1-s}/e^{r+1}$, such that

$$\alpha(\mathcal{G}_r) \geq |I_r| + \alpha(\mathcal{G}_{r+1}) \quad \text{and} \quad |I_r| \geq (0.99 \cdot n_{r-1} \cdot w_r)/(e \cdot t_{r-1})$$
$$|V_r| \geq (n_{r-1} \cdot (1 - \varepsilon))/e$$
$$d_{i;r}(v) \leq \binom{k-1}{i-1} \cdot (r+1)^{\frac{k-i}{k-1}} \cdot (t_r)^{i-1}$$

for each $v \in V_r$, where $d_{i;r}(v)$ is the degree for the i-element edges in $\mathcal{G}_r$ of v.

With $(1 + \epsilon)^n > 1 + \varepsilon \cdot n$, $1 + \varepsilon \leq e^\varepsilon$, $r \leq 10^{-2} \cdot \ln T$ and $\varepsilon = 10^{-6}/\ln T$ we have

$$\frac{n_r}{t_r} \geq \frac{(3/4) \cdot N \cdot (1 - \varepsilon)^{r+1-s}/e^{r+1}}{T \cdot (1 + \varepsilon)^{r+1-s}/e^{r+1}} \geq \frac{(3/4) \cdot N}{T} \cdot \frac{(1 - \varepsilon)^r}{(1 + \varepsilon)^r} \geq 0.74 \cdot \frac{N}{T}.$$

Hence, with $w_r = (r + 1)^{1/(k-1)} - r^{1/(k-1)}$ and $s = 10^{-3} \cdot \ln T$, we obtain for some constant $C_k > 0$ an independent set $I = I_s \cup \cdots \cup I_{(\ln T)/10^2}$ in $\mathcal{G}$ with

$$\alpha(\mathcal{G}) \geq |I| = \sum_{r=s}^{(\ln T)/10^2} |I_r| \geq 0.99 \cdot \frac{0.74}{e} \cdot \frac{N}{T} \cdot \sum_{r=s}^{(\ln T)/10^2} w_r \geq$$
$$\geq \frac{0.73}{e} \cdot \frac{N}{T} \cdot \sum_{r=s}^{(\ln T)/10^2} ((r+1)^{\frac{1}{k-1}} - r^{\frac{1}{k-1}}) \geq C_k \cdot \frac{N}{T} \cdot (\ln T)^{\frac{1}{k-1}},$$

which gives the lower bound (1) in Theorem 1. The time bound for the corresponding deterministic algorithm can be estimated as follows: Lemma 2 is applied in time $O(|V| + \sum_{i=2}^{k} |\mathcal{E}_i|)$ and all applications of Lemma 1 are done in time $O(\sum_{r=(\ln T)/10^3}^{(\ln T)/10^2} ((N/e^r) \cdot (T \cdot (1 + \varepsilon)^{r+1-s}/e^{r+1})^{4(k-1)})) = o(N \cdot T^{4(k-1)})$, compare Lemma 1, hence we have the time bound $o(N \cdot T^{4(k-1)})$. □

In [9] it has been shown that one may relax in Theorem 3 the assumptions: it suffices to have a linear hypergraph. Similarly, one can show:

Theorem 4. Let $k \geq 3$ be a fixed integer. Let $\mathcal{G} = (V, \mathcal{E}_3 \cup \cdots \cup \mathcal{E}_k)$ be a linear hypergraph with $|V| = N$ such that the average degrees $t_i^{i-1} := i \cdot |\mathcal{E}_i|/|V|$ for the i-element edges satisfy $t_i^{i-1} \leq c_i \cdot T^{i-1} \cdot (\ln T)^{(k-i)/(k-1)}$ for some number $T \geq 1$, where $c_i > 0$ are constants with $c_i < 1/32 \cdot \binom{k-1}{i-1}/(10^{(3(k-i))/(k-1)} \cdot k^6)$, $i = 3, \ldots, k$.

Then, for some constant $C_k > 0$, one can find deterministically in time $O(N \cdot T^{4k-2})$ an independent set $I \subseteq V$ such that $|I| = \Omega((N/T) \cdot (\ln T)^{1/(k-1)})$.

3   Areas of the Convex Hull of j Points

For distinct points $P, Q \in [0,1]^2$ let PQ denote the *line* through P and Q, and let $[P, Q]$ be the *segment* between P and Q. Let dist (P, Q) denote the *Euclidean distance* between the points P and Q. For points $P_1, \ldots, P_l \in [0,1]^2$ let area $(P_1, \ldots, P_l)$ be the area of the convex hull of $P_1, \ldots, P_l$. A *strip* centered at the line PQ of width w is the set of all points in $\mathbb{R}^2$, which are at Euclidean distance at most $w/2$ from the line PQ. We define a lexicographic order $\leq_{lex}$ on the unit square $[0,1]^2$: for points $P = (p_x, p_y) \in [0,1]^2$ and $Q = (q_x, q_y) \in [0,1]^2$ let $P \leq_{lex} Q :\Longleftrightarrow (p_x < q_x)$ or $(p_x = q_x$ and $p_y < q_y)$.

Lemma 3. *(a) Let $P_1, \ldots, P_l \in [0,1]^2$, $l \geq 3$, be points. If area $(P_1, \ldots, P_l) \leq A$, then for any distinct points P_i, P_j every other point P_k, $k \neq i, j$, is contained in a strip centered at the line $P_i P_j$ of width $4 \cdot A / \text{dist}(P_i, P_j)$.*

(b) Let $P, R \in [0,1]^2$ be distinct points with $P \leq_{lex} R$. Then all points $Q \in [0,1]^2$ with $P \leq_{lex} Q \leq_{lex} R$ and area $(P, Q, R) \leq A$ are contained in a parallelogram of area $4 \cdot A$.

In the following we prove Theorem 2.

Proof. Let $k \geq 3$ be a fixed and let $n \geq k$ be any integer. For a constant $\beta > 0$, which will be specified later, we select uniformly at random and independently of each other $N := n^{1+\beta}$ points $P_1, \ldots, P_N$ in $[0,1]^2$. Set $D_0 := N^{-\gamma}$ for a constant γ with $0 < \gamma < 1$ and let $A_3, \ldots, A_k > 0$ be numbers, which will be fixed later. We form a random hypergraph $\mathcal{G} = (V, \mathcal{E}_2 \cup \cdots \cup \mathcal{E}_k)$ with vertex-set $V = \{1, \ldots, N\}$, where vertex $i \in V$ corresponds to the random point $P_i \in [0,1]^2$, and with edges of cardinality at most k. Let $\{i_1, i_2\} \in \mathcal{E}_2$ if and only if $\text{dist}(P_{i_1}, P_{i_2}) \leq D_0$. Moreover, for $j = 3, \ldots, k$, let $\{i_1, \ldots, i_j\} \in \mathcal{E}_j$ if and only if area $(P_{i_1}, \ldots, P_{i_j}) \leq A_j$ and $\{i_1, \ldots, i_j\}$ does not contain any edges from $\mathcal{E}_2$.

We want to find a large independent set $I \subseteq V$ in $\mathcal{G}$, as I yields a subset $P(I) \subseteq [0,1]^2$ of size $|I|$ such that the area of the convex hull of each j distinct points, $j = 3, \ldots, k$, from $P(I)$ is bigger than A_j. To do so, first we estimate the expected numbers $E[|\mathcal{E}_j|]$ of j-element edges and $E[s_{2;(g,i,j)}(\mathcal{G})]$ of $(2; (g, i, j))$-cycles in $\mathcal{G}$, and we prove that these numbers are not too big. Then we show the existence of a certain induced, linear subhypergraph $\mathcal{G}^* = (V, \mathcal{E}_3^* \cup \cdots \cup \mathcal{E}_k^*)$ (no 2-element edges anymore) of $\mathcal{G}$, which satisfies the assumptions of Theorem 4, and then we obtain a large independent set.

Lemma 4. *For $j = 3, \ldots, k$, there exist constants $c_j > 0$ such that*

$$E[|\mathcal{E}_j|] \leq c_j \cdot A_j^{j-2} \cdot N^j. \tag{5}$$

Proof. For integers $i_1, \ldots, i_j$ with $1 \leq i_1 < \cdots < i_j \leq N$ we estimate the probability Prob (area $(P_{i_1}, \ldots, P_{i_j}) \leq A_j$). We may assume that $P_{i_1} \leq_{lex} \cdots \leq_{lex} P_{i_j}$. Then area $(P_{i_1}, \ldots, P_{i_j}) \leq A_j$ implies area $(P_{i_1}, P_{i_g}, P_{i_j}) \leq A_j$ for $g = 2, \ldots, j - 1$. The points P_{i_1} and P_{i_j} with $P_{i_1} \leq_{lex} P_{i_j}$ may be anywhere in $[0,1]^2$. Given $P_{i_1}, P_{i_j} \in [0,1]^2$, by Lemma 3(b) all points P_{i_g}, $g = 2, \ldots, j - 1$, are contained in a parallelogram of area $4 \cdot A_j$, which happens with probability at most $(4 \cdot A_j)^{j-2}$. As there are $\binom{N}{j}$ choices for j out of N points, for some constants $c_j > 0$, $j = 3, \ldots, k$, we obtain $E[|\mathcal{E}_j|] \leq c_j \cdot A_j^{j-2} \cdot N^j$.   □

Next we estimate the expected numbers $E[s_{2;(g,i,j)}(\mathcal{G})]$ of $(2;(g,i,j))$-cycles, $2 \leq g \leq i \leq j \leq k$ but $g < j$, in $\mathcal{G} = (V, \mathcal{E}_2 \cup \cdots \cup \mathcal{E}_k)$.

Lemma 5. *Let* $2 \leq g \leq i \leq j \leq k$ *with* $i \geq 3$ *and* $g < j$, *and let* $0 < A_3 \leq \cdots \leq A_k$. *Then, there exist constants* $c_{2;(g,i,j)} > 0$ *such that for* $D_0^2 \geq 2 \cdot A_j$ *it is*

$$E[s_{2;(g,i,j)}(\mathcal{G})] \leq c_{2;(g,i,j)} \cdot A_i^{i-2} \cdot A_j^{j-g} \cdot N^{i+j-g} \cdot (\log N)^3. \qquad (6)$$

Proof. We estimate the probability that $(i + j - g)$ points, which are chosen uniformly at random and independently of each other in $[0,1]^2$, form sets of i and j points with areas of the convex hulls at most A_i and A_j, respectively, conditioned on the event that distinct points have Euclidean distance bigger than D_0. Both sets have g points in common, say $P_1, \ldots, P_g$, where $P_1 \leq_{lex} \cdots \leq_{lex} P_g$. Let the sets of i and j points be $P_1, \ldots, P_i$ and $P_1, \ldots, P_g, Q_{g+1}, \ldots, Q_j$ with area $(P_1, \ldots, P_i) \leq A_i$ and area $(P_1, \ldots, P_g, Q_{g+1}, \ldots, Q_j) \leq A_j$, respectively.

The point P_1 may be anywhere in $[0,1]^2$. Given $P_1 \in [0,1]^2$, we have Prob $(r \leq \text{dist}(P_1, P_g) \leq r + dr) \leq \pi \cdot r \, dr$. Given $P_1, P_g \in [0,1]^2$ with dist $(P_1, P_g) = r$, by Lemma 3(b) all points $P_2, \ldots, P_{g-1}$ are contained in a parallelogram with area $4 \cdot A_i$, which happens at most with probability $(4 \cdot A_i)^{g-2}$.

Given $P_1, \ldots, P_g \in [0,1]^2$ with dist $(P_1, P_g) = r$, by Lemma 3(a) all points $P_{g+1}, \ldots, P_i$ are contained in a strip S_i of width $w = 4 \cdot A_i/r$, and all points $Q_{g+1}, \ldots, Q_j$ are contained in a strip S_j of width $w = 4 \cdot A_j/r$, where both strips are centered at the line $P_1 P_g$. Set $S_i^* := S_i \cap [0,1]^2$ and $S_j^* := S_j \cap [0,1]^2$, which have areas at most $4 \cdot \sqrt{2} \cdot A_i/r$ and $4 \cdot \sqrt{2} \cdot A_j/r$, respectively.

For the convex hulls of $P_1, \ldots, P_i$ and $P_1, \ldots, P_g, Q_{g+1}, \ldots, Q_g$ we denote their *extremal* points by P', P'' and Q', Q', respectively, i.e., $P', P'' \in \{P_1, \ldots, P_i\}$ and $Q', Q'' \in \{P_1, \ldots, P_g, Q_{g+1}, \ldots, Q_j\}$ and, say $P' \leq_{lex} P''$ and $Q' \leq_{lex} Q''$, it is $P' \leq_{lex} P_1, \ldots, P_i \leq_{lex} P''$ and $Q' \leq_{lex} P_1, \ldots, P_g, Q_{g+1}, \ldots, Q_j \leq_{lex} Q''$.

Given $P_1 \leq_{lex} \cdots \leq_{lex} P_g$, there are three possibilities each for the convex hulls of $P_1, \ldots, P_i$ and $P_1, \ldots, P_g, Q_{g+1}, \ldots, Q_j$: extremal are (i) P_1 and P_g, or (ii) only one point, P_1 or P_g, or (iii) none of P_1 and P_g.

Consider the convex hull of the points $P_1, \ldots, P_i$. In case (i), given $P_1, \ldots, P_g \in [0,1]^2$ with dist $(P_1, P_g) = r$, as in the proof of Lemma 4 we infer

$$\text{Prob (area }(P_1, \ldots, P_i) \leq A_i \mid P_1, \ldots, P_g \text{ ; case (i))} \leq (4 \cdot A_i)^{i-g}. \qquad (7)$$

In case (ii), either P_1 or P_g is extremal for the convex hull of $P_1, \ldots, P_i$. By Lemma 3(a), the second extremal point is contained in the set S_i^*, which happens with probability at most $4 \cdot \sqrt{2} \cdot A_i/r$. Given both extremal points $P', P'' \in [0,1]^2$, by Lemma 3(b) all points $P_{g+1}, \ldots, P_i \neq P', P''$ are contained in a parallelogram of area $4 \cdot A_i$, hence, with dist $(P_1, P_g) = r$ we infer

$$\text{Prob(area}(P_1, \ldots, P_i) \leq A_i \mid P_1, \ldots, P_g \text{ ; case (ii))} \leq ((4 \cdot A_i)^{i-g} \cdot \sqrt{2})/r. \qquad (8)$$

In case (iii) neither point P_1 nor P_g is extremal for the convex hull of $P_1, \ldots, P_i$. With area $(P_1, \ldots, P_i) \leq A_i$, by Lemma 3(a) both extremal points P' and P'', say $P' \leq_{lex} P_1 \leq_{lex} P_g \leq_{lex} P''$, are contained in the strip S_i of width $4 \cdot A_i/r$, which is centered at the line $P_1 P_g$. Given $P_1 \in [0,1]^2$, the probability that dist $(P_1, P') \in [s, s+ \mathrm{d}s]$ is at most the difference of the areas of the balls with center P_1 and with radii $(s+ \mathrm{d}s)$ and s, respectively, intersected with the strip S_i. Since distinct points have Euclidean distance bigger than D_0, we have $r, s > D_0$. A circle with center P_1 and radius $s > D_0$ intersects both boundaries of the strip S_i of width $4 \cdot A_i/r$ in four points $R \leq_{lex} R'$ and $R'' \leq_{lex} R'''$, where R, R' are on one boundary of the strip S_i and R'', R''' are on the other boundary. To see this, notice that $s > 2 \cdot A_i/r$ follows from $r, s > D_0$ and $D_0^2 > 2 \cdot A_j \geq 2 \cdot A_i$. Let $\varepsilon(s)$ be the angle between the lines $P_1 R$ and $P_1 R''$. Then, by using $\varepsilon/2 \leq \sin \varepsilon$ for $\varepsilon \leq \pi/2$ and $\sin(\varepsilon(s)/2) = 2 \cdot A_i/(r \cdot s) < 2 \cdot A_i/D_0^2 \leq 1$, we infer

$$\text{Prob (dist } (P_1, P') \in [s, s+ \mathrm{d}s] \mid P_1) \leq ((2 \cdot \varepsilon(s)))/(2 \cdot \pi) \cdot 2 \cdot \pi \cdot s \, \mathrm{d}s \leq$$
$$\leq 8 \cdot \sin(\varepsilon(s)/2) \cdot s \, \mathrm{d}s = (16 \cdot A_i/r) \, \mathrm{d}s .$$

Given $P' \in [0,1]^2$ with dist $(P_1, P') = s$, the second extremal point $P'' \in [0,1]^2$ is contained in a strip centered at the line $P_1 P'$ of width $4 \cdot A_i/s$, which occurs with probability at most $4 \cdot \sqrt{2} \cdot A_i/s$. Given both points P', P'', by Lemma 3(b) all points $P_{g+1}, \ldots, P_i \neq P', P''$ are contained in a parallelogram of area $4 \cdot A_i$. Hence, given $P_1, \ldots, P_g \in [0,1]^2$, with $s > D_0 = N^{-\gamma}$ and $\gamma > 0$, we infer:

$$\text{Prob (area } (P_1, \ldots, P_i) \leq A_i \mid P_1, \ldots, P_g \text{ ; case (iii)})$$
$$\leq (4 \cdot A_i)^{i-g} \cdot \int_{D_0}^{\sqrt{2}} \frac{4 \cdot \sqrt{2}}{r \cdot s} \, \mathrm{d}s = \sqrt{32} \cdot (4 \cdot A_i)^{i-g} \cdot \frac{\ln \sqrt{2} + \gamma \cdot \ln N}{r} . \qquad (9)$$

Summarizing cases (i–iii) with (7)–(9), and $r \leq \sqrt{2}$ and $0 < \gamma < 1$ we obtain:

$$\text{Prob (area } (P_1, \ldots, P_i) \leq A_i \mid P_1, \ldots, P_g)$$
$$\leq (4 \cdot A_i)^{i-g} \cdot \frac{\sqrt{8} + \sqrt{8} \cdot (\ln 2 + 2 \cdot \gamma \cdot \ln N)}{r} \leq (4 \cdot A_i)^{i-g} \cdot \frac{11 \cdot \ln N}{r}. \qquad (10)$$

Similarly, it follows Prob (area $(P_1, \ldots, P_g, Q_{g+1}, \ldots, Q_g) \leq A_j \mid P_1, \ldots, P_g) \leq ((4 \cdot A_j)^{j-g} \cdot 11 \cdot \ln N)/r$ holds. Hence, we obtain for constants $c^*_{2;(g,i,j)} > 0$:

$$\text{Prob } (P_1, \ldots, P_i \text{ and } P_1, \ldots, P_g, Q_{g+1}, \ldots, Q_j \text{ is a } (2; (g,i,j))\text{-cycle}) \leq$$
$$\leq \int_{D_0}^{\sqrt{2}} (4 \cdot A_i)^{g-2} \cdot \left((4 \cdot A_i)^{i-g} \cdot \frac{11 \cdot \ln N}{r} \right) \cdot \left((4 \cdot A_j)^{j-g} \cdot \frac{11 \cdot \ln N}{r} \right) \cdot \pi \cdot r \, \mathrm{d}r$$
$$\leq c^*_{2;(g,i,j)} \cdot A_i^{i-2} \cdot A_j^{j-g} \cdot (\log N)^3 \qquad \text{as } D_0 = N^{-\gamma}, \gamma > 0 \text{ is constant}. \qquad (11)$$

There are $\binom{N}{i+j-g}$ choices for $i + j - g$ out of N points, hence for constants $c_{2;(g,i,j)} > 0$, $j = 2, \ldots, k-1$, we get with (11) the upper bound:

$$E[s_{2;(g,i,j)}(\mathcal{G})] \leq c_{2;(g,i,j)} \cdot A_i^{i-2} \cdot A_j^{j-g} \cdot N^{i+j-g} \cdot (\log N)^3. \qquad \square$$

For distinct points $P, Q \in [0,1]^2$, it is Prob (dist $(P,Q) \leq D_0) \leq \pi \cdot D_0^2$. With $D_0 = N^{-\gamma}$ we infer $E[|\mathcal{E}_2|] \leq \binom{N}{2} \cdot \pi \cdot D_0^2 \leq c_2 \cdot N^{2-2\gamma}$ for some constant $c_2 > 0$. By Markov's inequality, using this and the estimates (5) and (6) there exist N points $P_1, \ldots, P_N \in [0,1]^2$ such that the resulting hypergraph $\mathcal{G} = (V, \mathcal{E}_2 \cup \cdots \cup \mathcal{E}_k)$ with $|V| = N$ satisfies for $2 \leq g \leq i \leq j \leq k$ but $g < j$:

$$|\mathcal{E}_2| \leq k^3 \cdot c_2 \cdot N^{2-2\gamma} \quad \text{and} \quad |\mathcal{E}_j| \leq k^3 \cdot c_j \cdot A_j^{j-2} \cdot N^j \tag{12}$$

$$s_{2;(g,i,j)}(\mathcal{G}) \leq k^3 \cdot c_{2;(g,i,j)} \cdot A_i^{i-2} \cdot A_j^{j-g} \cdot N^{i+j-g} \cdot (\log N)^3. \tag{13}$$

For suitable constants $c_j' > 0$, $j = 3, \ldots, k$, which will be fixed later, we set

$$A_j := (c_j' \cdot (\log n)^{1/(j-2)})/n^{(j-1)/(j-2)}. \tag{14}$$

Lemma 6. *For fixed $\gamma > 1/2$ it is $|\mathcal{E}_2| = o(|V|)$.*

Proof. Using (12) and $|V| = N$, we have $|\mathcal{E}_2| = o(|V|)$ provided that $N^{2-2\gamma} = o(N) \Longleftrightarrow N^{1-2\gamma} = o(1)$, which holds for $\gamma > 1/2$. □

Lemma 7. *For fixed $2 \leq g \leq i \leq j \leq k$ but $g < j$ and for fixed constant β with $0 < \beta < (j-g)/((j-2) \cdot (i+j-g-1))$ it is $s_{2;(g,i,j)}(\mathcal{G}) = o(|V|)$.*

Proof. By using (13) and (14) and $|V| = N = n^{1+\beta}$ with fixed $\beta > 0$ we have $s_{2;(g,i,j)}(\mathcal{G}) = o(|V|)$ for $j = 2, \ldots, k-1$, provided that

$$A_i^{i-2} \cdot A_j^{j-g} \cdot N^{i+j-g} \cdot (\log N)^3 = o(N)$$

$$\Longleftrightarrow (\log n)^{4+\frac{j-g}{j-2}} \cdot n^{(1+\beta)(i+j-g-1)-(i-1)-\frac{(j-g)(j-1)}{j-2}} = o(1)$$

$$\Longleftrightarrow (1+\beta) \cdot (i+j-g-1) < i-1 + ((j-g) \cdot (j-1))/(j-2),$$

which holds for $\beta < (j-g)/((j-2) \cdot (i+j-g-1))$. □

Fix $\beta := 1/(2 \cdot k^2)$ and $\gamma := k/(2 \cdot (k-1))$. Then, with (14) and $D_0 = N^{-\gamma}$ and $N = n^{1+\beta}$ all assumptions in Lemmas 5–7 are fulfilled. We delete one vertex from each 2-element edge $E \in \mathcal{E}_2$ and each $(2; (g,i,j))$-cycle, $2 \leq g \leq i \leq j \leq k$ but $g < j$, in $\mathcal{G}$. Let $V^* \subseteq V$ be the set of remaining vertices. The induced subhypergraph $\mathcal{G}^* = (V^*, \mathcal{E}_3^* \cup \cdots \cup \mathcal{E}_k^*)$ of $\mathcal{G}$ with $\mathcal{E}_j^* := \mathcal{E}_j \cap [V^*]^j$, $j = 3, \ldots, k$, is linear, and by (12), and Lemmas 6 and 7 fulfills $|V^*| \geq N/2$ and $|\mathcal{E}_j^*| \leq k^3 \cdot c_j \cdot A_j^{j-2} \cdot N^j$. By (14), the hypergraph $\mathcal{G}^*$ has average degree

$$t_j^{j-1} = j \cdot |\mathcal{E}_j^*|/|V^*| \leq 2 \cdot k^3 \cdot j \cdot c_j \cdot (c_j')^{j-2} \cdot N^{j-1} \cdot \log n/n^{j-1} =: (t_j(1))^{j-1}$$

for the j-element edges. Fix a constant $c' > 0$ such that $C_k/(2 \cdot c') \cdot \beta^{1/(k-1)} > 1$ and set $T := c' \cdot (N/n) \cdot (\log n)^{1/(k-1)}$. Then fix constants $c_j' > 0$, $j = 3, \ldots, k$, in (14) such that

$$(t_j(1))^{j-1} = (2 \cdot k^3 \cdot j \cdot c_j \cdot (c_j')^{j-2} \cdot N^{j-1} \cdot \log n)/n^{j-1} \leq$$

$$\leq 1/32 \cdot \binom{k-1}{j-1}/(10^{(3(k-j))/(k-1)} \cdot k^6) \cdot T^{j-1} \cdot (\log T)^{(k-j)/(k-1)}.$$

Then, the assumptions in Theorem 4 are satisfied for $\mathcal{G}^*$, and its independence number $\alpha(\mathcal{G}^*)$ satisfies for some constant $C_k > 0$:

$$\alpha(\mathcal{G}) \geq \alpha(\mathcal{G}^*) \geq C_k \cdot (|V^{**}|/T) \cdot (\log T)^{\frac{1}{k-1}} \geq C_k \cdot (N/(2 \cdot T)) \cdot (\log T)^{\frac{1}{k-1}} \geq$$
$$\geq \frac{C_k \cdot n}{2 \cdot c' \cdot (\log n)^{\frac{1}{k-1}}} \cdot \left(\log(n^\beta)\right)^{\frac{1}{k-1}} \geq n.$$

The vertices of an independent set I with $|I| = n$ yield n points among the N points $P_1, \ldots, P_N \in [0,1]^2$, such that for $j = 3, \ldots, k$ the area of the convex hull of any j distinct points of these n points is $\Omega((\log n)^{1/(j-2)}/n^{(j-1)/(j-2)})$ as desired. $\qquad\square$

References

1. M. Ajtai, J. Komlós, J. Pintz, J. Spencer and E. Szemerédi, *Extremal Uncrowded Hypergraphs*, Journal of Combinatorial Theory Ser. A, 32, 1982, 321–335.
2. G. Barequet, *A Lower Bound for Heilbronn's Triangle Problem in d Dimensions*, SIAM Journal on Discrete Mathematics 14, 2001, 230–236.
3. G. Barequet and J. Naor, *Large k−D Simplices in the D-Dimensional Unit Cube*, Proceedings '17th Canadian Conference on Computational Geometry', 2005, 30–33.
4. G. Barequet and A. Shaikhet, *The On-Line Heilbronn's Triangle Problem in d Dimensions*, Proceedings '12th Annual Computing and Combinatorics Conference COCOON'06', Springer Verlag, LNCS 4112, 2006, 408–417.
5. C. Bertram–Kretzberg and H. Lefmann, *The Algorithmic Aspects of Uncrowded Hypergraphs*, SIAM Journal on Computing 29, 1999, 201–230.
6. C. Bertram-Kretzberg, T. Hofmeister and H. Lefmann, *An Algorithm for Heilbronn's Problem*, SIAM Journal on Computing 30, 2000, 383–390.
7. P. Brass, *An Upper Bound for the d-Dimensional Analogue of Heilbronn's Triangle Problem*, SIAM Journal on Discrete Mathematics 19, 2005, 192–195.
8. B. Chazelle, *Lower Bounds on the Complexity of Polytope Range Searching*, Journal of the American Mathematical Society 2, 1989, 637–666.
9. R. A. Duke, H. Lefmann and V. Rödl, *On Uncrowded Hypergraphs*, Random Structures & Algorithms 6, 1995, 209–212.
10. A. Fundia, *Derandomizing Chebychev's Inequality to find Independent Sets in Uncrowded Hypergraphs*, Random Structures & Algorithms, 8, 1996, 131–147.
11. J. Komlós, J. Pintz and E. Szemerédi, *On Heilbronn's Triangle Problem*, Journal of the London Mathematical Society, 24, 1981, 385–396.
12. J. Komlós, J. Pintz and E. Szemerédi, *A Lower Bound for Heilbronn's Problem*, Journal of the London Mathematical Society, 25, 1982, 13–24.
13. H. Lefmann, *Distributions of Points in the d Dimensions and Large k-Points Simplices*, Proceedings '11th Annual Computing and Combinatorics Conference COCOON'05', Springer Verlag, LNCS 3595, eds. Lusheng Wang, 2005, 514–523.
14. H. Lefmann, *Distributions of Points in the Unit-Square and Large k-Gons*, Proceedings '16th ACM-SIAM Symposium on Discrete Algorithms SODA', ACM and SIAM, 2005, 241–250.
15. K. F. Roth, *On a Problem of Heilbronn*, Journal of the London Mathematical Society 26, 1951, 198–204.

16. K. F. Roth, *On a Problem of Heilbronn, II, and III,* Proc. of the London Mathematical Society (3), 25, 1972, 193–212, and 543–549.

17. K. F. Roth, *Estimation of the Area of the Smallest Triangle Obtained by Selecting Three out of n Points in a Disc of Unit Area,* Proc. of Symposia in Pure Mathematics, 24, 1973, AMS, Providence, 251–262.

18. K. F. Roth, *Developments in Heilbronn's Triangle Problem,* Advances in Mathematics, 22, 1976, 364–385.

19. W. M. Schmidt, *On a Problem of Heilbronn,* Journal of the London Mathematical Society (2), 4, 1972, 545–550.

20. J. Spencer, *Turán's Theorem for k-Graphs,* Discrete Mathematics, 2, 1972, 183–186.

Optimal *st*-Orientations for Plane Triangulations

Huaming Zhang[1,*] and Xin He[2,**]

[1] Computer Science Department
University of Alabama in Huntsville
Huntsville, AL, 35899, USA
hzhang@cs.uah.edu
[2] Department of Computer Science and Engineering
SUNY at Buffalo
Buffalo, NY, 14260, USA
xinhe@cse.buffalo.edu

Abstract. For a plane triangulation G with n vertices, it has been proved that there exists a plane triangulation G with n vertices such that for any st-orientation of G, the length of the longest directed paths of G from s to t is $\geq \lfloor \frac{2n}{3} \rfloor$ [18]. In this paper, we prove the bound $\frac{2n}{3}$ is optimal by showing that every plane triangulation G with n-vertices admits an st-orientation with the length of its longest directed paths bounded by $\frac{2n}{3} + O(1)$. In addition, this st-orientation is constructible in linear time. A by-product of this result is that every plane graph G with n vertices admits a visibility representation with height $\leq \frac{2n}{3} + O(1)$, constructible in linear time, which is also optimal.

1 Introduction

st-orientations (also known as *st-numberings* or *bipolar orientations*) of undirected graphs satisfy certain criteria. They define no cycles and have exactly one source s and one sink t. From which, st-orientations for planar graphs play key roles in many graphs algorithms. For example, starting with an undirected biconnected graph $G = (V, E)$, many graph drawing algorithms, such as *hierarchical drawings* [1], *visibility representations* [16,14,5,6,9,18,19] and *orthogonal drawings* [13], use an st-orientation of G in order to compute a drawing of G. Therefore, the importance of st-orientations in Graph Drawing is evident.

Given a biconnected undirected graph $G = (V, E)$, with n vertices and m edges, and two nodes s, t, an st-orientation of G is defined as an orientation of its edges such that a directed acyclic graph with exactly one source s and exactly one sink t is produced. An st-orientation of an undirected graph can be easily computed using an st-numbering [8] of the respective graph G and orienting the edges of G from *low* to *high*. An st-numbering of G is a numbering of its vertices such that s receives number 1, t receives number n and every other node except s, t is adjacent to at least one lower-numbered vertex and at least one higher-numbered vertex.

---

* Corresponding author.
** Research supported in part by NSF Grant CCR-0309953.

M.-Y. Kao and X.-Y. Li (Eds.): AAIM 2007, LNCS 4508, pp. 296–305, 2007.
© Springer-Verlag Berlin Heidelberg 2007

st-numberings were first introduced in 1967 in [7], where it was proved (together with an $O(nm)$ time algorithm) that given any edge (s,t) of a biconnected undirected graph G, we can construct an st-numbering. Later in the year 1976, Even and Tarjan proposed an algorithm that computes an st-numbering of an undirected biconnected graph in $O(n+m)$ time [8]. Ebert [3] proposed a slightly simpler algorithm for the computation of such a numbering, which was further simplified by Tarjan [17]. The planar case has been extensively investigated in [14], where a linear time algorithm was presented which may reach any st-orientation of a planar graph. However, no information regarding the bound of the length of the longest directed paths resulted from the st-orientation was given, except the trivial bound $(n-1)$. In [10], a parallel algorithm was described. An overview of the work concerning bipolar orientations was presented in [4,11].

However, all algorithms mentioned above compute an st-numbering without expecting any specific properties of the oriented graph. Recently, by using *canonical ordering trees* and *Schnyder's realizers* for plane triangulations, Zhang and He [19] proved that for any plane triangulation G with n vertices, there is an st-orientation of G, constructible in linear time such that the length of the longest directed paths is at most $\frac{2n}{3} + 2\lceil\sqrt{n}\rceil$, which is nearly optimal because there is a plane triangulation G with n vertices such that the length of the longest directed paths for any st-orientations of G is $\geq \lfloor\frac{2n}{3}\rfloor$ [18]. There is still a $O(\sqrt{n})$ gap between the lower bound and the current nearly optimal algorithm. In this paper, we close this gap by using three trees of a Schnyder's realizer of a plane triangulation and by utilizing each such tree in a somewhat thorough way. As a direct by-product of this result, we prove that any plane graph G with n vertices admits a visibility representation, constructible in linear time, with height at most $\frac{2n}{3} + O(1)$, which is also optimal. For empirical algorithms on shortening the length of the longest directed paths, we refer readers to [12]. It introduced parameterized st-orientations, trying to control the length of the longest path of the resulting directed acyclic graph.

The present paper is organized as follows. Section 2 introduces preliminaries. Section 3 presents the construction of an st-orientation for a plane triangulation G with the length of its longest directed paths $\leq \frac{2n}{3} + O(1)$. The result on visibility representation with optimal height for a plane graph G is also presented in this section.

2 Preliminaries

We begin this section by mentioning some motivations of investigating st-orientations. Many algorithms in Graph Drawing use st-orientations as a first step. More importantly, the length of the longest paths from s to t of the specific st-orientation determines certain aesthetics of the drawing. For example:

Hierarchical Drawings: One of the most common algorithms in hierarchical drawing is the longest path laying [1]. This algorithm applies to directed acyclic graphs (which are necessarily planar). The height of such a drawing is always equal to the length of the longest paths of the directed acyclic graph, l. If we want to visualize an undirected graph G using this algorithm, we must first *st*-orient G. The height of the produced drawing will be equal to the length of the longest path l of the produced *st*-orientation.

Visibility Representations: In order to compute visibility representations of planar graphs, we first assign unit-weights to the edges of the graph and compute the longest path to each one of its vertices from source s. The y-coordinate of each vertex u in the visibility representation is equal to the length of the longest path from s to u [16]. Therefore, we have the following Lemma [14,16]:

Lemma 1. *Let G be a 2-connected plane graph with an st-orientation. A visibility representation of G can be obtained from an st-orientation $\mathcal{O}$ of G in linear time. Furthermore, the height of the visibility representation equals the length of the longest directed path in $\mathcal{O}$.*

Therefore, for the above graph drawing problems, it is crucial to find better *st*-orientations such that the length of the longest directed paths is as small as possible. Many researches have been done on shortening the length of the longest directed paths. From the empirical algorithms' side, [12] introduced parameterized *st*-orientations, trying to control the length of the longest paths of the resulting directed acyclic graph. From the theoretical algorithms' side, using canonical ordering trees and Schnyder's realizer for plane triangulations, Zhang and He [19] proved that for any plane triangulation G with n vertices, there is an *st*-orientation of G, constructible in linear time such that the length of the longest directed paths is at most $\frac{2n}{3} + 2\lceil\sqrt{n}\rceil$, which is nearly optimal because there is a plane triangulation G with n vertices such that the length of the longest directed paths for any *st*-orientations of G is at least $\lfloor\frac{2n}{3}\rfloor$ [18]. There is still a $O(\sqrt{n})$ gap between the lower bound and the current nearly optimal algorithm. We are going to close this gap in this paper.

Next, we are going to give definitions and preliminary results. Definitions not mentioned here are standard.

A *planar graph* is a graph $G = (V, E)$ such that the vertices of G can be drawn in the plane and the edges of G can be drawn as non-intersecting curves. Such a drawing is called an *embedding*. The embedding divides the plane into a number of connected regions. Each region is called a *face*. The unbounded face is called *external face*. The other faces are *internal faces*. A *plane graph* is a planar graph with a fixed embedding. A *plane triangulation* is a plane graph where every face is a triangle (including the external face). A subgraph $G' = (V', E')$ of $G = (V, E)$ is called a spanning subgraph of G if $V' = V$.

Let $\mathcal{O}$ be an orientation (or a numbering) of a graph G. We will use length($\mathcal{O}$) to denote the length of the longest directed paths of G. (Note that, we do not require that $\mathcal{O}$ is necessarily an *st*-orientation.)

An *ordered list* $\mathcal{O}$ consisting of elements $a_1, a_2, \ldots, a_k$ is written as $\mathcal{O} =< a_1, a_2, \ldots, a_k >$. For two elements a_i and a_j, if a_i appears before a_j in $\mathcal{O}$, we write $a_i \prec_{\mathcal{O}} a_j$. The reverse of an ordered list $\mathcal{O} =< a_1, a_2, \ldots, a_k >$ is the ordered list $< a_k, \ldots, a_2, a_1 >$, which is going to be denoted by $\mathcal{O}^r$. The concatenation of two ordered lists $\mathcal{O}_1$ and $\mathcal{O}_2$ is written as $\mathcal{O}_1\mathcal{O}_2$.

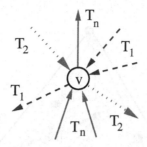

Fig. 1. Edge directions around an internal vertex *v*

Definition 1. Let G be a plane triangulation of n vertices with three external vertices v_1, v_2, v_n in counterclockwise order. A *realizer* (also called a *Schnyder's realizer*) $\mathcal{R} = \{T_1, T_2, T_n\}$ of G is a partition of its internal edges into three sets T_1, T_2, T_n of directed edges such that the following hold:

- For each $i \in \{1, 2, n\}$, the internal edges incident to v_i are in T_i and directed toward v_i.
- For each internal vertex of G, v has exactly one edge leaving v in each of T_1, T_2, T_n. The counterclockwise order of the edges incident to v is: leaving in T_1, entering in T_n, leaving in T_2, entering in T_1, leaving in T_n, and entering in T_2 (See Fig. 1). Each entering block could be empty.

Figure 2 show a realizer of a plane triangulation G. The dashed lines (dotted lines and solid lines, respectively) are the edges in T_1 (T_2 and T_n, respectively).

Schnyder showed in [15] that every plane triangulation G has a realizer which can be constructed in linear time. It is also shown that each set T_i of a realizer is a tree rooted at the vertex v_i. For each T_i of a realizer, we denote by $\overline{T}_i$ the tree composed of T_i augmented with the two edges of the external face incident to the root v_i of T_i. Obviously $\overline{T}_i$ is a spanning tree of G.

Let $i = 1, j = 2$ and $k = n$. (Or $i = 2, j = n$ and $k = 1$, or $i = n, j = 1$ and $k = 2$, respectively.) Let u be an internal vertex of G. Consider the tree $\overline{T}_i$. Let w_j and w_k be the parent of u in $\overline{T}_j$ and $\overline{T}_k$ respectively. Then we have the following simple observation [2]:

Observation 1: w_j precedes u in the counterclockwise postordering of $\overline{T}_i$. w_k precedes u in the clockwise postordering of $\overline{T}_i$. w_j succeeds u in the clockwise postordering of $\overline{T}_i$. w_k succeeds u in the counterclockwise postordering of $\overline{T}_i$.

For example, in Figure 2, $\overline{T}_n$ is T_n (the tree in thick solid lines) augmented with edges (v_1, v_n) and (v_2, v_n), which is a spanning tree of G. Consider the case for $i = n, j = 1$ and $k = 2$. Consider the tree $\overline{T}_n$, for the vertex $u = 7$, its parent in $\overline{T}_1$ is $w_j = 6$, its parent in $\overline{T}_2$ is $w_k = 9$. They certainly hold the properties stated in Observation 1.

Let $T = (V, E)$ be a tree drawn in the plane, a *balanced partition* [19] of T is the partition of V into three ordered subsets A, B, C such that: Let a_i

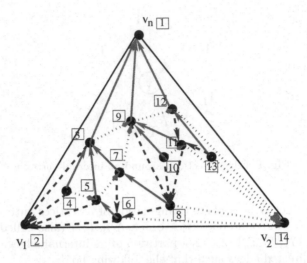

Fig. 2. A plane triangulation G and a realizer $\mathcal{R}$ of G

be the ith vertex of T in counterclockwise postordering, and b_i be the ith vertex of T in clockwise postordering. Marking the vertices of T in the order $a_1, b_1, a_2, b_2, \ldots, a_i, b_i, \ldots$, continue this process as long as the next pair of the vertices a_{i+1}, b_{i+1} have not been marked. This process stops when either $a_{k+1} = b_{k+1}$ or b_{k+1} is already marked. This vertex is called the *merge vertex* of T. When the marking process stops, the un-marked vertices of T form a single path from the merge vertex a_{k+1} to the root of T. This path is called the *leftover* path of T. Then $A = <a_1, a_2, \ldots, a_k>$, $B = <b_1, b_2, \ldots, b_k>$, and C is the leftover path ordered from the merge vertex to the root of T.

In the above balanced partition, the subgraph induced by vertices in $A \cup B$ defines a subgraph of a *ladder graph* of order k. Ladder graph is defined as follows [19]:

Definition 2. Let $L = (V_L, E_L)$ be a plane graph. If the vertex set V_L is partitioned into two ordered lists $A = <a_1, a_2, \ldots, a_k>$ and $B = <b_1, b_2, \ldots, b_k>$. For edges, if $E_L = E_A \cup E_B \cup E_{cross}$, where: $E_A = \{(a_i, a_{i+1}) | 1 \leq i < k\}$; $E_B = \{(b_i, b_{i+1}) | 1 \leq i < k\}$; and E_{cross} consists of edges between a vertex $a_i \in A$ and a vertex $b_j \in B$. Then any spanning subgraph of L is called a *ladder graph* of order k. The edges in E_{cross} are called *cross edges*.

A numbering $\mathcal{O}$ of the vertices of a ladder graph $L = (A \cup B, E_L)$ is *consistent with respect to L* if for any $i < j$, $a_i \prec_{\mathcal{O}} a_j$ and $b_i \prec_{\mathcal{O}} b_j$.

The following lemma was proved in [19]:

Lemma 2. Let $L = (A \cup B, E_L)$ be a ladder graph of order k. Then L has a consistent numbering $\mathcal{O}$ such that $\text{length}(\mathcal{O}) \leq k + 2\lceil \sqrt{k} \rceil - 1$. $\mathcal{O}$ can be constructed in linear time.

3   Optimal *st*-Orientations for Plane Triangulations

Let G be a plane triangulation with n vertices, v_1, v_2, v_n be its three external vertices in counterclockwise order, $\mathcal{R} = \{T_1, T_2, T_n\}$ be a realizer of G. For any internal vertex u of G, consider the three paths $P_1(u), P_2(u), P_n(u)$ from u to the roots of T_1, T_2, T_n respectively, they only intersect at the vertex u. In addition, they separate the interior regions of G into three subregions $R_{12}(u), R_{2n}(u), R_{n1}(u)$ ($R_{ij}(u)$ is also denoted by $R_{ji}(u)$), in which the subregion $R_{ij}(u)$ is enclosed by the paths $P_i(u), P_j(u), i, j \in \{1, 2, n\}, i \neq j$. (Note that the regions specified here don't include their boundaries.) We have the following technical lemma, its proof is omitted due to space limitation:

Lemma 3. *Let G be a plane triangulation with n vertices, v_1, v_2, v_n be its three external vertices in counterclockwise order, $\mathcal{R} = \{T_1, T_2, T_n\}$ be a realizer of G. We have the following:*

1. *For any $i, j \in \{1, 2, n\}, i \neq j$, the subgraph of $\overline{T_i}$ induced by the vertices in $R_{ij}(u) \cup P_i(u) - \{u\}$ is a subtree of $\overline{T_i}$.*
2. *For any $i \in \{1, 2, n\}$, the subgraph of $\overline{T_i}$ induced by the vertices in $R_{ij}(u) \cup R_{ik}(u) \cup P_i(u) - \{u\}$ is a subtree of $\overline{T_i}$.*

We are going to denote the subtree in Lemma 3 (1) by $\overline{T_i^j}(u)$, the subtree in Lemma 3 (2) by $\overline{T_i^{jk}}(u)$ in the remaining of the paper. For example, in Fig. 2, consider the vertex $u = 7$. $\overline{T_1^n}(u)$ is the tree with vertices $\{2, 4, 5, 6\}$ in $\overline{T_1}$. $\overline{T_2^{1n}}(u)$ is the tree with vertices $\{8, 9, 10, 11, 12, 13, 14\}$ in $\overline{T_2}$.

Before we proceed to next lemma, we would like to point out another simple observation:

Observation 2: Let T be a tree rooted at v. T' be a subtree of T, also with v as its root. Let u, w be two vertices in T'. Then omitting vertices which are not in T' from the counterclockwise (clockwise, respectively) postordering of T produces a counterclockwise (clockwise, respectively) postordering of T'. This ordering of T' will be called the induced counterclockwise (clockwise, respectively) postordering from T to T'.

Lemma 4. *let $G = (V, E)$ be a plane triangulation with n vertices, v_1, v_2, v_n be three external vertices in counterclockwise order, $\mathcal{R} = \{T_1, T_2, T_n\}$ be a realizer of G. If there is a path in any of $\overline{T_i}, i = 1, 2, n$ with length $\geq \frac{n}{3}$, then G has an st-orientation $\mathcal{O}$, constructible in linear time with length$(\mathcal{O}) \leq \frac{2n}{3} + O(1)$.*

Proof. We will prove for the case $i = n$, the other cases are similar. Assume that there is a path $P_n(v)$ in $\overline{T_n}$, starting from a vertex v to the root v_n, whose length is $> \frac{n}{3}$. (If the path does not end at v_n, extend it to the root v_n. Its length is still $\geq \frac{n}{3}$). The vertices of G can be decomposed into three sets: $A = \{$vertices in $\overline{T_1^n}(v)\}$, $B = \{$vertices in path $P_n(v)\}$, and $C = \{$vertices in $\overline{T_2^{1n}}(v)\}$. Obviously $V = A \cup B \cup C$. We are going to order A, B, C into ordered lists as the following:

1. For the vertices in A, let $\mathcal{O}_A$ be the induced clockwise postordering from $\overline{T_1}$ to $\overline{T_1^n(v)}$.
2. For the vertices in B, they are all on the path $P_n(v)$. When traversing the path from v to v_1, we insert the first vertex to the very end, the second vertex to the very front, recursively insert the remaining vertices into the next available end or next available front of the ordered list. So the third vertex in the path actually sits on the second last position in the ordered list, the fourth vertex actually sits on the second position in the ordered list, and so on. Denote this order by $\mathcal{O}_B$.
3. For the vertices in C, let $\mathcal{O}_C$ be the induced counterclockwise postordering from $\overline{T_2}$ to $\overline{T_2^{1n}(v)}$

Claim: The concatenation of $\mathcal{O}_A^r \mathcal{O}_B \mathcal{O}_C$ defines an st-orientation $\mathcal{O}$ for G.

Proof of Claim

1. v_1 is the first in $\mathcal{O}$, v_2 is the last in $\mathcal{O}$.
2. v_n succeeds v_1, but it proceeds v_2 in $\mathcal{O}$.
3. For any vertex $u \neq v_1$ in A, its parent in $\overline{T_1^n(v)}$ proceeds u in $\mathcal{O}$. According to Observation 1, the parent w of u in $\overline{T_n}$ proceeds u in the clockwise postordering in $\overline{T_1}$. So if w is also in $\overline{T_1^n(v)}$, then according to Observation 2, w also proceeds u in the clockwise postordering $\mathcal{O}_A$, since it is the induced clockwise postordering. Thus w succeeds u in $\mathcal{O}_A^r$, and hence in $\mathcal{O}$. On the other hand, if w is not in $\overline{T_1^n(v)}$, then w is either in B or in C. Either way, w succeeds u in $\mathcal{O}$.
4. For any vertex $u \in B$, its parent w_1 in $\overline{T_1}$ is in A, so w_1 proceeds u in $\mathcal{O}$. Its parent w_2 in $\overline{T_2}$ is in C, so w_2 succeeds u in $\mathcal{O}$.
5. For any vertex $u \neq v_2 \in C$, its parent in $\overline{T_2^{1n}(v)}$ succeeds it in $\mathcal{O}$. Again, according to Observation 1, the parent w of u in $\overline{T_n}$ proceeds u in the counterclockwise postordering in $\overline{T_2}$. So if w is also in $\overline{T_2^{1n}(v)}$, then according to Observation 2, w also proceeds u in the counterclockwise postordering $\mathcal{O}_C$, since it is the induced counterclockwise postordering. Therefore, w proceeds u in $\mathcal{O}_C$, and hence in $\mathcal{O}$. On the other hand, if w is not in $\overline{T_2^{1n}(v)}$, then w is either in A or in B. Either way, w proceeds u in $\mathcal{O}$.

Therefore, $\mathcal{O}$ is an st-orientation for G, since every vertex other than the source v_1 or the sink v_2 has one vertex proceeding it and one vertex succeeding it in the ordering.

End of Proof of Claim

Obviously, $\mathcal{O}$ can be constructed in linear time.

Consider a longest directed path P of $\mathcal{O}$. It can only passes through at most 2 vertices on path $P_n(v)$. Therefore, the total number of vertices it can pass through is at most $\frac{2n}{3} + 2$. Hence, $\text{length}(\mathcal{O}) \leq \frac{2n}{3} + O(1)$.

Lemma 5. *let $G = (V, E)$ be a plane triangulation with n vertices, v_1, v_2, v_n be three external vertices in counterclockwise order, $\mathcal{R} = \{T_1, T_2, T_n\}$ be a realizer of G. If every path in $\overline{T_i}, i = 1, 2, n$ has length $< \frac{n}{3}$, then G has an st-orientation $\mathcal{O}$, constructible in linear time with $\text{length}(\mathcal{O}) \leq \frac{2n}{3} + O(1)$.*

Proof. Let l be the length of the longest paths in all of $\overline{T_i}, i = 1, 2, n$. Hence, $l < \frac{n}{3}$. Let $l = \frac{n}{3} - 2k$. Therefore $k > 0$.

Consider the balanced partition A, B, C of $\overline{T_n}$, in which C is the leftover path. Therefore $|C| = r \leq l$. Because it is a balanced partition, we have $|A| = |B| = \frac{n-r}{2} \geq \frac{n-l}{2} = \frac{n}{3} + k$.

Claim: The subgraph of $\overline{T_1}$ induced by A is a subtree of $\overline{T_1}$, denote it by T_1^A. The subgraph of $\overline{T_2}$ induced by B is a subtree of $\overline{T_2}$, denote it by T_2^B.

Proof of Claim: We will only show the case for A. The other case is similar. When we do balanced partition of $\overline{T_n}$, we use counterclockwise postordering to accumulate the set A. For any vertex u in A, its parent w in $\overline{T_1}$ comes before u in this counterclockwise postordering according to Observation 1. Therefore, both u and its parent w are in A. Hence, A induces a subtree of $\overline{T_1}$.

End of Proof of Claim

For the tree T_1^A, we can further consider its balanced partition, denote it by A', B', C', in which C' is the leftover path. Consider C', its length is $\leq l$. Therefore, we have $|A'| = |B'| = \frac{|A| - |C'|}{2} \geq \frac{\frac{n}{3} + k - l}{2} = \frac{3k}{2}$. Since both $|A'|$ and $|B'|$ are necessarily integers, therefore $|A'| = |B'| \geq \lceil \frac{3k}{2} \rceil$. For the tree T_2^B, we can further consider its balanced partition, denote it by A'', B'', C'', in which C'' is the leftover path. Similarly, $|A''| = |B''| \geq \lceil \frac{3k}{2} \rceil$.

We are going to order A, B, C into ordered lists as the following:

1. For the vertices in $A = A' \cup B' \cup C'$, they form the tree T_1^A. Apply Lemma 2 to the subgraph induced by $A' \cup B'$ (which is a ladder graph) to obtain an ordering $\mathcal{O}_{A'B'}$, then to the leftover path C' (towards to its end vertex v_1, i.e. v_1 is the last vertex in this ordering.) to obtain an ordering $\mathcal{O}_{C'}$. Let $\mathcal{O}_A$ be the concatenation of $\mathcal{O}_{A'B'}\mathcal{O}_{C'}$.
2. For the vertices in C, they are all on one path. When traversing the path from one end v to the other end v_n, we insert the first vertex to the very end, the second vertex to the very front, recursively insert the remaining vertices into the next available end or next available front of the ordered list. So the third vertex in the path actually sits on the second last position in the ordered list, the fourth vertex actually sits on the second position in the ordered list, and so on. Denote this order by $\mathcal{O}_C$.
3. For the vertices in $B = A'' \cup B'' \cup C''$, they form the tree T_2^B. Apply Lemma 2 to the subgraph induced by $A'' \cup B''$ (which is a ladder graph) to obtain an ordering $\mathcal{O}_{A''B''}$, then to the leftover path C'' (towards to its end vertex v_2, i.e. v_2 is the last vertex in this ordering.) to obtain an ordering $\mathcal{O}_{C''}$. Let $\mathcal{O}_B$ be the concatenation of $\mathcal{O}_{A''B''}\mathcal{O}_{C''}$.

Claim: The concatenation of $\mathcal{O}_A^r\mathcal{O}_C\mathcal{O}_B$ defines an st-orientation $\mathcal{O}$ for G.
Proof of Claim

1. v_1 is the first in $\mathcal{O}$, v_2 is the last in $\mathcal{O}$.
2. v_n succeeds v_1, but it proceeds v_2 in $\mathcal{O}$.

3. For any vertex $u \neq v_1$ in A, its parent in $\overline{T}_1^A$ proceeds u in $\mathcal{O}$. If $u \in A'$, consider its parent w in $\overline{T}_2$, according to Observation 1, w proceeds u in the counterclockwise postordering of $\overline{T}_1$. According to Observation 2, if w is also in $\overline{T}_1^A$, then w proceeds u in the counterclockwise postordering of $\overline{T}_1^A$, since it is the induced counterclockwise postordering. Because A' is accumulated counterclockwisely from the balanced partition of $\overline{T}_1^A$, so w is also in A' and w proceeds u in $\mathcal{O}_{A'B'}$, and hence in $\mathcal{O}_A$. Therefore, w succeeds u in $\mathcal{O}_A^r$, and hence in $\mathcal{O}$. On the other hand, if w is not in $\overline{T}_1^A$, then w is either in B or in C. Either way, w succeeds u in $\mathcal{O}$. Similarly, if $u \in B'$, its parent in $\overline{T}_n$ succeeds u in $\mathcal{O}$. If $u \in C'$, then both its parents in $\overline{T}_2$ and $\overline{T}_n$ succeed u in $\mathcal{O}$.

4. The case for a vertex $u \neq v_2$ in B is similar to the above case for $u \neq v_1$ in A.

5. For any vertex $u \in C$, its parent w_1 in $\overline{T}_1$ is in A, so w_1 proceeds u in $\mathcal{O}$. Its parent w_2 in $\overline{T}_2$ is in B, so w_2 succeeds u in $\mathcal{O}$.

Therefore, $\mathcal{O}$ is an st-orientation for G, since every vertex other than the source v_1 or the sink v_2 has one vertex proceeding it and one vertex succeeding it in the ordering.

End of Proof of Claim

Obviously, $\mathcal{O}$ can be constructed in linear time.

Now we need to show that length$(\mathcal{O})$ is $\leq \frac{2n}{3} + O(1)$.

Similar to Lemma 4, any longest directed path P of $\mathcal{O}$ can pick at most two vertices from C. Applying Lemma 2, within $\mathcal{O}_A^r$, since the ladder graph $A' \cup B'$ is of order at least $p = \lceil \frac{3k}{2} \rceil$, therefore, $\mathcal{O}_A^r$ can pick at most $p + 2\lceil \sqrt{p} \rceil - 1 + 1 = p + 2\lceil \sqrt{p} \rceil$ vertices from $A' \cup B'$. (The plus 1 is because here we consider number of vertices instead of number of edges.). Similarly, for the ladder graph $A'' \cup B''$, it is of order at least p, therefore, $\mathcal{O}_B$ can pick at most $p + 2\lceil \sqrt{p} \rceil$ vertices from $A'' \cup B''$. So even if P picks all the vertices from C' and C'' (Both are $\leq l$), the total length of P is still at most $l + l + 2 + 2(p + 2\lceil \sqrt{p} \rceil) = \frac{2n}{3} - 4k + 2(p + 2\lceil \sqrt{p} \rceil) + 2$. Because $p = \lceil \frac{3k}{2} \rceil$, it is easy to see that $-4k + 2(p + 2\lceil \sqrt{p} \rceil) + 2 \leq O(1)$. Therefore, the length$(\mathcal{O}) \leq \frac{2n}{3} + O(1)$.

Now we can have our main theorem:

Theorem 1. *1. Let G be a plane triangulation with n vertices. Then G admits an st-orientation $\mathcal{O}$, constructible in linear time with length$(\mathcal{O}) \leq \frac{2n}{3} + O(1)$.*

2. Let G be a plane graph with n vertices. Then G admits a visibility representation, constructible in linear time with height $\leq \frac{2n}{3} + O(1)$.

Proof. 1. It comes directly from Lemma 4 and 5.

2. If G is not a plane triangulation, we add dummy edges to triangulate it into a plane triangulation G'. Then according to (1) and Lemma 1, G' admits a visibility representation with height $\leq \frac{2n}{3} + O(1)$, which is obviously constructible in linear time. After deleting corresponding vertical line segments for these added dummy edges, we have a visibility representation for G with the same height, which is $\leq \frac{2n}{3} + O(1)$.

Note that, according to a result in [18], both above bounds are optimal.

References

1. G. di Battista, P. Eades, R. Tammassia, and I. Tollis, Graph Drawing: Algorithms for the Visualization of Graphs, Princeton Hall, 1998
2. Y.-T. Chiang, C.-C. Lin and H.-I. Lu, Orderly spanning trees with applications to graph encoding and graph drawing, in *Proc. of the 12th Annual ACM-SIAM SODA*, pp. 506-515, ACM Press, New York, 2001.
3. J. Ebert, *st*-ordering the vertices of biconnected graphs, *Computing* Vol. 30 no. 1 (1983), pp. 19-33.
4. H.D. Fraysseix, P.O. de Mendez, and P. Rosenstiehl, Bipolar orientations revisited, *Discrete Applied Mathematics*, 56 (1995), pp. 157-179.
5. G. Kant, A more compact visibility representation. *International Journal of Computational Geometry and Applications* 7 (1997), 197-210.
6. G. Kant and X. He, Regular edge labeling of 4-connected plane graphs and its applications in graph drawing problems. *Theoretical Computer Science* 172 (1997), 175-193.
7. A. Lempel, S. Even and I. Cederbaum, An algorithm for planarity testing of graphs, in *Theory of Graphs (Proc. of an International Symposium, Rome,July 1966)*, pp. 215-232, Rome, 1967.
8. S. Even and R. Tarjan, Computing an *st*-numbering, *Theoretical Computer Science*, 2 (1976), 339-344.
9. C.-C. Lin, H.-I. Lu and I-F. Sun, Improved compact visibility representation of planar graph via Schnyder's realizer, in: *SIAM Journal on Discrete Mathematics*, 18 (2004), 19-29.
10. Y. Maon, B. Schieber, U. Vishkin, Parallel ear decomposition search (eds) and *st*-numbering in graphs, *theoretical Computer Science* Vol. 47, (1986) 277-298.
11. P. Ossona de Mendez, Orientations bipolaires, PhD thesis, Ecole des Hautes Etudes en Sciences Sociales, Paris, 1994.
12. C. Papamanthou and I. G. Tollis, Applications of parameterized *st*-orientations in graph drawings, in: *Proc. of 13th International Symposium on Graph Drawing*, Lecture Notes in Computer Science, Vol. 3843, pp. 355-367.
13. A. Papakostas and I.G. Tollis, Algorithms for area-efficient orthogonal drawings, *Computational Geometry: Theory and Applications*, 9 (1998), 83-110.
14. P. Rosenstiehl and R. E. Tarjan, Rectilinear planar layouts and bipolar orientations of planar graphs. *Discrete Comput. Geom.* 1 (1986), 343-353.
15. W. Schnyder, Planar graphs and poset dimension. *Order* 5 (1989), 323-343.
16. R. Tamassia and I.G.Tollis, An unified approach to visibility representations of planar graphs. *Discrete Comput. Geom.* 1 (1986), 321-341.
17. R. Tarjan, Two streamlined depth-first search algorithms, *Fundamentae Informatica*, Vol. 9 (1986), pp. 85-94.
18. H. Zhang and X. He, *New Theoretical Bounds of Visibility Representation of Plane Graphs*, in: Proc. GD'2004, Lecture Notes in Computer Science, Vol. 3383, (Springer-Verlag, 2005) pp. 425-430.
19. H. Zhang and X. He, Nearly Optimal Visibility Representations of Plane Graphs, in: Proc. ICALP'06, Lecture Notes in Computer Science, Vol. 4051, (Springer-Verlag, 2006) pp. 407-418.

Minimum Spanning Tree with Neighborhoods*

Yang Yang, Mingen Lin, Jinhui Xu, and Yulai Xie

Department of Computer Science and Engineering
University at Buffalo, the State University of New York
Buffalo, NY 14260, USA
{yyang6,mlin6,jinhui,xie}@cse.buffalo.edu

Abstract. We consider a natural generalization of the classical minimum spanning tree problem called *Minimum Spanning Tree with Neighborhoods (MSTN)* which seeks a tree of minimum length to span a set of 2D regions called neighborhoods. Each neighborhood contributes exact one node to the tree, and the MSTN has the minimum total length among all possible trees spanning the set of nodes. We prove the NP-hardness of this problem for the case in which the neighborhoods are a set of disjoint discs and rectangles. When the regions considered are a set of disjoint 2D unit discs, we present the following approximation results: (1) A simple algorithm that achieves an approximation ratio of 7.4; (2) Lower bounds and two 3-approximation algorithms; (3) A PTAS for this problem. Our algorithms can be easily generalized to higher dimensions.

1 Introduction

Finding minimum spanning trees (MST) in graphical or geometric settings is one of the most fundamental problems in computer science and has received a great deal of attentions in the past [1]. In geometric settings, optimal solution has been achieved in 2D using Delaunay triangulation. In higher dimensions, Agarwal *et. al.* designed a method to compute MSTs in time $O(n^{2-2/(\lceil d/2 \rceil+1)+\epsilon})$, where d is the dimension. More efficient approximation algorithms were also investigated. Har-Peled [2] showed that it is possible to construct an ϵ-approximation of the Voronoi diagram in $O(\frac{n}{\epsilon^d}(\log n) \log \frac{n}{\epsilon})$ time in d-dimensional space, and thus can be used to construct an approximate MST. Later, Arya and Mount *et. al.* [3] presented an improved algorithm for computing the ϵ-approximation of the Voronoi diagram, which runs in $O(n\epsilon^{\frac{d-1}{2}}\gamma^{\frac{3(d-1)}{2}} \log \gamma)$ time, where $2 \le \gamma \le 1/\epsilon$.

In this paper, we consider a generalization of the geometric MST problem called *Minimum Spanning Tree with Neighborhoods (MSTN)*. In the MSTN problem, each point in the Euclidean MST problem becomes a region called *neighborhood*. The objective is to identify a representative point from each neighborhood so that the MST of the set of representative points has the minimum total length among all possible spanning trees. The version of MSTN studied in this paper (i.e., each neighborhood is a unit disc) is motivated by applications in biology.

---

* The research of this work was supported in part by an NSF CARRER Award CCF-0546509.

M.-Y. Kao and X.-Y. Li (Eds.): AAIM 2007, LNCS 4508, pp. 306–316, 2007.

In cell biology, an important problem is to determine the DNA replication and transcription networks using a set of replication and transcription sites (i.e., points) in some microscopic images [4]. Since the transcription sites are often labeled RNA scattering along their corresponding transcribing DNA, they are presumably close to the replication network for a possible switch of genomic function between replication and transcription. Thus the replication network needs not only to span the replication sites but also to pass the neighborhood of each transcription site, and therefore can be computed by solving 3D MSTN.

The concept of neighborhoods has been investigated in other geometric optimization problems. The most notable one is the *Traveling Salesman Problem with Neighborhoods (TSPN)*, in which a minimum-length tour of the neighborhoods is sought. As a generalization of the classical TSP problem, the TSPN problem is clearly NP-hard. The problem was first studied by Arkin and Hassin [5]. They gave an $O(1)$-approximation algorithm for some special cases such as regions with parallel diameter segments and comparable diameters. Mata and Mitchell [6] obtained an $O(\log n)$-approximation algorithm for the general case of connected polygonal regions, based on the Guillotine rectangular subdivisions, and runs in $O(m^5)$ time, where m is the total complexity of the n regions. The time bound was later improved by Gudmundsson and Levopoulos [7]. For any fixed $\epsilon > 0$, their algorithm either outputs an $O(\log n)$-approximation tour in time $O(n \log n + m)$, or a $(1 + \epsilon)$-approximation tour in time $O(m^3)$. Dumitrescu and Mitchell [8] also designed a constant approximation algorithm for the case of arbitrary connected neighborhoods having comparable diameters, and a PTAS for the case of unit disc neighborhoods. When the neighborhoods are a set of disjoint convex fat objects, a polynomial-time constant-approximation algorithm was given by M. de Berg *et. al.* [9]. Recently, Mitchell [10] showed the first PTAS result of TSPN for fat regions (using a very weak notion of "fat").

To our best knowledge, there is no previous hardness result on MSTN. In Section 2, by reducing from the planar 3-satisfiability problem, we show that it is NP-hard to solve MSTN when the neighborhoods are a set of 2D disjoint unit discs and rectangles. As for the approximation results, it is likely that the constant-approximation algorithm for TSPN given by M. de Berg in [9] can be generalized here, but the algorithm is not so practical due to its large constant factor (In the paper they showed a factor of 93 for the supposedly better case when input is a set of disjoint axis-aligned squares). In Section 3.1, we present a simple approximation algorithm with asymptotic approximation ratio of 7.4 for MSTN of disjoint unit discs. Later on, we improve the approximation factor by establishing two lower bounds on the length of MST of a set of disjoint discs in Section 3.2. Two near linear time 3-approximation algorithms are presented in Section 3.3, both easy to implement. It seems also possible to generalize the PTAS for TSPN given by Mitchell in [10], which is based on a novel extension of the m–guillotine method, to the MSTN problem. In Section 3.4, we introduce a polynomial time approximation scheme for MSTN of disjoint unit discs, which is based on an interesting generalization of Arora's framework for Euclidean TSP and other related problems [11], and runs in $O(n(\log n)^{O(1/\epsilon)})$ time.

The m–guillotine based method [10] is quite involved and generally has higher runnning time comparing to our PTAS. Moreover, it is not clear whether the m–guillotine based method can be extended to higher dimension, whereas all our approximation algorithms (including the PTAS) can be easily extended to higher dimensions. Our algorithms can also be extended to the cases in which the discs have comparable radii and/or slightly overlap. Due to the space limit, we omit many details of our algorithms and some details of the hardness proof from this version.

2 Hardness Result of the MSTN Problem

Problem 1. **Minimum   Spanning   Tree   of   Discs   and   Rectangles (MSTDR):**
Given: A set D of disjoint discs, and a set R of disjoint rectangles in a plane.
Objective: A minimum spanning tree of D and R.

To show the NP-hardness of MSTDR, we reduce from the planar 3-satisfiability.

Problem 2. **Planar 3-Satisfiability (P3SAT):**
Given: A set L of literals, and a collection C of clauses over L with $|c| = 3$ for all $c \in C$. Furthermore the bipartite graph $G = (V, E)$ is planar, where $V = L \cup C$ and $E = \{\{x, c\} : x \text{ or } \bar{x} \text{ occurs in } c\}$.
Problem: Is there a satisfying truth assignment for C?

The NP-completeness proof of P3SAT was given in [12]. We use the planarity of the underlying graph of P3SAT to construct a planar instance in which we can easily determine MSTDR. We start with a planar embedding of a given instance (L, C) of P3SAT, and then extend the literal and clause vertices to form geometric components consisting of discs and rectangles. The constructed instance contains literal components and clause components that reflects the relation between the satisfiability of clauses and the existence of a MST. Figure 1 illustrates three literals and one clause containing these three literals.

As shown in Figure 1, we use a shadowed rectangle with rounded corners to represent a sequence of slightly disjoint unit discs. We call such rectangles as *disc sticks*. By "slightly disjoint", we mean that the distance between two consecutive discs can be ignored compared to the total length of MST. The centers of all discs in one disc stick are collinear and the supporting line of all centers is called the *center line* of the disc stick. Let r be the radius of unit discs, and δ be any number $\gg r$. We construct literal components and clause components as follows.

Literal components

1. For each literal x (including its negation) appearing in a total of m_x clauses, create m_x vertical disc sticks (called *positive disc sticks* or x *disc sticks*) to represent x and and m_x vertical disc sticks (called *negative disc sticks* or $\bar{x}$ *disc sticks*) to represent $\bar{x}$. These x and $\bar{x}$ disc sticks are arranged alternately and spaced by a distance of $4\delta - 2r$. See Figure 1.

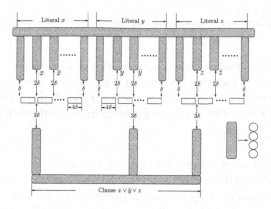

Fig. 1. An illustration of the hardness proof

2. Put two longer vertical disc sticks (called *forcing disc sticks*) to start and end the array of disc sticks of literal x. See Figure 1. The starting forcing disc stick which is neighboring to an x disc stick is called a negative forcing disc stick, and the ending disc stick which is neighboring to an $\bar{x}$ disc stick is called a positive forcing disc stick.
3. Create one horizontal disc stick (called an L *disc stick*) to "connect" all literals together. See Figure 1.
4. Put a sequence of slightly disjoint rectangles, each of length 4δ and width $\ll \delta$, below the vertical disc sticks of each literal. All the rectangles are aligned on the same horizontal slap, and the center lines of the vertical sticks pass through the "gaps" between neighboring rectangles. Thus for a literal x appearing in m_x clauses, there are $2m_x + 1$ rectangles for x.
5. The distance from an x (or $\bar{x}$) disc stick to the nearest rectangle is 2δ.
6. The distance from a forcing disc stick to the nearest rectangle is δ.

Clause components

1. For each clause c containing three literals, construct three vertical disc sticks (each corresponding to a literal in c) and one horizontal disc stick, as shown in Figure 1. The four disk sticks together are called a c *disc stick*.
2. Each vertical disc stick in a c disc stick is aligned with a vertical disc stick of the corresponding literal. The three literal disc sticks aligned with the vertical disc sticks of the c disc stick are called c's *selected literal disc sticks*.
3. No two clauses share a selected literal disc stick.
4. The center lines of the vertical disc stick pass through the gap between neighboring rectangles.
5. The distance from a vertical disc stick of c to the nearest rectangle in the literal component is 3δ.
6. The disc sticks of the clauses are placed according to the embedding of the underlying planar graph. No disc sticks from two clauses intersect each other.

7. The two horizontal disc sticks of any two distinct clauses are vertically separated by a distance of 4δ.

In the above construction, the radius of the discs and the width of the rectangles are $\ll \delta$, therefore can be ignored when considering the length of MST.

Next we introduce edges to connect these constructed components, and show that a MST of the instance can be computed from a subset of these edges.

Spanning edges

- **Type 1:** Edges between neighboring discs in a disc stick. Since $r \ll \delta$, such edges are part of MST, and their total length can be viewed as a constant.
- **Type 2:** Edges between a positive (negative) disc stick to its nearest rectangles. Since the center line of such a disc stick passes through the gap between two neighboring rectangles, we create a vertical edge from the last type 1 edge of the disc stick to simultaneously connect the two nearest neighboring rectangles. We call such an edge a *positive edge* (or *negative edge*). The length of a positive (negative) edge is 2δ, according to the construction.
- **Type 3:** Edges between a vertical disc stick of a c disc stick to its nearest rectangles. Similar to a type 2 edge, each such edge connect a type 1 edge to two neighboring rectangles (or a type 2 edge). This type of edges are called *truth assignment edges* for clause c, and the length of each such edge is 3δ.

Now we consider the MST. Given a literal x, let P_x be the portion of x in the literal component, and l_x be the total length of type 1 edges in P_x.

Lemma 1. *Any MST of P_x contains either all positive edges and the positive forcing edge of x or all negative edges and the negative forcing edge of x.*

Proof. Suppose this is not true. Let T_x be such a MST which fails the lemma.

First, notice that the set of positive (negative) edges and the positive (negative) forcing edge, together with the type 1 edges of P_x, form a MST T'_x. The length of T'_x is $l_x + (2m_x + 1)\delta$, where m_x is the occurrences of x in the clauses.

Since T_x fails the lemma, one of the following two cases must be true. (Assuming that all type 2 edges in P_x are sorted into a sequence based on x-coordinates.)

1. T_x contains three consecutive type 2 edges. This cannot be true, since one can remove the edge in the middle to obtain a shorter spanning tree.
2. T_x misses at least two consecutive type 2 edges. Let $\bar{x}_i$ and x_{i+1} be two consecutive type 2 edges missing from T_x. Let x_i and $\bar{x}_{i+1}$ be the two type 2 edges immediately before $\bar{x}_i$ and immediately after x_{i+1}. To span the three rectangles, say r_l, r_m, r_r, between edge x_i and $\bar{x}_{i+1}$, there must be at least three edges (i.e. x_i, $\bar{x}_{i+1}$ and an edge traversing r_l (or r_r)) of a total length of 8δ to connect the three rectangles to the rest of the T_x. But the total length of $\bar{x}_i$ and x_{i+1} is at most 4δ. Thus by replacing the edges traversing r_l (or r_r) with $\bar{x}_i$ (or x_{i+1}) result in a tree with shorter length.

Since both cases lead to contradictions, the lemma follows. $\qquad\square$

Let m_x be the number of occurrences of a literal x in the clauses and m be the total number of clauses. We have the following lemma regarding the truth assignment (type 3) edges and the total length of a MST of the instance.

Lemma 2. *Any MST T of the constructed components contains at most one truth assignment edge for each clause, and has a total length of at least $l + \sum_{x \in L} (2m_x + 1)\delta + 3m\delta$, with the minimum achieved when it contains exactly one truth assignment edge for each clause connecting to a type 2 edge of T.*

Proof. Suppose the lemma is not true. Assume that T contains two truth assignment edges t_x and t_y for a clause c. Let $\theta(x)$ and $\theta(y)$ be the truth assignments represented by t_x and t_y, there are two cases about the connectivity of T:

1. Either all $\theta(x)$ (type 2) edges or all $\theta(y)$ (type 2) edges are contained in T (by Lemma 1). Suppose without loss of generality that all $\theta(x)$ edges are in T. Since literals x and y are connected by horizontal type 1 edges, removing t_y from T results in a shorter spanning tree. A contradiction.
2. None of the $\theta(x)$ and $\theta(y)$ (type 2) edges are contained in T. Then T must traverse one rectangle in the literal components of either x or y to connect the clause component of c. Again, one of the two truth assignment edges t_x and t_y can be removed to obtain a shorter spanning tree. A contradiction.

For the total length of T, note that the length of a truth assignment edge is 3δ, the length of a rectangle is 4δ, and the minimum distance between two clause disc sticks is 4δ. Hence, the way of connecting a clause c's component to the literal components with the least cost is to add one truth assignment edge for c that can directly connect to a type 2 edge contained in T. Therefore for each clause c, at least one truth assignment edge with length of 3δ is needed for T. Thus by Lemma 1, the total length of the MST is at least $l + \sum_{x \in L} (2m_x + 1)\delta + 3m\delta$. The minimum is achieved when T contains exactly one truth assignment edge for each clause connecting to a type 2 edge of T.   □

With Lemmas 1, 2 we are ready to show the equivalence of Problem 2 and 1.

Theorem 1. *There exists a MST of D and R with length of $l + \sum_{x \in L} (2m_x + 1)\delta + 3m\delta$, if and only if the set of clauses C of (L, C) is satisfiable.*

Proof. **Sufficiency:** Suppose that C is satisfiable, and let θ be a satisfiable truth assignment. We construct the spanning tree T as follows:

1. T contains all the type 1 edges.
2. T contains one and only one type 3 edge for each clause, which can be chosen according to θ.
3. For each literal $x \in L$, T contains all positive and the positive forcing edges of x if the assignment $\theta(x)$ is *true*, or all negative and the negative forcing edges of x if otherwise.

By Lemma 1 and the properties of all types of edges, T is a spanning tree of D and R with length of $l + \sum_{x \in L} (2m_x + 1)\delta + 3m\delta$. By Lemma 2, T is minimum.

Necessity: By Lemma 2, we know that the length of any MST is at least $l + \sum_{x \in L} (2m_x + 1)\delta + 3m\delta$, and the minimum is achieved when it contains exactly one truth assignment edge for each clause connecting to a type 2 edge of the MST. Thus, if there exists a tree T with a total length of $l + \sum_{x \in L} (2m_x + 1)\delta + 3m\delta$, we can have a unique truth assignment θ by setting $\theta(x)$ to be *true* for each $x \in L$ if T contains all positive and positive forcing edges of x, and setting $\theta(x)$ to be *false* for all other literals. Since T contains one truth assignment edge for each clause, θ satisfies all clauses. $\square$

3   MST of Disjoint Unit Discs

3.1   A Simple Constant Algorithm

Property 1. In a MST of n disjoint discs, the number of internal nodes $\geq \lceil \frac{n-1}{5} \rceil$.

We call a disc represented by an internal node in T as *an internal disc*.

Proposition 1. *Let $\mathcal{D}$ be a set of n discs of radius r. Let T and L be a MST of $\mathcal{D}$ and its length, respectively. Then, $n \leq \frac{10}{\pi r}L + 1$.*

Our algorithm simply constructs a MST of the center points of the n discs [1]. We call this tree T_c as a *minimum center tree*.

Lemma 3. *The length L_c of a minimum center tree T_c is at most $(1 + \frac{20}{\pi})L + 2r$.*

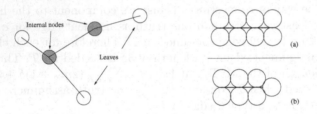

Fig. 2. The tree of the centers   **Fig. 3.** Put as many as discs along a segment

3.2   Two Lower Bounds of the MST of Discs

We define distance d_{ij} between two discs d_i and d_j to be the minimum Euclidean distance between any two points $p_i \in d_i$ and $p_j \in d_j$. The segment s_{ij} between the closest pair is called the distance segment of d_i and d_j.

Let S be the set of $n^2 - n$ distance segments of $\mathcal{D}$, called the *distance set*. We say two discs d_i and d_j is connected in a tree if their distance segment is included. A *minimum connecting tree* T_n of $\mathcal{D}$ is defined as a subset of S which connects all the discs and with the minimum total length $\sum\limits_{s_{ij} \in T_n} d_{ij}$. Below we show that the length of a minimum connecting tree is a lower bound of a MST.

---

[1] This algorithm follows the spirit of the constant approximation for TSP of discs [8].

Lemma 4. *Let $\mathcal{D}$ be the set of discs, and T and L be a MST of $\mathcal{D}$ and its length respectively. Let T_n and L_n be a minimum connecting tree of $\mathcal{D}$ and its length respectively. $L_n \leq L$.*

Next, we show another lower bound which investigates the relationship between L, r, and n. To determine a good lower bound for L, we consider the "dual" question of "What is the minimum length of a tree spanning a set of n discs?", that is, "How many disjoint discs can a tree of length L intersect?". This can be viewed as a disc "packing" problem along line segments, as the tree can be decomposed into a set of line segments.

Lemma 5. *Let s be a segment of length L, and n is the maximum number of disjoint discs which intersect s. Then, $L \geq (n - 2)r$ when $n \geq 4$ is an even number, and $n \geq 3$, $L \geq (n - 4 + \sqrt{3})r$ when $n \geq 3$ is an odd number .*

Proof. Assume s is horizontal. If n is an even number, to have s intersect as many disjoint discs as possible, the best way is to put $n/2$ discs on top of s (side by side), and the other $n/2$ beneath s, as shown in Figure 3. If n is odd, we just add the "extra" one to either end of s, as shown in Figure 3. □

3.3 Two 3-Approximation Algorithms

Algorithm 1. A 3-approximation algorithm

Construct a minimum connecting tree T_n from the distance set. Let L_n be the length of T_n.

for every internal disc d_i in T_n **do**

  Among all the distance segments incident to d_i, arbitrarily pick one of the distance segment's endpoint p_{i1} on the boundary of d_i as the representative point of d_i.
  Connect the endpoints of all other distance segments of d_i, $p_{i2}, \ldots, p_{ik}$, to p_{i1} by adding intra-disc segments.

end for

Output the new T_n'.

Theorem 2. *Let T_n' and L_n' be the spanning tree generated by Algorithm 1 and its length, respectively. Let L be the length of a MST. Then,*

$$L_n' \leq \begin{cases} 3L, & \text{if } n \text{ is even and } n \geq 4 \\ 3L + (4 - 2\sqrt{3})r, & \text{if } n \text{ is odd and } n \geq 3. \end{cases}$$

Consider the minimum center tree algorithm mentioned in Section 3.1. We could improve its length bound by a more careful analysis.

Theorem 3. *Let T_c be a minimum center tree. Cut off the intra-disc subsegments of all leaves in T_c. Let L_c' be the length of the resulting tree. Then,*

$$L_c' \leq \begin{cases} 3L + 2r, & \text{if } n \text{ is even and } n \geq 4 \\ 3L + (6 - 2\sqrt{3})r, & \text{if } n \text{ is odd and } n \geq 3. \end{cases}$$

3.4   A Polynomial Time Approximation Scheme

Our PTAS follows the framework of Arora's scheme for the Euclidean TSP problem [13,11]. Below we first sketch the main steps of Arora's technique.

1. Perturb the set of input points such that they are well separated and rounded to integral points;
2. Recursively partition the set of rounded points and build a randomly shifted quadtree such that each leaf contains at most one point;
3. In a bottom-up fashion, for each node of the quadtree decomposition, use dynamic programming to compute the shortest (m, r)-light path, where $r = O(1/\epsilon)$ and $m = O(\log n/\epsilon)$.

When the geometric objects change from points to discs, a number of new challenges occur. To overcome the additional difficulties, we adopt the following modifications according to the special properties of our problem (See Figure 4):

1. Recursively partition the perturbed problem instance into a randomly shifted quadtree structure, based on the positions of the disc centers, such that each leaf contains at most one disc center;
2. The "boundary" of a subproblem is the "outer envelope" of the followings:
   (a) The boundary of a square of the quadtree decomposition.
   (b) The boundaries of a set of discs whose centers are inside this square;
3. A set of portals is evenly distributed on the boundary of each subproblem:
   (a) Place portals on the straight line segment part (contributed by the square edges) of a boundary segment;
   (b) Place portals on the arc portion (contributed by the boundary of the discs) of a boundary segment;
   (c) The minimal distance between any two portals are bounded by l_s/m, where l_s is the size of the square and m is a constant whose value will be determined later on.

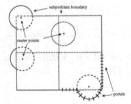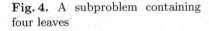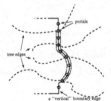

Fig. 4. A subproblem containing four leaves

Fig. 5. An illustration of the patching lemma. The segments added during the patching process are shown as the "dash-dot-dot-dot" line.

We define the size of a subproblem boundary as $1/4$ of the total length of the boundary associated with the subproblem. The next lemma upper-bounds the boundary size of a subproblem.

Lemma 6. *Let l_s be the size of a square and l be the size of the subproblem boundary. Then $l \leq (3\pi/8 + 1/2)l_s$.*

A "bended tree" is a tree that visits all the input discs and some subset of portals. A bended tree is (m, r)-light with respect to the shifted quadtree decomposition if it crosses each side of the subproblem-boundary for at most r times and always at portals. A bended tree could contain curve segments as its "bended" edges, which can be straightened at the end of the algorithm to get a spanning tree without increasing the cost.

We place a constant number (i.e., $O(1/\epsilon^2)$) of evenly distributed grid points in each disc and make them the candidates for the representative point of each disc. Thus, the subproblem of finding a (m, r)-light tree in each leaf square can be solved in a constant time.

Lemma 7. *(Patching Lemma) Let Π be any spanning tree of $\mathcal{D}$ and b be any boundary edge of length l that Π crosses at least 3 times, there exists a set of segments whose total length is at most $2l$ and whose addition to Π changes it into another spanning tree crossing b only once.*

The next lemma is useful in the analysis, and can be easily proved by using the arguments given in [11].

Lemma 8. *Let T be an optimal MST with length L, and $t(T, b)$ be the number of crossings by T on all the boundaries. Then $t(T, b)$ is bounded by $\sqrt{2}L$.*

Theorem 4. *Let $\mathcal{D}$ be a set of n disjoint unit discs contained in a minimal bounding box of size $B = O(n/\epsilon)$, and T be its MST of length L. For a random (a, b)-shift, $0 \leq a, b \leq B$, and $0 < \epsilon < 1$, there exists an algorithm computing a $(1 + \epsilon)$-approximation of T in $O(n(\log n)^{O(1/\epsilon)})$ time with probability $\geq 1/2$.*

References

1. Cormen, T.H., Leiserson, C.E., Rivest, R.L., Stein, C.: Introduction to Algorithms (Second Edition). MIT Press and McGraw-Hill (2001)
2. Har-Peled, S.: A replacement for voronoi diagrams of near linear size. In: FOCS. (2001) 94–103
3. Arya, S., Malamatos, T., Mount, D.M.: Space-efficient approximate voronoi diagrams. In: STOC '02: Proceedings of the thiry-fourth annual ACM symposium on Theory of computing, New York, NY, USA, ACM Press (2002) 721–730
4. Wei, X., Samarabandu, J., Devdhar, R., Siegel, A., Acharya, R., Berezney, R.: Segregation of transcription and replication sites into higher order domains.
5. Arkin, E.M., Hassin, R.: Approximation algorithms for the geometric covering salesman problem. Discrete Applied Mathematics **55**(3) (1994) 197–218
6. Mata, C.S., Mitchell, J.S.B.: Approximation algorithms for geometric tour and network design problems (extended abstract). In: Symposium on Computational Geometry. (1995) 360–369
7. Gudmundsson, J., Levcopoulos, C.: A fast approximation algorithm for tsp with neighborhoods. Nord. J. Comput. **6**(4) (1999) 469–488

8. Dumitrescu, A., Mitchell, J.S.B.: Approximation algorithms for tsp with neighborhoods in the plane. In: Proc. 12th ACM-SIAM Sympos. Discrete Algorithms (SODA). (2001) 38–46
9. de Berg, M., Gudmundsson, J., Katz, M., Levcopoulos, C., Overmars, M., van der Stappen, F.: Constant factor approximation algorithms for tspn with fat objects.
10. Mitchell, J.S.B.: A ptas for tsp with neighborhoods among fat regions in the plane. In: ACM-SIAM Symposium on Discrete Algorithms. (2007)
11. Arora, S.: Polynomial time approximation schemes for euclidean traveling salesman and other geometric problems. J. ACM **45**(5) (1998) 753–782
12. Mansfield, A.: Determining the thickness of graphs is np-hard. Math. Proc. Cambridge Philos. Soc. (1983) 9–23
13. Arora, S.: Polynomial time approximation schemes for euclidean TSP and other geometric problems. In: FOCS. (1996) 2–11

An Almost Linear Time 2.8334-Approximation Algorithm for the Disc Covering Problem*

Bin Fu[1], Zhixiang Chen[1], and Mahdi Abdelguerfi[2]

[1] Dept. of Computer Science, University of Texas - Pan American
TX 78539, USA
{chen,binfu}@cs.panam.edu
[2] Dept. of Computer Science, University of New Orleans, LA 70148, USA
mahdi@cs.uno.edu

Abstract. The disc covering problem asks to cover a set of points on the plane with a minimum number of fix-sized discs. We develop an $O(n(\log n)^2(\log \log n)^2)$ deterministic time 2.8334-approximation algorithm for this problem. Previous approximation algorithms [7,3,6], when used to achieve the same approximation ratio for the disc covering problem, will have much higher time complexity than our algorithms.

1 Introduction

The disc covering problem is to find a minimum number of discs of a prescribed radius r to cover a given set of points on the plane. This problem has many applications in areas such as image processing, wireless communication and patten recognition. It was proved to be NP-hard [4]. The first approximation algorithm was derived by Hochbaum and Maass [7]. Their algorithm has computational time $n^{2\lfloor l\sqrt{2}\rfloor^2}$ for approximation ratio $(1 + \frac{1}{l})^2$, where $l > 0$ is an integer accuracy control parameter. This approximation algorithm was further improved to $n^{6\lfloor l\sqrt{2}\rfloor}$ in [3,6]. The high computational complexity of those polynomial time approximation algorithms make them impractical for implementation in practice. Therefore, it is interesting, but challenging, to design faster polynomial time approximation algorithms for the disc covering problem.

In this paper we try to find faster approximation algorithms for the disc covering problem with some reasonably small approximation ratios. We derive an almost linear time approximation algorithm for the disc covering problem. This algorithm runs in $O(n(\log n)^2(\log \log n)^2)$ time with a 2.8334-approximation ratio. Previous approximation algorithms [7,3,6] will have much higher time complexity than our algorithms, when they are used to achieve the same 2.8334-approximation ratio for the disc covering problem. An $O(n)$ time $3(1 + \beta)$-approximation algorithm for the two-dimensional disc covering problem was shown in [5]. We generalize this linear time approximation algorithm to any

* This research is supported by Louisiana Board of Regents fund under contract number LEQSF(2004-07)-RD-A-35, and in part by NSF Grant CNS-0521585.

M.-Y. Kao and X.-Y. Li (Eds.): AAIM 2007, LNCS 4508, pp. 317–326, 2007.

fixed d-dimensional space by using the concept of Borsuk number, which is the minimum number of d-dimensional balls of radius r to fill a d-dimensional ball of a radius that is slightly larger than r.

We develop some novel method to cover the points in the local region, which is roughly occupied by one disc. Instead of covering each local region by three discs like [5], we let two local regions share one disc in some cases. Our method involves the nontrivial algorithm by Chan [1] for covering points with two fixed size discs, and another interesting algorithm by Meggido and Supowit [9] for covering points with one fixed size disc.

2   Notations and Shifting Strategy

Given a set of input points P and a radius r, Let $o(P)$ denote the minimum number of discs of radius r to cover all the points in P. For any given approximation algorithm A, which outputs $A(P)$ many discs of radius r to cover P, we shall have $A(P) \geq o(P)$. The approximation ratio of the algorithm A is defined as $max_P \frac{A(P)}{o(P)}$. Let $C_r(p)$ be a disc with a radius r and centered at the point p.

For two points p_1, p_2 in the d-dimensional Euclidean space R^d, $dist(p_1, p_2)$ is the Euclidean distance between p_1 and p_2. For a set $A \subseteq R^d$, $dist(p_1, A) = min_{q \in A} dist(p_1, q)$.

A point in R^d is a grid point if all of its coordinates are integers. For a d-dimensional point $p = (i_1, i_2, \cdots, i_d)$ and $a > 0$, define $grid_a(p)$ to be the set $\{(x_1, x_2, \cdots, x_d) | i_j - \frac{a}{2} \leq x_j < i_j + \frac{a}{2}, j = 1, 2, \cdots, d\}$, which is a half open and half close a^d-volume d-dimensional cubic region. For $a_1, \cdots, a_d > 0$, a $(a_1, \cdots, a_d)$-grid point is a point $(i_1 a_1, \cdots, i_d a_d)$ for some integers $i_1, \cdots, i_d$. For a ball B in R^d, let $r(B)$ denote the radius of B, $center(B)$ denote the center of B, and $extend_\delta(B)$ be the ball with the same center as B but with a larger radius $(1 + \delta) r(B)$ for $\delta > 0$.

We will use the shifting method developed by Hochbaum and Maass [7] to handle some subcases of our algorithms. For completeness, we give the description of the shifting method to deal with the disc covering problem. Let $l > 0$ be the integer parameter to control the accuracy of approximation. Assume that all the points in the input set P are in a region B, and discs of radius r are used to cover P. The region B is partitioned into vertical strips of width $2r$, $B_1, B_2, \cdots, B_k$. Without loss of generality, we assume that the union of every two consecutive strips intersects P (otherwise, the covering problem can be decomposed into two independent covering problems). This indicates that the number of strips is $O(|P|)$. Group every l consecutive strips into a wider strip of width $2rl$. In other words, each wider strip is $L_i = B_i \cup B_{i+1} \cdots \cup B_{i+l-1}$ for $i = 1, \cdots, k - l + 1$, and $L_i = B_i \cup B_{i+1} \cdots \cup B_k$ for $i = k - l + 2, \cdots, k$. We also define $L_i^0 = B_1 \cup B_2 \cdots B_{i-1}$, which is the union of first $i - 1$ blocks. The i-th shifted case has a set of wider strips $P_i = \{L_i^0, L_i, L_{i+l}, L_{i+2l}, \cdots, L_{i+t_i l}\}$, forming a partition for B ($B = L_i^0 \cup L_i \cup L_{i+l} \cup \cdots L_{i+tl}$).

Define $opt_P(D)$ to be the set of the discs in an optimal solution for covering the points of the set P in the region D. Let $d_i = \sum_{L \in P_i} |opt_P(L)|$. The crucial property of the shifting method [7] is that $\sum_{i=1}^{l} d_i \leq (1+l)|opt_P(B)|$. This implies that $\min_{1 \leq i \leq l} d_i \leq (1+\frac{1}{l})|opt_P(B)|$. Assume we have a local algorithm A for solving each local area L_i with approximation ratio AP_A. The solution of the algorithm A for the partition P_i is $s_i = \sum_{L \in P_i} A(L) \leq AP_A \cdot d_i$. The shifting method SA applies the algorithm A for each partition P_i, $1 \leq i \leq l$, and outputs the result $SA(B) = \min_{i=1}^{l} s_i$. Therefore, $SA(B) \leq (1+\frac{1}{l}) \cdot AP_A \cdot opt_P(B)$.

Theorem 1 ([7]). *Assume that a local algorithm A has an approximation ratio AP_A for the disc covering problem. Then, the approximation ratio AP_{SA} of the shifting method utilizing A satisfies $AP_{SA} \leq (1+\frac{1}{l})AP_A$.*

For the d-dimensional ball covering problem, repeating the shifting method at the directions of d-axis, we can get the following result:

Theorem 2 ([7]). *Assume that a local algorithm A has an approximation ratio AP_A for the disc covering problem in the region of volume l^d. Then, the approximation ratio AP_{SA} of the shifting method utilizing A satisfies $AP_{SA} \leq (1+\frac{1}{l})^d$. Furthermore, its computational time is $O(ndlT_d(l))$, where $T_d(l)$ is the computational time for the optimal solution in a local d-dimensional cubic region of volume $(2rl)^d$.*

3   Borsuk Number and Disc Covering

For any given dimension $d > 0$ and any given radius $r > 0$, let the Borsuk number $B(d)$ be the minimum number of d-dimensional balls of radius r in R^d that can fill a d-dimensional ball of radius $r + \delta$ in R^d for some $\delta > 0$. It is well-known that $B(2) = 3$ and $B(3) = 4$. Given a set P of points in R^d, the d-dimensional disc (or ball) covering problem is to find a minimum number of d-dimensional discs (or balls) of radius r to cover all the points in P.

Lemma 1. *For any given dimension $d > 0$ and any fixed parameter $\delta > 0$, there is an $O(n)$ time algorithm that, given a set of n points P in R^d, returns a set of points $Sketch_\delta(P) \subseteq P$ such that for every $(\delta, \cdots, \delta)$-grid point q, $grid_\delta(q) \cap P \neq \emptyset$ iff $|grid_\delta(q) \cap Sketch_\delta(P)| = 1$.*

Proof. Set $Sketch_\delta(P) = \emptyset$. Unmark all the $(\delta, \delta, \cdots, \delta)$-grid points. For each point p in P, find the $(\delta, \delta, \cdots, \delta)$-grid point q such that $p \in grid_\delta(q)$. If q is not marked, add p to $Sketch_\delta(P)$ and mark q. This takes $O(n)$ time.

Theorem 3. *Given a fixed dimension $d > 0$, a constant $\beta > 0$ and a radius $r > 0$, for any set P of n points in R^d, we have two algorithms for covering $Sketch_\delta(P)$ for some constant $\delta > 0$ and P, respectively:*

1. *There exists an $O(n)$ time $(1+\beta)$-approximation algorithm for covering all the points in $Sketch_\delta(P)$ with discs of radius r.*
2. *There exists an $O(n)$ time $B(d)(1+\beta)$-approximation algorithm for covering all the points in P with discs of radius r.*

Proof. Select an integer l such that $(1 + \frac{1}{l})^d \leq 1 + \beta$. Assume that $\delta_1 > 0$ is the constant such that a d-dimensional ball of radius $r(1 + \delta_1)$ can be filled by $B(d)$ many d-dimensional balls of radius r. Let $\delta = \frac{r\delta_1}{\sqrt{d}}$. Let Q be the set $Sketch_\delta(P)$ derived from Lemma 1. Apply the shifting method to find the $(1+\beta)$-approximation to the minimum number of balls to cover all the points in Q. By Theorem 2, we can get the $(1 + \beta)$-approximation for covering the points of Q in $O(nlT(l))$ steps, where $T(l)$ is the time in the $(2rl)^d$ region that has at most $(\frac{2rl}{\delta} + 1)^d$ $(\delta, \cdots, \delta)$-grid points. Therefore, it has at most $(\frac{2rl}{\delta} + 3)^d$ points in Q. We use d points to determine the position of a ball in d-dimensional space. Finding the optimal covering for the points of Q in a $(\frac{2rl}{\delta})^d$ region can be done in $O((\frac{2rl}{\delta} + 3)^{2d}) = O((\frac{2\sqrt{d}l}{\delta_1} + 3)^{2d})$ steps for fixed d. This completes the proof for first part of the theorem.

To prove the second the part of the theorem, we continue with the set of balls, denoted by S, obtained by the algorithm for the first part for covering $Sketch_\delta(P)$. By the construction of $Sketch_\delta(P)$, every point p in P is either covered by a ball in S, or it is not covered but is within distance $\sqrt{d}\delta$ to some ball S. We replace each ball D in S by a ball D' of radius $r + \sqrt{d}\delta$ and centered at $center(D)$, i.e., $D' = extend_{\sqrt{d}\delta}(D)$. Let S' denote the new set of those larger balls. Obviously, balls in S' covers P. By the choice of δ, $r + \sqrt{d}\delta = r(1 + \delta_1)$. Thus, every ball in S' can be filled with $B(d)$ balls of radius r. Therefore, replacing each ball D' in S' with $B(d)$ ball of radius r that fill D' yields a set of balls of radius r for covering all the points in P. This completes the proof for the second part of the part of the theorem.   □

4   A 2.8334-Approximation Algorithm for 2D Covering

We present our main result in this section. We derive a 2.8334 approximation algorithmfor the 2D disc covering problem with almost linear computational time. We will use the linear time algorithm for finding the minimum disc to cover a set of points by Meggido and Supowit [9]. We also use the $O(n(\log n)^2(\log \log n)^2)$-time algorithm developed by Chan, who improved the previous $O(n(\log n)^9)$-time deterministic algorithm by Sharir [10] to check if a set of points on the plane can by covered by two discs. An $O(n(\log n)^2)$time randomized algorithm to check if a set of points on the plane can by covered by two discs was developed by Epstein [2].

Lemma 2. *Let $r > 0$ be a real number. For any constant $1 \geq \alpha > 0$, there are three constants $\alpha \geq \alpha_1, \alpha_2, \alpha_3 > 0$ such that for every disc D_1 of radius $r' = (1 + \beta)r$ with $\beta \leq \alpha_1$ and every disc D_2 of radius r, if $2r' - \frac{r}{2} \leq dist(center(D_1), center(D_2)) \leq 2r' - \alpha_2 r$, then the line through their intersection points has distance at most $r' - \alpha_3 r$ to $center(D_1)$.*

Proof. We first compute the two intersection points of the two discs D_1 and D_2. Without loss of generality, we assume that the center of D_1 is at the origin $(0, 0)$ and the center of D_2 is at x-axis $(d, 0)$, where $d = dist(center(D_1), center(D_2)) \leq$

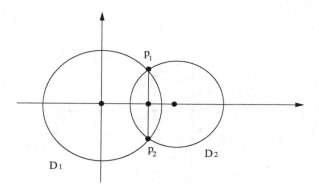

Fig. 1. Two Discs with Intersection

$2r' - \alpha_2 r$. See Figure 1 for an illustration. The two intersection points are at $p_1 = (x, y)$ and $p_2 = (x, -y)$. It is easy to see that $r'^2 - x^2 = r^2 - (d - x)^2$. Thus, we have $x = \frac{r'^2 - r^2 + d^2}{2d}$. It is easy to see that x is maximal when $d = 2r' - \alpha_2 r$. Thus, x is at most

$$\frac{r'^2 - r^2}{2d} + \frac{d}{2} = \frac{(r' - r)(r' + r)}{2d} + \frac{d}{2} \tag{1}$$

$$= \frac{\beta r(2 + \beta) r}{2(2(1 + \beta)r - \alpha_2 r)} + \frac{2(1 + \beta)r - \alpha_2 r}{2} \tag{2}$$

$$= (1 + \frac{\beta(2 + \beta)}{2(2(1 + \beta) - \alpha_2)} + \beta - \frac{\alpha_2}{2})r \tag{3}$$

$$\leq (1 + \frac{3\beta}{2} + \beta - \frac{\alpha_2}{2})r. \tag{4}$$

We use that fact $0 < \beta, \alpha_2 \leq 1$ in the transition from (3) to (4), which gives that $\frac{\beta(2+\beta)}{2(2(1+\beta)-\alpha_2)} \leq \frac{\beta(2+1)}{2(2(1+0)-1)} = \frac{3\beta}{2}$. Assign $\alpha_1 = \frac{\alpha}{4}$ and $\alpha_2 = \alpha$, and $\alpha_3 = \frac{\alpha}{8}$. Thus we have $x \leq (1 + \beta + \frac{3\alpha_1}{2} - \frac{\alpha_2}{2})r \leq (1 + \beta - \frac{\alpha}{8})r = r' - \alpha_3 r$. □

Lemma 3. *Let $r > 0$ and $\beta > 0$. There exist constants ϵ and δ with $\beta \geq \delta > \epsilon > 0$ such that if D is a disc of radius $r' = (1 + \epsilon)r$ and L is a line of the distance $d \leq r' - \delta r$ to the center of D, then the larger part of D on one of two sides of L can be covered by two discs of radius r.*

Proof. We select positive constants ϵ, δ and t that satisfy the conditions below:

$$t = 5 \tag{5}$$

$$\delta = 12t^2 \epsilon^{\frac{2}{3}} \tag{6}$$

$$\frac{t}{2} > t^2 \epsilon + (12t^2)^2 \epsilon^{\frac{1}{6}} \tag{7}$$

$$\min(\beta, \frac{1}{4}) > \delta, \epsilon \tag{8}$$

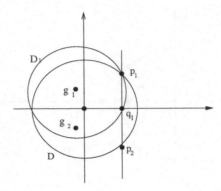

Fig. 2. A Disc with an Intersection Line

It is easy to see the existence of those three constants. Without loss of generality, assume that the center of D is at the origin $(0,0)$ and line L is parallel to y-axis. Let $p_1 = (x, y)$ and $p_2 = (x, -y)$ be the intersection points of disc D and line L (see Figure 2). Since the center of D has distance at most $r' - \delta r$ to line L,

$$x \le r' - \delta r. \tag{9}$$

We put a disc D_1 of radius r with center at point $g_1 = (-t\epsilon r, \delta r)$. We put the second disc D_2 of radius r with center at $g_2 = (-t\epsilon r, -\delta r)$.

We use $L_{\le x}$ to denote the half plane on left side of line L. We will prove that for every point q in the area of $D \cap L_{\le x}$ has either $dist(q, g_1) \le r$ or $dist(q, g_2) \le r$. Let $q = (x_1, y_1)$ be a point on the boundary of D with $-r' \le x_1 \le x$ and $y_1 \ge 0$. Clearly, $x_1^2 + y_1^2 = r'^2$.

$$dist(q, g_1)^2 = (x_1 + t\epsilon r)^2 + (y_1 - \delta r)^2 \tag{10}$$
$$= x_1^2 + 2t\epsilon r x_1 + (t\epsilon r)^2 + y_1^2 - 2(\delta r)y_1 + (\delta r)^2 \tag{11}$$
$$= r'^2 + 2t\epsilon r x_1 + (t\epsilon r)^2 + (\delta r)^2 - 2(\delta r)y_1 \tag{12}$$
$$= (r + \epsilon r)^2 + 2t\epsilon r x_1 + (t\epsilon r)^2 + (\delta r)^2 - 2\delta r y_1 \tag{13}$$
$$= r^2 + 2\epsilon r^2 + \epsilon^2 r^2 + 2t\epsilon r x_1 + (t\epsilon r)^2 + (\delta r)^2 - 2\delta r y_1 \tag{14}$$

Case 1. $-\frac{r}{2} < x_1 \le x \le r' - \delta r$. This condition implies that $r' + x_1 > \frac{r}{2}$ and $r' - x_1 \ge r' - x \ge \delta r$. Therefore,

$$y_1 = \sqrt{r'^2 - x_1^2} = \sqrt{(r' - x_1)(r' + x_1)} \ge \sqrt{\frac{\delta}{2}}r. \tag{15}$$

Now we prove that the distance between q and g_1 is bounded by r. By (14),

$$dist(q, g_1)^2 = r^2 + 2\epsilon r^2 + \epsilon^2 r^2 + 2t\epsilon r x_1 + (t\epsilon r)^2 + (\delta r)^2 - 2\delta r y_1 \tag{16}$$
$$\le r^2 + 2\epsilon r^2 + \epsilon^2 r^2 + 2t\epsilon r x_1 + (t\epsilon r)^2 + (\delta r)^2 - 2\sqrt{\frac{1}{2}}\delta^{\frac{3}{2}}r^2 \tag{17}$$

$$\leq r^2 + 2\epsilon r^2 + \epsilon^2 r^2 + 2t\epsilon r^2 + (t\epsilon r)^2 + (\delta r)^2 - \sqrt{2}\delta^{\frac{3}{2}} r^2 \tag{18}$$

$$\leq r^2 + (2\epsilon + \epsilon^2 + 2t\epsilon + (t\epsilon)^2 + \delta^2 - \sqrt{2}\delta^{\frac{3}{2}})r^2 \tag{19}$$

$$\leq r^2 + (6t^2\epsilon - \frac{\sqrt{2}\delta^{\frac{3}{2}}}{2})r^2 \tag{20}$$

$$< r^2 \tag{21}$$

Transition from (16) to (17) is from (15). In transition (19) to (20), we use the fact thats $\epsilon, \epsilon^2, t\epsilon, t^2\epsilon^2 \leq t^2\epsilon$ and $\delta^2 \leq \frac{\delta^{\frac{3}{2}}}{2} \leq \frac{\sqrt{2}\delta^{\frac{3}{2}}}{2}$. Those are from conditions (6) to (8). Transition (20) to (21) is from (6).

Case 2. $-r' \leq x_1 \leq -\frac{r}{2}$. We have that $2t\epsilon r x_1 \leq -t\epsilon r^2$. By (14),

$$dist(q, g_1)^2 \leq r^2 + 2\epsilon r^2 + \epsilon^2 r^2 + 2t\epsilon r x_1 + (t\epsilon r)^2 + (\delta r)^2 - 2\delta r y_1 \tag{22}$$

$$\leq r^2 + 2\epsilon r^2 + \epsilon^2 r^2 - t\epsilon r^2 + (t\epsilon r)^2 + (\delta r)^2 \tag{23}$$

$$\leq r^2 + (t^2\epsilon^2 + \delta^2 - \frac{t\epsilon}{2})r^2 \tag{24}$$

$$\leq r^2 + (t^2\epsilon^2 + (12t^2)^2\epsilon^{\frac{4}{3}} - \frac{t\epsilon}{2})r^2 \tag{25}$$

$$\leq r^2 + \epsilon(t^2\epsilon + (12t^2)^2\epsilon^{\frac{1}{3}} - \frac{t}{2})r^2 \tag{26}$$

$$< r^2 \tag{27}$$

In transition (23) to (24), we use that fact that $2 + \epsilon \leq \frac{t}{2}$, which is derived from (8). Transition (24) to (25) follow from (6). (26) to (27) is from the condition (7).

Case 3. $q_1 = (x, 0)$. Notice that x satisfies (9). We have that

$$dist(q_1, g_1) = (x + t\epsilon r)^2 + (\delta r)^2 \tag{28}$$

$$= x^2 + 2t\epsilon r x + (t\epsilon r)^2 + (\delta r)^2 \tag{29}$$

$$\leq (r' - \delta r)^2 + 2t\epsilon r x + (t\epsilon r)^2 + (\delta r)^2 \tag{30}$$

$$\leq r'^2 - 2\delta r r' + \delta^2 r^2 + 2t\epsilon r^2 + (t\epsilon r)^2 + (\delta r)^2 \tag{31}$$

$$\leq (1 + \epsilon)^2 r^2 - 2\delta r^2 + \delta^2 r^2 + 2t\epsilon r^2 + (t\epsilon r)^2 + (\delta r)^2 \tag{32}$$

$$\leq r^2 + 2\epsilon r^2 + (\epsilon r)^2 - 2\delta r^2 + 2t\epsilon r^2 + (t\epsilon r)^2 + 2(\delta r)^2 \tag{33}$$

$$\leq r^2 + (6t^2\epsilon - \delta)r^2 \tag{34}$$

$$< r^2. \tag{35}$$

For transition (33) to (34), we use the facts that $\epsilon, \epsilon^2, t\epsilon, t^2\epsilon^2 \leq t^2\epsilon$ and $2\delta^2 \leq \delta$, which are derived from conditions (6) to (8). Transition (34) to (35) is due to condition (6).

Let $B(D \cap L_{\leq x})$ be the set of all the points (x_1, y_1) on the boundary of the top half of disc D with $y_1 \geq 0$ and $-r' \leq x_1 \leq x$. Notice $q_1 = (x, 0)$ as in Case 3. For every point $p = (u, v)$ in the top half of $D \cap L_{\leq x}$, i.e., $u^2 + v^2 \leq r'^2$, $-r' \leq u \leq x$, and $0 \leq v$, we have

$$dist(p, g_1) \leq \max(\max_{q \in B(D \cap L_{\leq x})}(dist(q, g_1), dist(q_1, g_1)) \leq r.$$

Thus, we have proved that D_1 covers the top half of $D \cap L_{\leq x}$. Similarly, we can also prove that D_2 covers the bottom half of $D \cap L_{\leq x}$. Therefore, D_1 and D_2 completely cover $D \cap L_{\leq x}$. $\qquad\square$

Lemma 4. *Let $r > 0$ be any given real number. For every constant $\alpha \in (0,1)$, there exist constants ϵ and ρ in the interval $[0,\alpha]$ such that for every two discs D_1, D_2 of radius $(1+\epsilon)r$, if $dist(center(D_1), center(D_2)) \leq 4(1+\epsilon)r - \rho r$, then they can be covered by five discs of radius r.*

Proof. We prove the lemma by Lemmas 2 and 3. By Lemma 2, we have three constants α_1, α_2 and α_3 such that for every $0 < \epsilon_1 < \alpha_1$, for any disc A_1 of radius $(1+\epsilon_1)r$ and any disc A_2 of radius r, if the distance of their centers is within the range $[2(1+\epsilon_1)r - \frac{r}{2}, 2(1+\epsilon_1)r - \alpha_2 r]$, the line through the intersection points between A_1 and A_2 has distance at most $(1+\epsilon_1)r - \alpha_3 r$ to the center of A_1. Let $\beta = min\{\frac{1}{20}, \alpha_1, \alpha_3\}$ and $\rho = 2\alpha_2$. By Lemma 3, there are constants ϵ and δ such that $\beta \geq \delta > \epsilon > 0$ satisfy the condition in Lemma 3.

Case 1: The distance of the two centers of D_1 and D_2 is at least $4(1+\epsilon)r - r$. Let p be the middle point on the line through the centers of D_1 and D_2. By the condition of the lemma, $dist(p, center(D_1)) \leq 2(1+\epsilon)r - \frac{\rho}{2}r \leq 2(1+\epsilon)r - \alpha_2 r$. So, $dist(p, center(D_1)) \in [2(1+\epsilon)r - \frac{r}{2}, 2(1+\epsilon)r - \alpha_2 r]$. Let B be the disc of radius r and of center at the middle point p. By Lemma 2, the line through the intersection points between D_1 and B has distance at most $(1+\epsilon)r - \alpha_3 r \leq (1+\epsilon)r - \delta r$ to the center of D_1. By Lemma 3, $D_1 - B$ can be covered by two discs of radius r. Similarly $D_2 - B$ can also be covered by two discs of radius r. Therefore, $D_1 \cup D_2$ can be covered by at most five discs.

Case 2: The distance of the two centers of D_1 and D_2 is less than $4(1+\epsilon)r - r$. We consider two subcases:

Subcase 1: Disc D_1 and disc D_2 have intersection points p_1 and p_2 and the center of D_1 has distance at most $r' - \delta r$ to the line L through p_1 and p_2. By Lemma 3, the larger part of D_1 at one side of L can be covered by two discs of radius r. The part of D_2 at other side of L can be also covered by two discs of radius r. Therefore, $D_1 \cup D_2$ can be covered by four discs.

Subcase 2: Disc D_1 and disc D_2 have no intersection or the center of D_1 has distance $\geq r' - \delta r$ to the line L through the intersection points of D_1 and D_2. We put disc D with its center at the median on the line segment connecting the centers of D_1 and D_2. Disc D has enough intersection with both D_1 and D_2 (the center of D to the line through the intersection points between D and D_1 (D_2) will be small enough) so that by Lemma 3, both $D_1 - D$ can be covered by two discs of radius r and $D_2 - D$ can be covered by two discs of radius r. Therefore, $D_1 \cup D_2$ can be covered by five discs. $\qquad\square$

Theorem 4. *There exists an $O(n(\log n)^2(\log\log n)^2)$-time 2.8334-approximation algorithm for the disc covering problem on the plane.*

Proof. Let r be the radius of discs for the covering problem. Let γ be a positive real constant such that every disc of radius $\leq (1+\gamma)r$ can be covered by three discs of radius r. Let $\alpha = \gamma$. Select ϵ and ρ according to Lemma 4. We set $a = \frac{1}{3}$

and $\eta = \frac{\epsilon r}{\sqrt{2}}$. Select the constant $\beta > 0$ small enough such that $\frac{6-a}{2}(1 + \beta) \leq$ 2.8334 and $2 + \frac{5a}{2}(1+\beta) \leq 2.8334$. Such a β exists because $\frac{6-a}{2} = 2 + \frac{5a}{2} < 2.8334$. Let r' be equal to $r(1 + \epsilon)$.

Algorithm
Input: A set of n points P on the plane.
(1) With the parameter β, use the algorithm of the first part of Theorem 3 to find a set of discs S of radius r to cover $Sketch_\eta(P)$ (see Lemma 1).
(2) Let $T = \emptyset$ and $U = S$.
(3) For each disc $D \in U$
(4) begin
(5)   if (there is a disc $D' \in S$ with
            $dist(center(D), center(D')) \leq 4r' - \rho r$) then
(6)   begin
(7)   let $U = U - \{D, D'\}$, and
(8)   let $T = T \cup \{(D, D')\}$.
(9)   end (if)
(10) end (for)
(11) Let V be the set of discs in the pairs of T.
(12) For each $(D, D') \in T$, cover $extend_\eta(D)$ and $extend_\eta(D')$ with at most 5 discs of radius r.
(13) For each disc in $D \in U$, cover all the points in $extend_\eta(D) \cap P$ with a minimal number of discs (at most 3 discs are needed).
End of Algorithm

Let m be the total number of discs in the set S obtained by the algorithm in the first part of Theorem 3 for covering $Sketch_\eta(P)$. Let S, T and U be the sets after running the algorithm above. Recall in section 2 that we define $o(A)$, for any set A of points, as the minimal number of discs of radius r for covering all the points in A. As $Sketch_\eta(P)$ is a subset of P and $\frac{m}{o(Sketch_\eta(P))} \leq 1 + \beta$, we have $o(Sketch_\eta(P)) \leq o(P)$ and

$$o(P) \geq \frac{m}{1 + \beta}. \tag{36}$$

Let t be the number of pairs of discs of distance $4r' - \rho r$ that have been identified by the algorithm. Those pairs are put into T.

Case 1. $2t \geq am$. In other words, $|V| \geq am$. The algorithm outputs at most $\frac{5}{2}am + 3(1 - a)m = \frac{6-a}{2}m$ discs of radius r to cover P. The approximation ratio is $\leq \frac{6-a}{2}m / \frac{m}{(1+\beta)} = \frac{6-a}{2}(1 + \beta) \leq 2.8334$.

Case 2. $2t < am$. In other words, $|V| < am$. So, $|U| \geq (1-a)m$. Notice that any two discs in U has distance $> 4r' - \rho r$. Let $m_i (i = 1, 2, 3)$ be the number of discs D in U such that $extend_\eta(D) \cap P$ requires i discs of radius r to cover. Over all, the algorithm needs at least $\frac{m_1+2m_2+3m_3}{2}$ discs of radius r to cover all $extend_\eta(D) \cap P$ for every D in U, since each disc can be shared by at most two adjacent regions ($extend_\eta(D_1)$ and $extend_\eta(D_2)$ for discs D_1 and D_2 in U). On the other hand, we can cover all points in P with $(m_1 + 2m_2 + 3m_3) + \frac{5}{2}am$ discs of radius r, and

these many discs have been identified by the algorithm. Combining with (36), we have $o(P) \geq \max(\frac{m}{1+\beta}, \frac{m_1+2m_2+3m_3}{2})$. The approximation ratio of the algorithm is at most $\frac{(m_1+2m_2+3m_3)+\frac{5}{2}am}{o(P)} \leq \frac{(m_1+2m_2+3m_3)+\frac{5}{2}am}{\max(\frac{m}{1+\beta}, \frac{m_1+2m_2+3m_3}{2})} \leq \frac{(m_1+2m_2+3m_3)}{\frac{m_1+2m_2+3m_3}{2}} + \frac{\frac{5}{2}am}{\frac{m}{1+\beta}} \leq$
$2 + \frac{5a}{2}(1+\beta) \leq 2.8334$.

By Theorem 3, the shifting part at step (1) in the algorithm for covering $Sketch_\eta(P)$ takes $O(n)$ time. We can find the smallest circle to cover a set of points in linear time [9]. We need $O(z(\log z)^2(\log \log z)^2)$ time to check if a set of z points on the plane can be covered by two discs of radius r [1]. It is easy to see that each disc only intersects with $O(1)$ other discs. This shows that steps (3) to (10) in the algorithm takes $O(|S|) = O(n)$ time. In summary, for each disc D in U, it takes at most $O(z(\log z)^2(\log \log z)^2)$ time to find a minimal number of discs of radius r to cover $extend_\eta(D) \cap P$, where $z = |extend_\eta(D) \cap P|$. Since each point stays in $O(1)$ discs in S, the total time of step (9) is $O(n(\log n)^2(\log \log n)^2)$. $\square$

Acknowledgements. We are grateful to two anonymous reviewers of SODA 2007 and Binhai Zhu for telling us Chan's paper [1].

References

1. T. M. Chan, More planar two-center algorithms, Computational Geometry Theory and Applications, 13, 1999, pp. 189-198.
2. D. Eppstein, Faster construction of planar two-centers, Proc. 8th ACM and SIAM Symp. Discrete Algorithms, Jan 1997, pp. 131138.
3. T. Feder and D. Greene, Optimal algorithms for approximate clustering, Proceedings of the 20th annual ACM symposium on the theory of computing, 1988, pp. 434-444.
4. R. J. Fowler, M. S. Paterson, and S. L. Tanimoto, Optimal packing and covering in the plane are NP-complete, Information processing letters, 3(12), 1981, pp. 133-137.
5. M. Franceschetti, M. Cook, and J. Bruck , A geometric theorem for network design, IEEE transaction on computers, 53(4), 2004. pp. 483-489.
6. T. Gonzalez, Covering a set of points in the multidimensional space, Information processing letters, 4(40) 1991, pp. 181-188.
7. D. S. Hochbaum, and W. Maass, Approximation schemes for covering and packing problems in image processing and VLSI, Journal of the ACM,1(32), 1985, pp. 130-136.
8. N. Meggido and K. Supowit, On the complexity of some common geometric location problems, SIAM journal on computing, v. 13, 1984, pp. 1-29.
9. N. Meggido, Linear-time algorithms for linear programming in R^3 and related problems, SIAM journal on computing, 1291983), pp. 759-776.
10. M. Sharir, A near-linear algorithm for the planar 2-center problem, Discrete & Computational Geometry, 1997, 18(2), pp. 125 - 134
11. D. W. Wong and Y. S. Kuo, A study of two geometric location problems, Information processing letters, v. 28, 1988, pp. 281-286.

Optimal Field Splitting with Feathering in Intensity-Modulated Radiation Therapy*

Xiaodong Wu[1,2] and Xin Dou[1]

[1] Department of Electrical and Computer Engineering
[2] Department of Radiation Oncology
The University of Iowa
Iowa City, IA 52242, USA
{xiaodong-wu,xin-dou}@uiowa.edu

Abstract. In this paper, we study an interesting geometric partition problem, called *optimal field splitting with feathering* (OFSF), which arises in *Intensity-Modulated Radiation Therapy* (IMRT). In current clinical practice, a multi-leaf collimator (MLC) with a *maximum field size* is used to deliver the prescribed intensity maps (IMs). However, the maximum field size of an MLC may require to split a large intensity map into several overlapping sub-IMs each being delivered separately, which may result in sacrificed treatment quality. Few IM splitting techniques reported in the literature have addressed the issue of treatment delivery accuracy for large IMs. We develop a new algorithm for solving the OFSF problem while minimizing the total delivery error. Our basic idea is to formulate the OFSF problem as computing a d-link shortest path in a directed acyclic graph, which expresses a special "layered" structure. The edge weights of the graph satisfy the Monge property, which enables us to solve this d-link shortest path problem by examining only a small portion of the graph, yielding an optimal linear time algorithm for the OFSF problem.

Keywords: Intensity map splitting, shortest paths, Monge property, Algorithms, IMRT.

1 Introduction

In this paper, we study an interesting geometric partition problem, called *optimal field splitting with feathering* (OFSF), which arises in *Intensity-Modulated Radiation Therapy* (IMRT) [11]. IMRT, a state-of-the-art radiation therapy technique for cancer treatments, aims to deliver a highly conformal radiation dose to a target tumor while sparing the surrounding normal tissues. The quality of IMRT crucially depends on the ability to accurately and efficiently deliver

* This research was supported in part by an NIH-NIBIB research grant R01-EB004640, and in part by a seed grant award from the American Cancer Society through an Institutional Research Grant to the Holden Comprehensive Cancer Center, the University of Iowa, Iowa City, Iowa, USA.

M.-Y. Kao and X.-Y. Li (Eds.): AAIM 2007, LNCS 4508, pp. 327–336, 2007.

the prescribed dose distributions of radiation, commonly called *intensity maps* (IMs). An intensity map is specified by a set of nonnegative integers on a 2-D grid. The number in a grid cell indicates the amount (in unit) of radiation to be delivered to the body region corresponding to that cell. The delivery of an IM is carried out by a set of cylindrical radiation beams orthogonal to the IM grid.

One of the most advanced tools today for delivering IMs is the *multileaf collimator* (MLC) [11]. An MLC has multiple pairs of tungsten leaves of the same rectangular shape and size. The two opposite leaves of each pair are aligned to each other. The leaves can move up and down to form (say) an x-monotone rectilinear region (i.e., monotone to the x-axis), called an *MLC-aperture*. The cross-section of a cylindrical radiation beam (generated by a radiotherapy machine) is shaped by this MLC-aperture to deliver certain units of radiation to (a portion of) an IM. The mechanical design of the MLCs restricts what kinds of beam-shaping regions are allowed [11]. A common constraint is called the *maximum field size*: Due to the limitation on the fixed number of MLC leaf pairs and the over-travel distance of the leaves, an MLC cannot enclose an IM of a too large size (called the *field size*). During the delivery of an IM, the isocenter O of the MLC is always aligned with the center of the IM.

One popular IMRT approach for delivering IMs using an MLC is the "step-and-shoot" technique [4,11,13]. Mathematically, the "step-and-shoot" delivery planning can be viewed as the following segmentation problem: Given an intensity map A defined on a 2-D $m \times n$ grid, decompose A into the form of $A = \sum_{k=1}^{\kappa} \alpha_k S_k$, where S_k is a special 0-1 matrix specifying an MLC-aperture, α_k is the amount of radiation delivered through S_k, and κ is the number of MLC-apertures used to deliver A. The reader is referred to [13,11] for more details on the step-and-shoot IMRT technique.

Two key criteria are used to measure the quality of an IMRT treatment. (1) The **delivery error** (accuracy): Due to the special geometric shapes of the MLC leaves [2,11,13] (i.e., the "tongue-and-groove" interlock feature), an MLC-aperture cannot be delivered perfectly. Instead, there is a *delivery error* between the planned dose and actual delivered dose (called the "*tongue-and-groove*" *error* in medical literature [11,13]). Chen *et al.* [2] showed that for an IM $B = (b_{i,j})_{m' \times n'}$ of size $m' \times n'$ (no larger than the maximum allowed field size), the minimum amount of error for delivering B is captured by the following formula (note that B contains only nonnegative integers): $Err(B) = \sum_{j=1}^{n'} \left\{ b_{1,j} + \sum_{i=1}^{m'-1} |b_{i,j} - b_{i+1,j}| + b_{m',j} \right\}$. Minimizing the delivery error is important because according to a recent study [5], the maximum delivery error can be up to 10%, creating underdose/overdose spots in the target region. (2) The **treatment time** (efficiency): Minimizing the treatment time is crucial since it not only lowers the treatment cost for each patient but also increases the patient throughput of the hospitals; in addition, it reduces the risk associated with the uncertainties in the treatment.

In current clinical radiation therapy, large intensity maps frequently occur [7,6,10]. Due to the maximum field size constraint of the MLC design, a large IM needs to be split into several sub-IMs each being delivered separately using the step-and-shoot delivery technique. However, such splitting may result in prolonged beam-on time and increased delivery error, and thus compromise the treatment quality. The *field splitting problem*, roughly speaking, is to split an IM of a large size into multiple sub-IMs whose sizes are all no larger than a threshold size, such that the treatment quality is optimized.

One simple way to split a large IM is to use straight lines, yielding *abutting* sub-IMs. One of the problems associated with this field splitting method is the field mismatching problem that occurs in the field junction region due to the uncertainties in setup and organ motion [9,10]. If the borders of two abutting sub-IMs do not precisely align each other, it may result in *hotspots* or *coldspots*. To alleviate the field mismatching problem, a commonly used medical practice is to apply a so-called *field feathering* technique [9,6,10]. Using this technique, a large IM A is split into a set of sub-IMs, $A_1, A_2, \ldots, A_\kappa$, such that each sub-IM A_k is subject to the maximum field size constraint, and any two adjacent sub-IMs overlap over a central *feathering region*. Note that in the former splitting method, each IM cell belongs to exactly one sub-IM; but in the latter method, each cell of the feathering region can belong to two adjacent sub-IMs, with non-negative intensity value in both sub-IMs.

A few field splitting algorithms have been recently reported in the literature to address the issue of treatment delivery efficiency for large IMs. To our best knowledge, Kamath *et al.* [8] first gave an $O(mn^2)$ time algorithm to split an $m \times n$ IM using vertical lines into *at most three* sub-IMs (thus restricting the maximum width of a large IM) while minimizing the total beam-on time. The beam-on time of a treatment is the time while a patient is exposed to actual irradiation [11], which is closely related to the total treatment time. Wu [12] formulated the field splitting problem for an arbitrary field width using vertical lines as a k-link shortest path problem and developed an $O(mnw)$ time algorithm, where w is the maximum allowed field width. Kamath *et al.* recently studied the field splitting with feathering [9]. However, their algorithm is optimal only for the case that the input IM has one row and the width of the IM is $\leq 3w$. Chen and Wang [3] further developed an $O(mn + m\xi^{d-2})$ time algorithm for optimally splitting an IM of size $m \times n$, where d is the number of resulting sub-IMs and ξ is the remainder of n divided by w. Although it is useful to consider field splitting with feathering to minimize the delivery error, to our best knowledge, no field splitting algorithms are known aiming to minimizing the total delivery error.

In this paper, we study the following **optimal field splitting with feathering (OFSF)** problem: Given an IM $A = (a_{i,j})_{m \times n}$ of size $m \times n$, a maximum allowable field size $l \times w$ with $m \leq l$ and $n > w$, and the width range $[\delta .. \Delta]$ of each feathering region $(0 < \delta < \Delta < w)$, split A using vertical lines into a sequence of $d = \lceil \frac{n-\delta}{w-\delta} \rceil$ (≥ 2) sub-IMs $\mathcal{S} = \{S_1, S_2, \ldots, S_d\}$ (from left to right), such that: (1) the width of each sub-IM S_k is w; (2) any two neighboring sub-IMs in the sequence overlap each other and the width of the overlapping (feathering) region ranges from δ to Δ; (3) no sub-IM overlaps completely with its neighboring sub-IM(s); and (4) the total delivery error of these d sub-IMs is minimized. Note that d is the minimum number of sub-IMs needed for delivering A. We may use more sub-IMs. But, that could significantly increase the total treatment time, which is undesirable. We thus assume that each sub-IM has a maximum width w since we can introduce columns filled with 0's to the sub-IM without increasing its delivery error. It is also reasonable here to assume that $m \leq l$ (i.e., we do not need to split the IM along the rows of A) since the splitting along the rows of A does not affect the delivery error.

We present an optimal linear time algorithm for solving the above OFSF problem. In our algorithm, we first model the OFSF problem as a d-link shortest path problem in a directed acyclic graph (DAG) of $O(n)$ vertices and $O(n(\Delta - \delta))$ edges. The computation of each edge weight takes pseudo-polynomial time. Interestingly, this DAG has a special "layered" structure, which consists of d layers with any two adjacent ones inducing a bipartite graph. We are able to calculate each edge weight in $O(m)$ time after a certain preprocessing. Moreover, the edge weights of the DAG satisfy the Monge property [1]. Thus, we can solve this d-link shortest path problem by examining only a small portion of the graph, and our algorithm runs in an optimal $O(mn)$ time.

2   The d-Link Shortest Path Model

In this section, we present a d-link shortest path model for solving the OFSF problem.

Denote by $A[j]$ the j-th column of IM A, i.e., $A[j] = \{a_{1,j}, a_{2,j}, \ldots, a_{m,j}\}$, and $A[j .. k]$ consists of all rows of A from column j to column k. Since the width of each sub-IM is fixed as w, d vertical lines $\{j_1, j_2, \ldots, j_d\}$ are needed to determine the starting column of each sub-IM in the splitting (including the first vertical line which is always corresponding to the first column of A, i.e., $j_1 = 1$). The k-th feathering region F_k consists of multiple columns of A starting from Column j_{k+1} to Column $j_k + w - 1$, denoted by $A[j_{k+1} .. j_k + w - 1]$. F_k is somehow decomposed into $F_k^{(0)}$ and $F_k^{(1)}$ such that $F_k = F_k^{(0)} + F_k^{(1)}$ (i.e., the value of every element in F_k is partitioned into two non-negative integers, one in $F_k^{(0)}$ and the other in $F_k^{(1)}$). Then, a feasible splitting $\mathcal{S} = \{S_1, S_2, \ldots, S_d\}$ of A is defined, as follows. For each $k = 1, 2, \ldots, d$, $S_k = F_{k-1}^{(1)} \,\|\, A[j_{k-1} + w .. j_{k+1} - 1] \,\|\, F_k^{(0)}$, where $\|$ is a concatenation operator, $F_0^{(1)} = F_d^{(0)} = \emptyset$, $j_0 = -w + 1$ and $j_{d+1} = n + 1$. The decomposition of each feathering region F_k may increase the total delivery error. Let $\Delta_{err}(j)$ denote the minimum *increase* of the delivery error resulting from the decomposition of F_k. Two issues need to be resolved in order to solve the OFSF problem: (1) how to choose a set of d vertical splitting line $\{j_k\}_{k=1}^{d}$; and (2) how to decompose each feathering region such that the increased delivery error is minimized.

To address the first issue, we model the OFSF problem as a d-link shortest path problem in a weighted directed acyclic graph $G = (V, E)$, which is defined as follows.

The vertices of G are defined as $V = \{t\} \cup \{v_j \mid 1 \leq j \leq n\}$. Each column $A[j]$ $(j = 1, 2, \ldots, n)$ of A corresponds to exact one vertex v_j in G. Note that if $j_k = j$

$(k < d)$, then $j + w - \Delta \leq j_{k+1} \leq j + w - \delta$. Thus, for every $j \in \{1, 2, \ldots, n - \delta\}$, vertex v_j has a directed edge in E to each vertex in $\{v_k \mid j + w - \Delta \leq k \leq j + w - \delta$ and $k \leq n\}$. That is, every edge $(v_j, v_{j'}) \in E$ $(j < j')$ specifies a feathering region $A[j' .. j + w - 1]$. The *weight* $c(v_j, v_{j'})$ equals to the minimum increase Δ_{err} of the delivery error resulting from the decomposition of this feathering region. Meanwhile, the only possible starting column of the last sub-IM S_d of a feasible splitting is in the range of $[n - w + 1 .. n - \delta + 1]$. Hence, for each vertex v_j $(j = n - w + 1, \ldots, n - \delta + 1)$, we put in E a directed edge from v_j to t. The cost of edge (v_j, t) is set to be 0.

Our algorithm then computes in G a d-link shortest path p from v_1 to t: $v_1 \rightarrow v_{j_2^*} \rightarrow v_{j_3^*} \rightarrow \ldots \rightarrow v_{j_d^*} \rightarrow t$. Obviously, the sequence of vertical lines defined by $\{1, j_2^*, \ldots, j_d^*\}$ specifies a feasible splitting S^* of A. The total increase of the delivery error due to the splitting of S^* equals the total weight $c(p)$ of the path p, which is minimized. Thus, S^* is an optimal splitting of A. The graph G has $O(n)$ vertices and $O(n(\Delta - \delta))$ edges. It takes $O(n(\Delta - \delta)d)$ time to compute a d-link shortest path in G after the graph is constructed. Note that there are $O(n(\Delta - \delta))$ possible feathering regions. Assume that it takes a time T for computing the minimum increase Δ_{err} of the delivery error induced from the decomposition of each feathering region. The total running time is $O(n(\Delta - \delta)(d + T))$.

Lemma 1. *The OFSF problem can be solved in $O(n(\Delta - \delta)(d + T))$ time by computing a d-link shortest path in a DAG G.*

We next show how to efficiently compute Δ_{err} for each possible feathering regions, actually, in $O(m)$ time after an $O(mn)$ time preprocessing, and the decomposition to achieve the optimal Δ_{err} can be obtained in linear time. In Section 4, we exploit the Monge property of the graph G to speed up the d-link shortest path computation, yielding an optimal linear $O(mn)$ time algorithm for the OFSF problem.

3   Linear Time Algorithm for Optimal Feathering Region Decomposition

In this section, we characterize the increase of the delivery error due to the decomposition of a feathering region. Then, a linear time algorithm for the optimal feathering region decomposition is developed.

3.1   Characterizing the Increase of the Delivery Error

Consider a feathering region $F_k = A[l .. r]$ with $r = l + \delta - 1$ $(1 < l < r < n)$, which is the overlapping region of sub-IMs S_{k-1} and S_k. Assume that the decomposition of F_k is $F_k^{(0)} = (y_{i,j})_{m \times \delta}$ plus $F_k^{(1)} = (x_{i,j})_{m \times \delta}$. Then, S_{k-1} ends with $F_k^{(0)}$ and S_k starts with $F_k^{(1)}$. The contribution $R(F_k)$ of the feathering region F_k $(A[l .. r])$ to the delivery error of IM A is $\sum_{i=1}^{m} \sum_{j=l}^{r+1} |a_{i,j} - a_{i,j-1}|$.

While the contribution $R(F_k^{(0)})$ to the delivery error of S_{k-1} (i.e., toward the total sum of the delivery errors of the sub-IMs in the splitting $\mathcal{S}$ of A) is

$$\sum_{i=1}^{m} \left(|y_{i,l} - a_{i,l-1}| + \sum_{j=l+1}^{r} |y_{i,j} - y_{i,j-1}| + |0 - y_{i,r}| \right),$$

and the contribution $R(F_k^{(1)})$ to the delivery error of S_k is

$$\sum_{i=1}^{m} \left(|x_{i,l} - 0| + \sum_{j=l+1}^{r} |x_{i,j} - x_{i,j-1}| + |a_{i,r+1} - x_{i,r}| \right).$$

Thus, the *increase* of the delivery error due to the decomposition (i.e., $F_k^{(0)}$ and $F_k^{(1)}$) of the feathering region F_k is $R(F_k^{(0)}) + R(F_k^{(1)}) - R(F_k)$. We next develop a linear time algorithm for an optimal decomposition of an feathering region to minimizing the increase of the delivery error.

Observing that the decomposition of F_k can be performed row by row, we define the following **optimal vector decomposition (OVD)** problem. Define the *complexity* $\mathcal{C}(\mathbf{z})$ of a vector $\mathbf{z} = (z_1, z_2, \ldots, z_N)$ as $\sum_{j=2}^{N} |z_j - z_{j-1}|$. Given a non-negative integer vector $\mathbf{b} = (b_1, b_2, \ldots, b_N)$, decompose $\mathbf{b}$ into two non-negative vectors $\mathbf{x} = (x_1, x_2, \ldots, x_N)$ and $\mathbf{y} = (y_1, y_2, \ldots, y_N)$, such that (1) $x_1 = 0$ and $x_N = b_N$; (2) for each $j = 1, 2, \ldots, N$, $b_j = x_j + y_j$, and (3) the total complexity of $\mathbf{x}$ and $\mathbf{y}$ (i.e., $\mathcal{C}(\mathbf{x}) + \mathcal{C}(\mathbf{y})$) is minimized.

Then, for a given feathering region $F_k = A[l..r]$, we may view each *extended* row $(a_{i,l-1}, a_{i,l}, \ldots, a_{i,r}, a_{i,r+1})$ as a vector $\mathbf{a}_i$. Applying the OVD algorithm, $\mathbf{a}_i$ is decomposed into two vectors, $\mathbf{x}_i = (0, x_{i,l}, \ldots, x_{i,r}, a_{i,r+1})$ and $\mathbf{y}_i = (a_{i,l-1}, y_{i,l}, \ldots, y_{i,r}, 0)$, with $\mathcal{C}(\mathbf{x}_i) + \mathcal{C}(\mathbf{y}_i)$ being minimized. Clearly, $(\mathbf{x}_i)_{i=1}^{m}$ and $(\mathbf{y}_i)_{i=1}^{m}$ can used to specify a decomposition of F_k (i.e., $F_k^{(0)}$ and $F_k^{(1)}$). Note that $R(F_k) = \sum_{i=1}^{m} \mathcal{C}(\mathbf{a}_i)$, $R(F_k^{(0)}) = \sum_{i=1}^{m} \mathcal{C}(\mathbf{x}_i)$, and $R(F_k^{(1)}) = \sum_{i=1}^{m} \mathcal{C}(\mathbf{y}_i)$. Thus, $F_k^{(0)}$ and $F_k^{(1)}$ is an optimal decomposition of F_k and the minimum increase $\Delta_{err}(F_k)$ of the delivery error is $\sum_{i=1}^{m} [\mathcal{C}(\mathbf{x}_i) + \mathcal{C}(\mathbf{y}_i) - \mathcal{C}(\mathbf{a}_i)]$.

We next presents our optimal $O(N)$ time OVD algorithm for computing an optimal decomposition of a given vector $\mathbf{b} = (b_1, b_2, \ldots, b_N)$. The OVD problem is first modeled as computing a shortest path in a directed acyclic graph of pseudo-polynomial size. By exploiting the convexity of the edge weight functions of the graph, we show that the OVD problem can optimally be solved in linear time. Chen and Wang [3] also studied an optimal vector decomposition problem, but optimized a different criterion.

3.2   The Graph Model of the OVD Problem

Given a non-negative integer vector $\mathbf{b} = (b_1, b_2, \ldots, b_N)$, we define an edge-weighted DAG $G' = (V', E')$ for the OVD problem, as follows.

The element b_1 (resp., b_N) of $\mathbf{b}$ corresponds to exactly one vertex $v_1(0)$ (resp., $v_N(b_N)$) in V', briefly called the *source* (resp., *sink*) vertex s (resp., t) of G'. For every other element b_j $(j = 2, 3, \ldots, N - 1)$, there is a set $Col(j)$ of $b_j + 1$ vertices in G' corresponding to b_j; $Col(j) = \{v_j(h) \mid h = 0, 1, \ldots, b_j\}$, namely the b_j-*column* of G'. Intuitively, the vertices in $Col(j)$ give all possible distinct ways to decompose b_j into two non-negative integers (i.e., each vertex $v_j(h)$ corresponds to decomposing b_j into h and $b_j - h$). We also say that column $Col(1)$ (resp., $Col(N)$) consists of only one vertex $v_1(0)$ (resp., $v_N(b_N)$). For any two adjacent columns $Col(j)$ and $Col(j + 1)$ $(j = 1, 2, \ldots, N - 1)$, each vertex $v_j(h) \in Col(j)$ has a directed edge e to every vertex $v_{j+1}(h')$, with an edge *weight* $w(e) = |h' - h| + |(b_{j+1} - h') - (b_j - h)|$.

Consider any s-t path p in G', with $p = v_1(h_1) \to v_2(h_2) \to \ldots \to v_{N-1}(h_{N-1})$ $\to v_N(h_N)$, where $h_1 = 0$ and $h_N = b_N$ (i.e., $v_1(h_1)$ is the source s and $v_N(h_N)$ is the sink t). Let $\mathbf{x}(p) = (x_1, x_2, \ldots, x_N)$ and $\mathbf{y}(p) = (y_1, y_2, \ldots, y_N)$ be two non-negative integer vectors defined from the path p, in the following way: for each $j = 1, 2, \ldots, N$, $x_j = h_j$ and $y_j = b_j - h_j$. Note that each s-t path p in G' actually define a feasible decomposition of $\mathbf{b}$, i.e., $\mathbf{b} = \mathbf{x}(p) + \mathbf{y}(p)$. The total complexity of $\mathbf{x}(p)$ and $\mathbf{y}(p)$, $\mathcal{C}(\mathbf{x}(p)) + \mathcal{C}(\mathbf{y}(p))$, equals to the total sum of the weights of the edges on p, i.e., $w(p) = \mathcal{C}(\mathbf{x}(p)) + \mathcal{C}(\mathbf{y}(p))$. Hence, a shortest s-t path in G', which can be computed in $O(|V| + |E|)$ time, specifies an optimal decomposition of $\mathbf{b}$.

This is a pseudo-polynomial time algorithm for the OVD problem, which may not be efficient enough, especially when the elements of $\mathbf{b}$ are large. However, this DAG model lays down a base for further exploiting the intrinsic structures of the OVD problem.

3.3   Our Optimal OVD Algorithm

Our OVD algorithm hinges on the piecewise linearity and convexity of the edge weight functions of G'. For each $j = 1, 2, \ldots, N - 1$, based on b_j and b_{j+1}, we define a function $f_j(\Delta h)$: $\mathbf{Z} \to \mathbf{Z}^+$, as follows.

(1) If $b_j \leq b_{j+1}$,

$$f_j(\Delta h) = \begin{cases} 2\Delta h + (b_{j+1} - b_j), & \text{if } \Delta h > 0 \\ b_{j+1} - b_j, & \text{if } b_j - b_{j+1} \leq \Delta h \leq 0 \\ -2\Delta h + (b_j - b_{j+1}), & \text{if } \Delta h < b_j - b_{j+1} \end{cases} \qquad (*)$$

(2) If $b_j > b_{j+1}$,

$$f_j(\Delta h) = \begin{cases} 2\Delta h + (b_{j+1} - b_j), & \text{if } \Delta h > b_j - b_{j+1} \\ b_j - b_{j+1}, & \text{if } 0 \leq \Delta h \leq b_j - b_{j+1} \\ -2\Delta h + (b_j - b_{j+1}), & \text{if } \Delta h < 0 \end{cases} \qquad (**)$$

Note that $f_j(\Delta h)$ is piecewise linear and convex with respect to Δh. For any edge $(v_j(h), v_{j+1}(h'))$ between two adjacent columns $Col(j)$ and $Col(j + 1)$, Lemma 2 reveals the relation between the edge weight $w(v_j(h), v_{j+1}(h'))$ and the function $f_j(\Delta h)$.

Lemma 2. $w(v_j(h), v_{j+1}(h')) = f_j(h - h')$.

We next consider a shortest path from s to a specific vertex $v_{\hat{j}}(\hat{h}) \in Col(\hat{j})$ in G', denoted by $s \leadsto v_{\hat{j}}(\hat{h})$. We first define the following series $\{h_{\bar{j}}^-\}_{j=1}^{\hat{j}}$.

$$
h_j^- = \begin{cases} \hat{h}, & j = \hat{j} \\ \max\left\{0, h_{j+1}^- - \max\{0, b_j - b_{j+1}\}\right\}, & 1 < j < \hat{j} \\ 0, & j = 1 \end{cases}
$$

Intuitively, for any $j < \hat{j}$, $v_j(h_j^-)$ is the "bottom-most" vertex on $Col(j)$ such that the edge $(v_j(h_j^-), v_{j+1}(h_{j+1}^-))$ has the minimum edge weight among all edges connecting a vertex in $Col(j)$ to $v_{j+1}(h_{j+1}^-)$. Let $sw_j(h)$ denote the shortest path weight from s to $v_j(h)$.

Lemma 3. (1) The path $s \leadsto v_{\hat{j}}(\hat{h})$ defined by the series $\{h_j^-\}_{j=1}^{\hat{j}}$, with $s \leadsto v_{\hat{j}}(\hat{h})$ $= v_1(h_1^-) \to v_2(h_2^-) \to \cdots \to v_{\hat{j}}(h_{\hat{j}}^-)$, is a shortest path from s to $v_{\hat{j}}(\hat{h})$.

(2) The weight $w(s \leadsto v_{\hat{j}}(\hat{h}))$ of the path $s \leadsto v_{\hat{j}}(\hat{h})$ is $sw_2(h_2^-) + \sum_{j=3}^{\hat{j}} |b_j - b_{j-1}|$.

Furthermore, $sw_2(h_2^-) = h_2^- + |b_2 - h_2^- - b_1| = 2\max\{h_2^-, b_2 - b_1\} - (b_2 - b_1)$, and $h_2^- = \max\{0, \hat{h} - \sum_{j=3}^{\hat{j}} \max\{0, b_j - b_{j-1}\}\}$ from the definition of the series $\{h_j^-\}_{j=1}^{\hat{j}}$. For a given instance of the OVD problem, $\hat{h} = b_N$ and obviously, the series $\{h_j^-\}_{j=1}^{N}$ can be computed in an optimal $O(N)$ time.

Lemma 4. *Given a non-negative integer vector* $\mathbf{b} = (b_1, b_2, \ldots, b_N)$, *an optimal decomposition* $(\mathbf{x}, \mathbf{y})$ *of* $\mathbf{b}$ *(*$\mathbf{b} = \mathbf{x} + \mathbf{y}$*) minimizing the total complexity* $C(\mathbf{x}) + C(\mathbf{y})$ *can be computed in* $O(N)$ *time, and the total complexity of the optimal solution is* $2\max\{h^*, b_2 - b_1\} - (b_2 - b_1) + \sum_{j=3}^{N} |b_j - b_{j-1}|$, *where* $h^* = \max\{0, b_N - \sum_{j=3}^{N} \max\{0, b_j - b_{j-1}\}\}$.

3.4   Computing the Optimal Decomposition of a Feathering Region

As analyzed in Section 3.1, for a given feathering region $F_k = A[l .. r]$, we can view each extended row $(a_{i,l-1}, a_{i,l}, \ldots, a_{i,r}, a_{i,r+1})$ as a vector $\mathbf{a}_i$. Applying the OVD algorithm to decompose each $\mathbf{a}_i$ into two vectors, $\mathbf{x}_i$ and $\mathbf{y}_i$, we have the minimum increase $\Delta_{err}(F_k)$ of the delivery error for F_k is $\sum_{i=1}^{m} [C(\mathbf{x}_i) + C(\mathbf{y}_i) - C(\mathbf{a}_i)]$. Based on Lemma 4, $\Delta_{err}(F_k) = \sum_{i=1}^{m} [2\max\{h_i^*, a_{i,l} - a_{i,l-1}\} - (a_{i,l} - a_{i,l-1}) - |a_{i,l} - a_{i,l-1}|]$, where $h_i^* = \max\{0, a_{i,r+1} - \sum_{j=l+1}^{r+1} \max\{0, a_{i,j} - a_{i,j-1}\}\}$. We thus can introduce an additional matrix $B = (b_{i,j})_{m \times n}$ such that $b_{i,j} = \sum_{k=1}^{j+1} \max\{0, a_{i,k} - a_{i,k-1}\}\}$ ($a_{i,0} = a_{i,n+1} = 0$). The matrix B can be computed in $O(mn)$ time. Hence, we have the following lemma.

Lemma 5. *After an* $O(mn)$ *time preprocessing, the minimum increase* $\Delta_{err}(F_k)$ *of the delivery error for any feathering region* F_k *can be computed in* $O(m)$ *time, and the decomposition of* F_k *to achieve the optimal* $\Delta_{err}(F_k)$ *can be obtained in linear time.*

4 Speeding Up the d-Link Shortest Path Computation in G

In this section, we exploit the special "layered" structure of the graph G defined in Section 2, and further show that G has the Monge property [1], which enables us to compute a d-link shortest path in G in $O(mn)$ time.

First, let us exam the possible starting column j_k of the k-th (assume that $k > 1$ since j_1 is always 1) sub-IM S_k in a feasible splitting of the IM A. Note that there are $k-1$ sub-IMs to the left and $d-k$ sub-IMs to the right of S_k. Each sub-IM has a fixed width of w and the minimum width of a feathering region is δ. We thus have $n - [(d-k)(w-\delta)+w]+1 \le j_k \le (k-1)(w-\delta)+1$, denoted by C_k. Interestingly, the possible starting columns of any two overlapping sub-IMs are disjoint.

Lemma 6. *For any $k = 1, 2, \ldots, d-1$, $C_k \cap C_{k+1} = \emptyset$ ($C_1 = \{1\}$).*

To reflect this disjointness property, we redefine the graph $G = (V, E)$, as follows. G consists of d layers of vertices, with each layer k corresponding to all possible starting columns of the k-th sub-IM and each vertex corresponding to a column of A. More precisely, the first layer L_1 contains one vertex v_1 and the k-th layer L_k ($1 < k \le d$) contains vertices $\{v_j \mid n - [(d-k)(w-\delta)+w]+1 \le j \le (k-1)(w-\delta)+1\}$ (i.e., $j \in C_k$). For any vertices $v_j \in L_k$ and $v_{j'} \in L_{k+1}$ ($k = 1, 2, \ldots, d-1$), if $w - \Delta \le j' - j \le w - \delta$, we put a directed edge $e(v_j, v_{j'})$ in E from v_j to $v_{j'}$. The weight $c(v_j, v_{j'})$ is set to be the minimum increase of the delivery error for the corresponding feathering region $A[j' .. j + w - 1]$. We introduce a dummy vertex t. Each vertex in Layer L_d has a directed edge to t with a weight of 0. Due to the layered structure of G, the computation of a d-link shortest v_1-to-t path becomes computing a shortest v_1-to-t path. We next show the weights of the edges between two adjacent layers, L_k and L_{k+1} ($k = 2, 3, \ldots, d-1$), has the Monge property.

Lemma 7. *Given four vertices $v_{j'}, v_{j'+1} \in L_k$ and $v_{j''}, v_{j''+1} \in L_{k+1}$ in G with $2 \le k < d$, $c(v_{j'}, v_{j''}) + c(v_{j'+1}, v_{j''+1}) \le c(v_{j'}, v_{j''+1}) + c(v_{j'+1}, v_{j''})$.*

sThe Monge property as shown in Lemma 7 indicates that there always exists a set of shortest paths from v_1 to every vertex on Layer L_k, such that no two paths in the set "cross" each other. This is due to the fact that we can always replace two crossing edges, if any, on two such shortest paths with two non-crossing edges, and obtain two new paths whose weight sum is no larger than that of the two original shortest paths. It follows that the two new paths thus generated must also be the shortest.

For every vertex v_j in the k-th layer L_k, let $sw_k(j)$ denote the shortest path length from v_1 to $v_j \in L_k$ in G. Clearly, $sw_k(j) = \min\{sw_{k-1}(j') + c(v_{j'}, v_j) \mid v_{j'} \in L_{k-1}$ and $w - \Delta \le j - j' \le w - \delta\}$. Recall that an edge $(v_{j'}, v_j) \in E$ if and only if $v_{j'} \in L_{k-1}$, $v_j \in L_k$, and $w - \Delta \le j - j' \le w - \delta$. Hence, the set of all outgoing edges of each vertex $v_{j'}$ and the set of all incoming edges of each v_j can be represented implicitly (such that we can access any edge of G in $O(1)$ time and compute its weight in $O(m)$ time as shown in Section 3.4). Note that the Monge property is normally defined on a matrix [1]. Lemma 7 actually shows the Monge property of the matrix containing the path weight $sw_{k-1}(j') + c(v_{j'}, v_j)$ for every edge $(v_{j'}, v_j)$ between the vertices on two consecutive layers L_k and L_{k+1} of G, with $1 < k < d$. Thus, applying the matrix-searching technique in [1], it takes $O(m(w - \delta))$ time to compute all shortest paths from v_1 to all vertices on the k-th layer while knowing all $sw_{k-1}(j')$'s of Layer L_{k-1}. Hence, a shortest v_1-to-t path in G can be obtained in $O(dm(w - \delta)) = O(mn)$ time.

Theorem 1. *Given an IM A of size $m \times n$, the OFSF problem on A is solvable in $O(mn)$ time.*

References

1. A. Aggarwal, M.M. Klawe, S. Moran, P. Shor, and R. Wilber, Geometric Applications of a Matrix-Searching Algorithm, *Algorithmica*, 2(1987), pp. 195-208.
2. D.Z. Chen, X.S. Hu, S. Luan, C. Wang, and X. Wu, Mountain Reduction, Block Matching, and Medical Applications, *Proc. of the 21st Annual ACM Symposium on Computational Geometry (SoCG)*, Pisa, Italy, June 6–8, 2005, pp. 35–44.
3. D.Z. Chen and C. Wang, Field Splitting Problems in Intensity-Modulated Radiation Therapy, *Lecture Notes in Computer Science, Springer-Verlag, Proc. 17th Int. Symp. on Algorithms and Computation (ISAAC)*, Vol. 4288, pp. 701-711, 2006.
4. J. Dai and Y. Zhu, Minimizing the Number of Segments in a Delivery Sequence for Intensity-Modulated Radiation Therapy with Multileaf Collimator, *Med. Phys.*, 28(10)(2001), pp. 2113-2120.
5. J. Deng, T. Pawlicki, Y. Chen, J. Li, S.B. Jiang, and C.-M. Ma, The MLC Tongue-and Groove Effect on IMRT Dose Distribution, *Physics in Medicine and Biology*, 46(2001), pp. 1039-1060.
6. N. Dogan, L.B. Leybovich, A. Sethi, and B. Emami, Automatic Feathering of Split Fields for Step-and-Shoot Intensity Modulated Radiation Therapy, *Phys. Med. Biol.*, 48(2003), pp. 1133-1140.
7. L. Hong, A. Kaled, C. Chui, T. LoSasso, M. Hunt, S. Spirou, J. Yang, H. Amols, C. Ling, Z. Fuks, and S. Leibel, IMRT of Large Fields: Whole-Abdomen Irradiation, *Int. J. Radiat. Oncol. Biol. Phys.*, 54(2002), pp. 278-289.
8. S. Kamath, S. Sahni, S. Ranka, J. Li, and J. Palta, Optimal Field Splitting for Large Intensity-Modulated Fields, *Medical Physics*, 31(12)(2004), pp. 3314-3323.
9. S. Kamath, S. Sahni, J. Li, J. Palta, and S. Ranka, A Generalized Field Splitting Algorithm for Optimal IMRT Delivery Efficiency, *The 47th Annual Meeting and Technical Exhibition of the American Association of Physicists in Medicine (AAPM)*, 2005. Also, *Med. Phys.*, 32(6)(2005), pp. 1890.
10. Q. Wu, M. Arnfield, S. Tong, Y. Wu, and R. Mohan, Dynamic Splitting of Large Intensity-Modulated Fields, *Phys. Med. Biol.*, 45(2000), pp. 1731-1740.
11. S. Webb, *Intensity-Modulated Radiation Therapy*, Institute of Cancer Research and Royal Marsden NHS Trust, Jan. 2001.
12. X. Wu, Efficient Algorithms for Intensity Map Splitting Problems in Radiation Therapy, *Lecture Notes in Computer Science, Springer-Verlag, Proc. 11th Annual International Computing and Combinatorics Conference (COCOON)*, volume 3595, pp. 504-513, 2005.
13. P. Xia and L.J. Verhey, MLC Leaf Sequencing Algorithm for Intensity Modulated Beams with Multiple Static Segments, *Med. Phys.*, 25(1998), pp. 1424-1449.

Approximating the Maximum Independent Set and Minimum Vertex Coloring on Box Graphs

Xin Han[1], Kazuo Iwama[1], Rolf Klein[2], and Andrzej Lingas[3]

[1] School of Informatics, Kyoto University, Kyoto 606-8501, Japan
{hanxin,iwama}@kuis.kyoto-u.ac.jp
[2] University of Bonn, Institute of Computer Science I, D-53117 Bonn, Germany
rolf.klein@informatik.uni-bonn.de
[3] Department of Computer Science, Lund University, 221 00 Lund, Sweden
Andrzej.Lingas@cs.lth.se

Abstract. A box graph is the intersection graph of a finite set of orthogonal rectangles in the plane. The problem of whether or not the maximum independent set problem (MIS for short) for box graphs can be approximated within a substantially sub-logarithmic factor in polynomial time has been open for several years. We show that for box graphs on n vertices which have an independent set of size $\Omega(n/\log^{O(1)} n)$ the maximum independent set problem can be approximated within $O(\log n/\log\log n)$ in polynomial time. Furthermore, we show that the chromatic number of a box graph on n vertices is within an $O(\log n)$ factor from the size of its maximum clique and provide an $O(\log n)$ approximation algorithm for minimum vertex coloring of such a box graph. More generally, we can show that the chromatic number of the intersection graph of n d-dimensional orthogonal rectangles is within an $O(\log^{d-1} n)$ factor from the size of its maximum clique and obtain an $O(\log^{d-1} n)$ approximation algorithm for minimum vertex coloring of such an intersection graph.

1 Introduction

A box graph is an intersection graph of a finite set of rectangles. Computing maximum or large independent sets (MIS) in box graphs has application in efficient automated map labeling in geographic information systems (GIS), data mining, VLSI design, image processing and multi-dimensional point location [1,4,10]. For example, in the map labeling application, each label may correspond to an orthogonal rectangle of fixed size and position, and the task is to place as many disjoint labels as possible. The problem of minimum number vertex coloring of box graphs is also relevant for the aforementioned applications.

For a graph G, let $\alpha(G)$ and $\omega(G)$ denote the size of maximum independent set and the maximum clique in G, respectively. Next, let $\chi(G)$ denote the chromatic number of G, i.e., the minimum number of colors sufficient to color all vertices of G in such a way that no two adjacent vertices share the same color.

The NP-completeness of MIS for box graphs follows from [8]. The problem of whether or not MIS for box graphs can be approximated within a substantially

M.-Y. Kao and X.-Y. Li (Eds.): AAIM 2007, LNCS 4508, pp. 337–345, 2007.
© Springer-Verlag Berlin Heidelberg 2007

sub-logarithmic factor in polynomial time has been open for several years. The $O(\log n)$ approximation algorithm from [1] has been improved to $O(\log_{k+1} n)$, for any positive fixed integer k, and generalized to include higher dimensions in [4] (see also [5]) [1]. In [14], an algorithm with optimum-sensitive approximation factor $1 + \log_2(\alpha(G))$ has been given. On the other hand, polynomial-time approximation schemes are known for the special cases where the rectangles have unit height [1], are squares or have constant aspect ratio [7]. Furthermore, a subexponential time algorithm for MIS in general box graphs has been presented in [12]. Recently, the APX-hardness, and hence the non existence of PTAS, for the generalization of MIS for box graphs to include d-dimensional boxes, $d \geq 3$, has been established in [6]. As for minimum vertex coloring, already in the 60s Asplund and Grünbaum proved that the chromatic number is always upper bounded by four times the square of the clique number in any box graph [3]. The NP-completeness of the k-colorability problem for box graphs has been established in [9,10]. Furthermore, it has been proved in [13] that the chromatic number of a box graph is linear in the maximum clique size if the heights of the rectangles are within a constant factor.

No other published results on chromatic number and minimum vertex coloring for box graphs are known to the authors.

Our contributions for a box graph G on n vertices are as follows. (i) We show that if $\alpha(G) = \Omega(n/\log^{O(1)} n)$ then MIS of G can be approximated within $O(\log n/\log \log n)$ in polynomial time [2] (ii) Furthermore, we show that $\chi(G) = O(\omega(G)\log n)$ and provide an $O(\log n)$ approximation algorithm for minimum vertex coloring of G. We also generalize the latter results to the intersection graph G of n d-dimensional orthogonal rectangles, showing $\chi(G) = O(\omega(G)\log^{d-1} n)$ and providing $O(\log^{d-1} n)$ approximation for minimum vertex coloring of G.

2   Approximating Large MIS

The *overlap* of a finite set R of orthogonal rectangles is the maximum number of rectangles in R overlapping (containing) the same point [3]. Note that the overlap of R is equal to the size of a maximum clique in the box graph induced by R (cf. [9]). Since it turned out that our following key lemma has been already mentioned in [5], we moved its proof to the appendix.

Lemma 1. *If the overlap of a set of n orthogonal rectangles is $O(\log^{O(1)} n)$ then an $O(\log n/\log \log n)$ approximation of the maximum independent set of the corresponding box graph can be computed in polynomial time.*

---

[1] In the manuscript [2], a β^{-1}-approximation of MIS is given for box graphs G on n vertices satisfying $\alpha(G) \geq \beta n$. This improves the logarithmic approximation factor only in case $\alpha(G) = \Omega(n/\log n)$.

[2] We have been informed by a referee that the key lemma in the proof of this result had been already observed by T. Chan in [5].

[3] The equivalent terms of *depth* [5], *thickness*, and *ply* are in use in the literature.

Lemma 1 yields an $O(\log n/\log\log n)$ approximation of MIS for box graphs with large MIS.

Theorem 1. *Let G be the box graph induced by a set of n rectangles. If $\alpha(G) = \Omega(n/\log^{O(1)} n)$ then an $O(\log n/\log\log n)$ approximation of MIS for G can be computed in polynomial time.*

Proof. Let $cn/\log^l n$ be a specification of $\Omega(n/\log^{O(1)} n)$, and let m be the size of MIS of G. Consider the grid formed by the vertical and horizontal straight-lines overlapping with the edges of the rectangles. Iterate the following step: whenever there is a grid point contained in at least $2\log^l n/c$ rectangles then remove all the rectangles. Note that the remaining set R' of rectangles has the following properties: (1) the overlap of R' is smaller than $2\log^l n/c$, and (2) the MIS of the box graph induced by R' has size at least $m/2$ which is at least $cn/2\log^l n$. The property (2) immediately follows from the following three facts: the MIS of original box graph has cardinality not less than $cn/\log^l n$, each rectangle iteration can decrease the MIS of the original box graph by one, the number of the iterations is at most $cn/2\log^l n$. It follows in particular that since R' has at least $cn/2\log^l n$ rectangles it has a polylogarithmic in its number of rectangles overlap. Now, it is sufficient to apply Lemma 1 to R' to obtain the theorem by the property (2).

3   Relationship Between $\chi(G)$ and $\omega(G)$

For a box graph G, its chromatic number $\chi(G)$ clearly cannot be smaller than its maximum clique size $\omega(G)$, which in turn is equivalent to the overlap of the set R of rectangles that induces G by Lemma 5 in [9]. We shall show that up to a logarithmic factor a reverse relationship holds.

Recall the rectilinear grid formed by the straight-lines passing through the edges of the input rectangles. We may without loss of generality normalize the grid by setting the distance between each pair of neighboring grid lines to 1 since such transformation does not change the intersection relationships. Thus, we may assume without loss of generality that the n input rectangles lie on an $O(n) \times O(n)$ integer grid.

Divide the rectangles into classes R_i, $i = 1, ..., O(\log n)$, where a rectangle r is in R_i iff the width of r is in $[2^i, 2^{i+1})$.

Lemma 2. *For each $i = 1, ..., O(\log n)$, the subgraph of G induced by R_i can be colored with $3t_i$ colors, where t_i is the overlap of R_i.*

Proof. Distinguish the vertical grid lines L whose x coordinates are divisible by 2^{i+1} (see Fig. 2). Note that any rectangle in R_i can intersect at most one such line L. We can easily color the set of all rectangles in R_i intersecting a given line L with t_i colors such that no two rectangles of the same color intersect. We simply sweep a horizontal line and stop at the horizontal edges of rectangles intersecting L in order to color newly scanned rectangles and/or release colors of those disappearing. It follows that all rectangles in R_i intersecting all the vertical lines L can be colored with $2t_i$ colors (by using t_i colors for the rectangles intersecting odd L lines and another t_i colors for those intersecting even L lines). Of course, there might also remain rectangles in R_i that fall between the lines L. To color them, let us move horizontally the lines L by 2^i. Now, each of the remaining rectangles intersects exactly one of the newly distinguished vertical lines. Also, no pair of remaining rectangles intersecting different newly distinguished lines can have a non-empty intersection since otherwise at least one of the rectangles would intersect one of the lines L. Hence, we can argue similarly to show that additional t_i colors are sufficient to color the remaining rectangles.

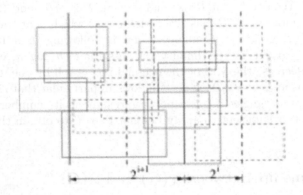

Fig. 1. Coloring the rectangles in R_i

By the definition of the classes R_i and Lemma 2, we obtain immediately our upper bound on the chromatic number in terms of the overlap parameter, or equivalently, the maximum clique size.

Theorem 2. *For any box graph G on n vertices, $\chi(G) = O(\omega(G) \log n)$ holds.*[4]

4   Approximating Minimum Vertex Coloring

Note that in the proof of Theorem 2, a construction of an $O(\omega(G) \log n)$ coloring of G is given. The construction can be easily implemented by using sorting and standard geometric sweeping techniques (e.g., see [15]) in time $O(n \log n)$. Hence, we obtain the following approximation result.

---

[4] It has been pointed to us that a similar upper bound can be also derived from the aforementioned result on the linearity of the chromatic number in the maximum clique size in case the heights of the rectangles are within a constant factor [13]. Indeed, the standard trick of partitioning the rectangles into a logarithmic number of classes satisfying the latter requirements yields a similar logarithmic upper bound in the general case. However, our method yields a better constant factor.

Theorem 3. *Given a set of n orthogonal rectangles, we can approximate minimum vertex coloring of the box graph induced by the rectangles within an $O(\log n)$ factor in time $O(n \log n)$.*

Just in case of constant overlap, one can generalize the divide and conquer k-line technique for maximum independent set (see Section 2) to include minimum vertex coloring in order to obtain an $O(\log_{k+1} n)$ approximation factor. On the other hand, one can easily generalize Theorem 3 to include minimum vertex coloring of d-dimensional box graphs, i.e., the intersection graphs of d-dimensional rectangles.

Theorem 4. *Let G be the intersection graph of a set of n orthogonal d-dimensional rectangles. $\chi(G) = O(\omega(G) \log^{d-1} n)$ holds and an $O(\log^{d-1} n)$ approximation of minimum vertex coloring of G can be computed in time $O(n \log n)$.*

Proof. It is sufficient to prove by induction on d that G can be colored with $O(\omega(G) \log^{d-1} n)$ colors in time $O(n \log n)$. For $d = 2$, this holds by the proof of Theorems 2 and 3. For $d > 2$, consider and normalize the $O(n^d)$ grid composed of the $d - 1$ dimensional hyperplanes including the $d - 1$ dimensional faces of the input d-dimensional rectangles. Generalizing the proof of Theorems 2 and 3, divide the rectangles into classes C_i, $i = 1, ..., O(\log n)$ such that a rectangle r is in C_i iff the size of r along the first axis is in $[2^i, 2^{i+1})$. Next, distinguish these $d - 1$-dimensional hyperplanes orthogonal to the first axis whose first coordinate is divisible by 2^{i+1} and consider the rectangles in C_i that intersect them. Note that each of the rectangles in C_i can intersect at most one of these hyperplanes. Hence, by the inductive assumption, we can color all the rectangles in C_i intersecting these hyperplanes with at most twice $O(\omega(G) \log^{d-2} n)$ colors. To color the remaining rectangles in C_i we again proceed analogously as in the proof of Theorems 2 and 3. That is, we just shift the aforementioned hyperplanes by 2^i along the first axis, and now each of the remaining in C_i rectangles has to intersect exactly one of the shifted hyperplanes. Thus, an additional portion of $O(\omega(G) \log^{d-2} n)$ colors is sufficient to color the remaining in C_i rectangles. Since there are $O(\log n)$ classes C_i, we have to use $O(\omega(G) \log^{d-1} n)$ colors totally.

The claimed time complexity follows from the inductive hypothesis and the fact that the n input d-dimensional rectangles can be divided into the C_i classes in time $O(n \log n)$.

5   Final Remarks

Our result on MIS for box graphs with large MIS does not seem to admit any straightforward generalization to include the intersection graphs of d-dimensional orthogonal rectangles where $d > 2$. On the other hand, it can be easily generalized to include the weighted version of MIS. It also partially complements the aforementioned result of Nielsen which yields substantially sub-logarithmic factor for box graphs with very small MIS [14].

Acknowledgement

The authors are grateful for valuable comments to the referees and to Martin Wahlén for technical help.

References

1. P.K. Agarwal, M. van Kreveld, and S. Suri. Label placement by maximum independent set in rectangles, *Computational Geometry: Theory and Applications*, 11, pp. 209-218, 1998.
2. P.K. Agarwal and N.H. Mustafa. Independent set of intersection graphs of convex objects in 2D. *Computational Geometry: Theory and Applications*, 34, pp. 83-95, 2006.
3. E. Asplund and B. Grünbaum. On coloring problem. *Mathematica Scandinavica*, 8, pp. 181-188, 1960.
4. P. Berman, B. DasGupta, S. Muthukrishnan, and S. Ramaswami. Efficient approximation algorithms for tiling and packing problems with rectangles. *Journal of Algorithms*, 41, pp. 443-470, 2001.
5. T.M. Chan. A note on maximum independent sets in rectangle intersection graphs. *Information Processing Letters*, 89, pp. 19-23, 2004.
6. M. Chlebík, J. Chlebiková. Approximation Hardness of Optimization Problems in Intersection Graphs of d-dimensional Boxes. Proc. SODA 2005, pp. 267-276.
7. T. Erlebach, K. Jansen, and E. Seidel. Polynomial-Time Approximation Schemes for Geometric Graphs. Proc. SODA 2001, pp. 671-679.
8. R.J. Fowler, M.S. Paterson, and S.L. Tanimoto. Optimal packing and covering the plane are NP-complete. *Information Processing Letters* 12 (1981), pp. 133-137.
9. H. Imai and T. Asano. Finding the connected components and a maximum clique of an intersection graph of rectangles in the plane. *Journal of Algorithms*, 4, pp. 310-323, 1983.
10. D.T. Lee and J.Y.-T. Leung. On the 2-dimensional channel assignment problem. *IEEE Trans. Comput.* 33 (1984), pp. 2-6.
11. L. Lewin-Eytan, J. Naor and A. Orda. Routing and Admission Control in Networks with Advance Reservations. Proc. 5th International Workshop on Approximation Algorithms for Combinatorial Optimization, APPROX 2002, LNCS 2462 (Springer), pp. 215-228.
12. A. Lingas and M. Wahlen. A note on maximum independent set and related problems on box graphs. *Information Processing Letters* 93(2005), pp. 169-171.
13. E. Malesinska. Graph-Theoretical Models for Frequency Assignment Problems. PhD Thesis, TU Berlin, 1997.
14. F. Nielsen. Fast stabbing of boxes in high dimensions. *Theoretical Computer Science* 246 (2000), pp. 53-72.
15. Preparata, F. and Shamos, M., Computational Geometry – an Introduction Springer Verlag, New York, NY, 1985.

Appendix: The Proof of Lemma 1

Proof. We generalize the divide and conquer method of Agarval et al. from [1] by slicing the set of input rectangles with k equal distant vertical straight-lines instead of one (cf. [4]) and returning the maximum of the union of the $k + 1$ solutions for the MIS of the $k + 1$ box graphs induced by the $k + 1$ slices and the MIS of the box graph induced by rectangles that intersect at least one of the k vertical straight-lines. By substituting $\log_{k+1} n$ for $\log_2 n$ and $n/(k + 1)$ for $n/2$ in the analysis from [1], we obtain an $O(\log_{k+1} n)$ approximation factor for this method (which agrees with the results in [4] presented in a more general setting).

In order to keep the proof more self-contained, we present the generalized algorithm and the generalized approximation analysis here.

Let R be a set of n orthogonal rectangles in the plane. Following [1], we preprocess R by sorting the horizontal edges of R by their y-coordinates and their vertical edges by their x-coordinates. The preprocessing takes $O(n \log n)$ time. Our refined recursive algorithm depending on a fixed positive integer k is as follows.

1. If $n \le k$ then determine a maximum independent set of rectangles in R by a brute force method.
2. For $i = 1, ..., k$, determine the $\lfloor i2n/(k + 1) \rfloor$ different element x_i in the the sorted list of $2n$ x-coordinates.
3. Partition the rectangles of R into $k + 2$ groups $R_1, R_2, ..., R_{k+1}$ and R^* such that R_1 contains rectangles lying to the left of the line $x = \lfloor 2n/(k + 1) \rfloor$, for $i = 2, ..., k$, R_i contains the rectangles lying between the lines $x = \lfloor (i - 1)2n/(k + 1) \rfloor$ and $\lfloor i2n/(k + 1) \rfloor$, R_{k+1} contains rectangles lying to the right of the line $x = \lfloor k2n/(k + 1) \rfloor$, and finally R^* consists of all rectangles intersecting the k aforementioned vertical lines.
4. Compute a maximum independent set I^* of R^*. Recursively compute approximate independent sets I_i for R_i, for $i = 1, ..., k + 1$ respectively.
5. If $|I^*| \ge \sum_{i=1}^{k+1} |I_i|$ then return I^* otherwise return $\bigcup_{i=1}^{k+1} I_i$.

The disjointness of the sets R_1 through R_{k+1} guarantees that the generalized algorithm always returns an independent set of rectangles in R.

Our proof of the asserted approximation factor is a natural generalization of the proof of the $O(\log_2 n)$ approximation factor for the simple algorithm from [1]. For a set of rectangles Q let $mis(Q)$ denote the maximum size of an independent set of rectangles in Q. We prove by induction on the number of input rectangles that the independent set returned by generalized recursive algorithm is of size not less than $mis(R)/\max\{k, \log_{k+1} n\}$. By step 1 of this algorithm, we may assume w.l.o.g $n > k$. Suppose that induction hypothesis is true for all $n' < n$. Let I be a maximum independent set in R. Since the set I^* computed by the generalized algorithm is a maximum independent set of rectangles in R^*, we have $|I^*| \ge |I \cap R^*|$.) By the induction hypothesis, we have for $i = 1, ..., k + 1$,

$$|I_i| \ge \frac{mis(R_i)}{\log_{k+1}(n/(k + 1))}.$$

Hence, we obtain $\max\{|I^*|, \sum_{i=1}^{k+1}|I_i|\} \geq$

$$\max\{|I \cap R^*|, \frac{\sum_{i=1}^{k+1}|I \cap R_i|}{\log_{k+1} n - 1}\} \geq \max\{|I \cap R^*|, \frac{|I| - |I \cap R^*|}{\log_{k+1} n - 1}\}.$$

If $|I \cap R^*| \geq |I|/\log_{k+1} n$, the induction step follows. Otherwise, we have the inequality:

$$\frac{\sum_{i=1}^{k+1}|I \cap R_i|}{\log_{k+1} n - 1} \geq \frac{|I| - |I|/\log_{k+1} n}{\log_{k+1} n - 1} = \frac{|I|}{\log_{k+1} n}.$$

By proving that the maximum independent set I^* of rectangles in R^* can be computed in time $n^{O(k)}$, one can immediately obtain a polynomial-time performance of this algorithm for $k = O(1)$ (cf. [4]).

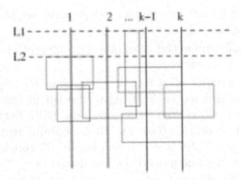

Fig. 2. The k vertical lines intersected by the horizontal line L

To show that I^* can be still computed in polynomial time for $k = O(\log n/\log\log n)$ under the overlap assumption (see Fig. 1), we make the following key observation.

For any horizontal straight-line L the set S_L of the input rectangles which can intersect L and simultaneously at least one of the k vertical lines has cardinality $O(k \log^{O(1)} n)$.

It follows that the family of at most $\sum_{i=1}^{k}\binom{O(k \log^{O(1)} n)}{i}$ sets of non-overlapping rectangles which are subsets of S_L is of polynomial cardinality for $k = O(\log n/\log\log n)$. Consequently, I^*, which is a MIS of the box graph induced by the input rectangles that intersect at least one of the $O(\log n/\log\log n)$ vertical straight-lines, can be computed in polynomial time as follows [5].

Sweep the horizontal line L top-down stopping whenever L overlaps with an edge of a rectangle in R. At each stop, inductively compute for each subset S of S_L with no overlapping rectangles an independent set including S and maximum number of rectangles above L intersecting one of the k vertical lines.

---

[5] One can also follow the method of Lemma 15 in [4] with the aforementioned subsets corresponding to the so called cuts in [4].

More precisely, if L' stands for the the preceding position of L and $S' = S \cap S_{L'}$ then the aforementioned independent set can be inductively obtained by merging $S \setminus S'$ with a largest among independent sets composed of a set S'', where $S'' \subset S_{L'}$ and $S' \subset S''$, and maximum number of rectangles above L' intersecting one of the k vertical lines.

We conclude that the generalized algorithm can be implemented in polynomial time for $k = O(\log n / \log \log n)$ under the overlap assumption. This combined with the proved $O(\log_{k+1} n)$ approximation factor of the algorithm yields the $O(\log_{\log n / \log \log n} n) = O(\log n / \log \log n)$ approximation in polynomial time.

BMA^*: An Efficient Algorithm for the One-to-Some Shortest Path Problem on Road Maps

Dan He

Department of Computer Science, University of Vermont
Burlington, VT 05405, USA

Abstract. The best known algorithm for the one-to-all shortest path problem is Dijkstra's algorithm, which can achieve time complexity $O(|E|+|V|\log(|V|))$ by the implementation of data structures like buckets. While for one-to-some shortest path problem, no matter how small the "some" is, the time complexity of Dijkstra's algorithm remains $O(|E|+|V|\log(|V|))$ and it often still needs to explore a large part of the graph and thus is not efficient. This paper proposes a novel algorithm which computes the shortest paths bidirectionally with A^* algorithm multiple times to solve the one-to-some shortest path problem on road maps efficiently, where the size of the destination set is much smaller than the total number of vertices in the graph. The experiments on both randomly generated graphs and real road maps show that our algorithm is more space and time efficient than Dijkstra's algorithm with buckets, one of the most efficient algorithm for one-to-some shortest path problem.

1 Introduction

Single source shortest path problem ($SSSP$) over varies graphs is a fundamental problem which has lots of applications in many areas, such as the network routing problem, the vehicle routing problem and Geographical Information Systems (GIS). The single destination shortest path problem ($SDSP$), which leads to more applications, can be considered as the complement of the $SSSP$: we just need to reverse the shortest paths computed from the destination to all the sources. Therefore an efficient algorithm for these problems is of great importance.

The most widely used algorithm for one-to-all shortest path problem, which aims to find the shortest paths from a single source to all the other vertices in the graph, is $Dijkstra$ algorithm [1]. With smart implementations such as using the data structure of buckets [14], Dijkstra's algorithm solves the one-to-all shortest path problem in $O(|E| + |V|\log(|V|))$ time on a directed graph $G = (V, E)$ with no negative edges. Lots of work has been done to improve the naive implementation of Dijkstra's algorithm by varies data structures [6,7,8,9]. For a long time Dijkstra's algorithm has been believed to be the best algorithm for computing shortest paths from single source to multiple destinations.

M.-Y. Kao and X.-Y. Li (Eds.): AAIM 2007, LNCS 4508, pp. 346–357, 2007.

While Dijkstra's algorithm is efficient and simple for computing shortest paths from single source to all the other vertices in the graph, it's often not efficient to solve the one-to-some shortest path problem, which aims to find shortest paths from a single source to a set of destinations $T \subset V$, where $|T| \ll |V|$ and $|V|$ is the total number of vertices in the graph. This is because Dijkstra's algorithm always expands the vertex in best-first order, namely the vertex closest to the source, no matter how far it is to the destination. Therefore if one destination is far from the source, Dijkstra's algorithm may need to explore a large part of the graph to find the shortest path to that destination. No matter how small the "some" is, the time complexity of Dijkstra's algorithm for one-to-some shortest path problem still remains $O(|E| + |V| \log(|V|))$. For single source single destination problem ($SSSD$), or one-to-one problem, many heuristic search algorithms have been developed. One of the most famous algorithms is the A^* algorithm [2], which uses a heuristic estimator to improve the efficiency of the Dijkstra's algorithm.

In this paper, we present a novel application of A^* algorithm, for the one-to-some shortest path problem on road maps where the destination set is much smaller than the vertex set of the graph. By applying A^* algorithm bidirectionally multiple times, we can achieve significant improvements on both time and space requirements. Our experiments on both randomly generated graphs and real road map show that our algorithm is more efficient than the implementation of Dijkstra's algorithm with buckets [14], one of the most efficient algorithm for the one-to-some shortest path problem.

The outline of this paper is as follows: In Section 2 we summarize the Dijkstra's algorithm, A^* algorithm, bidirectional algorithm and the algorithms from Shibuya. In Section 3 we introduce our novel algorithms. We show our experiment results in Section 4 and we include our final remarks in Section 5.

2 Preliminaries

The original Dijkstra's algorithm [1] aims to compute the shortest paths from a single source to all the other vertices in the graph. It can be modified easily to compute shortest paths from single source to a set of destinations.

Let $G = (V, E)$ be a directed graph with no negative edges, $s \in V$ be the source, $T = \{t_1, t_2, ..., t_m\}$ be the set of destinations, and $l(u, v)$ be the length of the edge $(u, v) \in E$. The Dijkstra's algorithm expands the vertex in best-first order, updates the score of v for each vertex v accordingly and stops if all the destinations are expanded. It is obvious that Dijkstra's algorithm always expands the vertex closest to the source first no matter how far the vertex is to the destination. Therefore if one of the destinations is far from the source, Dijkstra's algorithm may need to explore a large part of the graph to reach that destination.

Lots of work has been done on improving the performance of Dijkstra's algorithm [6,7,8,9]. Zhan [4] showed that one of the currently fastest algorithms for one-to-some shortest path problem is the implementation of Dijkstra's algorithm with buckets [14]. In the original Dijkstra's algorithm, searched vertices are treated as a non-ordered list. In order to find the right vertex to be expanded next, at each iteration we need to go though the whole explored vertices list, which turns out to be a bottleneck operation. This problem can be handled well by using buckets to maintain the explored vertices in a sorted fashion according to their current found distances from the source. The explored vertices with the same current found distances are kept in the same bucket and the vertices in each bucket are in FCFS manner. The time complexity of Dijkstra's algorithm with buckets (Dial's implementation [14]) is $O(|E| + |V|C))$, where C is the number of buckets.

The A^* algorithm [2] extends the Dijkstra's algorithm by taking into a heuristic function $h(u)$ from every vertex u in the graph to the destination t, which is a lower bound on the distance of shortest path from u to t. And the score for each vertex u is not simply the distance of u to the source $g(u)$, but the score of $f(u) = g(u) + h(u)$. Therefore the A^* algorithm prefers vertices close to both source and destination and can find the shortest path more efficiently than simple Dijkstra's algorithm. The vertices on the search frontier are stored in an *Open* list, and the already-expanded vertices are stored in a *Closed* list. The A^* algorithm can then build a shortest path tree from s to t, by updating the backtrack-pointer of each vertex to its latest parent. The shortest path from each vertex v on the tree to s can be obtained by tracing back from v to s along the tree. And by reversing the shortest path from v to s, we can also obtain the shortest path from s to v.

The bidirectional algorithm [3] applies the Dijkstra algortithm simultaneously from both the source s and the destination t. The algorithm terminates if the forward and backward explorations meet each other. Then the shortest distance is the minimum value of $l(f, b) + p_s(f) + p_t(b)$, where f is a vertex in the forward exploration vertex set p_s and b is a vertex in the backward exploration vertex set p_t such that $(f, b) \in E$ and $l(f, b) + p_s(f) + p_t(b)$ is minimized, $l(f, b)$ is the length of the edge (f, b), $p_s(f)$ is the shortest distance from s to f, and $p_t(b)$ is the shortest distance from t to b. The shortest path can be obtained by combining the s-f shortest path, edge (f, b) and the b-t shortest path.

One trivial way of combining the ideas of A^* algorithm and the bidirectional algorithm is to run the A^* algorithm in two directions simultaneously and stop as soon as they meet. This does not work, however, if the heuristic function for the forward A^* search $h_f()$ and for the backward A^* search $h_b()$ are not consistent [10]. The $h_f()$ and $h_b()$ are consistent if for any arc $e(u, v)$, $h_f(u, v)$ is equal to $h_b(v, u)$, where $h_f(u, v)$ is the heuristic estimator from u to v in the original graph and $h_b(v, u)$ is the heuristic estimator from v to u in the reverse graph. This actually indicates that $h_f(v) + h_b(v) =$const for every vertex v. Therefore if we use the same heuristic function for both the forward and backward A^* searches, the optimal shortest path is guaranteed.

The algorithms from Sanders and Schultes [13] and Goldberg et al. [11] achieve significant improvements over Dijkstra's algorithm by preprocessing the road network. They show that using preprocessing, the query time can be improved greatly. However, their algorithms are special for $SSSD$ problem and what's more, they assume preprocessing is allowed, which does not meet the requirement of this paper.

For $SSSD$ problem, or one-to-one shortest path problem, A^* algorithm and bidirectional algorithm often perform much better than simple Dijkstra's algorithm, but for a long time they are believed to be only applicable to one-to-one shortest path problem but not to one-to-some, one-to-all, or all pairs shortest path problems. Direct application of these two algorithms to the above problems often results in poor performances. Shibuya [5] first introduced 2 modified A^* algorithms to the $n \times m$ shortest path problem on real road networks: the *Euclidean-distance-based A^* algorithm* and the *bidirectional-method-based A^* algorithm*. Due to the space limit, we refer the readers to [5] for the detail of these algorithms.

3   New Applications of A* Algorithm to One-to-Some Shortest Path Problem on Road Maps

3.1 BMA^* Algorithm

In our algorithm, we first pick one destination t and use ordinary A^* algorithm to find the shortest path from the source s to t (we call *forward A^* search*). We can thus obtain a shortest path tree T_s. Then we apply A^* algorithm multiple times from each destination $t_i \in T/\{t\}$, to the source s (we call *backward A^* search*). And we know with monotonic heuristic estimators, if v is an expanded vertex on the shortest path tree T_s built by A^* algorithm from source s to the destination t, the path from s to v along the tree T_s must be the shortest path from s to v in the graph G [10]. Once the backward A^* search expands a vertex v which has been expanded already by the forward A^* search, we can stop immediately and obtain the shortest path from s to t_i by combining the path from s to v along the tree T_s and the path from v to t_i along the shortest path tree built by the backward A^* search. This is similar to the bidirectional A^* algorithm for Shibuya [5], with the differences that we do forward A^* search and backward A^* search separately instead of doing them simultaneously, and we do backward A^* searches multiple times. However, bidirectional A^* algorithm is designed for $n \times m$ shortest paths problem where n, m are large. It is not suitable for one-to-some shortest paths problem. Due to space limit, we put the detailed analysis of bidirectional A^* algorithm in the journal version of this paper. Since this algorithm searches in two directions with A^* algorithm multiple times, we call it the *Bidirectional-Multiple A^** algorithm (BMA^*). And since we use Euclidean distance as our heuristic estimator for both forward and backward A^* searches, the optimal shortest paths are guaranteed to be found correctly.

A key problem of our algorithm is how to find the destination on which we are going to apply forward A^* search. We hope to find the destination whose corresponding shortest path tree built by forward A^* search expands many vertices so that the backward A^* searches are highly possible to expand those vertices and then stop quickly. We take a simple way here: choose the destination with the highest heuristic value, by the assumption that the higher the heuristic value of the destination is, the more vertices the corresponding forward A^* search would probably expand. For example, for the shortest path problem on real road networks where we can use the euclidian distance as the heuristic estimator, we apply the forward A^* search on the destination with the greatest euclidian distance to the source. Our experiments on both randomly generated graphs and real road map show that this strategy is simple but surprisingly successful.

For all the backward A^* searches, during the search processes we should store the heuristic estimator for each newly explored vertex so that the same heuristic estimator of this vertex can be reused by all backward A^* searches.

When the source and the destinations are in two distant clusters and when the destinations are close to each other, BMA^* algorithm often explores much less vertices than Dijkstra's algorithm does. Because in these cases the backward A^* searches are highly possible to meet the vertices expanded by the forward A^* search and then stop quickly. While if the destinations are sparsely distributed around the source, simple application of BMA^* algorithm can perform worse than Dijkstra's algorithm, because the possibilities of the backward A^* searches to meet the vertices expanded by the forward A^* search are comparatively low. An extreme case can be shown in the real road network. If the destination with the highest heuristic value, namely the greatest euclidian distance to the source, is on one side of the source while all the other destinations are on the opposite side, the backward A^* searches nearly couldn't meet any vertices expanded by the forward A^* search.

3.2   Improved BMA^* Algorithms

In this subsection we propose two strategies to improve the performances of BMA^* algorithm and make it suitable to both situations where the source and the destinations are in distant clusters (we call "*distant*" case) and where the destinations are sparsely distributed around the source (we call "*sparse*" case).

First, not just choose one destination for the forward A^* search, but choose a forward search set F which consists of multiple destinations. Our algorithm is still based on the assumption that the farther the destination to the source is, the more vertices the corresponding forward A^* search would probably expand. And for destinations t_i and t_j, without loss of generality, we assume $h(t_i) > h(t_j)$, where $h()$ is the heuristic function. Then we first apply forward A^* search from source s to t_i and if the angle between radials st_i and st_j is less than a threshold d, the backward A^* search from t_j would probably meet the vertices expanded by forward A^* search from s to t_i, or else we'd better use forward A^* searches on both t_i and t_j. Our experiments show that when we set the threshold as 45 degree, the forward feature set strategy works well.

For the shortest path problem on road maps, the above algorithm cuts the 2-dimensional space into several pie-slice sectors centered at source s, each containing a cluster of destinations. We then apply BMA^* algorithm on each sector, namely for each $t_i \in F$, we apply forward A^* search from the source s to t_i, store the expanded vertices for all the forward searches and we store each vertex only once. Then we apply backward A^* search on each $t_j \in T/F$ and store the heuristic estimator for each vertex for the purpose of reuse. And because $|F|$ can be at most $360/d$, this algorithm takes $O(360m/d + m \log m)$ time given the size of the destination set as m, where the $O(m \log m)$ factor is the time for sorting the destinations. Since $m \ll |V|$, and we set d as 45 degree, the time for this preprocessing step is of no importance and therefore can be ignored.

Second, the success of BMA^* algorithm depends on whether the forward A^* searches can expand many vertices which may be reexpanded by the backward A^* searches. If we can enlarge the search space of the forward A^* searches, it is more possible for the backward A^* searches to meet the vertices expanded by the forward A^* searches and stop as soon as possible. One simple way to enlarge the number of vertices expanded by forward A^* searches without loss of heuristic power is to simply divide the heuristic estimator h by a factor $\varepsilon > 1$.

Theorem 1. *If h is a monotonic estimator for the A^* algorithm in graph G, h/ε is also a monotonic estimator, for some $\varepsilon \geq 1$.*

Proof. A heuristic function $h()$ is monotonic if for any nodes u and v, $l(u,v) + h(v) \geq h(u)$. Since h is monotonic, we can derive the inequality $l(u,v) + h(v) \geq h(u)$, where u, v are vertices in graph G, $l(u,v)$ is the length of edge (u,v) in graph G and $l(u,v) \geq 0$.

$$l(u,v) + h(v) \geq h(u) \Rightarrow l(u,v) \geq h(u) - h(v)$$
$$\Rightarrow l(u,v) \geq (h(u) - h(v))/\varepsilon, \quad \text{for } \epsilon \geq 1$$
$$\Rightarrow l(u,v) + h(v)/\varepsilon \geq h(u)/\varepsilon$$

Therefore h/ε ($\varepsilon \geq 1$) is also monotonic and can guarantee an optimal solution.

Since $h(v)$ is a lower bound for the real distance of vertex v to destination t, a higher value of $h(v)$ stands for a tighter lower bound and therefore can reduce the search space and a lower value of $h(v)$ stands for a looser lower bound and will enlarge the search space. So $h(v)/\varepsilon$ for some $\varepsilon > 1$ is a looser lower bound compared with the original $h(v)$ and will enlarge the search space. Notice we only modify the heuristic estimators for the forward A^* searches and keep the heuristic estimators for backward A^* searches the same as before. We call the BMA^* algorithm with heuristic estimators as h/ε ($\varepsilon \geq 1$) for the forward A^* searches the BMA^*-ε algorithm.

The new heuristic estimator $h(v)/\varepsilon$ for some $\varepsilon > 1$ is a tradeoff since although it makes the backward searches more efficient, the forward searches are going to search more vertices. Our experiments show that this method behaves well on big destination sets where backward searches benefit more by picking a looser heuristic function for forward searches.

4   Experiments

4.1   Experiments on Random Generated Graphs

We do all the experiments on randomly generated graphs where we can use Euclidean distance as the heuristic estimator. And we try to generate random

#destinations	#searched vertices	#loaded vertices
10	$45,780/8,324$	$9,439/3,180$
20	$45,994/10,249$	$9,594/3,536$
30	$48,453/16,262$	$9,763/4,002$
40	$48,897/30,058$	$9,777/4,229$
50	$49,962/40,250$	$9,850/4,837$

Dijkstra/BMA*

Fig. 1. The numbers of searched vertices and loaded vertices of Dijkstra's algorithm and BMA^* algorithm respectively on randomly generated graphs with 10,000 vertices, 25,000 arcs. The l-to-e ratio is set as ≤ 2. The source and destinations are in distant clusters.

graphs close to the real road networks. According to the experiments in [4] and [5], we set the arc-to-node ratios as 2.5. We build our graph in a $10,000 \times 10,000$ two dimensional grid and set the location for each vertex as a pair of (x, y) coordinates in Euclidean space. The coordinates for each dimension is from 0 to 10,000. We randomly generate graphs with 10,000 vertices and 25,000 arcs on this $10,000 \times 10,000$ two dimensional grid. We set the euclidian distance between the two terminals of each arc be no more than 1,500, and according to [4], we try to eliminate the irregularity of the graph that two vertices are "close" to each other in their locations but "far" from each other in their real distance. We set the ratio of arc length to euclidian distance (we call *l-to-e ratio*) be no more than 2. And the number of destinations $|T|$ in our experiments is always much less than the total number of vertices $|V|$. We use Euclidean distance as heuristic function for both forward and backward A^* searches. All experiments are done on an Intel Xeon 2.4GHZ processor with 2GB memory.

4.1.1 Source and Destinations Are in Distant Clusters

We randomly pick a vertex as the source in the circle centered at (3000,3000) with radius as 500. And we randomly pick the destinations in the circle centered at (7000,7000), with radius as 1000. All the vertices including source and destinations are randomly generated. The size of destination set varies from 10 to 50. And for each destination set size, we do our experiments on ten randomly generated graphs and show the average performance of each algorithm. Since this is "distant" case, we do not consider BMA^*-ε algorithm and we only compare the number of searched vertices and the number of loaded vertices for BMA^* algorithm and the Dijkstra's algorithm. #searched vertices is the total number of vertices searched by all searches, both forward and backward. #loaded vertices is the total number of vertices loaded into memory. And notice that the searched and loaded vertices for simple Dijkstra's algorithm and Dijkstra's algorithm with buckets are exactly the same.

In Figure 1, we show our experiment results for 10, 20, 30, 40, 50 destinations, respectively. We can see that for only 10 destinations, the Dijkstra's algorithm searches already as many as $45,780$ vertices and loads most of the vertices in the

#destinations	#searched vertices	#loaded vertices
40	$35,101/26,621/17,052$	$8,039/5,271/5,757$
50	$39,153/53,109/20,671$	$8,611/6,174/6,216$
80	$47,999/45,494/21,112$	$9,752/5,817/6,107$
90	$38,589/72,692/27,174$	$8,090/5,619/5,676$

$$Dijkstra/BMA^*/BMA^*\text{-}10$$

Fig. 2. The numbers of searched vertices and loaded vertices of Dijkstra's algorithm, BMA^* algorithm and BMA^*-10 algorithm, respectively on randomly generated graphs with 10,000 vertices, 25,000 arcs. The l-to-e ratio is set as ≤ 2. The destinations are sparsely distributed around the source.

graph. And BMA^* algorithm searches only 1/6th vertices and loads 1/3th vertices of those by Dijkstra's algorithm. With the increase of the size of destination set, the numbers of searched vertices and loaded vertices by Dijkstra's algorithm increase slowly but these by BMA^* algorithm increase fast. Therefore the advantage of BMA^* algorithm over Dijkstra's algorithm on space requirement is more significant for comparatively small destination set size. This is quite reasonable since the larger the destination set is, the more vertices the BMA^* algorithm explores and loads. Our experiments show that for destination set size under 50, our BMA^* algorithm loads only 1/3th to 1/2th of the number of vertices loaded by Dijkstra's algorithm.

4.1.2 Destinations Are Sparsely Distributed Around the Source

We randomly pick a vertex as the source in the circle centered at (3000,3000) with radius as 500. And we randomly pick the destinations in the circle centered also at (3000,3000), with radius as 5000. We set the size of the destination set as 40, 50, 80, 90, respectively. And we set the threshold in the algorithm for computing the forward feature set as 45 degree. We compare the numbers of searched vertices and loaded vertices by Dikstra algorithm, BMA^* algorithm and BMA^*-10 algorithm, respectively. Both the BMA^* algorithm and BMA^*-10 algorithm use forward search set method. And we set the l-to-e ratio as ≤ 2. For each destination set we do independent tests on ten randomly generated graphs and show the average performance of each algorithm.

As shown in Figure 2, the results are quite similar to those shown in Figure 1. With the increase of the destination set size to be more than 50, the number of searched vertices increases quickly to be larger than that by Dijkstra's algorithm. While by dividing the heuristic estimator by 10, as the BMA^*-10 algorithm does, we can reduce the number of searched vertices to nearly half of the original. And we can also see that the number of loaded vertices by BMA^* algorithm are close to that by BMA^*-10 algorithm, both of which are much less than that by Dijkstra's algorithm.

We also compare the execution time in both cases for simple Dijkstra's algorithm, Dijkstra's algorithm with buckets, the BMA^* algorithm, the BMA^*-10 algorithm. As stated before, we do not consider BMA^*-ε algorithm for "distant" case. As shown in Figure 3, the simple implementation of Dijkstra's algorithm

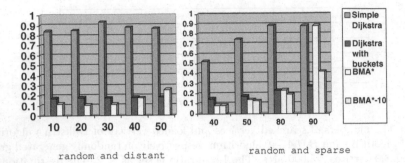

random and distant   random and sparse

Fig. 3. The execution times (sec.) for simple implementation of Dijkstra's algorithm, the implementation of Dijkstra's algorithm with buckets, the BMA^* algorithm and the BMA^*-10 algorithm respectively, on randomly generated graphs with 10,000 vertices, 25,000 arcs. In "distant" case, we do not consider BMA^*-10 algorithm.

spends much more time than the other three algorithms. For small destination sets, the execution times of BMA^* algorithm are quite close to those of BMA^*-10 algorithm. While for comparatively large destination sets, e.g. the destination set size of 40, 50 in "distant" case, the destination set sizes of 80, 90 in "sparse" case, the BMA^*-10 algorithm runs faster than BMA^* algorithm. Since the number of searched vertices for Dijkstra's algorithm does not change much with the increase of the destination set size, its execution times remain close for different destination set sizes and therefore it's inefficient for comparatively small destination set. While for BMA^* algorithm and BMA^*-10 algorithm, their execution times are largely affected by the destination set size. Therefore for small destination set sizes they can be much faster than Dijkstra's algorithm with buckets but as the number of destinations increases, they will spend more and more time on computing heuristic estimators and finally be slower than Dijkstra's algorithm with buckets.

4.2   Experiments on Real Road Network

We apply our BMA^* algorithm on real road map for Chittenden County, Vermont. This is a two-dimensional map with 2727 vertices and 3038 arcs in the bounding box with $(-29861.0\text{ft}, -103025.6\text{ft})$ and $(111551.6\text{ft}, 102624.5\text{ft})$ as the left-bottom and right-top corners. The location of each vertex is a pair of coordinates (x,y) in Euclidean space. We use Euclidean distance as heuristic function for both forward and backward A^* searches. All experiments are done on an Intel Xeon 2.4Ghz processor with 2GB memory. Since this real road map is much smaller than those randomly generated graphs, we set ε as 2 for the BMA^*-2 algorithm. Since we have already showed that simple Dijkstra's algorithm performs much worse than the other algorithms, we do not consider it here.

4.2.1   Source and Destinations Are in Distant Clusters

We randomly pick up the source vertex in a circle centered at $(75000\,ft, 75000\,ft)$ with radius $2000\,ft$. And we randomly choose destination vertices in a circle

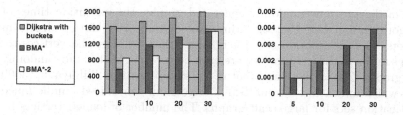

#loaded VS #destination   time(S.) VS #destination

Fig. 4. The numbers of loaded vertices and the execution times (sec.) of Dijkstra's algorithm with buckets, BMA^* algorithm and BMA^*-2 algorithm, respectively, on real road map Chittenden County, Vermont, with 2727 vertices and 3038 arcs. The source and destinations are in distant clusters.

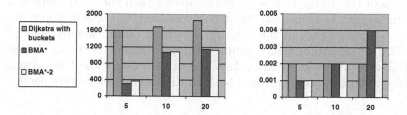

#loaded VS #destination   time(S.) VS #destination

Fig. 5. The numbers of loaded vertices and the execution times (sec.) of Dijkstra's algorithm with buckets, BMA^* algorithm and BMA^*-2 algorithm, respectively, on real road map Chittenden County, Vermont, with 2727 vertices and 3038 arcs. The destinations are sparsely distributed around the source.

centered at $(10000ft, 10000ft)$ with radius $40000ft$. We then compare the numbers of searched vertices and loaded vertices for our BMA^* algorithm, BMA^*-2 algorithm and the Dijkstra's algorithm with Buckets, and the execution time for each algorithm. We show the average result of 10 tests for each destination set size. The experiment results are shown in Figure 4.

4.2.2 Destinations Are Sparsely Distribute Around the Source

We randomly pick up the source vertex in a circle centered at $(10000ft, 10000ft)$ with radius $2000ft$. And we randomly choose destination vertices in a circle centered also at $(10000ft, 10000ft)$ with radius $30000ft$. We compare the numbers of searched vertices and loaded vertices for our BMA^* algorithm, BMA^*-2 algorithm and the Dijkstra's algorithm with Buckets, and the execution time for each algorithm. We show the average results of 10 tests for each destination set size. The experiment results are shown in Figure 5.

As shown in Figure 4 and Figure 5, we can see that on this real road map, the experiment results for both "distant" and "sparse" cases are similar to those on randomly generated graphs. Since the numbers of vertices and arcs are much smaller than the previous randomly generated graphs, the execution times of BMA^* algorithm and BMA^*-2 algorithm exceeds that of Dijkstra's algorithm with buckets when destination set size is larger than 30. But the vertices-to-destinations ratios $|V|/|T|$ (70 to 400) are even better than that in our randomly generated graphs (200 to 1000) and that in the experiments of Shibuya [5] (3000 to 10000), which indicates that our BMA^* algorithm can indeed handle much larger destination sets for large scale graphs. The number of loaded vertices by Dijkstra's algorithm is always much larger than that by BMA^* algorithm and BMA^*-2 algorithm.

5   Conclusion

We have proposed a novel algorithm for one-to-some shortest path problem, where the size of the destination set is much smaller than the total number of vertices in the graph. The experiments on both randomly generated graphs and real road map revealed that by applying A^* algorithm bidirectionally multiple times, our BMA^* algorithm makes significant improvements on the implementation of Dijkstra's algorithm with buckets, one of the most efficient algorithms for one-to-some shortest path problem, both on time and space requirements.

References

1. E. Dijkstra. A note on two problems in connection with graphs. *Numerical Mathematics*, 1, 1959, pp. 395-142.
2. P.Hart, N.Nilsson and B.Raphael. A formal basis for the heuristic determination of minimum cost paths. *IEEE Trans. Syst. Sci. Cybernet.*, 4(2): pp. 100-107,1968.
3. D.Champeaus. Bidirectional Heuristic Search Again. *J. ACM*, vol. 30, 1983, pp. 22-32.
4. F.B. Zhan and C. E. Noon. Shotest Path Algorithms: An Evaluation using Real Road Networks. *Transportation Sciences*, Vol. 32, No. 1, February 1998.
5. T. Shibuya. Computing the $n \times m$ Shortest Paths Efficiently. *J. ACM of Experimental Algorithmics*, ISSN 1084-6654, Vol. 5, No. 9, 2000.
6. B.V. Cherkassky, A. V. Goldberg and C. Silverstein. Buckets, heaps, lists and monotone priority queues. *Proc. 8th ACM-SIAM Symposium on Discrete Algorithm*, pp. 83-92, 1997.
7. M.L. Fredman and D.E. Willard. Trans-dichotomous algorithms for minimum spanning trees and shortest paths. *J. Comp. Syst. Sc.*, 48, pp. 533-551, 1994.
8. R. Raman. Priority queues: small monotones, and trans-dichotomous. *Proc. ESA'96, LNCS 1136*, pp. 121-137, 1996.
9. M. Thorup. On RAM priority queues. *Proc. 7th ACM-SIAM symposium on Discrete Algorithms*, pp. 59-67,1996.
10. A.V. Goldberg, C. Harrelson. Computing the Shortest Path: A* Search Meets Graph Theory. *16th Annual ACM-SIAM Symposium on Discrete Algorithms (SODA '05)*, Vancouver, Canada, 2005.
11. A.V. Goldberg, H. Kaplan, and R. Werneck. Reach for A*: Efficient Point-to-Point Shortest Path Algorithms. *Technical Report MSR-TR-2005-132, Microsoft Research, 2005.*

12. R. Gutman. Reach-based Routing: A New Approach to Shortest Path Algorithms Optimized for Road Net- works. *In Proc. 6th International Workshop on Algorithm Engineering and Experiments*, , pp. 100-111. SIAM, 2004.

13. P. Sanders and D. Schultes. Highway Hierarchies Hasten Exact Shortest Path Queries. *In Proc. 13th Annual European Symposium Algorithms*, , volume 3669 of LNCS, pp. 568-579. Springer, 2005.

14. R. B. Dial. Algorithm 360: Shortest Path Forest with Topological Ordering. *Communications of the ACM*, 12, pp. 632-633.

Strip Packing vs. Bin Packing

Xin Han[1], Kazuo Iwama[1], Deshi Ye[2], and Guochuan Zhang[3,*]

[1] School of Informatics, Kyoto University, Kyoto 606-8501, Japan
{hanxin,iwama}@kuis.kyoto-u.ac.jp
[2] Department of Computer Science, The University of Hong Kong, Hong Kong
yedeshi@cs.hku.hk
[3] Department of Mathematics, Zhejiang University, China
zgc@zju.edu.cn

Abstract. In this paper we establish a general algorithmic framework between bin packing and strip packing, with which we achieve the same asymptotic bounds by applying bin packing algorithms to strip packing. More precisely we obtain the following results: (1) Any offline bin packing algorithm can be applied to strip packing maintaining the same asymptotic worst-case ratio. Thus using FFD (First Fit Decreasing Height) as a subroutine, we get a practical (simple and fast) algorithm for strip packing with an upper bound 11/9. (2) A class of Harmonic-based algorithms for bin packing can be applied to online strip packing maintaining the same asymptotic competitive ratio. It implies online strip packing admits an upper bound of 1.58889 on the asymptotic competitive ratio. This significantly improves the previously best bound of 1.6910 and affirmatively answers an open question posed in [5].

1 Introduction

In strip packing a set of rectangles with widths and heights both bounded by 1, is packed into a strip with width 1 and infinite height. Rectangles must be packed such that no two rectangles overlap with each other and the sides of the rectangles are parallel to the strip sides. Rotations are not allowed. The objective is to minimize the height of the strip to pack all the given rectangles. If we know all rectangles before constructing a packing, then this problem is *offline*. In contrast in *online* strip packing rectangles are coming one by one and a placement decision for the current rectangle must be done before the next rectangle appears. Once a rectangle is packed it is never moved again.

It is well known that strip packing is a generalization of bin packing. Namely if we restrict all input rectangles to be of the same height, then strip packing is equivalent to bin packing. Thus any negative results for bin packing still hold for strip packing. More precisely, strip packing is NP-hard in the strong sense and the lower bound 1.5401 [15] is valid for online strip packing.

---

* Research supported by NSFC (10231060).

M.-Y. Kao and X.-Y. Li (Eds.): AAIM 2007, LNCS 4508, pp. 358–367, 2007.
© Springer-Verlag Berlin Heidelberg 2007

Previous results. For the offline version Coffman et al. [4] presented algorithms NFDH (Next Fit Decreasing Height) and FFDH (First Fit Decreasing Height), and showed that the respective asymptotic worst-case ratios are 2 and 1.7. Golan [6] and Baker et al. [2] improved it to 4/3 and 5/4, respectively. Using linear programming and random techniques, an asymptotic fully polynomial time approximation schemes (AFPTAS) was given by Kenyon and Rémila [9]. In the online version Baker and Schwarz [3] introduced an online strip packing algorithm called a shelf algorithm. A shelf is a rectangular part of the strip with width one and height at most one so that (i) every rectangle is either completely inside or completely outside of the shelf and (ii) every vertical line through the shelf intersects at most one rectangle. Shelf packing is an elegant idea to exploit bin packing algorithms. By employing bin packing algorithms *Next Fit* and *First Fit* Baker and Schwarz [3] obtained the asymptotic competitive ratios of 2 and 1.7, respectively. This idea was extended to the Harmonic shelf algorithm by Csirik and Woeginger [5], obtaining an asymptotic competitive ratio of $h_\infty \approx 1.6910$. Moreover it was shown that h_∞ is the best upper bound a shelf algorithm can achieve, no matter what online bin packing algorithm is used. Note that there were already several algorithms for online bin packing that have asymptotic competitive ratios better than h_∞ in late 80s and early 90s [10,11,12,16]. However, it was shown that whether these online algorithms can be applied to the online strip packing problem is an open question [5].

The core of shelf packing is reducing the two-dimensional problem to the one-dimensional problem. Basically shelf algorithms consist of two steps. The first one is *shelf design* which only takes the heights of rectangles into account. One shelf can be regarded as a bin with a specific height. The second step is *packing into a shelf*, where rectangles with similar heights are packed into the same shelves. This step is done by employing some bin packing algorithm to pack the rectangles with a total width bounded by one into a shelf. Clearly, to maintain the quality of bin packing algorithms in shelf packing we must improve the first step. Along this line we make the following contributions.

Our contributions. We propose a batch packing strategy and establish a general algorithmic framework between bin packing and strip packing. It is shown that any offline bin packing algorithm can be used for offline strip packing maintaining the asymptotic worst-case ratio. As an example, the well known bin packing algorithm FFD can approximate strip packing with an asymptotic worst-case ratio of 11/9. A simple AFPTAS can easily be derived from [8].

For online strip packing, we affirmatively answers the above question by showing that a class of online bin packing algorithm based on Super Harmonic algorithm [13] can be used in online strip packing maintaining the same asymptotic competitive ratio. This result implies that the known Harmonic based bin packing algorithms [10,11,12,13] can be converted into online strip packing algorithms without changing their asymptotic competitive ratios (better than h_∞). Note that the current champion algorithm for online bin packing is Harmonic++ by Seiden [13], which has an asymptotic competitive ratio of 1.58889. Hence strip packing admits an online algorithm with the same upper bound of 1.58889.

Main ideas. Recall that the strip packing problem becomes the bin packing problem if all rectangles have the same height. This motivates us to convert the strip packing problem into the bin packing problem by constucting a set of *new* rectangles called *slips* with the same height by bundling a subset of items. Then we just call the algorithm in the bin packing problem to pack the generated slips into the strip. More precisely, in the offline case, we pack in batch the rectangles with similar width into rectangular boxes (slips) of pre-specified height of c, where $c > 1$ is a sufficiently large constant. Then we obtain a set of new rectangles (slips) of the same height. The next step is to use bin packing algorithms on the new set. In the on-line case the strategy is slightly different. We divide the rectangles into two groups according to their widths, to which we apply the above batching strategy and the standard shelf algorithms respectively.

Asymptotic worst-case (competitive) ratio. To evaluate an approximation (online) algorithms for strip packing and bin packing we use the standard measure defined as follows.

Given an input list L and an approximation (online) algorithm A, we denote by $OPT(L)$ and $A(L)$, respectively, the height of the strip used by an optimal (offline) algorithm and the height used by (online) algorithm A for packing list L.

The *asymptotic worst-case (competitive) ratio* R_A^∞ of algorithm A is defined by

$$R_A^\infty = \lim_{n \to \infty} \sup_L \{A(L)/OPT(L) | OPT(L) = n\}.$$

2   The Offline Problem

Given a rectangle R, throughout the paper, we use $w(R)$ and $h(R)$ to denote its width and height, respectively.

Fractional strip packing. A fractional strip packing of L is a packing of any list L' obtained from L by subdividing some of its rectangles by horizontal cuts: a rectangle (w, h) can be replaced by a sequence $(w, h_1), (w, h_2), ..., (w, h_k)$ of rectangles such that $h = \sum_{i=1}^k h_i$.

Homogenous lists. Let L and L' be two lists where any rectangle of L and L' takes a width from q distinct numbers $w_1 > w_2 > \cdots > w_q$. List L is r-*homogenous* to L' where $r \geq 1$ if for $i = 1, 2, \ldots, q$,

$$\sum_{w(R')=w_i, R' \in L'} h(R') \leq \sum_{w(R)=w_i, R \in L} h(R) \leq r \cdot \sum_{w(R')=w_i, R' \in L'} h(R').$$

The following lemma is from Section 3.1 of the paper [9].

Lemma 1. *There is an approximation algorithm such that the cost by the algorithm for packing a list L of rectangles is bounded by*

$$\max\{(1 + \epsilon)s(L), OPT_{FSP}(L)\} + O(\epsilon^{-2}),$$

where $s(L)$ is the total area of all the rectangles in L and $OPT_{FSP}(L)$ is the optimal value of fractional strip packing for the list L.

The following lemma is an implicit byproduct of Lemma 1.

Lemma 2. *For an input list L of rectangles and $\epsilon > 0$, we have $OPT(L) \leq (1 + \epsilon)OPT_{FSP}(L) + O(\epsilon^{-2})$, where $OPT(L)$ is the optimal value for the list L.*

Proof. For every list L, we have $s(L) \leq OPT_{FSP}(L)$. So by Lemma 1, this lemma follows.

The next lemma shows a useful property of *homogenous* lists.

Lemma 3. *Given two lists L and L', if L is r-homogenous to L', we have $OPT_{FSP}(L') \leq OPT_{FSP}(L) \leq r \cdot OPT_{FSP}(L')$, where $r \geq 1$.*

Proof. If $r = 1$, it is easy to see that any fractional strip packing of L is a fractional packing of L' and vice versa. The conclusion thus follows immediately.

Now we consider the case that $r > 1$. By adding some rectangles to L' we can get a new list L'_1 which is 1-homogenous to L. We have

$$OPT_{FSP}(L') \leq OPT_{FSP}(L'_1) = OPT_{FSP}(L).$$

On the other hand we obtain another list L'_2 by prolonging in height all rectangles of L', i.e., if $(w, h) \in L'$, then $(w, rh) \in L'_2$. Clearly

$$OPT_{FSP}(L'_2) \leq r \cdot OPT_{FSP}(L').$$

Moreover, $OPT_{FSP}(L) \leq OPT_{FSP}(L'_2)$. The lemma holds.

Theorem 1. *Given two lists L and L', if L is r-homogenous to L', then for any $\epsilon > 0$*

$$OPT(L) \leq r(1 + \epsilon)OPT(L') + O(\epsilon^{-2}).$$

Proof. By Lemma 2,

$$OPT(L) \leq (1 + \epsilon)OPT_{FSP}(L) + O(\epsilon^{-2}).$$

By Lemma 3,

$$OPT_{FSP}(L) \leq r \cdot OPT_{FSP}(L').$$

Moreover $OPT_{FSP}(L') \leq OPT(L')$. Hence we have this theorem.

In the following we are ready to present our approach for offline strip packing. Given an input list $L = \{R_1, \ldots, R_n\}$ such that $w_1 \geq w_2 \geq \cdots \geq w_n$, where $R_i = (w_i, h_i)$, and a constant $c > 1$, we construct an offline algorithm $B\&P_A$ using some bin packing algorithm A as a subroutine. Basically the strategy consists of two stages.

Stage 1 - *Batching.* Pack $R_1, \ldots, R_i$ by NF (Next Fit) algorithm in the vertical direction into a slip $S_1 = (w_1, c)$, where $\sum_{j=1}^{i} h_j \leq c < \sum_{j=1}^{i+1} h_j$ and pack $R_{i+1}, \ldots, R_k$ into a slip $S_2 = (w_{i+1}, c)$, and so on, until all items are packed, shown as Figure 1. (Note that except for the last slip, all slips have the packed heights at least $(c - 1)$.)

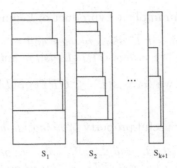

Fig. 1. Packing rectangles into slips

Stage 2 - *Packing*. Except for the last slip, pack all slips into the strip by algorithm A, since all slips have the same heights c. Then append the last slip on the top of the strip.

We present the main result for the offline case. In terms of the asymptotic worst case ratio, strip packing is essentially the same as bin packing.

Theorem 2. *The asymptotic worst-case ratio $R^\infty_{B\&P_A} = R^\infty_A$ for any bin packing algorithm A.*

Proof. Assume that $R^\infty_A = \alpha$. After the first stage of algorithm $B\&P_A$, we get a series of slips $S_1, \ldots, S_k, S_{k+1}$, shown as Figure 1. We then round up every item (w_j, h_j) in slip S_i to $(w(S_i), h_j)$ and obtain a new list $\bar{L}$, where $w(S_i)$ is the width of slip S_i. On the other hand, we obtain another list $\underline{L}$ by rounding down every item (w_j, h_j) in slip S_i to $(w(S_{i+1}), h_j)$ (here we set $w(S_{k+2}) = 0$). We have

$$OPT(\underline{L}) \leq OPT(L) \leq OPT(\bar{L}) \tag{1}$$

Let L_1 and L_2 be two sets of slips such that $L_1 = \{S_1, \ldots, S_k\}$ and $L_2 = \{S_2, \ldots, S_k\}$. Then

$$OPT(L_2) \leq OPT(L_1) \leq OPT(L_2) + c. \tag{2}$$

We can treat S_i as a one-dimensional item ignoring its height since $h(S_i) = c$ for $i = 1, 2, \ldots, k$. Let $I(L_1)$ be the corresponding item set for bin packing induced from the list L_1, i.e, $I(L_1) = \{w(S_1), w(S_2), \ldots, w(S_k)\}$. And $OPT(I(L_1))$ is the minimum number of bins used to pack $I(L_1)$. It follows that $OPT(L_1) = c \cdot OPT(I(L_1))$.

Note that L_2 is $c/(c-1)$-homogenous to $\underline{L}$, by Theorem 1, we have

$$OPT(L_2) \leq \frac{c}{c-1}(1+\epsilon)OPT(\underline{L}) + O(\epsilon^{-2}). \tag{3}$$

Now we turn to algorithm $B\&P_A$. After Stage 1 the list L becomes $L_1 \cup \{S_{k+1}\}$. At Stage 2 we deal with a bin packing problem: pack $k+1$ items with size of $w(S_i)$ into the minimum number of bins. The bin packing algorithm A is applied to $I(L_1)$ while S_{k+1} occupies a bin itself. Thus $B\&P_A(L) \le c \cdot A(I(L_1)) + c$. Since $R_A^\infty = \alpha$, we have $A(I(L_1)) \le \alpha OPT(I(L_1)) + O(1)$. Then

$$B\&P_A(L) \le c \cdot A(I(L_1)) + c \le \alpha \cdot c \cdot OPT(I(L_1)) + O(c) = \alpha \cdot OPT(L_1) + O(c).$$

Combining with (2),(3), (1), we have

$$B\&P_A(L) \le \alpha OPT(L_2) + O(c) \tag{4}$$

$$\le \frac{\alpha c}{(c-1)}(1 + \epsilon)OPT(\underline{L}) + O(\alpha \epsilon^{-2} + c\alpha) \tag{5}$$

$$\le \frac{\alpha c}{(c-1)}(1 + \epsilon)OPT(L) + O(\alpha \epsilon^{-2} + c\alpha). \tag{6}$$

As c goes to infinite and ϵ goes to zero, this theorem follows.

By Theorem 2, any offline bin packing algorithm can be transformed into an offline strip packing algorithm without changing the asymptotic worst case ratio. If the well known algorithm FFD ([1] [7][17]) is used in our approach, then we get a simple and fast algorithm $B\&P_{FFD}$ for strip packing and have the following result from Theorem 2.

Corollary 1. *Given constants $\epsilon > 0$ and $c > 1$, for any strip packing instance L, $B\&P_{FFD}(L) \le \frac{11c}{9(c-1)}(1 + \epsilon)OPT(L) + O(\epsilon^{-2} + c)$, where $c \le \epsilon OPT(L)$, and the time complexity is $O(n\log n)$.*

Remarks: If we batch all the rectangle by the approach in Stage 1 to generate a set of slips, then apply the bin packing algorithm [8] to pack slips into the strip. By the Theorem 2, this is an AFPTAS for the strip packing problem. Note that the above algorithm is quiet simple and only relies on the algorighm used for the bin packing problem.

3   The Online Problem

In this section we consider online strip packing. In the online case we are not able to sort the rectangles in advance because of no information on future items. Due to this point we cannot reach a complete matching between bin packing algorithms and strip packing algorithms generated from the former. However we can deal with a class H of Super Harmonic algorithms [13], which includes all known online bin packing algorithms based on Harmonic. Such an algorithm can be used in online strip packing without changing its asymptotic worst-case ratio.

A general algorithm of Super Harmonic algorithms has the following characteristics.

- Items are classified into $k + 1$ groups by their sizes, where k is a constant integer.
- Those items in the same group are packed by the same manner.

3.1   An Online Algorithm $G\&P_A$

Let A be any algorithm of Super Harmonic algorithm. Our approach $G\&P_A$ is presented below.

Grouping: A rectangle is *wide* if its width is at least ϵ; otherwise it is *narrow*, where $\epsilon > 0$ is a given small number. We further classify *wide* rectangles into k classes, where k is a constant, as Algorithm A does. Let $1 = t_1 > t_2 > \cdots > t_k > t_{k+1} = \epsilon$. Denote I_j to be the interval $(t_{j+1}, t_j]$ for $j = 1, ..., k$. A rectangle is of type-i if its width $w \in I_i$.

Packing narrow rectangles: Apply the standard shelf algorithm NF_r [3] to *narrow* rectangles $R = (w, h)$, where $0 < r < 1$ is a parameter. Round h to r^s if $r^{s+1} < h \leq r^s$. If R cannot be packed into the current open shelf with height of r^s, then close the current one and open a new one with height r^s and pack R into it, otherwise just pack R into the current one by NF.

Packing wide rectangles: We pack *wide* rectangles into bins of $(1, c)$, where $c = o(OPT(L)) > 1$ is a constant. Similarly as the offline case we batch the items of the same type and pack them into a slip. Here we specify the width of the slip by values t_i for $i < k + 1$ and name a slip (t_i, c) of type-i. Suppose that the incoming rectangle R is of type i ($w \in (t_{i+1}, t_i]$). If there is a slip of type-i with a packed height less than $c - 1$, then pack R into it by algorithm NF in the vertical direction. Otherwise create a new empty slip of type-i with size (t_i, c) and place R into the new slip by NF algorithm in the vertical direction. As soon as a slip is created, view it as one dimensional item and pack it by algorithm A into a bin of $(1, c)$. Figure 2(b) shows an illustration.

3.2   The Analysis for $G\&P_A$

The weighting function technique introduced by Ullman [14] has been widely used in performance analysis of bin packing algorithms [5][10][13]. Roughly speaking, the weight of an item indicates the maximum portion of a bin that

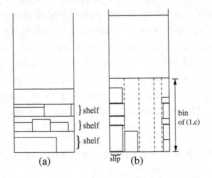

Fig. 2. Shelf packing vs our packing

the item occupies. Then, Seiden generalized the idea of weighting function and proposed a weighting system which can be used to analyze Harmonic, Refined Harmonic, Modified Harmonic, Modified Harmonic 2, Harmonic+1 and Harmonic++. The basic ideas in his paper are below,

- First show that the total cost of his algorithm is upper bounded by the total weights of all items.
- Then consider all the patterns of packing items into one bin and give an upper bound of the total weight over all the patterns.
- Finally, the asymptotic competitive ratio is implied.

Here, we first introduce some definitions and results in the Super Harmonic algorithm [13], then give a weighting function for our algorithm and prove that our definition is a geralization of the one in [13].

Weighting Systems. Let $\mathbb{R}$ and $\mathbb{N}$ be the sets of real numbers and non-negative integers, respectively. A *weighting system* for algorithm A is a tuple $(\mathbb{R}^m, \mathbf{w}_A, \xi_A)$. $\mathbb{R}^m$ is a vector space over the real numbers with dimension m. The function $\mathbf{w}_A : (0, 1] \mapsto \mathbb{R}^m$ is called the *weighting function*. The function $\xi_A : \mathbb{R}^m \mapsto \mathbb{R}$ is called the *consolidation function*. Seiden defined a $2K + 1$ dimensional weighting system for Super Harmonic, where K is a parameter of Super Harmonic algorithm. Real numbers $\alpha_i, \beta_i, \gamma_i, \epsilon$ and functions $\phi(i), \varphi(i)$ are defined in Super Harmonic algorithm. The unit basis vectors of the weighting system are denoted by

$$\mathbf{b}_0, \mathbf{b}_1, ..., \mathbf{b}_K, \mathbf{r}_1, ..., \mathbf{r}_K.$$

The weighting function for an item with size x is defined as below:

$$\mathbf{w}_A(x) = \begin{cases} (1 - \alpha_i)\frac{\mathbf{b}_{\phi(i)}}{\beta_i} + \alpha_i \frac{\mathbf{r}_{\varphi(i)}}{\gamma_i} & \text{if } x \in I_i \text{ with } i \leq k, \\ x\frac{\mathbf{b}_0}{1-\epsilon} & \text{if } x \in I_{k+1}. \end{cases}$$

The consolidation function is defined as below:

$$\xi_A(\mathbf{x}) = \mathbf{x} \cdot \mathbf{b}_0 + \max_{1 \leq j \leq K+1} \min \left\{ \sum_{i=j}^{K} \mathbf{x} \cdot \mathbf{r}_i + \sum_{i=1}^{K} \mathbf{x} \cdot \mathbf{b}_i, \sum_{i=1}^{K} \mathbf{x} \cdot \mathbf{r}_i + \sum_{i=1}^{j-1} \mathbf{x} \cdot \mathbf{b}_i \right\}.$$

Lemma 4. *[13] For all sequences of bin packing* $\delta = (p_1, ..., p_n)$,

$$cost_A(\delta) \leq \xi_A \left(\sum_{i=1}^{n} \mathbf{w}_A(p_i) \right) + O(1).$$

This means that the cost of Super Harmonic Algorithm is bounded by the total weight of the items. We can obtain a similar result in Lemma 5 by defining our weighting function as follows,

$$\mathbf{w}_A(P) = y \cdot \mathbf{w}_A(x),$$

where P is a rectangle of size (x, y).

Lemma 5. *For any sequence of rectangles* $L = (P_1, ..., P_n)$, *the cost by* $G\&P_A$ *is*

$$cost_A(L) \leq \max\{\frac{c}{c-1}, \frac{1}{r}\}\xi_A\Big(\sum_{i=1}^{n} \mathbf{w}_A(P_i)\Big) + O(1).$$

Due to page limitation, the proof of Lemma 5 is left to the full paper. (Refer to the page: www.lab2.kuis.kyoto-u.ac.jp/~hanxin/strip.ps)

For bin packing, a *pattern* is a tuple $q = \langle q_1, ..., q_k \rangle$ over $\mathbb{N}$ such that

$$\sum_{i=1}^{k} q_i t_{i+1} < 1,$$

where q_i is the number of items of type i contained in the bin. Intuitively, a pattern describes the contents of a bin. The weight of pattern q is

$$\mathbf{w}_A(q) = \mathbf{w}_A\Big(1 - \sum_{i=1}^{k} q_i t_{i+1}\Big) + \sum_{i=1}^{k} q_i \mathbf{w}_A(t_i).$$

Define $\mathcal{Q}$ to be the set of all patterns q. Note that $\mathcal{Q}$ is necessarily finite.

A *distribution* is a function $\chi : \mathcal{Q} \mapsto \mathbb{R}_{\geq 0}$ such that

$$\sum_{q \in \mathcal{Q}} \chi(q) = 1.$$

Given an instance of bin packing δ, Super Harmonic uses $cost(\delta)\chi(q)$ bins containing items as described by the pattern q.

Lemma 6. *[13] For any distribution* χ, *if we set* A *as Harmonic++ then*

$$\xi_A\Big(\sum_{q \in \mathcal{Q}} \chi(q)\mathbf{w}_A(q)\Big) \leq 1.58889.$$

By Lemmas 5 and 6, we have the folloing thorem. Its proof is left to the full paper.

Theorem 3. *If we set algorithm* A *to Harmonic++, then the asymptotic competitive ratio of algorithm* $G\&P_A$ *is* 1.58889, *where* c *is a constant..*

4   Concluding Remarks

Although strip packing is a generalization of the one dimensional bin packing problem, we show from the point of algorithmic view that it is essentially the same as bin packing. In terms of asymptotic performance we give a universal method to apply the algorithmic results for bin packing to strip packing maintaining the solution quality. However our approach cannot be applied to strip packing in terms of absolute performance. Note that algorithm FFD has an absolute worst-case ratio of $3/2$ which is the best possible unless $P = NP$. It is challenging to prove or disprove the existence of a $3/2$-approximation algorithm for offline strip packing.

References

1. B.S. Baker, A new proof for the first-fit decreasing bin-packing algorithm. *J. Algorithms* 6, 49-70, 1985.
2. B.S. Baker, D.J. Brown, and H.P. Katseff, A 5/4 algorithm for two-dimensional packing. *J. Algorithms* 2, 348-368, 1981.
3. B.S. Baker and J.S. Schwarz, Shelf algorithms for two-dimensional packing problems, *SIAM J. Comput.* 12, 508-525, 1983.
4. E.G. Coffman, M.R. Garey, D.S. Johnson, and R.E. Tarjan, Performance bounds for level oriented two dimensional packing algorithms, *SIAM J. Comput.* 9, 808-826, 1980.
5. J. Csirik and G.J. Woeginger, Shelf algorithm for on-line strip packing, *Information Processing Letters* 63, 171-175, 1997.
6. I. Golan, Performance bounds for orthogonal, oriented two-dimensional packing algorithms, *SIAM J. Comput.* 10, 571-582, 1981.
7. D.S. Johnson, Near-optimal bin-packing algorithms, *doctoral thesis, M.I.T., Cambridge, Mass.*, 1973.
8. N. Karmarkar and R.M. Karp, An efficient approximation scheme for the one-dimensional bin-packing problem, In *Proc. 23rd Annual IEEE Symp. Found. Comput. Sci.*, 312-320, 1982.
9. C. Kenyon and E.Rémila, A near-optimal solution to a two-dimensional cutting stock problem, *Mathematics of Operations Research* 25, 645-656, 2000.
10. C.C. Lee and D.T. Lee, A simple on-line bin-packing algorihtm, *J. ACM* 32, 562-572, 1985.
11. P.V. Ramanan, D.J. Brown, C.C. Lee, and D. T. Lee, On-line bin packing in linear Time, *J. Algorithms* 10, 305-326, 1989.
12. M.B. Richey, Improved bounds for harmonic-based bin packing algorithms, *Discrete Appl. Math.* 34, 203-227, 1991.
13. S.S. Seiden, On the online bin packing problem, *J. ACM* 49, 640-671, 2002.
14. J.D. Ullman, The performance of a memory allocation algorithm. *Tech. Rep. 100, Princeton University, Princeton, N.J.,Oct.*, 1971.
15. A. van Vliet, An improved lower bound for on-line bin packing algorithms, *Inform. Process. Lett.* 43, 277-284,1992.
16. A.C.-C. Yao, New Algorithms for Bin Packing, *J. ACM* 27, 207-227, 1980.
17. M. Yue, A simple proof of the inequality FFD(L) $\leq$ 11/9OPT(L) +1, $\forall L$ for the FFD bin-packing algorithm, *Acta mathematicae applicatae sinica* 7, 321-331, 1991.

Probe Matrix Problems:
Totally Balanced Matrices

David B. Chandler[1], Jiong Guo[2,*], Ton Kloks[3,**], and Rolf Niedermeier[2]

[1] Department of Mathematical Sciences, University of Delaware,
Newark, Delaware 19716
davidbchandler@gmail.com
[2] Institut für Informatik, Friedrich-Schiller-Universität Jena, Ernst-Abbe-Platz 2,
D-07743 Jena, Germany
guo@minet.uni-jena.de, niedermr@minet.uni-jena.de
[3] School of Computing, University of Leeds, Leeds LS2 9JT, UK

Abstract. Let $\mathcal{M}$ be a class of 0/1-matrices. A 0/1/$\star$-matrix A where the $\star$s induce a submatrix is a *probe matrix* of $\mathcal{M}$ if the $\star$s in A can be replaced by 0s and 1s such that A becomes a member of $\mathcal{M}$. We show that for $\mathcal{M}$ being the class of totally balanced matrices, it can be decided in polynomial time whether A is a probe totally balanced matrix. On our route toward proving this main result, we also prove that so-called partitioned probe strongly chordal graphs and partitioned probe chordal bipartite graphs can be recognized in polynomial time.

1 Introduction

With this paper, we bring together two lines of research. On the one hand, we consider totally balanced matrices and the closely related strongly chordal graphs, and, on the other hand, we study sandwich and, more specifically, probe problems. We provide first positive results on the recognizability of probe totally balanced matrices and, correspondingly, partitioned probe chordal bipartite and partitioned probe strongly chordal graphs.

Sandwich Problems. Sandwich problems are studied in graph and hypergraph theory as well as for matrix problems [10,13,14,16]. For a graph property Π, the corresponding sandwich problem is defined as follows: Let $G_1 = (V, E_1)$ and $G_2 = (V, E_2)$ be two graphs such that $E_1 \subseteq E_2$. Is there a graph $G = (V, E)$ such that G satisfies Π and $E_1 \subseteq E \subseteq E_2$? Similarly, in case of matrices one is given a matrix with entries from $\{0, 1, \star\}$, and one asks whether one can replace the $\star$s by 0s and 1s such that the matrix fulfills a given property (such as being totally balanced).

Sandwich problems can be seen as generalizations of recognition and completion problems; for instance, graph completion problems allow the arbitrary

---

* Supported by DFG, Emmy Noether research group "PIAF", NI 369/4.
** Partially supported by the National Science Council of Taiwan, grant NSC94-2627-B-007-001. During the final stage supported by DFG, project "OPAL", NI 369/2.

M.-Y. Kao and X.-Y. Li (Eds.): AAIM 2007, LNCS 4508, pp. 368–377, 2007.

addition of edges whereas in sandwich problems the addition of certain edges (those not in E_2) is disallowed. Unfortunately, as a rule, sandwich problems are notoriously hard. For instance, in their classic paper, Golumbic, Kaplan, and Shamir [14, Figure 3] have pointed out the NP-completeness of sandwich problems for many subclasses of perfect graphs. Very recently, Faria *et al.* [12] announced the NP-completeness of the sandwich problem for strongly chordal graphs. Analogous results hold for matrix sandwich problems [13,16]. For instance, considering the class of matrices with the consecutive ones property (see for example [4]), the corresponding sandwich problem is NP-complete [16].[1]

With the *probe* concept, motivated by applications in computational biology, a new, seemingly more tractable[2] sandwich concept entered the stage. It already received considerable attention in the graph-algorithms community; for example, see [3,5,15]. Given a class of graphs $\mathcal{G}$, a graph G is a *probe graph of* $\mathcal{G}$ if its vertices can be partitioned into two sets $\mathbb{P}$ (the *probes*) and $\mathbb{N}$ (the *nonprobes*), where $\mathbb{N}$ is an independent set, such that G can be *embedded* into a graph of $\mathcal{G}$ by adding edges between certain nonprobe vertices. Notice that this is a special version of graph sandwich problems with $G_1 = G$ and $G_2 = (V, E \cup E')$ where E' contains the edges between all vertex pairs from $\mathbb{N}$.

0/1-Matrix Problems. Interpreting 0/1-matrices as adjacency matrices, there is a direct connection between matrix and graph problems. In what follows, we will make extensive use of this close relationship, setting out our proofs in terms of graph theory rather than matrix theory. The original motivation for our work, however, comes from matrices and integer linear programming.

It is well-known that when the matrix of a linear program is balanced, totally balanced, or totally unimodular, then the corresponding integer linear programming problem can be solved in polynomial time. The general case is NP-hard (see, *e.g.*, [4,21]). The study of balanced 0/1-matrices, that is, matrices where no square submatrix of odd order contains exactly two 1s per row and per column, goes back to Berge [2]. In particular, he proved that a 0/1-matrix is balanced if and only if the corresponding bipartite graph has no induced cycle of length 2 mod 4. Later, Lovász suggested to study totally balanced 0/1-matrices, that is, matrices that correspond to bipartite graphs without any induced cycle of length more than 4. In this paper, we want to initiate a study of probe problems referring to these matrices. More precisely, we focus on the perhaps simplest case, that is, totally balanced matrices. This matrix class, a subclass of balanced matrices, finds applications in various contexts [1,4]. We employ the following, for our purposes most suitable definition of totally balanced matrices (due to Lovász). To this end, note that a 0/1-matrix uniquely corresponds to a bipartite graph where one color class stands for the rows and the other for the columns.

A 0/1-matrix is *totally balanced* if its corresponding bipartite graph is *chordal bipartite*, that is, if the bipartite graph has no chordless cycle of length more than

---

[1] The matrices with the consecutive ones property form a subclass of totally balanced matrices.

[2] For instance, whereas the sandwich problem for chordal graphs is NP-complete [14], the corresponding probe chordal problem is polynomial-time solvable [15,3].

four. Note that the recognition of chordal bipartite graphs has a long history, see Huang [18] for recent characterizations. The sandwich problem for totally balanced matrices is, given a matrix A with entries from $\{0, 1, \star\}$, try to replace the $\star$s with 0s and 1s such that A becomes totally balanced. Unfortunately, this problem turns out to be NP-complete [12]. Hence, instead we somewhat naturally "relax" the problem formulation, considering its probe version. Let A be a $0/1/\star$-matrix in which the $\star$s induce a *submatrix*. Then A is called *probe totally balanced* if the $\star$s in A can be replaced by 0s and 1s such that A becomes totally balanced.

Seen from a more general perspective, probe matrices stand in one-to-one correspondence with *partitioned* probe bipartite graphs; partitioned means that the partition of the vertices into probes and nonprobes is part of the input. More precisely, exactly those rows and columns that contain at least one $\star$-entry correspond to the nonprobe vertices.

The main result of this work in terms of matrices is to show that one can decide in polynomial time whether a given $0/1/\star$-matrix is probe totally balanced, and, if so, find a corresponding replacement of the $\star$s by 0s and 1s. This can also be considered as a step toward solving the corresponding recognition problems for probe balanced and probe totally unimodular matrices.

Due to the lack of space, some proofs are omitted.

2   Preliminaries

For notational convenience, for sets A and B and elements x we write $A + B$, $A - B$, $A + x$, and $A - x$ as shorthands for $A \cup B$, $A \setminus B$, $A \cup \{x\}$, and $A \setminus \{x\}$. Moreover, for a graph $G = (V, E)$ we denote by $G - x$ the induced subgraph $G[V - x]$. The *complement graph* $\bar{G} = (V, E')$ of a graph $G = (V, E)$ is given by $E' := \{\{u, v\} \mid u, v \in V, \{u, v\} \notin E\}$. For a vertex x we denote by $N_G(x)$ the set of its neighbors in graph G and we let $N_G[x] = N_G(x) + x$ be its closed neighborhood. For a subset $A \subseteq V$ we write $N_G(A) = \bigcup_{x \in A} N_G(x) - A$. Herein, we omit the subscript "G" when it is clear from the context. A subset I of vertices is called an *independent set* if the induced subgraph $G[I]$ has no edge, whereas a subset K of vertices is called a *clique* if $G[K]$ has all possible edges.

A *chord* in a (simple) cycle is an edge connecting two vertices of the cycle which are not adjacent in the cycle. A graph is *chordal* if it has no chordless cycle of length more than 3. A chord in an even cycle is *odd* if the distance between the endvertices along the cycle is odd.

The major technical contribution of this paper is based on the close relationship between the class of strongly chordal graphs and totally balanced matrices. The following is not the original definition of strong chordality but a characterization due to Farber [11] which best fits our purposes.

Definition 1 ([11]). *A graph is* strongly chordal *if it is chordal and every even cycle of length at least six has an odd chord.*

Polynomial-time recognition algorithms for strongly chordal graphs (based on doubly-lexical orderings) are due to Lubiw [19] and Paige and Tarjan [20].

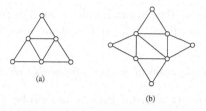

Fig. 1. A 3-sun (a) and a trampoline (b)

Strongly chordal graphs have a useful characterization based on the following notion.

Definition 2. *A simple vertex is a vertex* x *such that for every pair* $y, z \in N(x)$ *either* $N[y] \subseteq N[z]$ *or* $N[z] \subseteq N[y]$.

That is, the closed neighborhoods of the neighbors of a simple vertex form a chain under inclusion. Notice that a simple vertex is *simplicial*, *i.e.*, its neighborhood induces a clique.

Theorem 1 ([11]). *A graph is strongly chordal if and only if every induced subgraph has a simple vertex.*

The class of strongly chordal graphs can be characterized by forbidden induced subgraphs called suns. A *sun* is a graph obtained from an even cycle of length at least 6 by adding edges to make a maximum independent set into a clique. If the cycle is of length 2k, then the sun is called a k-*sun*. We call the set of vertices of degree 2 in a sun the *independent set of the sun*, and the set of vertices of degree at least 4, the *clique of the sun*. If some edges of the clique are missing but the graph is still chordal, it is called a *trampoline*. Figure 1 shows a 3-sun and a trampoline.

Theorem 2 ([6,11]). *A graph is strongly chordal if and only if it is chordal and has no induced sun.*

3 Partitioned Probe Strongly Chordal Graphs

Strongly chordal graphs are closely related to totally balanced matrices. As a basis for showing in Sect. 4 that probe totally balanced matrices can be recognized in polynomial time, here we show how to recognize partitioned probe strongly chordal graphs, or PP-strongly chordal graphs for short, in polynomial time.

In what follows, an *embedding* of a probe graph G into a graph class 𝒢 always means a graph contained in 𝒢 which is obtained from G by adding edges between the nonprobes in G. In our recognition algorithm the first two steps concentrate on finding an embedding of the partitioned probe graph into a chordal graph. In a third step we take care of the suns (cf. Theorem 2) in this chordal embedding.

To find a chordal embedding, we first prove that we only have to destroy chordless 4-cycles. More precisely, we show that probe strongly chordal graphs are *weakly chordal*. That is, they contain neither an induced *hole* nor an induced *antihole*. A hole is a chordless cycle of length at least 5. An antihole is the complement of such a cycle.

Lemma 1. *Probe strongly chordal graphs are weakly chordal.*

Proof. Since probe strongly chordal graphs are probe chordal, they are perfect [15] and hence they have no induced odd hole [7]. A probe chordal graph also contains no antihole. Suppose there is an antihole. It has at most two nonprobes, and the edge joining them must be added because an antihole is not chordal.[3] However, the resulting graph still contains at least the complement of a path induced by five vertices, and hence a 4-cycle, and is still not chordal. Thus, it remains to be shown that there are no even holes.

Consider a graph G containing an induced even hole of length 2k, $k \geq 3$. To make G chordal, one needs a set of k nonprobes. The even hole can be made into a sun by turning this set of nonprobes into a clique. However, if not all the edges between nonprobes are there, this even hole will be a trampoline. Since every trampoline contains a sun as an induced subgraph [6,11], any embedding of this even hole will have a sun. Thus, G is not probe strongly chordal.   □

Next we point out how to cope with chordless 4-cycles (the second step of our algorithm). To this end, we need some more notation and facts. A cycle in which probes and nonprobes alternate is called an *alternating cycle*.

Proposition 1 ([15]). *Let* $G = (\mathbb{P} + \mathbb{N}, E)$ *be a partitioned graph. Let* C *be a chordless 4-cycle in* G. *If* G *is probe chordal, then* C *is alternating and any chordal embedding of* G *must have the edge filled in between the two nonprobes of* C.

Enhanced graphs [15] play a central role for the recognition of probe chordal graphs [3,15] and are also a key concept in our recognition algorithm.

Definition 3 ([15]). *The* enhanced graph G^* *is obtained from a partitioned graph* $G = (\mathbb{P} + \mathbb{N}, E)$ *by adding all edges between nonprobes in alternating chordless 4-cycles of* G.

As stated in the following theorem [15], the enhanced graph is the desired chordal embedding for the second step of the algorithm.

Theorem 3 ([15]). *Let* $G = (\mathbb{P} + \mathbb{N}, E)$ *be a* PP-*chordal graph which is weakly chordal. Then the enhanced graph* G^* *is chordal.*

For the third and last step of our algorithm, by Theorem 2, it remains to destroy the induced suns in the enhanced graphs. As an example of destroying suns, take the 3-sun in which one independent set vertex and one clique vertex are

---

[3] An antihole is not chordal for the following reason. A length-5 antihole is the same as a length-5 hole. Every antihole with at least 6 vertices contains 4 vertices which induce a chordless cycle.

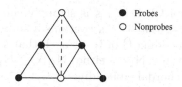

● Probes
○ Nonprobes

Fig. 2. A 3-sun in which one independent set vertex and one clique vertex are the only (nonadjacent) nonprobes. Adding the edge destroys the sun.

the only (nonadjacent) nonprobes (see Fig. 2). The enhanced graph is the 3-sun itself, which is not strongly chordal. A strongly chordal embedding is obtained by adding the edge between the two nonprobes. The final and most important concept for our strategy to destroy suns is the notion of a "probe simple vertex."

Definition 4. *Let* $G = (\mathbb{P} + \mathbb{N}, E)$ *be a partitioned graph. A vertex is* probe simple *if it can be made simple by adding some edges between nonprobes in* G.

Obviously, if G is PP-strongly chordal then, by Theorem 1, every induced subgraph has a probe simple vertex.

We summarize our findings in the following pseudo-code for recognizing PP-strongly chordal graphs. The correctness of the third and main step is heavily based on Theorem 4 shown subsequently.

ALGORITHM FOR RECOGNIZING PP-STRONGLY CHORDAL GRAPHS
Input: Partitioned probe graph $G = (\mathbb{P} + \mathbb{N}, E)$.
Output: YES if G is PP-strongly chordal; otherwise, NO.
1   **If** G is not weakly chordal **then** return NO;
2   **If** there is a chordless 4-cycle that is not alternating **then** return NO;
3   Construct the enhanced graph G^*;
    While G^* is not empty **do**
        If G^* has a probe simple vertex s **then**
            Insert an inclusion-minimal set of edges into G^* to make s simple;
            $G^* := G^* - s$
        else return NO;
    Return YES

In the following, given a graph $G = (V, E)$, a vertex $v \in V$, and a graph $H = (V_H, E_H)$ with $V_H \subseteq V \setminus \{v\}$, we use $H + v$ to denote the graph obtained by adding v to H and adding the edges between v and $N_G(v) \cap V_H$.

Theorem 4. *Let* s *be a probe simple vertex in an enhanced graph* G. *Let* G′ *be the graph obtained from* G *by adding an inclusion-minimal set of edges between nonprobes to* G *in order to make* s *simple. Then,* G *is* PP-*strongly chordal if and only if* G′ − s *is* PP-*strongly chordal.*

Proof. (Sketch) "⇐": We show this direction by constructing from a strongly chordal embedding H of G′ − s a strongly chordal embedding of G. Since s is simple in G', $H + s$ is chordal. If $H + s$ is strongly chordal, then we are done. Hence, we consider the case that $H + s$ contains a sun: We transform H into a new, strongly chordal embedding $\hat{H}$ of $G - s$ such that s is simple in $\hat{H} + s$; that is, for every two vertices $x, y \in N_G(s)$ either $N_{\hat{H}}[x] \subseteq N_{\hat{H}}[y]$ or $N_{\hat{H}}[y] \subseteq N_{\hat{H}}[x]$. Then $\hat{H} + s$ is a strongly chordal embedding of G.

To construct $\hat{H}$, we apply the following operation to all pairs $x, y \in N_G(s)$ with $N_{G'}[x] \subseteq N_{G'}[y]$ while $N_H[x] - N_H[y] \neq \varnothing$. Obviously, $N_H[x] - N_H[y] \subseteq N$. Consider removing all edges $\{x, z\}$ where $z \in N_H(x) - N_H[y]$. Assume this creates a chordless cycle C of length at least 4. Since H is chordal, $x \in C$. Then y is adjacent to the two neighbors of x in C and C contains a vertex z in $N_H(x) - N_H[y]$ which is not adjacent to y. It follows that the subgraph of H induced by $C - x + y$ contains a chordless cycle, which is a contradiction.

A similar argument shows that the resulting graph is *strongly* chordal. Repeated application of the above operation gives the graph $\hat{H}$.

"$\Rightarrow$": The basic idea of the proof of this direction is as follows. To show that $G' - s$ is PP-strongly chordal, we construct a strongly chordal embedding of $G' - s$ from the strongly chordal embedding H of G. More precisely, by adding and deleting edges, we will construct a graph $\hat{H}$ from H such that $\hat{H} - s$ is a strongly chordal embedding of $G' - s$. The central difficulty arising here is as follows. In order to make s simple in G', we have added edges to G to obtain G'. Among these edges there might be edges that are not present in H. Just adding these edges to H, however, does not necessarily guarantee that the resulting graph is strongly chordal. Thus, we will not only add these edges to H, but also delete and add some other edges when constructing $\hat{H}$ from H.

Without loss of generality let us assume the ordering $N_{G'}[s] \subseteq N_{G'}[x_1] \subseteq \cdots \subseteq N_{G'}[x_k]$ for $N_{G'}(s) = \{x_1, \ldots, x_k\}$. We write also $s = x_0$. The graph $\hat{H}$ shall have the following properties: $\hat{H}$ is strongly chordal (P1), $N_{\hat{H}}(s) = N_{G'}(s) = \{x_1, \ldots, x_k\}$ (P2), and $N_{\hat{H}}[s] \subseteq N_{\hat{H}}[x_1] \subseteq \cdots \subseteq N_{\hat{H}}[x_k]$ (P3).

Step 1. Apply the following operation for $i = 0, \ldots, k$: As long as x_i has a neighbor z in H for which there exists a probe vertex $y \in \{x_{i+1}, \ldots, x_k\} \cap \mathbb{P}$ with $z \notin N_H(y)$, remove the edge $\{x_i, z\}$ from H.

It can be shown that the resulting graph H_1 is strongly chordal.

Step 2. Turn $N_{H_1}[s]$ into a clique.

Let H_2 be the result. Assume H_2 has a chordless cycle C. Then C has exactly two vertices $x, y \in N_{H_1}(s)$. Then $C + s$ induces a chordless cycle in H_1 and this is a contradiction. Now let D be a component of $H_1 - N_{H_1}[s]$. Then any two vertices $x, y \in N_{H_1}(D)$ are adjacent and have no *private neighbors* in D. Herein, for two vertices x and y, a private neighbor of x is a vertex adjacent to x but not to y. It follows that a vertex in D which is simple in H_1 remains simple in H_2. Thus H_2 has a *simple elimination ordering*, i.e., there is a vertex ordering $(v_1, \ldots, v_n)$ of H_2 such that, for all $1 \leq i \leq n$, the vertex v_i is simple in the subgraph of H_2 induced by $\{v_i, \ldots, v_n\}$. Thus, H_2 is strongly chordal.

Step 3. For $i = 0, \ldots, k - 1$ and each $y \in \{x_{i+1}, \ldots, x_k\}$, make y adjacent to all vertices of $N_{H_2}[x_i]$.

Let H_3 be the resulting graph. Assume that making y adjacent to all vertices of $N_{H_2}[x_i]$ creates a chordless cycle C. Then for some $z \in N_{H_2}(x_i)$, $y, z \in C$. Now let z' be the other neighbor of y in C. Then $N_{H_2}(x_i) \cap C \subseteq \{y, z, z'\}$. It follows that $H_2[C + x_i]$ also contains a chordless cycle. Assume this step creates a "bad" cycle S, $i.e.$, an even cycle of length at least 6 without an odd chord. Let $y, z, z' \in S$ as above and let z'' be the other neighbor of z in S. Then x_i is adjacent to at least one of z', z'' and it follows that also H_2 has such a bad cycle, a contradiction.

Step 4. Remove the edges $\{s, z\}$ from H_3 for all $z \in N_{H_3}(s) - N_{G'}(s)$.

In the remaining graph $\hat{H}$, s has the same neighborhood as in G'. Notice that $\hat{H}$ is strongly chordal: Indeed, $H_3 - s = \hat{H} - s$, and so $\hat{H} - s$ is strongly chordal. The vertex s is simple in $\hat{H}$, thus also $\hat{H}$ is strongly chordal.   □

The correctness of the algorithm as given in the pseudo-code above follows from Lemma 1, Proposition 1, Theorem 3, and Theorem 4. Altogether, we obtain the following main result.

Theorem 5. *It can be decided in polynomial time if a partitioned graph G is PP-strongly chordal. If so, also an embedding of G can be found.*

4   Partitioned Probe Chordal Bipartite Graphs

Recall that our main goal is to devise a polynomial-time algorithm for the recognition of probe totally balanced matrices. To this end, we make use of the fact that a 0/1-matrix is totally balanced if and only if the corresponding bipartite graph is *chordal bipartite* [11].

In this section, we show that the recognition algorithm for PP-strongly chordal graphs can be used for recognizing PP-chordal bipartite graphs and indicate how this transfers to the recognition of probe totally balanced matrices.

One characterization of chordal bipartite graphs that is useful for our purposes is the following by Dahlhaus [9]. If $B = (X, Y, E)$ is a bipartite graph then we denote by $\mathsf{split}_X(B)$ the graph obtained from B by completing X into a clique.[4]

Theorem 6 ([9]). *A bipartite graph $B = (X, Y, E)$ is chordal bipartite if and only if $\mathsf{split}_X(B)$ is strongly chordal.*

From now on, let B be a partitioned probe bipartite graph. Without loss of generality, we assume that B is connected, since otherwise we may concentrate on the components individually. Obviously, we cannot simply apply the recognition algorithm for partitioned probe strongly chordal graphs to $\mathsf{split}_X(B)$, since the completion of X into a clique possibly adds edges between nonprobe vertices. Instead, we use the following trick: We add two probe vertices to Y, say α and ω, and make these adjacent to all vertices of X. Let $\alpha\omega(B)$ be the resulting bipartite graph. Next, for every probe vertex x in X, we add edges between x and all other

---

[4] The operation $\mathsf{split}_X(B)$ transforms B into a *splitgraph*, that is, a graph which has a partition of its vertex set into a clique and an independent set.

vertices in X. Since α and ω create a chordless 4-cycle with any pair of nonprobe vertices of X, any embedding of this new graph into a strongly chordal graph forces X into a clique. Let $\mathsf{PPsplit}_X(\alpha\omega(B))$ be the graph obtained from $\alpha\omega(B)$ by adding edges between vertices in X as described above. In the following, we present a "probe version" of Theorem 6. Afterwards, Theorem 8 justifies the application of the partitioned probe strongly chordal graph recognition algorithm from Section 3 to $\mathsf{PPsplit}_X(\alpha\omega(B))$.

Theorem 7. *A partitioned probe bipartite graph* $B = (X, Y, E)$ *is partitioned probe chordal bipartite if and only if* $\mathsf{PPsplit}_X(\alpha\omega(B)) = (X, Y \cup \{\alpha, \omega\}, E')$ *is partitioned probe strongly chordal.*

Combining Theorems 5 and 7, we arrive at the main result of this section.

Theorem 8. *Partitioned probe chordal bipartite graphs can be recognized in polynomial time.*

Proof. The algorithm works as follows: Given a partitioned probe bipartite graph $B = (X, Y, E)$, construct $\mathsf{PPsplit}_X(\alpha\omega(B)) = (X, Y \cup \{\alpha, \omega\}, E')$ as described above. The new graph is checked against being partitioned probe strongly chordal by the algorithm in Section 3. The correctness and the running time follow from Theorems 5 and 7.   □

Corollary 1. *Probe totally balanced matrices can be recognized in polynomial time.*

5   Future Work

With this paper we try to initiate research on special matrix sandwich problems, that is, probe matrix problems. These stand in close relationship with partitioned probe graph problems. As an important starting case, we settled the complexity of the recognition problem of the probe totally balanced matrices, thereby also showing the polynomial-time recognizability of partitioned probe strongly chordal graphs and of partitioned probe chordal bipartite graphs. Note that the corresponding sandwich versions are NP-complete [12]. As to future work, we face the following two challenges (refer to [4, Chapter 9] for definitions).
1. Show that probe balanced matrices can be recognized in polynomial time. In their seminal work, Conforti, Cornuéjols, and Rao [8] designed a polynomial-time algorithm for recognizing balanced matrices (also see [17,23]).
2. Show that probe totally unimodular matrices can be recognized in polynomial time. Here, Seymour's [22] famous decomposition result for totally unimodular matrices should be helpful (also see [21, Chapters 19–21]).

Further opportunities for future work include showing the polynomial-time recognizability of unpartitioned strongly chordal graphs and studying optimization versions of the probe problems considered. In the latter case, the natural task would be to minimize the number of edges added and the number of ⋆s turned into 1s, respectively.

References

1. V. L. Beresnev. An efficient algorithm for the uncapacitated facility location problem with totally balanced matrix. *Discrete Applied Mathematics*, 114:13–22, 2001.
2. C. Berge. Balanced matrices. *Mathematical Programming*, 2:19–31, 1972.
3. A. Berry, M. C. Golumbic, and M. Lipshteyn. Two tricks to triangulate chordal probe graphs in polynomial time. In *Proc. of 15th ACM–SIAM SODA*, pages 962–969. ACM–SIAM, 2004.
4. A. Brandstädt, V. B. Le, and J. P. Spinrad. *Graph Classes: a Survey*. SIAM Monographs on Discrete Mathematics and Applications, 1999.
5. G. Chang, A. J. J. Kloks, J. Liu, and S.-L. Peng. The PIGs full monty—a floor show of minimal separators. In *Proc. 22nd STACS*, volume 3404 of *LNCS*, pages 521–532. Springer, 2005.
6. G. J. Chang. k-*Domination and Graph Covering Problems*. PhD thesis, School of OR and IE, Cornell University, Ithica, NY, USA, 1982.
7. M. Chudnovsky, N. Robertson, P. D. Seymour, and R. Thomas. The strong perfect graph theorem. *Annals of Mathematics*, 164(1):51–229, 2006.
8. M. Conforti, G. Cornuéjols, and R. Rao. Decomposition of balanced matrices. *Journal of Combinatorial Theory, Series B*, 77:292–406, 1999.
9. E. Dahlhaus. Chordale Graphen im besonderen Hinblick auf parallele Algorithmen. Habilitationsschrift, Universität Bonn, Germany, 1991.
10. M. Farach, S. Kannan, and T. Warnow. A robust model for finding optimal evolutionary trees. *Algorithmica*, 13:155–179, 1995.
11. M. Farber. Characterizations of strongly chordal graphs. *Discrete Mathematics*, 43:173–189, 1983.
12. L. Faria, C. de Figueiredo, S. Klein, and R. Sritharan. On the complexity of the sandwich problem for strongly chordal graphs and chordal bipartite graphs. Submitted to *Theoretical Computer Science*, 2006.
13. M. C. Golumbic. Matrix sandwich problems. *Linear Algebra and its Applications*, 277:239–251, 1998.
14. M. C. Golumbic, H. Kaplan, and R. Shamir. Graph sandwich problems. *Journal of Algorithms*, 19:449–473, 1995.
15. M. C. Golumbic and M. Lipshteyn. Chordal probe graphs. *Discrete Applied Mathematics*, 143:221–237, 2004.
16. M. C. Golumbic and A. Wassermann. Complexity and algorithms for graph and hypergraph sandwich problems. *Graphs and Combinatorics*, 14:223–239, 1998.
17. R. Hayward and B. A. Reed. Forbidding holes and antiholes. In J. L. A. Ramírez and B. A. Reed, editors, *Perfect Graphs*. John Wiley & Sons, 2001.
18. J. Huang. Representation characterizations of chordal bipartite graphs. *Journal of Combinatorial Theory, Series B*, 96:673–683, 2006.
19. A. Lubiw. Doubly lexical orderings of matrices. *SIAM Journal on Computing*, 16:854–879, 1987.
20. R. Paige and R. E. Tarjan. Three partition refinement algorithms. *SIAM Journal on Computing*, 16:973–989, 1987.
21. A. Schrijver. *Theory of Linear and Integer Programming*. John Wiley & Sons, 1986.
22. P. D. Seymour. Decomposition of regular matroids. *Journal of Combinatorial Theory, Series B*, 28:305–359, 1980.
23. G. Zambelli. A polynomial recognition algorithm for balanced matrices. *Journal of Combinatorial Theory, Series B*, 95:49–67, 2005.

Efficiency of Data Distribution
in BitTorrent-Like Systems

Ho-Leung Chan[1,*], Tak-Wah Lam[2], and Prudence W.H. Wong[3]

[1] Department of Computer Science, University of Pittsburgh
hlchan@cs.pitt.edu
[2] Department of Computer Science, University of Hong Kong, Hong Kong
twlam@cs.hku.hk
[3] Department of Computer Science, University of Liverpool, UK
pwong@csc.liv.ac.uk

Abstract. BitTorrent (BT) in practice is a very efficient method to share data over a network of clients. In this paper we extend the recent work of Arthur and Panigrahy [1] on modelling the distribution of individual data blocks in BT systems, aiming at a better understanding of why BT can achieve a high degree of parallelism. In particular, we include in our study several new network features that BT systems are using, as well as different local heuristics for routing data blocks in each client. We conduct simulations to figure out to what extent the new network features and routing heuristics would affect the distribution efficiency. Our findings confirm that for the primitive network setting studied in [1], it does require $\Omega(b \log n)$ phases for n clients to download b data blocks. More interestingly, our work suggests that for the more realistic network setting, the heuristics Random and Rarest Block First both allow n clients to download b blocks in $b + O(\log n)$ phases. We believe that the latter bound better reflects the high degree of parallelism of BT observed in reality. It is also worth-mentioning that $b + \log n$ is the smallest possible number of phases needed; it is interesting to see that some simple local routing heuristics have a performance so close to the optimal.

1 Introduction

Let us consider the following problem. There are n clients (nodes) on a well-connected network. They want to download a file of b data blocks from a server in a cooperative and efficient, but distributed manner. The idea is to avoid each client directly downloading from the server. Instead the server uploads each data block to only one client, and let the clients distribute the block among themselves. Assume each client can upload one data block to only one neighbor in one phase. The key concern is whether there exists a good strategy for each client to determine in each phase which block and which neighbor to upload, so that most clients can have progress in each phase.

---

* Part of this research was done while the author was at University of Hong Kong.

M.-Y. Kao and X.-Y. Li (Eds.): AAIM 2007, LNCS 4508, pp. 378–388, 2007.

The above problem is based on the "flash crowd" scenario, where a large number of clients join the network almost simultaneously to download a data file (say, a soccer game). BitTorrent (BT) [3,7] has found to be a very efficient method for such a problem in practice; it does exploit the bandwidth among the clients, and it is often observed that using BT, most clients have swift progress in parallel. Arthur and Panigrahy [1] were the first to model the distribution of data blocks in BT systems mathematically, aiming at explaining the high degree of parallelism BT achieves. In particular, they consider the clients are connected via a directed BT-graph (definition given in Section 2), and in each phase each client can upload (send) as well as download (receive) one data block from its neighbors. Among others, they proved that using a random strategy, the n clients can download all b blocks in $O(b \log n)$ phases with high probability. Below we refer this number of phases as the *total distribution time*.

In this paper, we extend the model used in [1] to include other network features found in BT networks. First, traffic between two neighbors is usually bidirectional; i.e., a client can upload and download from each neighbor client. Next, we note that the size of a data block is chosen in such a way that in each phase, a client has sufficient bandwidth to upload a block to one of its neighbors, yet the download rate is usually a few times higher than the upload rate (this is imposed by some internet providers), and each client can receive several data blocks in each phase. Furthermore, each client should be able to make a better decision using a request-based protocol instead of a push-based protocol; the former protocol requires each client to listen to the requests from its neighbors before making a decision which block and which neighbor to upload. The first objective of this paper is to study to what extent these extended features of the model would affect the total distribution time.

Furthermore, we study three different local routing heuristics for each node to decide which block to upload to its neighbor in each phase, namely, Sequential, Random, and Rarest Block First. Sequential simply uploads blocks sequentially according to their order in the file. Random picks a random block to upload. Rarest Block First selects a block which is the rarest among the neighbors of a node; this is a strategy being used by BT [5,7]. Regardless of which heuristic a node uses, the node only uploads a block which the neighbor does not have.

Note that no matter which network setting and which heuristics we pick, the total distribution time is at least $b + \lceil \log n \rceil$. To see this lower bound, we observe that the server requires b phases to upload b blocks, and the last block can reach at most $2^i - 1$ clients after i phases. It is a challenging task to find a reasonable network setting and routing heuristic that can lead to a total distribution time close to $b + \log n$.

We have done a lot of simulation on different network settings and heuristics. We consider models with different combinations of the three network features mentioned above to see to what extent they affect the total distribution time. These features include directed vs undirected graph, single vs multiple receive, and push vs request protocol. For multiple receive, we further distinguish the cases where a limited number x (called receive-x) of blocks vs an unlimited number can be received by a node in each phase. For each combination, we study the three local routing heuristics. We also vary the number of nodes n and the number of blocks b. Our findings are summarized as follows.

- As expected, the undirected and request-based model admits a better data distribution.
- For the model used in [1] (i.e., directed_receive-one_push), we observe that the total distribution time does require $\Omega(b \log n)$.
- The most surprising result is related to the model using undirected BT-graph, multiple receives, and request-based protocol (i.e., undirected_receive-x_request). No matter random or rarest block first is used, our simulation shows that the total distribution time is in the order $b + O(\log n)$, very close to the lower bound. For example, when $b = 3000$ and $n = 2000$, the total distribution time is around 3050 on average (with a very small standard deviation of 5 over five different trails.)
- It is natural to expect that the total distribution time to decrease with maximum number of receives in each phase increases. The decrease is indeed very drastic when we vary the number from 1 to 2, but the effect is not visible once the number goes up to 3. In fact, we find no significant difference using a maximum of five receives and an unlimited number of receives.
- Rarest block first is a heuristic used by BT. It is most effective, but only a bit better than Random. On the other hand, using Random saves a lot of implementation overhead. We believe Random is a better choice.

Related work. Before the work of Arthur and Panigrahy [1], there have been some theoretical work, yet some assumptions were made which might not be true in practice. Qiu and Srikant [10] applied flow analysis to BT-like networks but assuming (1) the data blocks available in each client for download at each phase is random and independent; (2) constant arrival rates. The latter does not account for BT's strength in handling flash crowd scenario. Yang and de Vecianna [11] considered flash crowd scenario but assumed that distribution of one data block will not slow down the distribution of other data blocks, ignoring the possible interaction in the distribution. There is also empirical work attempting to demonstrate BT's routing policy work well in practice [8,9,2,6], but they do not consider the distribution of individual data block.

Organization of the paper. In Section 2, we review Arthur and Panigrahy's model and discuss the effects of various network features on data distribution time. In Section 3, the local routing heuristics are tested. Finally, we give some concluding remarks in Section 4.

2   Network Models and Distribution Time

In this section, we first review Arthur and Panigrahy's model and then discuss how different model features affect the total distribution time. These features include directed vs undirected graph, single vs multiple receive, and push vs request protocol. In the next section we will study the effect of different routing heuristics. The experiments in this section all assume the Random heuristic when a node decide which block to send/receive.

2.1 Arthur and Panigrahy's Model

Arthur and Panigrahy [1] model a BT-like network as a directed graph with n nodes, each representing a user. One of the nodes is the seed which holds a large file initially. The file is divided into b equal-size blocks. The remaining $n - 1$ nodes want to obtain the file and they have zero block of the file initially.

A BT system maintains a virtual network represented as a BT-C graph[1], for some integer C, which is constructed as follows. The BT-C graph starts with C nodes $v_1, v_2, \ldots, v_C$, with a directed edge from v_i to v_j if and only if $i < j$. While the total number of nodes is less than n, a new node is added. C existing nodes are selected at random from which directed edges are drawn to the new node. In most BT systems, C is set to 40 .

For the distribution of data blocks in the network, a node can send a block to a neighbor only if it already has the block, and a node obtains the whole file only if it has every block of the file. Time is divided into discrete time steps (*phases*). In each phase, each node can send at most one block to a neighbor along an outgoing edge. When multiple blocks are sent to a node in a phase, the node can receive at most one such block.

The efficiency of the system is measured by the *total distribution time*, which is the number of phases taken until all nodes obtain the file. Arthur and Panigrahy assume a simple protocol for sending blocks: In each phase, each node u randomly picks a neighbor v, and u sends to v a block that u already has but v does not. u is idle if no such block exists. They proved that the total distribution time is $O(b \log n)$ with high probability.

2.2 Identifying More Realistic Model Features

We observe that some features in Arthur and Panigrahy's model are too restrictive comparing to a real BT system. In particular, we focus on the following three important features and analyze how they affect the total distribution time.

1. **Directed vs Undirected graph.** When modelling the network as a directed graph, the connection is asymmetric, i.e., there are pairs of nodes, say u and v, connected by an edge for which u can send to v but not vice versa This is not the case in real-life BT systems (more precisely, the underlying Internet) in which connected nodes can send blocks in both directions. In our study, we analyze the effect of assuming directed vs undirected (bidirectional) edges.
2. **Single vs multiple receive.** This models the bandwidth limit of the nodes. When multiple blocks are sent to a node in a phase, Arthur and Panigrahy assume that the node randomly obtains one of the blocks. It corresponds to

---

[1] The network in which the users connect is actually the Internet. On the application layer, BT maintains a virtue network in the form of a BT-C graph.

Table 1. The most restrictive model DG-1-Ph (col. 1) performs the worst. The most relaxed models UG-5-Rq & UG-u-Rq (bolded) are the best. Request protocol improves distribution time (col. 3 vs 4; 5 vs 6). Multiple receive also outperforms single (col. 2 vs 4 & 6).

Distribution time	DG-1-Ph	UG-1-Rq	UG-5-Ph	UG-5-Rq	UG-u-Ph	UG-u-Rq
$n = 300, b = 1200$	3826	1913	1354	**1228**	1325	**1228**
$n = 2000, b = 1200$	5068	1935	1386	**1247**	1344	**1239**
$n = 300, b = 3000$	9915	4742	3300	**3034**	3212	**3030**
$n = 2000, b = 3000$	12866	4787	3368	**3047**	3243	**3044**

the situation that download bandwidth is limited to a similar extent as the upload bandwidth. In reality, the download bandwidth of a node is usually higher than the upload capacity. Thus, it is interesting to understand the effect when each node can receive $r \geq 2$ blocks in each phase. We will consider the cases of $r = 1$ vs $r = 5$ and r is unlimited.

3. **Push vs request protocol.** This refers to whether a node actively ask for a *missing* block that it does not have. When a node send blocks to the neighbors, it was assumed in [1] that the node randomly picks a neighbor and sends a block useful to that neighbor without prior communication. We call this a *push* protocol. In reality, a BT system uses a two-way protocol where at the beginning of each phase, each node sends requests to the neighbors asking for missing blocks, and then each node serves some of the requests it gets. We call this a *request* protocol. Intuitively, request protocol helps to avoid multiple neighbors sending the same block to a node.

2.3 Simulation and Findings

We consider models with different combinations of the above three features and see how they affect the total distribution time. We name the models using three fields: the first field is either DG (directed graph) or UG (undirected graph); the second field tells the maximum number of blocks a node can receive in one phase, with u means unlimited; the third field is Ph (push protocol) or Rq (request protocol). For example, DG-1-Ph refers to the model studied in [1]. On the other hand, UG-u-Rq is the most relaxed model, based on which we vary the features to also study models in between (see Table 1 for the list of models we study).

When the request protocol is used, we assume that a random routing heuristic is used, i.e., each node picks a random missing block in turn (until all missing blocks have been considered) and requests it from a random neighbor having it, with the restriction that no neighbor will be requested twice. Note that we will discuss other routing heuristics in Section 3. We also assume that after getting the requests, each node randomly serves one of the requests.

We perform simulations with $n = 300, 2000$ and $b = 1200, 3000$. Note that BT systems usually divide a file into pieces of $1/4$ MB each, $b = 1200$ and 3000 correspond to a file of 300MB and 700MB, respectively, the latter is about the

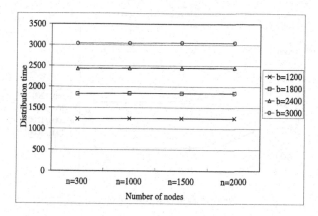

Fig. 1. The distribution time increases linearly with b (see the four different curves). The effect of n on the distribution time is small (see the flatness of each curve).

size of a CD. The choices of n allow us to see the effect of n being smaller than and larger than b. For each model and combination of n and b, we repeat the simulation for 5 times and record the average distribution time. The results are shown in Table 1. We have the following observations.

Near optimal distribution time. We observe that the total distribution time of models UG-5-Rq (col. 4) and UG-u-Rq (col. 6) are close to the theoretical lower bound of $b + O(\log n)$. It shows that the features of undirected graph, multiple receive and request protocols are very effective in improving the total distribution time. Note that a random heuristic is used in selecting blocks to send and receive. The experiments show that this is sufficient to obtain a good performance under an appropriate model.

To understand further the growth in distribution time under different combinations of n and b, we perform more experiments under the UG-5-Rq model, with $n = 300, 1000, 1500, 2000$ and $b = 1200, 1800, 2400, 3000$. We plot the total distribution time against n, with each value of b in a different curve. The results are shown in Figure 1. We can observe that the total distribution time grows very slowly as n increases. The distribution time is close to $b + k \log n$ where k is approximately 4.

Multiple receive is most effective. We observe that among the three features considered, allowing multiple receive per phase seems to be the most effective in improving the total distribution time. For example, comparing the models of UG-1-Rq (col. 2) and UG-5-Rq (col. 4), the total distribution time decreases by about 35%.

The effect of multiple receive can be explained as follows. Because each node decides on its own to whom it sends a block, it is common for multiple nodes sending blocks to the same target node. When each node can only receive one block per phase, bandwidth is lost, i.e., overall the number of data blocks received is smaller than that sent because some blocks have to be dropped. We call this

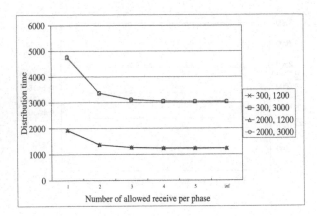

Fig. 2. The total distribution time drops significantly when the number of allowed receive increases from 1 to 2 and becomes stable beyond 3. Note that the two curves for $b = 1200$ (similarly, for $b = 3000$) are very close because the distribution time is dominated by the value of b.

receive collision. Multiple receive effectively reduces the bandwidth lost due to receive collision, thus leading to improved distribution time.

To obtain a better understanding of the effect of multiple receive, we vary the number of allowed receive from 1 to 5 and also unlimited. We perform experiments for $n = 300, 2000$ and $b = 1200, 3000$. The results are shown in Figure 2. We observe that the total distribution time improves greatly when number of allowed receive increases from 1 to 2, and has little effect beyond 3. It suggests that receive collision is common, but the number of blocks involved is usually small.

Undirected graph and request protocol are also effective. Undirected graphs have better distribution time than directed graphs (see col. 1 vs others) because of two reasons. Since the edges are bidirectional in an undirected graph, there are effectively double the possible connections among the nodes. Furthermore, the leaf nodes in directed graphs having no outgoing edges do not help to distribute the received blocks, so they reduce the efficiency by reducing the availability of the blocks.

Besides receive collision we mentioned before, there is another kind of collision that reduces the efficiency of distribution. Bandwidth is also lost when multiple copies of the same block are sent to the same node in a phase. We call this *send collision.* Request protocol avoids send collision because for each missing block, a node actively requests only one single neighbor to send the block.

Tightness of the analysis in [1]. Finally, we look at the results for the model DG-1-Ph [1] more closely. We observe that the total distribution time grows in the order of $b \log n$, it is roughly $0.4 \times b \log n$ (see col. 1). The experimental results suggest that data distribution does require $\Omega(b \log n)$ phases to finish in the DG-1-Ph model.

Table 2. Random and Rarest First have similar distribution time in the Model UG-5-Rq, which is close to the theoretical lower bound of $b + O(\log n)$

Distribution time	Random	Rarest First	Sequential
$n = 300, b = 1200$	**1228**	**1221**	3319
$n = 2000, b = 1200$	**1247**	**1233**	4678
$n = 300, b = 3000$	**3034**	**3024**	8494
$n = 2000, b = 3000$	**3047**	**3037**	11615

3 Performance of Simple Routing Heuristics

In BT systems, distribution of blocks is decided locally by each node, without a central coordination. Heuristics are used to determine which blocks are sent to which nodes at each phase. In this section, we consider both request and push protocols, and study how different heuristics affect the distribution time.

3.1 Routing Heuristics for Request Protocols

In models with request protocols, distribution of blocks is driven by what requests are sent by the nodes. We study three natural heuristics for deciding which blocks a node should request from its neighbors. To limit the communication overhead due to the requests, we restrict that at each phase, each node u can send at most one request for each block, and u can send at most one request to each of its neighbors.

1. **Random.** One simple heuristic for sending requests is by random. That is, at each phase, each node u repeatedly picks a random block it misses, and sends a request for it to a random unrequested neighbor having that block. This is the heuristic studied in the previous section, and it is observed that Random already achieves very good distribution time.

2. **Rarest First.** Real-life BT systems use the heuristic Rarest First to decide which blocks to request: At each phase, each node u counts the *availability* of each block it misses, where the availability of a block is the number of neighbors of u having that block. Then, starting from the rarest block (i.e., block with smallest availability), u sends requests for each block to a random unrequested neighbor having it.
   The motivation of Rarest First is to maintain balanced availability of each block in the network, so it avoids bad distribution time due to a small number of rare blocks.

3. **Sequential.** When a file is divided into blocks, each block corresponds to a different part of the file. With the heuristic Sequential, at each phase, each node sends requests for blocks sequentially according to their order in the file. The motivation for this heuristic is the ease of programming and it may lead to the good performance similar to that of Random. It may also support streaming of the file, i.e., the node can start using the file while the file is still being downloaded.

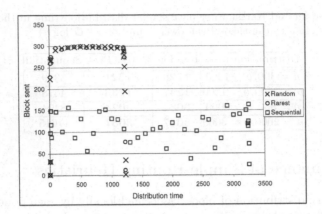

Fig. 3. The number of blocks sent in different phase during the file distribution, for $n = 300$ and $b = 1200$, in the Model UG-5-Rq. With Random and Rarest First, almost 300 blocks are sent per phase, while for Sequential, only 50 - 150 blocks are sent per phase.

Table 3. The distribution time of different heuristics in the Model UG-5-Ph. When a push protocol is used, Random has better distribution time than Rarest First in three out of the four cases (bolded).

	Random	Rarest First	Sequential
$n = 300, b = 1200$	**1354**	1356	8672
$n = 2000, b = 1200$	1386	1373	11365
$n = 300, b = 3000$	**3300**	3372	21594
$n = 2000, b = 3000$	**3386**	3407	28244

We assume that at each phase, each node randomly serves one of the requests it received. We study the performance of the three heuristics in the more realistic Model UG-5-Rq, i.e., undirected graph, at most 5 recevies per phase, and using request protocol. We perform simulations on different combinations of n and b. The results are shown in Table 2.

Random and Rarest First. We observe that the Random and Rarest First heuristics give very similar performance. Rarest First has slightly better distribution time than Random, but both can distribute the file in $b + O(\log n)$ phases.

To understand the reason for their efficiency, we investigate the case of $n = 300$ and $b = 1200$ and we measure the total number of blocks sent by the nodes in each phase throughout the distribution time. The result is shown in Figure 3. We observe that for both Random and Rarest First, in most phases during the distribution time, close to 300 blocks are sent in each phase. It shows that the uploading bandwidth among the nodes is very well utilized. We believe that Random and Rarest First both succeed in keeping the block content of the nodes heterogeneous, i.e., each node has content not similar to that of their neighbors.

Thus, each node can upload some block useful to its neighbors at each phase, and maintain the high utilization of bandwidth.

For the practical concern, Random can be implemented more easily than Rarest First, while maintaining similar performance. We believe that Random is a better choice than Rarest First in practice.

Sequential. Sequential has much worse distribution time than Random and Rarest First. We observe for the case of $n = 300$ and $b = 1200$, the number of blocks sent per phase is between 50 to 150 for most phases during the distribution time. The low usage of bandwidth is due to the fact that many nodes have similar content as their neighbors. Thus, no useful blocks can be uploaded to their neighbors.

3.2   Routing Heuristics for Push Protocols

Models with push protocols have the advantage that no overhead is needed for sending requests. In this subsection, we study the performance of the heuristics Random, Rarest First and Sequential for push protocols.

We first state clearly how the three heuristics are defined for push protocols. At each phase, each node u first randomly selects a neighbor v. Let S be the set of blocks that u has but v does not. Then, the three heuristics perform as follows.

1. **Random.** Send a random block in S to v.
2. **Rarest First.** Send the rarest block in S to v, where availability is measured according to u.
3. **Sequential.** Send the block in S corresponding to the earilest part of the file to v.

We study the performance of the three heuristics in the model UG-5-Push, with different combinations of n and b. The results are shown in Table 3.

We observe that Random and Rarest First have very similar performance in Push Protocols. In fact, in three out of the four cases tested, Random has smaller distribution time than Rarest first. The reason is that with Push protocols, there are send collisions where the same block is sent to a node from multiple neighbors in a phase. Rarest First makes send collisions more common, thus leading to a lost of efficiency. Sequential has even worse performance because send collision becomes very serious in this case.

4   Concluding Remarks

In this paper we extended [1] on modelling the distribution of data blocks in BT systems. We have studied new network features that BT systems are using and different local routing heuristics. Our simulation confirms that $\Omega(b \log n)$ phases are needed for the model in [1] while showing that the random and rarest block first heuristics under more realistic network setting lead to $b + O(\log n)$ total distribution time. An interesting open problem is to provide a mathematical analysis of these heuristics. Other interesting problems include modelling vary client bandwidth, and dynamic issues like new nodes joining, and nodes leaving the system.

References

1. D. Arthur and R. Panigrahy. Analyzing BitTorrent and Related Peer-to-Peer Networks. In *Proceedings of ACM-SIAM Symposium on Discrete Algorithms*, 2006, pages 961–969.
2. A. Bharambe, C. Herley, and V. Padmanabhan. Analyzing and Improving BitTorrent Performance. Technical Report MSR-TR-2005-03, Microsoft Research, 2005.
3. BitTorrent. http://www.bittorrent.com.
4. BitTorrent Accounts for 35% of Traffic. http://yro.slashdot.org/yro/04/11/04/1749257.shtml.
5. BitTorrent Protocol Specification v1.0.
   http://wiki.theory.org/BitTorrentSpecification.
6. C. Gkantsidis and P. Rodriguez. Network Coding for Large Scale Content Distribution. Technical Report MSR-TR-2004-80, Microsoft Research, 2004.
7. B. Cohen. Incentives Building Robustness in BitTorrent.
   http://www.bittorrent.org/bittorrentecon.pdf.
8. M. Izal, G. Urvoy-Keller, E. Biersack, P. Felber, A. Hamra and L. Garces-Erice. Dissecting BitTorrent: Five Months in a Torrent's Lifetime. In *Proceedings of Passive and Active Measurements*, 2004, pages 1–11.
9. J. Pouwelse, P. Garbacki, D. Epema and H. Sips. The BitTorrent P2P File-Sharing System: Measurements and Analysis, 2005, 205–126.
10. D. Qiu and R. Srikant. Modelling and Performance Analysis of BitTorrent-like Peer-top-Peer Networks. In *Proceedings of SIGCOMM*, 2004, pages 367–378.
11. X. Yang and G. de Vecianna. Service Capacity of Peer to Peer Networks. In *Proceedings of INFOCOM*, 2004, pages 2242–2252.

Design of a Fuzzy PI Controller to Guarantee Proportional Delay Differentiation on Web Servers

Ka Ho Chan and Xiaowen Chu

Department of Computer Science
Hong Kong Baptist University
Kowloon Tong
Hong Kong
{khchan, chxw}@comp.hkbu.edu.hk

Abstract. Proportional Integral (PI) controller has attracted research-ers in industrial control processes because of its simplicity and robust performance in a wide range of operating conditions. It has been used to provide proportional delay differentiation on web servers in previ-ous work. However, PI controller cannot achieve satisfactory results due to (1) the web server's non-linearity properties, and (2) the difficulty of building an accurate model for the web server. To address these is-sues, a nonlinear fuzzy PI controller is proposed in this paper, which has the advantage of fuzzy controller while maintaining the simplicity and robustness of PI controller. The proposed controller are self-tuned according to the periodical online performance measurement. The exper-imental results demonstrate that our fuzzy PI controller outperforms the PI controller in several aspects.

Keywords: QoS, Fuzzy PI Controller, Web servers.

1 Introduction

With the wide spread usage of all kinds of web applications, the access rates of popular web sites are growing rapidly. Besides, web servers experience an extreme variation in access demand: sometimes very lightly loaded, sometimes suffered from enormous connection requests. It is not economically feasible to design web servers for peak load, because even well-equipped web servers may still be overloaded. During overload period, not all requests can be served in a timely manner. However, it is possible to provide a better service to premium users. Performance-enhancing mechanisms that can achieve such QoS properties are very important.

Lu et al. have proposed a PI controller to guarantee delays ratio among dif-ferent classes [5]. They regarded the non-linear web server as a second order sys-tem and determined the system parameters by system identification technique. However, PI controller still cannot get satisfactory results on some performance metrics, such as settling time, and oscillation.

M.-Y. Kao and X.-Y. Li (Eds.): AAIM 2007, LNCS 4508, pp. 389–398, 2007.
© Springer-Verlag Berlin Heidelberg 2007

On the contrary, fuzzy controller is independent of accurate models and could be a good selection for providing delays guarantee [9] [11] [12] [13]. But it is hard to tune large amount of parameters. It is especially difficult to make initial approximate adjustment since there is no cookery book to do the job and the performance usually depends on the quality of expert knowledge.

In this paper, we first design a fuzzy PI controller to provide proportional delay differentiation for web servers. It combines the advantages of fuzzy controller and PI controller. Firstly, it shows shorter settling time with less oscillations, under the condition of very busty traffic. At the same time it performs as well as PI controller in steady state. Secondly, parameters determination is convenient. In addition, the simple fuzzy set definition is easy for the parameters adjustment.

The rest of the paper is organized as follows: Section II presents the background and related work. The architecture of the system is described in Section III. Section IV introduces the design of the fuzzy controller. Section V presents the experimental results. Finally, section VI concludes the paper.

2   Background

In this section, we first formally specify the term "delay" studied in this paper, and then compare the service delay guarantees.

2.1   Delays in Web Services

Apache [1] is typically structured as a pool of workers that handle HTTP requests. Our studies use release 1.3.9 in which a worker is simply a process. Our system can also apply to the case that a worker is a thread.

In Apache 1.3.9, when a request arrives, it enters the TCP Accept Queue and waits for a free worker. The number of worker processes is configurable but limited by system resources. If there is an available process, the incoming request can be served immediately. In HTTP/1.0, each request consumes one TCP connection. HTTP/1.1 introduced a new concept called persistent connection [3], which allows reuse of TCP connection for multiple requests from the same clients. Therefore, it is not necessary to establish and terminate TCP connection for every request. However, persistent connection gives rise to a peculiar server bottleneck: when there is no free process, any incoming TCP connection requests must wait until a process becomes available.

Let the *connection delay* denote the time interval between the arrival of connection and the time that he connection is established. Let the *processing delay* denote the time interval between the start of processing a request by web server and the end of the response transmission. The service time of a single connection may be very long for HTTP/1.1 which depends on the user behavior. If the web server is heavily loaded, which is the scenario that we consider in this paper, lots of connection requests are queued in the TCP listening queue. From queuing theory, the connection delay is related to the average queue length, the average service time, and the number of simultaneous connections allowed by the system. Generally, the connection delay is much longer than the processing delay during overload period. We simply use "*delay*" to represent connection delay in this paper.

Providing delay differentiated services to web system is popular in recent years. Admission control and scheduling strategies are used to provide differentiated services. In [7], a heuristic admission control method has been proposed to provide 90th percentile delay guarantee for the premium class in SEDA Web server by additive-increase multiplicative-decrease (AIMD) control mechanism. However, this approach cannot guarantee the QoS for a class and the performance of the algorithm is very sensitive to several control parameters. In [8], an admission control method based on a PI controller is proposed. However, they assume that web servers can be modeled by an M/G/1/PS queuing model. However, the exponential inter-arrival distribution is not accurate to characterize the web traffic [10]. Besides, resource scheduling by feedback control theory has been applied to web system, for example, the PI controller is used for differentiated services in [14]. Furthermore, the queuing theory has been combined to provide differentiated delay services in [5]. But the queuing theory can only work well in a long term scale.

Other than providing differentiated services, feedback control has been used to adjust the KeepAlive and MaxClient of Apache. The approach shows quick convergence and stability. However, it cannot directly address the important metrics for web server, like throughput and response time.

2.2 Semantics of Service Delay Guarantees

Lu et al. have introduced the concept of absolute delay guarantee and relative delay guarantee in [14].

- **Absolute Delay Guarantee**: It guarantees connection delays of specific high priority classes at the expense of longer delay of lower priority classes. [11] and [13] have proposed fuzzy controllers to guarantee absolute delay. However, if the system load grows arbitrarily high, it is impossible to satisfy the desired delays of all classes under overload conditions.
- **Relative Delay Guarantee**: It guarantees connection delay ratios between different classes. For example, if class 0 has a desired relative delay of 1.0, and class 1 has a desired relative delay of 3.0, it would like to guarantee that the connection delay of class 0 is one-third of that of class 1.

It is obvious that services for users of low priority classes cannot be guaranteed by *absolute delay guarantee*. Therefore, connections of low priority users may be reset. However, *relative delay guarantee* can provide fair services for users among different classes, and guarantee better services of high priority classes. Therefore, the fuzzy PI controller is proposed to provide relative delay guarantee.

3 Architecture

Our web server architecture is shown in Fig. 1, which includes three modules: Connection Scheduler, Delay Ratio Monitor, and Fuzzy PI Controller.

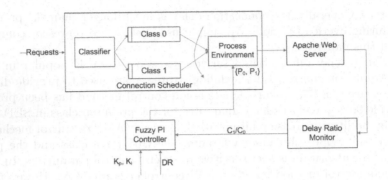

Fig. 1. The System Architecture

- **Connection Scheduler.** It listens to the well-known port and accepts in-coming requests. The Scheduler classifies requests into different classes according to IP addresses. The scheduler maintains a FIFO connection queue and a process counter for each class. A new request is allocated to a free process only if the number of consumed processes for the class is less than its process counter.
- **Delay Ratio Monitor.** It carries out the measurements of the proportional delay ratios experienced by requests of adjacent class k_i and class k_{i+1} periodically.
- **Fuzzy PI Controller.** It guarantees a preset delay ratios between different classes. The delay ratios are guaranteed by assigning suitable number of processes to handle requests from different classes.

At the mth sampling period, the controller computes the desired relative delays, $W_k|0 \leq k < N$, based on the measured delays, $C_k(m)|0 \leq k < N$, gathered by the Delay Ratio Monitor. At the beginning of every sampling period, $p_k|0 \leq k < N$, are re-computed. p_k is used to enforce the number of processes assigned to class k. The major objective of the controller is to keep the delay ratios between adjacent classes to the desired delay ratio (3.0) which we called it set point. The desired delay ratio between class k and $k-1$ is denoted by DR_k.

4   Design of the Proposed Fuzzy PI Controller

Feedback control provides a sound way to keep the delay ratios around the set point. The PI controller is widely used due to its simplicity. As earlier mentioned, PI controller cannot have the satisfactory results on several metrics, i.e., settling time and oscillation, due to nonlinearity of the web servers and the absence of an accurate model. To overcome the limited performance of PI controller, a fuzzy PI controller is proposed in this section. Control gains in the proposed fuzzy PI controller are self-tuned according to nonlinear functions of the inputs when compare with the fixed gains of PI controller. In this section, we briefly describe the mathematical principle of fuzzy PI controller, and then introduce how it

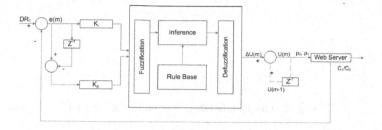

Fig. 2. The structre of the fuzzy PI controller for two classes

self-tines to system uncertainty. We first establish the three main steps in the design of fuzzy PI controller: fuzzification, inference, and defuzzification.

In our paper, two classes (premium and basic) are considered. The basic structure of the proposed fuzzy PI controller for two classes is presented as Fig. 2. There are two inputs for the proposed fuzzy PI controller. One is the product of a system parameter K_i and the delay error $(e(m))$, that is the difference between the set point DR_1 and the measured delay ratio $V(m) = C_1/C_0$. Another input is the product of system parameter K_p and the error rate, $e_v(m)$ which is defined as: $e_v(m) = e(m) - e(m-1)$. We have introduced a parameter $U(m)$ which equals p_0/p_1 to control the number of processes assigned to class 0 and class 1. The value of U, p_0 and p_1 is updated once for a sampling period. Therefore, ΔU, are used to adjust the output of the fuzzy PI controller, U. The two system parameters, K_p and K_i are the conventional proportional and integral gains, which can be obtained as follows:

Firstly, we consider a classic PI controller:

$$U(m) = K_p e(m) + K_i \sum_{n=0}^{m} e(n) \tag{1}$$

where $U(m) = p_0(m)/p_1(m)$, $V(m) = C_1(m)/C_0(m)$ and $e(m) = DR_1 - V(m)$; Then, it is easy to get:

$$\Delta U(m) = K_p(e(m) - e(m-1)) + K_i e(m) \tag{2}$$

Let $e_v(m) = e(m) - e(m-1)$, then the fuzzy PI controller output is:

$$\Delta U(m) = f_{fuzzy}(K_i e(m), K_p e_v(m)) \tag{3}$$

4.1 The Computation of K_p and K_i

To design the proportional and integral gains for the proposed controller, the model for web server should be constructed. Since no accurate model can be used for the web server, we use the model by system identification which has been defined in [14]. Due to space limitation, we only conclude the results here. The model for our specific system is:

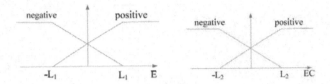

Fig. 3. The membership function of E and EC

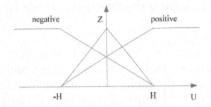

Fig. 4. The membership function of U

$$V(m) = 0.34V(m-1) - 0.08V(m-2) + 0.51U(m-1) + 0.41U(m-2) \quad (4)$$

where the coefficients are obtained by the least-squares estimator. The PI controller for our system used as in [14] is:

$$U(m) = U(m-1) + g(E(m) - rE(m-1)) \quad (5)$$

Combining (1) and (5), the proportional and integral gains as following:

$$K_p = g \cdot r, K_i = g - g \cdot r \quad (6)$$

Based on above model, the root locus method can be used to get the value of g and r. For our system, $g = 0.268$ and $r = 0.05$ are suitable.

4.2   The Proposed Fuzzy PI Controller

In this section, we describe the three steps of the fuzzy controller.

- **Fuzzification.** The membership functions for input and output were defined as Fig. 3 and Fig. 4. The linguistic variables for two numeric inputs and one output are E, EC and U respectively. They are used to handle uncertainties in computer systems.

  In Fig. 3, L_1 and L_2 are defined as the maximum value of $K_i e(m)$ and $K_p e_v(m)$. In our experiment, [-12.2, 12.2] is large enough to observe delay ratio error and the error rate. Therefore, we have:

  $$L_1 = 12.2K_i, \text{ and } L_2 = 12.2K_p$$

Note that $e(m)$ can be arbitrarily large due to busty traffic or the initial state. When this happen, the controller behaves like a classic PI controller according to the value of $e_v(m)$. In the membership function of output, H is set as 4. Numerous experiments verify that this setting is reasonable and acceptable.

- **Inference**: The actions of the fuzzy controller are guided by a set of IF-THEN rules. The rules established for our controller is relatively easy which can be concluded as following: *Rule 1*: If E is negative and EC is negative THEN U is negative;
Rule 2: If E is negative and EC is positive THEN U is zero;
Rule 3: If E is positive and EC is negative THEN U is zero;
Rule 4: If E is positive and EC is positive THEN U is positive;
For each rule, the IF part considers a position of current systems and THEN part indicates how the process ratio should be changed. Negative (decrease) and positive (increase) are the linguistic variables. In short, Rule 1 and 4 consider situations that the E and EC are of the same sign, so the output should be large to make the process quick converge. Conversely, Rule 2 and 3 describe situations that E and EC are of the opposite sign, so the output should be small to reduce overshoots and settling time.
- **Defuzzification**: To defuzzify the incremental control of fuzzy control law, the membership function for output is

$$U = \frac{\sum b_i \cdot \mu_i}{\sum \mu_i} \qquad (7)$$

where b_i is the output for Rule i and μ_i is the corresponding membership value for Rule i. μ_i is computed by the minimum value, i.e., $\mu_i = \mu_m(E) \wedge \mu_n(EC)$, where $\mu(M)$ is the membership degree of linguistic variable "M" for the linguistic value "m". In our controller, m can be "positive" or "negative".

According to equation 2 and 7, the control gains K_i and K_p are self-tuning. This is also the reason why the fuzzy PI controller performs better than the PI controller with fixed gains.

Finally, we remark that the overhead of this algorithm is small since only multiplications and additions are found in the algorithm. Furthermore, the controller only needs to adjust the process assignment once a sampling period.

5   Experiments

We have modified the source code of Apache 1.3.9 web server which runs on a Linux platform [2] to implement the adaptive architecture and fuzzy PI control algorithm. The sampling period was set to 30 seconds. The timeout of HTTP/1.1 connection was set to default value (15 seconds). The controller computes the number of processes assigned to each class.

All experiments were conducted on a test-bed includes a server with a 2.8GHz Pentium processor and 1 GB RAM running Linux-2.6.12 and a set of clients running

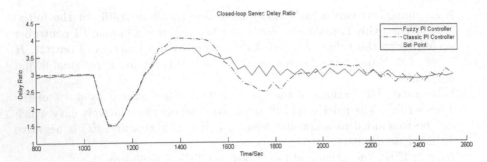

Fig. 5. The delay ratio of the PI controller and the fuzzy PI controller

Linux-2.4.18. The server and the clients were connected by a 1-Gbps Ethernet. The Surge workload generator [4] was used to generate web traffic. The percentage of base, embedded, and loner objects were 30%, 38%, and 32%, respectively.

5.1   Experiment 1

- **Server**: The total number of processes was configured to 128. It is a constant number throughout the experiment. The numbers of processes were initially the same which is 64 for both class 0 (premium) and class 1 (basic).
- **Client**: 300 class 0 and 200 class 1 clients are simulated at the beginning. At 1000th second, 200 new class 0 and 200 new class 1 clients are simulated.

Fig. 5 shows the behaviors of the PI controller and the fuzzy PI controller. The delay ratio of both controllers deviate from the set point seriously after load changing. After shorter period of time, the fuzzy PI controller reacted and re-converged to the set point at around 1700th second. However, it takes longer time for the PI controller to converge to the set point (at around 2200th second). The PI controller also shows more violent oscillations. In order to quantify the oscillation, the relative difference between the measured and desired delay from the initial state to the time when the server reached the steady state has been defined as following.

$$E_t = \sum_{i=1}^{n} T_i \cdot \mid Error_i \mid \tag{8}$$

where T_i represents the sampling point i and $Error_i$ represents the error between the delay ratio and the set point. Smaller E_t means better controller performance. We also define another metric $Diff$ to measure the performance improvement on oscillations over the PI controller.

Table 1. E_t of Classic PI controller and Fuzzy PI controller

	PI Controller	Fuzzy PI Controller
E_t	366.1	266.6

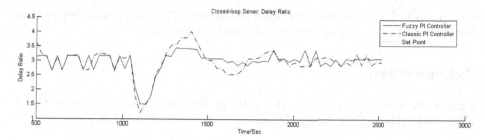

Fig. 6. The delay ratio of the PI controller and the fuzzy PI controller

Table 2. E_t of Classic PI controller and Fuzzy PI controller

	PI Controller	Fuzzy PI Controller
E_t	280.8	183.6

$$Diff = \frac{E_t(PI) - E_t(FuzzyPI)}{E_t(PI)} \qquad (9)$$

Table 1 shows the E_t of the fuzzy PI controller is much smaller, and the fuzzy PI controller improved the performance by $Diff = 27.2\%$.

5.2   Experiment 2

This experiment demonstrated the fuzzy PI controller performed much better than the PI controller under large load disturbance.

- **Server**: The server is configured as in experiment 1.
- **Client**: 200 class 0 and 300 class 1 clients are simulated at the beginning. At 1000th second, 200 new class 0 and 200 new class 1 clients are simulated.

The experimental results are shown in Fig. 6. At 1000th second, the delay ratios of both controllers decreased to a trough value about 1.5. The delay of class 1 users increased, since number of class 1 users increased in a sudden. In such circumstance, the delay ratio drops to a very small value. After 200 seconds, both controllers react and the delay ratios rise. The delay ratio of fuzzy PI controller converged at around 1550th seconds, and some low amplitude oscillations around the set point after 1550th seconds. However, the delay ratio of PI controller converged at 2000th seconds, and there are some oscillations around the set point after the convergent time.

Table 2 shows E_t of PI controller is also much larger than the fuzzy PI controller due to its large oscillation and long settling time. The fuzzy PI controller improved performance by $Diff = 34.6\%$.

6   Conclusions and Future Works

In this paper, we present a fuzzy PI controller which combines the advantages of classic PI controller and fuzzy controller. The self-tuning property makes it more robust. The experimental results demonstrate that it performs better than the classic PI controller on settling time and oscillations.

Acknowledgement

This work was supported by Hong Kong Baptist University under Grant FRG/05-06/II-34, Hong Kong, China.

References

1. Apache Software Foundation. http://www.apache.org/
2. Fedora Core 4 – Red Hat Linux. http://www.redhat.com/en_us/USA/fedora/
3. R. fielding, J. Gettys, J.Mogul, H. Frystyk, L. Masinter, P. Leach, and T. Berners-Lee: Hypertext Transfer Protocol-HTTP/1.1. IETF RFC 2616, June 1999.
4. P. Barford and M. Crovella: Generating Representative Web Workloads for Network and Server Performance Evaluation. In Proceedings ACM SIGMETRICS'98, Madison WI, 1998.
5. Y. Lu, T. F. Abdelzaher, C. Lu, L. Sha, and X. Liu: Feedback Control with Queuing-theoretic Prediction for Relative Delay Guarantees in Web Server. In Proceedings of IEEE Real-Time and Embedded Technology and Applications Symposium, pp.208-217, 2003.
6. Y. Diao, X. Lui, S. Froehlich, J. L. Hellerstein, S. Parekh, and L. Sha: On-line Response Time Optimization of an Apache Web Server. In Proceedings of International Workshop on Quality of Service, Monterey, Ca, June 2003.
7. M. Welsh and D. Culler: Adaptive Overload Control for Busy Internet Servers. In Proceedings of USITS, 2003.
8. A. Kamra, V. Misra, and E. Nahum: Yaksha: A Controller for Managing the Performance of 3-tiered Websites. In Proceedings of the Twelfth IEEE Workshop on Quality of Service (IWQoS 2004), Montreal, Canada, June, 2004.
9. B.-G. Hu, G.K.I. Mann, and R.G. Gosine: A Systematic Study of Fuzzy PID Controllers - Function-based Evaluation Approach. IEEE Transactions on Fuzzy Systems, Vol.9(5):699-712, 2001.
10. V. Paxson and S. Floyd: Wide Area Traffic: the Failure of Poisson Modeling. IEEE/ACM Transactions on Networking, 3(3):226-244, June 1995.
11. Jianbin Wei, and Cheng-Zhong Xu: A Self-tuning Fuzzy Control Approach for End-to-End QoS Guarantees in Web Servers. In Proceedings of IWQoS 2005, Lecture Notes in Computer Science, Volume 3552, pp: 123-135.
12. P. Pivonka: Comparative Analysis of Fuzzy PI/PD/PID Controller Based on Classical PID Controller Approach. In Proceedings of the 2002 IEEE World Congress on Computational Intelligence, USA, 2002, pp.541-546.
13. Y. Wei, C. Lin, X.-W. Chu, and T. Voigt: Fuzzy Control for Guaranteeing Absolute Delays in Web Servers. International Journal of High Performance Computing and Networking, to appear.
14. C. Lu, Y. Lu, T. F. Abdelzaher, J. A. Stankovic, and S. H. Son: Feedback Control Architecture and Design Methodology for Service Delay Guarantees in Web Servers. IEEE Transactions on Parallel and Distributed Systems 2006, vol. 17, no. 9, pp.1014-1027.

Improved Approximation Algorithms for Predicting RNA Secondary Structures with Arbitrary Pseudoknots

Minghui Jiang

Department of Computer Science, Utah State University, Logan, Utah 84322-4205, USA
mjiang@cc.usu.edu

Abstract. We study three closely related problems motivated by the prediction of RNA secondary structures with arbitrary pseudoknots: the MAXIMUM BASE PAIR STACKINGS problem proposed by Ieong et al. [19], the MAXIMUM STACKING BASE PAIRS problem proposed by Lyngsø [22], and the 2-INTERVAL PATTERN problem proposed by Vialette [36]. For MAXIMUM BASE PAIR STACKINGS and MAXIMUM STACKING BASE PAIRS, we present improved approximation algorithms that can incorporate covariance information from comparative analysis as explicit input of candidate base pairs. For 2-INTERVAL PATTERN, we present improved approximation algorithms on unitary and near-unitary input, and propose a new variant called LENGTH-WEIGHTED BALANCED 2-INTERVAL PATTERN, which is natural in the nearest-neighbor energy model that emphasizes base pair stacking.

1 Introduction

RNAs are versatile molecules: messenger RNAs carry genetic information and act as the intermediary agent between DNAs and proteins; ribosomal RNAs, transfer RNAs, and other non-coding RNAs play important structural, regulatory, and catalytic roles in cells. To understand fully the various functions of RNAs, we need to first understand their structures. The *primary structure* of an RNA is the sequence of nucleotides (that is, the four different bases A, C, G, and U) in its single-stranded polymer. An RNA folds into a three-dimensional structure by forming hydrogen bonds between pairs of complementary bases, such as the Watson-Crick pairs G-C and A-U and the wobble pair G-U, that are nonconsecutive in the sequence. The three-dimensional arrangement of the atoms in the folded RNA molecule is its *tertiary structure*; the collection of base pairs in the tertiary structure is the *secondary structure*. The secondary structure of an RNA is the scaffold of its tertiary structure; accurate prediction of secondary structures is a prerequisite for the more detailed structural analysis of RNAs.

Most early research on RNA secondary structure prediction adopts the thermodynamic approach: each possible secondary structure has a global free energy depending on the recursive decomposition of the structure and on a set of experimentally determined energy parameters; the optimal secondary structure corresponds to the minimum global free energy. When pseudoknots are excluded, the optimal secondary structure of an RNA can be computed by dynamic programming algorithms in $O(n^3)$ time and $O(n^2)$ space [27,39,38,31,24]. These algorithms are the basis of popular software packages for RNA secondary structure prediction such as mfold [25,37] and Vienna

M.-Y. Kao and X.-Y. Li (Eds.): AAIM 2007, LNCS 4508, pp. 399–410, 2007.

[17,16]. There has been considerable effort [29,35,2] on extending the dynamic programming algorithms to include pseudoknots. However, these extended algorithms typically have very high complexities, ranging from $O(n^4)$ to $O(n^6)$ in time and from $O(n^3)$ to $O(n^4)$ in space, which make them impractical even for RNA sequences of only a few hundred bases. Furthermore, as noted by Lyngsø and Pedersen [23] and also by Ieong et al. [19], these algorithms can handle only limited types of pseudoknots: the exact types are implicit in the algorithms and difficult to determine. On the other hand, if arbitrary pseudoknots may be included, then the prediction problem becomes exceedingly difficult, typically NP-hard or even APX-hard [2,23,19,22,4,36,9]. The apparent difficulty of the pseudoknot prediction problem naturally prompts researchers to explore heuristic approaches such as quasi-Monte-Carlo simulation [1], genetic algorithms [14,5,32], and, very recently, simulated annealing with pull moves on a 3D triangular lattice [21]. However, the lack of theoretical guarantees of these heuristics makes it impossible to bound how far a given prediction is from the optimal solution.

Hybrid methods for pseudoknot prediction, such as maximum weighted matching [10,33] and iterated loop matching [30], are based on a combination of thermodynamic and comparative [28] approaches. These methods can often achieve better prediction accuracies than algorithms based on the thermodynamic approach alone because, beside the RNA sequence, they also take as input a covariance matrix with one entry for each pair of bases in the sequence. The covariance matrix can be obtained from a multiple alignment of the input RNA sequence and its homologous sequences; it contains additional evolutionary evidence for the likelihood of individual base pairs in the RNA secondary structure that cannot be captured by the few energy parameters in the thermodynamic approach. The maximum weighted matching method [10,33], for example, then reduces the RNA secondary structure prediction problem to a graph-theoretical problem by interpreting the covariance matrix as the adjacency matrix of a weighted graph: each RNA base is a vertex, and each pair of bases is an edge weighted by the pair covariance.

One shortcoming of the maximum weighted matching method [10,33] is that it implicitly adopts an energy model in which the energy of each base pair is considered independently. In the more realistic nearest-neighbor energy model [2,23,19,22], the energy of each base pair depends not only on its two bases but also on the other adjacent base pairs. According to the Tinoco model [34], an RNA structure can be recursively decomposed into loops with independent free energy; the energy of each loop is an affine function in the number of unpaired bases and the number of interior base pairs. The only type of loops without unpaired bases are formed by base pair stacking; such loops have negative energy and stabilize the RNA structure. We next review some concepts related to base pair stacking.

Let S be an RNA sequence. Denote by $S[i]$ the i'th base of S and by $S[i,j]$ the subsequence of bases $S[i], S[i+1], \ldots, S[j]$. A *base pair* (i,j) of S is a pair of bases $S[i]$ and $S[j]$ such that $j \geq i+2$ (the two bases must be nonconsecutive). A *pseudoknot* is composed of two interleaving base pairs (i,j) and (k,l) with $i < k < j < l$. A *stacking loop* is a loop of four bases formed by two adjacent base pairs (i,j) and $(i+1, j-1)$. A set of m base pairs $\mathcal{S} = \{(i_1, j_1), \ldots, (i_m, j_m)\}$ is a valid *secondary structure* of S if the $2m$ indices i_k and j_k are all distinct, that is, each base participates

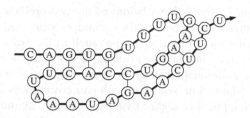

Fig. 1. Pseudoknot PKb of the upstream pseudoknot domain (UPD) of the 3'-UTR of beet soilborne virus RNA 1. The sequence of bases follows the thick curve; the base pairs are connected by thin lines.

in at most one base pair. For a secondary structure S, the number of *base pair stackings* is defined as

$$BPS(S) = \left|\{(i, j) \in S \mid (i + 1, j - 1) \in S\}\right|,$$

and the number of *stacking base pairs* is defined as

$$SBP(S) = \left|\{(i, j) \in S \mid (i + 1, j - 1) \in S \vee (i - 1, j + 1) \in S\}\right|.$$

Note that the number of base pair stackings is exactly the number of stacking loops in a secondary structure. Denote by $([i, i + q], [j, j - q])$ a *helix* of q consecutive stacking loops formed by $q + 1$ base pairs (i, j), $(i + 1, j - 1)$, ..., $(i + q, j - q)$. Given a maximal helix $\mathcal{H}$ of m base pairs, we have $BPS(\mathcal{H}) = m - 1$ for $m \geq 1$, $SBP(\mathcal{H}) = m$ for $m > 1$, and $SBP(\mathcal{H}) = 0$ for $m = 1$. For example, in the pseudoknotted secondary structure[1] shown in Figure 1, the number of base pairs is 10, the number of base pair stackings is 7, and the number of stacking base pairs is 9.

Ieong et al. [19] formulated the RNA secondary structure prediction problem as an optimization problem: given an RNA sequence, the MAXIMUM BASE PAIR STACK-INGS problem is to find a secondary structure with the maximum number of stacking loops (base pair stackings). Ieong et al. [19] showed that MAXIMUM BASE PAIR STACKINGS is NP-hard when the secondary structure is restricted to be planar. Their construction uses only the Watson-Crick base pairs A-U and C-G. Later, Lyngsø [22] demonstrated that MAXIMUM BASE PAIR STACKINGS is still NP-hard without the planar restriction, even for binary sequences with 0-1 base pairs. Lyngsø also showed that the related MAXIMUM STACKING BASE PAIRS problem, which is to find a secondary structure with the maximum number of stacking base pairs, is also NP-hard when the input sequence is over an unbounded alphabet Σ and the legal pair types form a subset of $\Sigma \times \Sigma$. Several algorithms have been proposed for the two problems. For MAXIMUM BASE PAIR STACKINGS over the canonical {A, C, G, U} alphabet with the Watson-Crick base pairs, Ieong et al. [19] presented an $O(n^3)$ time $O(n^2)$ space 2 approximation for the planar case and an $O(n)$ time 3 approximation for the general case. Lyngsø [22] presented a polynomial time exact algorithm for MAXIMUM BASE PAIR STACKINGS and a polynomial time approximation scheme for MAXIMUM STACKING BASE PAIRS, both over a fixed-size alphabet Σ with a subset $\mathcal{B} \subseteq \Sigma \times \Sigma$ of legal pair types.

---

[1] From the PseudoBase [6] at http://wwwbio.leidenuniv.nl/~batenburg/PKBase/PKB00116.html

In Ieong et al.'s and Lyngsø's formulations of the two problems [19,22], the set of possible base pairs that might appear in the secondary structures are determined implicitly by a prespecified set of legal pair types such as the Watson-Crick base pairs. Instead, the candidate base pairs may be given explicitly as input. For example, we can take advantage of the additional information from comparative analysis [28,10,33,30] by composing a candidate set of base pairs with pair covariances at least some threshold value. On explicit input of candidate base pairs, the two optimization problems are clearly NP-hard, since they generalize the previous formulations [19,22]. It is natural that we investigate their approximation algorithms. Lyngsø's algorithms [22], beside having very high complexities of $\Omega(n^{81})$ in time and $\Omega(n^{80})$ in space even for the canonical {A, C, G, U} alphabet, depend on a look-up table technique assuming that legal base pairs are implicitly determined by the pair types; therefore they cannot be adapted to explicit input of candidate base pairs. On the other hand, Ieong et al.'s 3 approximation for MAXIMUM BASE PAIR STACKINGS [19] can be easily adapted to input of candidate base pairs. In this paper, we improve Ieong et al.'s 3 approximation for MAXIMUM BASE PAIR STACKINGS to a 8/3 approximation, and present the first non-trivial 5/2 approximation for MAXIMUM STACKING BASE PAIRS, both on explicit input of candidate base pairs.

Vialette [36] proposed a geometric representation of the RNA secondary structure as a set of 2-intervals. Given a single-stranded RNA molecule, a subsequence of consecutive bases of the molecule can be represented as an interval on a single line, and a possible (stacked) pairing of two disjoint subsequences can be represented as a 2-interval, which is the union of two disjoint intervals. Given a candidate set of 2-intervals, a maximum pairwise-disjoint subset restricted to certain prespecified geometrical constraints gives a macroscopic approximation of the RNA secondary structure.

We review some definitions [36]. A 2-interval $D = (I, J)$ consists of two disjoint (closed) intervals I and J such that $I < J$, that is, I is completely to the left of J. Consider two 2-intervals $D_1 = (I_1, J_1)$ and $D_2 = (I_2, J_2)$. D_1 and D_2 are *disjoint* if the four intervals I_1, J_1, I_2, and J_2 are pairwise disjoint. Define three binary relations for disjoint pairs of 2-intervals:

Preceding: $D_1 < D_2 \iff I_1 < J_1 < I_2 < J_2$.
Nesting: $D_1 \sqsubset D_2 \iff I_2 < I_1 < J_1 < J_2$.
Crossing: $D_1 \between D_2 \iff I_1 < I_2 < J_1 < J_2$.

The two 2-intervals D_1 and D_2 are *R-comparable* for some $R \in \{<, \sqsubset, \between\}$ if either $(D_1, D_2) \in R$ or $(D_2, D_1) \in R$. (For example, D_1 and D_2 are $\sqsubset$-comparable if either $D_1 \sqsubset D_2$ or $D_2 \sqsubset D_1$.) Note that the set of binary relations $\{<, \sqsubset, \between\}$ is complete in the sense that any two disjoint 2-intervals are R-comparable for some $R \in \{<, \sqsubset, \between\}$. Given a *model* $\mathcal{R}$, which is a non-empty subset of $\{<, \sqsubset, \between\}$ (there are 7 such subsets), a set $\mathcal{D}$ of 2-intervals is $\mathcal{R}$-*structured* if any two distinct 2-intervals in $\mathcal{D}$ are R-comparable for some $R \in \mathcal{R}$. Given a set $\mathcal{D}$ of 2-intervals and a model $\mathcal{R}$, the 2-INTERVAL PATTERN problem is to find a maximum-size $\mathcal{R}$-structured subset of 2-intervals in $\mathcal{D}$.

Beside the various models $\mathcal{R}$, various restrictions can also be imposed on the input 2-interval set $\mathcal{D}$ for the 2-INTERVAL PATTERN problem. Define the *support* of a set $\mathcal{D}$

of 2-intervals, Support($\mathcal{D}$), as the set of intervals $\{I, J \mid (I, J) \in \mathcal{D}\}$. There are four common types of restrictions:

Unlimited: No restrictions.
Balanced: Every 2-interval in $\mathcal{D}$ consists of two intervals of equal length.
Unitary: Every interval in the support of $\mathcal{D}$ has a unit length.
Point: The intervals in the support of $\mathcal{D}$ are pairwise disjoint (therefore they can be considered as intervals of unit length, or, points).

The three types of restrictions, unlimited, unitary, and point, were originally introduced by Vialette [36]; the balanced restriction was later proposed by Crochemore et al. [12] because it is natural in the biological setting. In particular, we note that a helix ($[i, i + q], [j, j-q]$) of q consecutive stacking loops can be represented by a balanced 2-interval.

Table 1. (a) The complexities of the 2-INTERVAL PATTERN problem. $\mathcal{L} = \Theta(n^2)$ and $d = \Theta(n)$ in the worst case [11]. (b) The best approximation ratios for the 2-INTERVAL PATTERN problem. Our improvements are marked by "old → new".

	Unlimited	Balanced	Unitary	Point
$\{<, \sqsubset, \lozenge\}$	APX-hard [4]			$O(n\sqrt{n})$ [26]
$\{\sqsubset, \lozenge\}$	APX-hard [4]			$O(n \log n + \mathcal{L})$ [11]
$\{<, \lozenge\}$	NP-complete [9]			complexity unknown
$\{<, \sqsubset\}$	$O(n \log n + dn)$ [11]			
$\{\lozenge\}$	$O(n \log n + \mathcal{L})$ [11]			
$\{\sqsubset\}$	$O(n \log n)$ [9]			
$\{<\}$	$O(n \log n)$ [36]			

(a)

	Unlimited	Balanced	Unitary	Point
$\{<, \sqsubset, \lozenge\}$	4 [4]	4 [12]	3 [4] → 2 + ϵ	N/A
$\{\sqsubset, \lozenge\}$	4 [12]	4 [12]	3 [12] → 2 + ϵ	N/A
$\{<, \lozenge\}$	2 [20]	2 [20]	2 [20]	2 [12]

(b)

Since Vialette's pioneering work [36], the 2-INTERVAL PATTERN problem has been extensively studied. We summarize the complexities of the problem over its various models and restrictions in Table 1(a). Because the $\lozenge$ relation directly models the pseudo-knots in RNA secondary structures, it is not surprising that the 2-INTERVAL PATTERN problem is NP-hard or even APX-hard over the three models $\{<, \sqsubset, \lozenge\}$, $\{\sqsubset, \lozenge\}$, and $\{<, \lozenge\}$; these results [4,9] are compatible with the hardness results for the other models [2,23,19,22] and are consistent with our knowledge that RNA secondary structures with pseudoknots are difficult to predict in practice. Naturally, researchers have directed their attention to the design of efficient approximation algorithms. We refer to Table 1(b) for the best approximation ratios of polynomial time approximation algorithms for the 2-INTERVAL PATTERN problem. In this paper, we present improved approximation algorithms for the 2-INTERVAL PATTERN problem on unitary input (as marked in Table 1(b)) and on *near-unitary* input (to be introduced later).

The rest of the paper is organized as follows. In Section 2, we prove a lemma for 2-interval graphs. In Section 3, we present an improved approximation algorithm for the MAXIMUM BASE PAIR STACKINGS problem and the first non-trivial approximation algorithm for the MAXIMUM STACKING BASE PAIRS problem, on explicit input of candidate base pairs. In Section 4, we present improved approximation algorithms for the 2-INTERVAL PATTERN problem. In Section 5, we conclude with discussion.

2   A Graph-Theoretic Lemma for 2-Intervals

Let $\mathcal{D}$ be a set of 2-intervals. Without loss of generality, assume that each interval in Support($\mathcal{D}$) is a closed segment $[u, v]$ between two integer coordinates u and v. Define the *length* of an interval as the number of integer coordinates (which correspond to the individual bases) it contains: the length of $[u, v]$ is therefore $v - u + 1$. Denote by $\ell_{max}(\mathcal{D})$ and $\ell_{min}(\mathcal{D})$, respectively, the lengths of the longest and the shortest intervals in Support($\mathcal{D}$). Define the *2-interval graph* $G(\mathcal{D})$ as the undirected graph with a vertex for each 2-interval in $\mathcal{D}$ and with an edge between a pair of vertices if and only if the corresponding 2-intervals are not disjoint. In an undirected graph, a *d-claw* C is an induced subgraph $K_{1,d}$ that consists of an independent set T_C of d vertices, called *talons*, and a *center vertex* z_C that is connected to all the talons. A graph is *d-claw-free* if it has no d-claws. We observe an important property of 2-interval graphs:

Lemma 1. *For a set $\mathcal{D}$ of 2-intervals with $\ell_{min}(\mathcal{D}) = a$ and $\ell_{max}(\mathcal{D}) = b$, the 2-interval graph $G(\mathcal{D})$ is $(5 + 2\lfloor \frac{b-2}{a} \rfloor)$-claw-free.*

Proof. Let I be an interval in Support($\mathcal{D}$) and let $\mathcal{I}_I \subseteq$ Support($\mathcal{D}$) be a set of disjoint intervals that intersect I. All intervals in $\mathcal{I}_I$ are completely contained in I except the leftmost one and the rightmost one, which occupy at least two integer coordinates in I. Therefore the total number of intervals in $\mathcal{I}_I$ is at most $2 + \lfloor \frac{b-2}{a} \rfloor$. It follows that each 2-interval in $\mathcal{D}$ has at most $4 + 2\lfloor \frac{b-2}{a} \rfloor$ independent neighbors in the 2-interval graph $G(\mathcal{D})$.   □

3   Approximation Algorithms for MAXIMUM BASE PAIR STACKINGS and MAXIMUM STACKING BASE PAIRS

We first present a 8/3 approximation algorithm for MAXIMUM BASE PAIR STACKINGS. For an input sequence S of n bases and a candidate set $\mathcal{C}$ of m base pairs, let $\mathcal{S}$ be the set of stacking loops output by the algorithm. Initially, $\mathcal{S}$ is empty, and all bases in S are unmarked. Our algorithm consists four steps:

1. Repeatedly find the *leftmost* 5 consecutive stacking loops (that is, find the 2-interval $([u, u + 5], [v - 5, v])$ such that u is as small as possible) formed by base pairs in $\mathcal{C}$ with unmarked bases in S, add these stacking loops to $\mathcal{S}$, then mark all their bases.
2. Repeatedly find *any* 4 consecutive stacking loops formed by base pairs with unmarked bases, add them to $\mathcal{S}$, then mark all their bases.
3. Repeatedly find *any* 3 consecutive stacking loops formed by base pairs with unmarked bases, add them to $\mathcal{S}$, then mark all their bases.

4. For each single stacking loop or two consecutive stacking loops formed by base pairs with unmarked bases, add the corresponding 2-interval to $\mathcal{D}$. Construct the 2-interval graph $G(\mathcal{D})$ and assign each vertex a weight: 1 for a single stacking loop, and 2 for two consecutive stacking loops. Find a maximum weight independent set $\mathcal{I}$ in $G(\mathcal{D})$ using Berman's $5/2$ approximation algorithm [7] for MAXIMUM WEIGHT INDEPENDENT SET in 5-claw-free graphs. For each 2-interval in $\mathcal{I}$, add the corresponding stacking loops to $\mathcal{S}$.

We show that our algorithm indeed achieves a $8/3$ approximation for MAXIMUM BASE PAIR STACKINGS. The first three steps of our algorithm are similar to the first two steps of Ieong et al.'s *GreedySP* algorithm [19]: our first step is identical to Ieong et al.'s first step with parameter $i = 5$; our second and third steps are Ieong et al.'s second step with parameters $k = 4$ and $k = 3$. For completeness of exposition, we nevertheless incorporate a sketch of their analysis in our analysis in the following.

Let s_1, s_2, s_3, and s_4, respectively, be the numbers of stacking loops found by the first, second, third, and fourth steps of our algorithm. Let $\mathcal{S}^*$ be the set of stacking loops in an optimal secondary structure. Let s_1^*, s_2^*, and s_3^*, respectively, be the numbers of stacking loops in $\mathcal{S}^*$ that intersect the stacking loops found by the first, second, and third steps of our algorithm. Let s_4^* be the number of remaining stacking loops in $\mathcal{S}^*$, which are represented by 2-intervals in $\mathcal{D}$. We have $|\mathcal{S}| = s_1 + s_2 + s_3 + s_4$ and $|\mathcal{S}^*| = s_1^* + s_2^* + s_3^* + s_4^*$.

For each k consecutive stacking loops D found by the first three steps of our algorithm, it is clear that the number of stacking loops in $\mathcal{S}^*$ that intersect them is at most $2(k + 2)$, with $k + 2$ for each interval of the 2-interval D. By always choosing the *leftmost* 5 consecutive stacking loops D_5 in the first step, we can guarantee that the left interval of the 2-interval D_5 intersects at most $5 + 1$ stacking loops in $\mathcal{S}^*$. Suppose the contrary that the left interval of D_5 intersects 7 stacking loops in $\mathcal{S}^*$, then these 7 stacking loops must be consecutive, and the leftmost 5 of these stacking loops should have been chosen instead of D_5. We therefore have the following inequality:

$$\frac{s_1^*}{s_1} \leq \frac{5+1+5+2}{5} = 13/5 = 2.6. \tag{1}$$

With all 5 consecutive stacking loops found by the first step, we can guarantee that each interval of a 2-interval D_4 (consisting of 4 consecutive stacking loops) found by the second step of our algorithm intersects at most $4 + 1$ stacking loops in $\mathcal{S}^*$. Suppose the contrary that an interval of D_4 intersects 6 stacking loops in $\mathcal{S}^*$, then these 6 stacking loops must be consecutive and, consequently, must contain 5 consecutive stacking loops. This is a contradiction. We have the following inequality:

$$\frac{s_2^*}{s_2} \leq \frac{4+1+4+1}{4} = 10/4 = 2.5. \tag{2}$$

A similar analysis shows the following inequality for the third step:

$$\frac{s_3^*}{s_3} \leq \frac{3+1+3+1}{3} = 8/3 \approx 2.67. \tag{3}$$

In the fourth step, each 2-interval in $\mathcal{D}$ is balanced and corresponds to either a single stacking loop with interval length 2, or two consecutive stacking loops with interval length 3. Therefore we have $\ell_{\min}(\mathcal{D}) = 2$ and $\ell_{\max}(\mathcal{D}) = 3$. It follows from Lemma 1 that the 2-interval graph $G(\mathcal{D})$ is 5-claw-free. Berman's $5/2$ approximation algorithm [7] for MAXIMUM WEIGHT INDEPENDENT SET in 5-claw-free graphs guarantees that

$$\frac{s_4^*}{s_4} \le 5/2 = 2.5. \tag{4}$$

Combining inequalities (1), (2), (3), and (4), we have

$$\frac{|\mathcal{S}^*|}{|\mathcal{S}|} = \frac{s_1^* + s_2^* + s_3^* + s_4^*}{s_1 + s_2 + s_3 + s_4} \le 8/3.$$

We give an analysis of the complexities of our algorithm. Using an adjacency matrix representation of the candidate set of base pairs, the first three steps of our algorithm can be implemented in $O(n^2)$ time and space, which are optimal in the worst case. The fourth step of our algorithm is the dominating step. The 2-interval graph $G(\mathcal{D})$ has at most $O(m)$ vertices and $O(m^2)$ edges; the construction of the graph takes $O(n^2 + m^2)$ time and space. Berman's algorithm [7], in general, achieves only a $d/2 + \epsilon$ approximation for MAXIMUM WEIGHT INDEPENDENT SET in d-claw-free graphs: when the weights of the vertices are super-polynomial in the number of vertices, a rescaling procedure is necessary to ensure a polynomial running time at the price of an additional ϵ in the approximation ratio. However, we note that each of the $O(m)$ vertices in our 2-interval graph $G(\mathcal{D})$ has an integer weight of either 1 or 2. This implies that the number of iterations of Berman's local-improvement algorithm is at most $O(m)$. For our application, the rescaling procedure is therefore unnecessary and a 2.5 approximation can be obtained. Each of the $O(m)$ iterations of Berman's SQUAREIMP algorithm [7] on $G(\mathcal{D})$ runs in $O(m^4)$ time to find an improving 4-claw. The overall complexities of our algorithm are therefore $O(n^2 + m^2 + T_w(m))$ in time and $O(n^2 + m^2)$ in space, where $T_w(m) = O(m^5)$.

We next present a $5/2$ approximation algorithm for MAXIMUM STACKING BASE PAIRS. This algorithm is almost identical to our algorithm for MAXIMUM BASE PAIR STACKINGS except that the first three steps are omitted and the fourth step is modified to assign a weight of 2 for a single stacking loop and a weight of 3 for two consecutive stacking loops. A crucial observation here, as noted by Lyngsø [22], is that we only need to consider balanced 2-intervals of interval length (the number of stacking base pairs) either 2 or 3, since every balanced 2-interval of interval length more than 3 can be decomposed into several balanced 2-intervals of interval length either 2 or 3 with the same total weight. We have the following theorem:

Theorem 1. *Given a sequence of n bases and a candidate set of $m = O(n^2)$ base pairs,* MAXIMUM BASE PAIR STACKINGS *can be approximated with a ratio of $8/3$ and* MAXIMUM STACKING BASE PAIRS *can be approximated with a ratio of $5/2$ in $O(n^2 + m^2 + T_w(m))$ time and $O(n^2 + m^2)$ space, where $T_w(m) = O(m^5)$ is the time required to obtain a $5/2$ approximation for* MAXIMUM WEIGHT INDEPENDENT SET *in a 5-claw-free graph of m vertices with small integer weights.*

4   Approximation Algorithms for 2-INTERVAL PATTERN

The $\{<, \sqsubset, \between\}$-structured 2-INTERVAL PATTERN problem is essentially the MAXIMUM INDEPENDENT SET problem in 2-interval graphs. Halldórsson [15, Theorem 3.1 and Corollary 3.1] showed that, using a local-improvement heuristic[2], MAXIMUM INDE-PENDENT SET in $(k+1)$-claw-free graphs can be approximated with a ratio of $k/2 + \epsilon$ in $O(n^{\log_k 1/\epsilon})$ time, for any $k \geq 4$. Given a set $\mathcal{D}$ of n 2-intervals, we can construct the 2-interval graph $G(\mathcal{D})$ in $O(n^2)$ time, then find a $c + \epsilon$ approximation for MAXI-MUM INDEPENDENT SET in this $(2c+1)$-claw-free graph in $O(n^{O(\log_c 1/\epsilon)})$ time using Halldórsson's algorithm, where $c = 2 + \lfloor \frac{\ell_{\max}(\mathcal{D}) - 2}{\ell_{\min}(\mathcal{D})} \rfloor$. We have the following theorem:

Theorem 2. *For a set $\mathcal{D}$ of n 2-intervals, the $\{<, \sqsubset, \between\}$-structured 2-INTERVAL PAT-TERN problem can be approximated with a ratio of $c + \epsilon$, for any constant $\epsilon > 0$, in $O(n^2 + T_1(n))$ time, where $c = 2 + \lfloor \frac{\ell_{\max}(\mathcal{D}) - 2}{\ell_{\min}(\mathcal{D})} \rfloor$ and $T_1(n) = O(n^{O(\log_c 1/\epsilon)})$ is the time required to obtain a $c + \epsilon$ approximation for MAXIMUM INDEPENDENT SET in a $(2c+1)$-claw-free graph of n vertices.*

The following two corollaries, for the unitary and the near-unitary cases, respectively, are immediate:

Corollary 1. *For a set of unitary 2-intervals, the $\{<, \sqsubset, \between\}$-structured 2-INTERVAL PATTERN problem can be approximated with a ratio of $2 + \epsilon$, for any constant $\epsilon > 0$, in polynomial time.*

Corollary 2. *For a set $\mathcal{D}$ of 2-intervals, the $\{<, \sqsubset, \between\}$-structured 2-INTERVAL PAT-TERN problem can be approximated with a ratio of $3 + \epsilon$ when $\ell_{\max}(\mathcal{D}) \leq 2\ell_{\min}(\mathcal{D}) + 1$, and with a ratio of $2 + \epsilon$ when $\ell_{\max}(\mathcal{D}) \leq \ell_{\min}(\mathcal{D}) + 1$, for any constant $\epsilon > 0$, in polynomial time.*

As noted by Crochemore et al. [12], the $\{\sqsubset, \between\}$-structured 2-INTERVAL PATTERN prob-lem on input of n 2-intervals reduces to $O(n)$ $\{<, \sqsubset, \between\}$-structured 2-INTERVAL PAT-TERN problems. Therefore, with an extra $O(n)$ multiplicative factor in the running time, our algorithms for $\{<, \sqsubset, \between\}$-structured 2-INTERVAL PATTERN can be extended to ap-proximate $\{\sqsubset, \between\}$-structured 2-INTERVAL PATTERN with the same approximation ratios.

5   Discussion

In this paper, we presented improved approximation algorithms for three closely related optimization problems with application to the prediction of RNA secondary structures with arbitrary pseudoknots. Admittedly, although our algorithms achieve improved ap-proximation ratios, their theoretical time complexities are prohibitive due to heavy use of the local-improvement technique [15,7]. Local-improvement algorithms are expen-sive because of the exhaustive search of improving local graph structures. The search algorithms themselves, however, are very simple. It is quite possible that, with more

---

[2] Halldórsson [15] noted that Hurkens and Schrijver [18] had previously obtained similar results using the same local-improvement heuristic.

sophisticated search techniques (such as branch-and-bound) and careful algorithmic engineering, our approximation algorithms can be made practical after all. The empirical study of these algorithms is an interesting topic for future research.

The technique that we have used to obtain the improved approximation algorithms in this paper has been used earlier on another interesting problem in computational biology called NONOVERLAPPING LOCAL ALIGNMENT [3,7,8]. This problem is essentially the MAXIMUM WEIGHT INDEPENDENT SET problem for proper 2-union graphs as defined by Bar-Yehuda et al. [4], and is also related to the weighted version of the $\{\sqsubset, \emptyset\}$-structured 2-INTERVAL PATTERN problem. The current best approximation algorithm for the NONOVERLAPPING LOCAL ALIGNMENT problem was also obtained using the local-improvement technique in d-claw-free graphs [7]; it has an approximation ratio of $2.5 + \epsilon$ and a very high time complexity. Berman et al. [8] was able to design a simple $O(n \log n)$ time 3 approximation for this problem using the local-ratio technique.

The 2-INTERVAL PATTERN problem is the MAXIMUM INDEPENDENT SET problem in 2-interval graphs. A straight-forward extension, the WEIGHTED 2-INTERVAL PATTERN problem, is the MAXIMUM WEIGHT INDEPENDENT SET problem in 2-interval graphs. This extended problem has been studied by Bar-Yehuda et al. [4] in the general framework of t-interval graphs and, very recently, by Crochemore et al. [13]. As we noted earlier, q consecutive stacking loops formed by $q + 1$ stacking base pairs can be represented by a balanced 2-interval of interval length $q + 1$. The connection between balanced 2-intervals and stacking base pairs suggests the following natural variant of the WEIGHTED 2-INTERVAL PATTERN problem:

Definition 1. *Given a set of balanced 2-intervals with weight equal to the interval length, the* LENGTH-WEIGHTED BALANCED 2-INTERVAL PATTERN *problem is to find a maximum weight independent set in the corresponding 2-interval graph.*

The LENGTH-WEIGHTED BALANCED 2-INTERVAL PATTERN problem contains the unitary case of the 2-INTERVAL PATTERN problem, so it is also APX-hard [4]. Bar-Yehuda et al.'s $2t$ approximation for MAXIMUM WEIGHT INDEPENDENT SET in t-interval graphs [4] implies a polynomial time 4 approximation for WEIGHTED 2-INTERVAL PATTERN. For WEIGHTED 2-INTERVAL PATTERN on input of balanced 2-intervals, Crochemore et al. [13] designed a simpler and more efficient 4 approximation using the local-ratio technique. Since LENGTH-WEIGHTED BALANCED 2-INTERVAL PATTERN has a very special weight function, can we design a polynomial time algorithm for it with an approximation ratio less than 4?

Acknowledgment

The author thanks the anonymous referees for the two references [13,22] and helpful comments. This research was partially supported by Utah State University research funds A13501 and A14766.

References

1. J. P. Abrahams, M. van den Berg, E. van Batenburg, and C. Pleij. Prediction of RNA secondary structure, including pseudoknotting, by computer simulation. *Nucleic Acids Research*, 18(10):3035–3044, 1990.

2. T. Akutsu. Dynamic programming algorithms for RNA secondary structure prediction with pseudoknots. *Discrete Applied Mathematics*, 104(1-3):45–62, 2000.
3. V. Bafna, B. Narayanan, and R. Ravi. Nonoverlapping local alignments (weighted independent sets of axis-parallel rectangles). *Discrete Applied Mathematics*, 71:41–53, 1996.
4. R. Bar-Yehuda, M. M. Halldórsson, J. (S.) Naor, H. Shachnai, and I. Shapira. Scheduling split intervals. *SIAM Journal on Computing*, 36(1):1–15, 2006.
5. F. H. D. van Batenburg, A. P. Gultyaev, and C. W. A. Pleij. An APL-programmed genetic algorithm for the prediction of RNA secondary structure. *Journal of Theoretical Biology*, 174(3):269–280, 1995.
6. F. H. D. van Batenburg, A. P. Gultyaev, C. W. A. Pleij, J. Ng, and J. Oliehoek. Pseudobase: a database with RNA pseudoknots. *Nucleic Acids Research*, 28(1):201–204, 2000.
7. P. Berman. A $d/2$ approximation for maximum weight independent set in d-claw free graphs. *Nordic Journal of Computing*, 7:178–184, 2000.
8. P. Berman, B. DasGupta, and S. Muthukrishnan. Simple approximation algorithm for nonoverlapping local alignments. In *Proceedings of the 13th Annual ACM-SIAM Symposium on Discrete Algorithms (SODA'02)*, pages 677–678, 2002.
9. G. Blin, G. Fertin, and S. Vialette. New results for the 2-interval pattern problem. In *Proceedings of the 15th Annual Symposium on Combinatorial Pattern Matching (CPM'04)*, LNCS 3109, pages 311–322, 2004.
10. R. B. Cary and G. D. Stormo. Graph-theoretic approach to RNA modeling using comparative data. In *Proceedings of the 3rd International Conference on Intelligent Systems for Molecular Biology (ISMB'95)*, pages 75–80, 1995.
11. E. Chen, L. Yang, and H. Yuan. Improved algorithms for largest cardinality 2-interval pattern problem. *Journal of Combinatorial Optimization*, Special Issue on Bioinformatics, to appear.
12. M. Crochemore, D. Hermelin, G. M. Landau, and S. Vialette. Approximating the 2-interval pattern problem. In *Proceedings of the 13th Annual European Symposium on Algorithms (ESA'05)*, LNCS 3669, pages 426–437, 2005.
13. M. Crochemore, D. Hermelin, G. M. Landau, D. Rawitz, and S. Vialette. Approximating the 2-interval pattern problem. *Theoretical Computer Science*, to appear.
14. A. P. Gultyaev, F. H. D. van Batenburg, and C. W. A. Pleij. The computer simulation of RNA folding pathways using a genetic algorithm. *Journal of Molecular Biology*, 250(1):37–51, 1995.
15. M. M. Halldórsson. Approximating discrete collections via local improvements. In *Proceedings of the 6th Annual ACM-SIAM Symposium on Discrete Algorithms (SODA'95)*, pages 160–169, 1995.
16. I. L. Hofacker. Vienna RNA secondary structure server. *Nucleic Acids Research*, 31(13):3429–3431, 2003.
17. I. L. Hofacker, W. Fontana, P. F. Stadler, S. Bonhoeffer, M. Tacker, and P. Schuster. Fast folding and comparison of RNA secondary structures. *Monatshefte für Chemie*, 125(2):167–188, 1994.
18. C. A. J. Hurkens and A. Schrijver. On the size of systems of sets every t of which have an SDR, with an application to the worst-case ratio of heuristics for packing problems. *SIAM Journal on Discrete Mathematics*, 2(1):68–72, 1989.
19. S. Ieong, M.-Y. Kao, T.-W. Lam, W.-K. Sung, and S.-M. Yiu. Predicting RNA secondary structure with arbitrary pseudoknots by maximizing the number of stacking pairs. *Journal of Computational Biology*, 10(6):981–995, 2003.
20. M. Jiang. A 2-approximation for the preceding-and-crossing structured 2-interval pattern problem. *Journal of Combinatorial Optimization*, Special Issue on Bioinformatics, to appear.
21. M. Jiang, M. Mayne, and J. Gillespie. Delta: a toolset for the structural analysis of biological sequences on a 3D triangular lattice. In *Proceedings of the 2007 International Symposium on Bioinformatics Research and Applications (ISBRA'07)*, to appear.

22. R. B. Lyngsø. Complexity of pseudoknot prediction in simple models. In *Proceedings of the 31st International Colloquium on Automata, Languages and Programming (ICALP'04)*, pages 919–931, 2004.

23. R. B. Lyngsø and C. N. S. Pedersen. RNA pseudoknot prediction in energy-based models. *Journal of Computational Biology*, 7(3/4):409–427, 2000.

24. R. B. Lyngsø, M. Zuker, and C. N. S. Pedersen. Fast evaluation of interval loops in RNA secondary structure prediction. *Bioinformatics*, 15(6):440–445, 1999.

25. D. H. Mathews, J. Sabina, M. Zuker, and D. H. Turner. Expanded sequence dependence of thermodynamic parameters improves prediction of RNA secondary structure. *Journal of Molecular Biology*, 288(5):911–940, 1999.

26. S. Micali and V. V. Vazirani. An $O(\sqrt{|V|}|E|)$ algorithm for finding maximum matching in general graphs. In *Proceedings of the 21st Annual Symposium on Foundations of Computer Science (FOCS'80)*, pages 17–27, 1980.

27. R. Nussinov, G. Pieczenik, J. R. Griggs, and D. J. Kleitman. Algorithms for loop matching. *SIAM Journal on Applied Mathematics*, 35(1):68–82, 1978.

28. N. R. Pace, B. C. Thomas, and C. R. Woese. Probing RNA structure, function, and history by comparative analysis. In *The RNA World*, 2nd edition, pages 113–141, Cold Spring Harbor Laboratory Press, 1999.

29. E. Rivas and S. R. Eddy. A dynamic programming algorithm for RNA structure prediction including pseudoknots. *Journal of Molecular Biology*, 285:2053–2068, 1999.

30. J. Ruan, G. D. Stormo, and W. Zhang. An iterated loop matching approach to the prediction of RNA secondary structure with pseudoknots. *Bioinformatics*, 20(1):58–66, 2004.

31. D. Sankoff. Simultaneous solution of the RNA folding, alignment and protosequence problems. *SIAM Journal on Applied Mathematics*, 45(5):810–825, 1985.

32. B. A. Shapiro and J. C. Wu. Predicting RNA H-type pseudoknots with the massively parallel genetic algorithm *Computer Applications in the Biosciences*, 13(4):459–471, 1997.

33. J. E. Tabaska, R. B. Cary, H. N. Gabow, and G. D. Stormo. An RNA folding method capable of identifying pseudoknots and base triples. *Bioinformatics*, 14(8):691–699, 1998.

34. I. Tinoco, P. N. Borer, B. Dengler, M. D. Levine, O. C. Uhlenbeck, D. M. Crothers, and J. Gralla. Improved estimation of secondary structure in ribonucleic acids. *Nature New Biology*, 246:40–42, 1973.

35. Y. Uemura, A. Hasegawa, S. Kobayashi, and T. Yokomori. Tree adjoining grammars for RNA structure prediction. *Theoretical Computer Science*, 210(2):277–303, 1999.

36. S. Vialette. On the computational complexity of 2-interval pattern matching problems. *Theoretical Computer Science*, 312:223–249, 2004.

37. M. Zuker. Mfold web server for nucleic acid folding and hybridization prediction. *Nucleic Acids Research*, 31(13)3406–3415, 2003.

38. M. Zuker and D. Sankoff. RNA secondary structures and their prediction. *Bulletin of Mathematical Biology*, 46:591–621, 1984.

39. M. Zuker and P. Stiegler. Optimal computer folding of large RNA sequences using thermodynamics and auxiliary information. *Nucleic Acids Research*, 9(1):133–148, 1981.

A Heuristic Method for Selecting Support Features from Large Datasets[*]

Hong Seo Ryoo[**] and In-Yong Jang

Division of Information Management Engineering, Korea university
1, 5-Ka, Anam-Dong, Seongbuk-Ku, Seoul, 136-713, Korea
Phone: +82-2-3290-3394; Fax: +82-2-929-5888
{hsryoo,jjiy}@korea.ac.kr

Abstract. For feature selection in machine learning, set covering (SC) is most suited, for it selects support features for data under analysis based on the individual and the collective roles of the candidate features. However, the SC-based feature selection requires the complete pair-wise comparisons of the members of the different classes in a dataset, and this renders the meritorious SC principle impracticable for selecting support features from a large number of data.

Introducing the notion of implicit SC-based feature selection, this paper presents a feature selection procedure that is equivalent to the standard SC-based feature selection procedure in supervised learning but with the memory requirement that is multiple orders of magnitude less than the counterpart. With experiments on six large machine learning datasets, we demonstrate the usefulness of the proposed implicit SC-based feature selection scheme in large-scale supervised data analysis.

Keywords: feature selection, combinatorial optimization, supervised learning, large datasets.

1   Introduction

The classification of two types of data is a fundamental problem in machine learning and data mining and bears close resemblance to real-life problems (e.g., [1,2,3,4,5,6,7,8]). Furthermore, multicategory classification can be seen as successive binary classification (e.g., [9,10,11]). For convenience in presentation, therefore, we refer to the general classification of data as binary classification and denote the two types of data as 'positive' $(+)$ and 'negative' $(-)$ data in this paper.

Supervised learning to binary classification is aimed at discovering a classification theory on past (training) data in order to classify new (testing) observations in a manner consistent with the past classifications. In view that no decision rule is obtained in supervised learning without the process of 'learning (training)' and

---

[*] This work was supported by the Korea Research Foundation Grant funded by the Korean Government (MOEHRD) (KRF-2005-003-D00445).

[**] Corresponding author.

M.-Y. Kao and X.-Y. Li (Eds.): AAIM 2007, LNCS 4508, pp. 411–423, 2007.

that the nature of classification of data in general is difficult (e.g., [12,13,14]), the degree of difficulties associated with analyzing a set of data is more or less determined by the difficulties associated with solving the corresponding training problem, more specifically, the number of training observations in the dataset and the number of features describing them.

One problem that has proved most useful for identifying a minimal set of variables that collectively can explain the difference between the observations in one class from the other and vice versa is set covering (SC). SC is a well-known $\mathcal{NP}$−complete problem [15]. However, owing to having an array of practical applications, SC has attracted a number of efficient (meta-)heuristic procedures to be developed in the literature (e.g., [16,17,18,19,20,21]), and this has led the SC-based feature selection to be adopted as a standard feature selection procedure in supervised learning, for example, in the logical analysis of data (LAD) (e.g., [22,23]) and for probe selection in genomics (e.g., [4,24,25]).

Let us briefly review the standard SC-based feature selection scheme in supervised learning. Let $S^\bullet$ denote the index set of $m^\bullet$ observations of type $\bullet$, where $\bullet \in \{+,-\}$. Without loss of generality [22] and for ease of presentation and understanding, consider binary observations p_i, $i \in S^\bullet$, $\bullet \in \{+,-\}$. Let $N = \{1,\ldots,n\}$. Let a_j, $j \in N$, denote the j−th binary attribute and let p_{ij} denote the value of the j−th attribute in observation p_i, $i \in S^\bullet$. Furthermore, let

$$a_k^{(i,j)} = \begin{cases} 1, & \text{if } p_{ik} \neq p_{jk}, \\ 0, & \text{otherwise}, \end{cases} \tag{1}$$

for each pair of p_i and p_j, where $i \in S^+$ and $j \in S^-$. It is now seen that 1's in a cover (a feasible solution) $\mathbf{x}$ for the SC instance below identifies a set of support features that distinguishes the two types of data under analysis (e.g., [22,23,24])

$$\min \quad \sum_{k \in N} c_k x_k \tag{2}$$

$$\text{s.t.} \quad \sum_{k \in N} a_k^{(i,j)} x_k \geq 1, \quad i \in S^+, j \in S^- \tag{3}$$

$$x_k \in \{0,1\}, \qquad k \in N, \tag{4}$$

where usually $c_k = 1$ for all $k \in N$.

This SC-based feature selection is advantageous in at least two regards. First, the SC-based feature selection examines the collective as well as the individual roles of the candidate features in selecting support features. Second, as mentioned earlier, a number of efficient (meta-)heuristic procedures exists for efficient solution of SC.

Let us recall the most popular, textbook heuristic for SC (e.g., [26,27]). Denote by M the Cartesian product of S^+ and S^-, that is, $M := S^+ \times S^-$, and I_j the index set of rows in M that the column j can cover. Given a partial cover $\mathbf{x}$ that does not satisfy all of the cover inequalities in (3), denote by M_u the index set of those rows of M in (3) that are not covered by the partial cover at hand. The textbook heuristic builds a cover for SC by repeatedly selecting one variable at a time by a greedy rule

$$j \leftarrow \operatorname{argmin} \left\{ k \in N, \, x_k = 0, I_k \cap M_u \neq \emptyset : \frac{c_k}{|I_k \cap M_u|} \right\}$$

until all cover inequalities are covered. For convenience in presentation, we refer to the feature selection via the explicit formulation of the SC instance on the data at hand via (1)-(4) and its solution via the textbook heuristic as standard fs in this paper. We note that standard fs is the standard procedure in supervised learning, featured, for example, in the implementations of LAD of [22,23] for support feature selection and in the algorithm of [4] for oligonucleotide probe selection. [24] also uses the same explicit SC formulation-based approach for selecting short DNA probes but uses the SC heuristic from [17] for solving the SC instances.

Note from (1)-(4) that the SC-based feature selection requires the complete pairwise differencing between each pair of + and − observations of the dataset under analysis. This entails solving an SC instance with $m^+ m^-$ cover inequalities in n variables and renders feature selection impossible if the membership of the dataset is large. For example, consider the adult dataset available from the UC Irvine Repository of Machine Learning Databases [28]. Counting the observations without any missing attribute values, this dataset has 45,222 observations with about 24.78% of the data belonging to one class with income > \$50K and the remaining to the other class with income ≤ \$50K. Now, formulating the SC feature selection instance on the 30,162 training data of the dataset, we obtain an SC problem with about 1.7×10^8 cover inequalities and 1×10^4 binary variables. In order to solve this SC by the textbook heuristic, we need two double precision integer arrays, one for storing the information about the nonzero elements and the other for recording the row or column starts. Suppose now that about 30% of the elements in the cover coefficient matrix are nonzero (see Table 2 in Section 4 for the rationale). Then, noting that a double precision integer array requires 4 bytes of main memory per element, we obtain that the first integer array alone requires about 2.0×10^{12} bytes of main memory.

In this paper, we introduce the notion of implicit SC-based feature selection and present a memory-efficient feature selection procedure that is equivalent to standard fs. Specifically, assuming that $m^+ \leq m^-$, without loss of generality, standard fs requires $O(m^+ m^- n)$ of memory to store the corresponding SC feature selection instance but the proposed procedure requires $O(\max\{m^- n, m^+ m^-\})$ of main memory, which is multiple orders of magnitude less than that required by the standard procedure. The proposed feature selection scheme is SC-based, hence preserves the aforementioned merit of SC-based feature selection in supervised learning. Furthermore, owing to the efficient use of memory, the proposed procedure allows much larger datasets to be analyzed by supervised learning methodologies.

Briefly summarizing the organization of this paper, we develop in Section 2 an implicit SC-based feature selection procedure that is equivalent to standard fs. For reasons of space and readability, we omit proofs for a few mathematical

results in this section. Next, we illustrate the steps of the proposed feature selection procedure in Section 3 and demonstrate its efficiency in comparison with
the standard SC-based feature selection scheme in Section 4. In Section 5, we
test the proposed scheme in feature selection experiments with six large machine
learning datasets from [28,29]. These datasets are of the size that do not per-
mit their support features to be selected by the standard SC-based scheme, and
experimental results in this section illustrate the usefulness of the implicit SC-
based feature selection scheme in large-scale data analysis. Finally, we conclude
the paper with a summary in Section 6.

Before proceeding, we note that any other primal SC heuristic procedure from
the literature (e.g., [20,21]) can replace the role of the textbook heuristic in the
proposed feature selection procedure that we develop in the following section:
the primal-dual SC heuristics (e.g., [16,17,18]) are not suited for analyzing large
datasets because of their use of an excessive amount of additional memory for
storing the dual information. As it will become apparent by the end of Section 2,
the proposed procedure and **standard fs** are equivalent as long as the same SC
heuristic procedure is used in them. Our use of the textbook SC heuristic in
this paper simply owes to its popular usage for feature selection in the literature
(e.g., [22,23,30]) and ease of implementation.

2   Implicit SC-Based Feature Selection

Consider a set of binary (or binarized) data, described by n 0-1 binary/Boolean
variables. For $j \in N = \{1, \ldots, n\}$, let l_j be the literal associated with attribute
a_j that instructs to take or negate the value of the attribute in all observations
p_i in $S^+ \cup S^-$ via $l_j = a_j$ or $l_j = \bar{a}_j$, respectively. A term t is a conjunction of
literals. Let $N_t \subseteq N$ denote the index set of the literals included in a term t,
that is, $t := \bigwedge_{j \in N_t} l_j$. For a term t, let

$$C^\bullet(t) := \left\{ i \in S^\bullet : t(p_i) = \prod_{\substack{l_j = a_j, \\ j \in N_t}} p_{ij} \prod_{\substack{l_j = \bar{a}_j, \\ j \in N_t}} \bar{p}_{ij} = 1 \right\}$$

for $\bullet \in \{+, -\}$. As before, let I_j denote the index set of rows in (3) that setting
$x_j = 1$ for $j \in N$ and $x_k = 0$ for all $k \in N \setminus \{j\}$ satisfies.

Proposition 1. *For $j \in N$, $I_j = \left(C^+(a_j) \times C^-(\bar{a}_j) \right) \cup \left(C^+(\bar{a}_j) \times C^-(a_j) \right)$.*

Proposition 2. *For $t = \bigwedge_{j \in N_t} l_j$, where $N_t \subseteq N$ and $N_t \neq \emptyset$, $C^\bullet(t) = \bigcap_{j \in N_t} C^\bullet(l_j)$*

for $\bullet \in \{+, -\}$.

Let $\bar{t} := \bigwedge_{j \in N_t} \bar{l}_j$ in the following.

Lemma 1. *For* $t = \wedge_{j \in N_t} l_j$, *where* $N_t \subseteq N$,

$$M_u(t) = \left(C^+(t) \times C^-(t)\right) \cup \left(C^+(\bar{t}) \times C^-(\bar{t})\right)$$

gives the index set of the rows in (3) *that are not satisfied by the solution* $x_j = 1$ *for* $j \in N_t$ *and* $x_k = 0$ *for* $k \in N \setminus N_t$.

Corollary 1. *For some* $N_t \subseteq N$, $N_t \neq \emptyset$, *let* T *be the collection of all terms of the form* $t = \wedge_{j \in N_t} l_j$. *Then,*

$$M_u = \bigcup_{t \in T} M_u(t)$$

gives the index set of the rows in (3) *that are not satisfied by the solution* $x_j = 1$ *for* $j \in N_t$ *and* $x_k = 0$ *for* $k \in N \setminus N_t$.

For some $N_t \subseteq N$ and $N_t \neq \emptyset$, let

$$T = \left\{ t = \wedge_{j \in N_t} l_j : C^+(t) \neq \emptyset, C^-(t) \neq \emptyset \right\}. \tag{5}$$

If $T \neq \emptyset$, then for each $t \in T$, let

$$M'_u(t) = C^+(t) \times C^-(t).$$

Theorem 1. *For some* $N_t \subseteq N$, $N_t \neq \emptyset$, *let* T *be defined via* (5). *If* $T = \emptyset$, *then the solution obtained by setting* $x_j = 1$ *for* $j \in N_t$ *and* $x_k = 0$ *for* $k \in N \setminus N_t$ *satisfies the SC constraints in* (3). *If* $T \neq \emptyset$, *then*

$$M_u = \bigcup_{t \in T} M'_u(t) = \bigcup_{t \in T} C^+(t) \times C^-(t)$$

gives the index set of the rows in (3) *that are not satisfied by the solution* $x_j = 1$ *for* $j \in N_t$ *and* $x_k = 0$ *for* $k \in N \setminus N_t$.

Proof. If $T = \emptyset$, consider any term of the form $t = \wedge_{j \in N_t} l_j$. Then, $T = \emptyset$ implies that at least one of $C^\bullet(t) = \emptyset$ and at least one of $C^\bullet(\bar{t}') = \emptyset$ for $\bullet \in \{+, -\}$. This trivially yields the result.

Now, if $T \neq \emptyset$, we have

$$M_u = \bigcup_{t \in T} M_u(t) = \bigcup_{t \in T} M'_u(t) \cup M'_u(\bar{t}).$$

Here, $M'_u(\bar{t}) = C^+(\bar{t}) \times C^-(\bar{t})$ may or may not be empty. If empty, the result is immediate. If not empty, then $\bar{t}$ is also a member of T, hence $M'_u(\bar{t})$ needs not be considered here with t. This completes the proof.

Theorem 2. *For some* $N_t \subseteq N$ *and* $N_t \neq \emptyset$, *let* T *be defined via* (5). *If* $T \neq \emptyset$, *then for* $j \in N \setminus N_t$,

$$I_j \cap M_u = \bigcup_{t \in T} \left(C^+(t \wedge a_j) \times C^-(t \wedge \bar{a}_j) \right) \cup \left(C^+(t \wedge a_j) \times C^-(t \wedge a_j) \right) \tag{6}$$

gives the index set of the rows in (3) *that are not satisfied by the solution* $x_k = 1$ *for* $k \in N_t$ *and* $x_l = 0$ *for* $l \in N \setminus N_t$ *that setting* $x_j = 1$ *additionally covers.*

Proof. First, note that, barring infeasibility of the feature selection problem, $T \neq \emptyset$ implies that $N \setminus N_t \neq \emptyset$. Next, note for $t \in T$ and $j \in N \setminus N_t$ that

$$
\begin{aligned}
I_j \cap M_u'(t) &= \left\{ \left(C^+(a_j) \times C^-(\overline{a}_j) \right) \cup \left(C^+(\overline{a}_j) \times C^-(a_j) \right) \right\} \cap \left(C^+(t) \times C^-(t) \right) \\
&= \left\{ \left(C^+(a_j) \times C^-(\overline{a}_j) \right) \cap \left(C^+(t) \times C^-(t) \right) \right\} \cup \\
&\quad \left\{ \left(C^+(\overline{a}_j) \times C^-(a_j) \right) \cap \left(C^+(t) \times C^-(t) \right) \right\} \\
&= \left\{ \left(C^+(a_j) \cap C^+(t) \right) \times \left(C^-(\overline{a}_j) \cap C^-(t) \right) \right\} \cup \\
&\quad \left\{ \left(C^+(\overline{a}_j) \cap C^+(t) \right) \times \left(C^-(a_j) \cap C^-(t) \right) \right\} \\
&= \left(C^+(a_j \wedge t) \times C^-(\overline{a}_j \wedge t) \right) \cup \left(C^+(\overline{a}_j \wedge t) \times C^-(a_j \wedge t) \right) \\
&= \left(C^+(t \wedge a_j) \times C^-(t \wedge \overline{a}_j) \right) \cup \left(C^+(t \wedge \overline{a}_j) \times C^-(t \wedge a_j) \right).
\end{aligned}
$$

Finally, $I_j \cap M_u = I_j \cap \bigcup_{t \in T} M_u'(t) = \bigcup_{t \in T} \left(I_j \cap M_u'(t) \right)$ gives the desired result.

Corollary 2. *For some $N_t \subseteq N$ and $N_t \neq \emptyset$, let T be defined via (5). If $T \neq \emptyset$, then for $j \in N \setminus N_t$,*

$$
|I_j \cap M_u| = \sum_{t \in T} |C^+(t \wedge a_j)| \times |C^-(t \wedge \overline{a}_j)| + |C^+(t \wedge \overline{a}_j)| \times |C^-(t \wedge a_j)|, \quad (7)
$$

gives the number of the rows in (3) that are not satisfied by the solution $x_k = 1$ for $k \in N_t$ and $x_l = 0$ for $l \in N \setminus N_t$ that setting $x_j = 1$ additionally covers.

We use the results above to devise a memory-efficient heuristic procedure that we propose for selecting support features from large datasets.

procedure proposed fs
input: binary and contradiction free data p_i, $i \in S^+ \cup S^-$
output: N_t, an index set of support features
begin
    obtain $C^\bullet(a_j)$ for $\bullet \in \{+, -\}$ for $j \in N$.
    set $N_t = \{j \in N : I_j \text{ is maximal}\}$ and define T via (5).
    set $N_u = N \setminus \{T\}$ and $k = 1$.
    while $T \neq \emptyset$ **do** (iteration k)
        obtain $C^\bullet(t)$ for $\bullet \in \{+, -\}$ for each $t \in T$
        set $l \leftarrow \text{argmin} \left\{ j \in N_u : \frac{c_j}{|I_j \cap M_u|} \right\}$, where $|I_j \cap M_u|$ is calculated for
$j \in N_u$ via (7).
        set $N_t \leftarrow N_t \cup \{l\}$ and define T via (5).
        if $T \neq \emptyset$ **then** set $N_u \leftarrow N_u \setminus \{l\}$ and set $k \leftarrow k + 1$.
    end while
end

The following states that **proposed fs** and **standard fs** are equivalent in terms of the cover they build for the SC feature selection instance at hand.

Theorem 3. *For any SC feature selection instance defined by (1)-(4), **proposed fs** selects the same set of support features as **standard fs**.*

We close this section with the comparison of the memory requirements by standard fs and proposed fs. As we are concerned with the analysis of large datasets, we assume that at least one of m^+ and m^- is a large number and that $m^+ \leq m^-$, without loss of generality. Assume further that $m^+ m^- > n$, for otherwise, the dataset may be too large to be analyzed, especially when m^+ and m^- are both large numbers: in this case, we may first partition the column set into K smaller subsets of n_i columns, where $n_i < m^+ m^-$ for $i = 1, \ldots, K$ and then successively solve smaller SC feature selection instances and aggregate support features from (the) subproblems and remove the redundant ones to compose a set of support features for the dataset.

Note that the nonzero elements of the constraint matrix of an SC instance is 1. Hence, standard fs requires one double precision integer column array of length in the number of nonzero elements for storing the row indices of the nonzero elements. This yields that the amount of memory required by standard fs for support feature selection is $O(m^+ m^- n)$.

Now, proposed fs requires an integer array for $C^\bullet(a_j)$ for each $j \in N_u$ for $\bullet \in \{+, -\}$ and a different array for $C^\bullet(t)$ for each $t \in T$ for $\bullet \in \{+, -\}$. In the worst case, $C^-(a_j)$ can have m^- elements in it, T can include up to $\frac{1}{2}m^+$ elements, and $C^-(t)$ can be of length $\frac{1}{2}m^-$ for some $t \in T$. These yield that the memory requirement by proposed fs for storing an SC feature instance is $O(\max\{m^- n, m^+ m^-\})$, which is multiple orders of magnitude down from that required by standard fs.

3 Illustrative Example

For data in Table 1, we apply standard fs and proposed fs and demonstrate the equivalency between standard fs and proposed fs, as stated in Theorem 3, and also the steps of proposed fs.

When formulated on the data in Table 1, the constraint matrix of the feature selection problem becomes:

For this problem, standard fs identifies $N_t = \{1, 2, 3, 4, 5\}$ with the individual support features selected in order 5-1-3-2-4 (with the use of the lowest index first rule for breaking ties).

When applied, proposed fs also selects $\{a_1, a_2, a_3, a_4, a_5\}$ as support features. The following details the steps of proposed fs. For notational convenience, we use α_j for $|I_j \cap M_u|$ for $j \in N_u$ in this example.

Initialization. We first obtain $C^\bullet(a_j)$, $\bullet\{+, -\}$, for $j \in N$:

$$C^+(a_1) = \{1, 3, 4\}, \quad C^-(a_1) = \{1, 4, 5\}$$
$$C^+(a_2) = \{1, 2\}, \quad C^-(a_2) = \{5\}$$
$$C^+(a_3) = \{3\}, \quad C^-(a_3) = \{2, 5\}$$
$$C^+(a_4) = \{2, 4\}, \quad C^-(a_4) = \{3\}$$
$$C^+(a_5) = \{1, 2, 3, 5\}, \quad C^-(a_5) = \{1, 2\}$$
$$C^+(a_6) = \{3, 5\}, \quad C^-(a_6) = \{2, 3\}$$

Table 1. Example dataset 1

	i	a_1 p_{i1}	a_2 p_{i2}	a_3 p_{i3}	a_4 p_{i4}	a_5 p_{i5}	a_6 p_{i6}
	1	1	1	0	0	1	0
	2	0	1	0	1	1	0
S^+	3	1	0	1	0	1	1
	4	1	0	0	1	0	0
	5	0	0	0	0	1	1
	1	1	0	0	0	1	0
	2	0	0	1	0	1	1
S^-	3	0	0	0	1	0	1
	4	1	0	0	0	0	0
	5	1	1	1	0	0	0

$$\mathbf{A} = \{a_j^{(k,l)}, k \in S^+, l \in S^-, j \in N\} = \begin{bmatrix} 0 & 1 & 0 & 0 & 0 & 0 \\ 1 & 1 & 1 & 0 & 0 & 1 \\ 1 & 1 & 0 & 1 & 1 & 1 \\ 0 & 1 & 0 & 0 & 1 & 0 \\ 0 & 0 & 1 & 0 & 1 & 0 \\ 1 & \overline{1} & 0 & \overline{1} & 0 & 0 \\ 0 & 1 & 1 & 1 & 0 & 1 \\ 0 & 1 & 0 & 0 & 1 & 1 \\ 1 & 1 & 0 & 1 & 1 & 0 \\ 1 & 0 & 1 & 1 & 1 & 0 \\ 0 & 0 & \overline{1} & 0 & 0 & \overline{1} \\ 1 & 0 & 0 & 0 & 0 & 0 \\ 1 & 0 & 1 & 1 & 1 & 0 \\ 0 & 0 & 1 & 0 & 1 & 1 \\ 0 & 1 & 0 & 0 & 1 & 1 \\ 0 & 0 & 0 & \overline{1} & \overline{1} & 0 \\ 1 & 0 & 1 & 1 & 1 & 1 \\ 1 & 0 & 0 & 0 & 0 & 1 \\ 0 & 0 & 0 & 1 & 0 & 0 \\ 0 & 1 & 1 & 1 & 0 & 0 \\ \overline{1} & 0 & 0 & 0 & 0 & \overline{1} \\ 0 & 0 & 1 & 0 & 0 & 0 \\ 0 & 0 & 0 & 1 & 1 & 0 \\ 1 & 0 & 0 & 0 & 1 & 1 \\ 1 & 1 & 1 & 0 & 1 & 1 \end{bmatrix}$$

Next, we calculate $|I_j| = |C^+(a_j)| \times |C^-(\overline{a}_j)| + |C^+(\overline{a}_j)| \times |C^-(a_j)|$ for $j \in N$:

$$|I_1| = 3 \times 2\,(= m^- - |C^-(a_1)|) + 2\,(= m^+ - |C^+(a_1)|) \times 3 = 12$$

$$|I_2| = 2 \times 4 + 3 \times 1 = 11$$
$$|I_3| = 1 \times 3 + 4 \times 2 = 11$$
$$|I_4| = 2 \times 4 + 3 \times 1 = 11$$
$$|I_5| = 4 \times 3 + 1 \times 2 = 14$$
$$|I_6| = 2 \times 3 + 3 \times 2 = 12$$

With $|I_5| = 14$, $N_t \leftarrow \{5\}$ and $T \leftarrow \{t_1 = a_5, t_2 = \overline{a}_5\}$ via (5). Set $N_u \leftarrow \{1, 2, 3, 4, 6\}$ and $k = 1$.

Iteration 1. For $t \in T$, we have $C^+(t_1) = \{1, 2, 3, 5\}$, $C^-(t_1) = \{1, 2\}$, $C^+(t_2) = \{4\}$ and $C^-(t_2) = \{3, 4, 5\}$. In order to calculate $\alpha_1(= I_1 \cap M_u)$, we first obtain $|C^+(t_1 \wedge a_1)| = 2$, $|C^-(t_1 \wedge \overline{a}_1)| = 1$, $|C^+(t_1 \wedge \overline{a}_1)| = 2$ and $|C^-(t_1 \wedge a_1)| = 1$ for t_1 and $|C^+(t_2 \wedge a_1)| = 1$, $|C^-(t_2 \wedge \overline{a}_1)| = 1$ and $|C^-(t_2 \wedge \overline{a}_1)| = 0$ for t_2. As $|C^-(t_2 \wedge \overline{a}_1)| = 0$, we need not compute $|C^-(t_2 \wedge a_1)|$ here. Using these in (7), we next obtain $\alpha_1 = 5$. Likewise, we obtain $\alpha_2 = 5$, $\alpha_3 = 5$, $\alpha_4 = 4$ and $\alpha_6 = 5$. By the lowest index first rule for breaking ties, $N_t \leftarrow \{1, 5\}$ and $N_u \leftarrow \{2, 3, 4, 6\}$. T updates to $\{a_1 a_5, \overline{a}_1 a_5, a_1 \overline{a}_5\}$.

Iteration 2. We have $C^+(t_1) = \{1, 3\}$, $C^-(t_1) = \{1\}$, $C^+(t_2) = \{2, 5\}$, $C^-(t_2) = \{2\}$, $C^+(t_3) = \{4\}$ and $C^-(t_3) = \{4, 5\}$ and obtain $\alpha_2 = 3$, $\alpha_3 = 4$, $\alpha_4 = 3$ and $\alpha_6 = 2$. Hence, $N_t \leftarrow \{1, 3, 5\}$, $N_u \leftarrow \{2, 4, 6\}$ and $T \leftarrow \{a_1 \overline{a}_3 a_5, a_1 \overline{a}_3 \overline{a}_5\}$.

Iteration 3. We have $C^+(t_1) = \{1\}$, $C^-(t_1) = \{1\}$, $C^+(t_2) = \{4\}$ and $C^-(t_2) = \{4\}$ and calculate $\alpha_2 = 1$, $\alpha_4 = 1$ and $\alpha_6 = 0$. This updates N_t to $\{1, 2, 3, 5\}$ and N_u to $\leftarrow \{4, 6\}$. $T \leftarrow \{a_1 \overline{a}_2 \overline{a}_3 \overline{a}_5\}$.

Iteration 4. Similarly as above, we obtain $N_t \leftarrow \{1, 2, 3, 4, 5\}$ and $N_u \leftarrow \{6\}$. As $T \leftarrow \emptyset$, proposed fs terminates with selecting $\{a_1, a_2, a_3, a_4, a_5\}$ as a set of support features for this dataset.

4   Comparative Experiments with Adult Data

The adult dataset [28] is a two-class (income over \$50K or below) dataset with 45,220 observations that are defined in terms of 14 non-binary attributes. To demonstrate the difficulty associated with selecting support features via the standard SC-based approach, we used the first m_t observations of the dataset, with the value of m_t varied from 1,000 to 8,000 in the increment of 1,000, and calculated the size of the corresponding SC feature selection instance in the first six columns of Table 2. Specifically, going from left to right, we list in Table 2 the number of rows (m), the number of columns (n), the number of nonzero elements (ne) and the density of the nonzero elements in the SC constraint matrix and then provide the amount of space required by standard fs to store the nonzero elements in a single double precision integer array of length ne in order to select support features on the m_t adult data. In the last two columns of Table 2, we summarize the computational times required by standard fs and proposed fs in selecting support features on the m_t adult data. We performed

Table 2. Comparison of feature selection results on adult data by standard fs and proposed fs

$m_t^\S$	standard fs						proposed fs
	$m = m^+ m^-$ (in 10^6)	n (in 10^3)	$ne^\dagger$ (in 10^8)	density‡ (in %)	mem. req.* (in Gb)	feature selection time^a	feature selection time^a
1,000	0.18	0.58	0.30	28.3	0.12	0.9	0.1
2,000	0.76	1.00	2.28	29.7	0.91	8.24	0.7
3,000	1.73	1.38	7.26	30.4	2.90	26.8	2.2
4,000	3.01	1.73	16.01	30.7	6.4	462.4	5.5
5,000	4.69	2.07	30.30	31.3	12.1	1,021.3	10.6
6,000	6.75	2.41	51.42	31.6	20.6	-	18.3
7,000	9.14	2.74	80.02	31.9	32.0	-	27.6
8,000	11.87	3.10	117.73	31.9	47.1	-	37.6

$\S$: The number of training data: the first m_t observations of adult dataset.
$\dagger$: The number of nonzero $a_k^{(i,j)}$'s (1's) in (3).
$\ddagger$: Calculated by $\frac{ne}{mn} \times 100\%$.
$*$: Calculated via 4 bytes per double precision integer array element.
a: Total time required for feature selection in CPU seconds.
-: Did not solve due to large memory requirement by standard fs.

these experiments on an Intel 2.66GHz Linux PC with 512Mb of main memory and 80Gb of hardisk space and with using the Intel Fortran 90 compiler.

Recalling that standard fs and proposed fs select the same set of support features on any given set of data, the comparative feature selection results in Table 2 illustrate the efficiency of the implicit SC-based feature selection well.

5   Experiments on Large Datasets

For testing the proposed feature selection scheme, we obtained six larger size machine learning datasets from [28,29]. In Table 3, we summarize the information on the datasets analyzed and the feature selection results on them by proposed fs. More specifically, from left to right, we provide the number of classes, the number of attributes before data binarization, and the number of training observations in each of the six datasets analyzed. Then, we provide the number of binary features before feature selection, the number of support features selected and the total time spent for feature selection by proposed fs. For four datasets with more than two types of data, we selected support features in the $k - 1$ successive classification setup of 'one type against the remaining,' where k is the numbers of classes in those datasets. Again, we used the Intel Fortran 90 compiler and an Intel 2.66GHz Linux PC with 512Mb of main memory and 80Gb of hardisk space for these experiments.

Noting the size of the training datasets and in conjunction with the illustration made in the previous section, the results summarized in Table 3 clearly illustrate the usefulness of the proposed SC-based feature selection scheme in large-scale supervised data analysis without the need for further discussions.

Table 3. Feature selection results on larger datasets by `proposed fs`

Database	Number of			Number of Features		Feature Selection
	classes	attributes	training data	total[1]	support[2]	time (CPU sec.)
Adult	2	14	30,161	9,654	139	1,692.3
Ann-thyroid	3	21	3,772	580	12	0.4
Face Detection	2	361	6,977	87,552	17	8,184.7
Forest Covertype	7	54	11,340	5,676	43	264.3
Letter Recognition	26	16	16,000	168	19	29.9
Statlog Satellite Image	6	36	4,435	1,637	23	10.6

[1]: The number of binary features in the dataset.
[2]: The number of support features selected.

6   Conclusion

Feature selection in machine learning is a combinatorics and optimization natured problem that holds a key to successful analysis of large datasets. Although SC is most suited for feature selection, SC-based feature selection requires the complete pair-wise comparisons among the members of the different classes in a dataset under analysis, and, as the result, the size of an SC instance grows rapidly in the number of observations in a dataset. This in turn makes the meritorious SC principle impractical to use for feature selection on a large number of data.

Introducing the notion of implicit SC-based feature selection, we presented in this paper a memory-efficient SC procedure that is equivalent to the standard SC-based feature selection procedure in supervised learning. With the efficient use of memory, the proposed feature selection procedure allows larger datasets to be successfully analyzed by a supervised learning methodology. With experiments on six large machine learning dataset, we illustrated the usefulness of the proposed implicit SC-based feature selection scheme in large-scale data analysis.

References

1. Apté, C., Weiss, S., Grout, G.: Predicting defects in disk drive manufacturing: A case study in high-dimensional classification. In: Proceedings of the 9th Conference on Artificial Intelligence for Applications, Orlando, Florida (1993) 212–218
2. Bhandari, I., Colet, E., Parker, J., Pines, Z., Pratap, R., Ramanujam, K.: Advanced scout: Data mining and knowledge discovery in nba. Data Mining and Knowledge Discovery 1 (1997) 121–125
3. Carter, C., Catlett, S.: Assessing credit card applications using machine learning. IEEE Expert (1987) 71–79
4. Kim, K., Ryoo, H.: A lad-based method for selecting short oligo probes for genotyping applications. OR Spectrum: Special Issue on OR and Biomedical Informatics (2006) accepted for publication.
5. Osuna, E., Freund, R., Girosi, F.: Training support vector machines: an application to face detection. In: IEEE Conference on Computer Vision and Pattern Recognition, Puerto Rico (1997) 130–136

6. Rahmann, S.: Fast large scale oligonucleotide selection using the longest common factor approach. Journal of Bioinformatics and Computational Biology **1**(2) (2003) 343–361

7. Wang, X., Seed, B.: Selection of oligonucleotide probes for protein coding sequences. Bioinformatics **19**(7) (2003) 796–802

8. Wolberg, W., Mangasarian, O.: Multisurface method of pattern separation for medical diagnosis applied to breast cytology. Proceedings of the National Academy of Sciences **87** (1990) 9193–9196

9. Cortes, C., Vapnik, V.: Support vector networks. Machine Learning **20** (1995) 273–297

10. Ullman, J.: Pattern Recognition Techniques. Crane, London (1973)

11. Vapnik, V.: Statistical Learning Theory. Wiley-Interscience (1998)

12. Bennett, K., Mangasarian, O.: Robust linear programming discrimination of two linearly inseparable sets. Optimization Methods and Software **1** (1992) 23–34

13. Falk, J., Lopez-Cardona, E.: The surgical separation of sets. Journal of Global Optimization **11** (1997) 433–462

14. Megiddo, N.: On the complexity of polyhedral separability. Discrete and Computational Geometry **3** (1988) 325–337

15. Garey, M., Johnson, D.: Computers and Intractability: A Guide to the Theory of $\mathcal{NP}$-Completeness. Freeman, New York (1979)

16. Balas, E., Carrera, M.: A dynamic subgradient-based branch-and-bound procedure for set covering problem. Operation Research **44**(6) (1996) 875–890

17. Caprara, A., Fischetti, M., Toth, P.: A heuristic method for the set covering problem. Operations Research **47**(5) (1999) 730–743

18. Ceria, S., Nobili, P., Sassano, A.: A lagrangian-based heuristic for large-scale set covering problems. Mathematical Programming **81**(2) (1998) 215–228

19. Fisher, M., Kedia, P.: Optimal solution of set covering/partitioning problems using dual heuristics. Management Science **36** (1990) 674–688

20. Vasko, F., Wilson, G.: An efficient heuristic for large set covering problem. Naval Research Logistics Quarterly **31** (1984) 163–171

21. Vasko, F., Wilson, G.: Hybrid heuristics for minimum cardinality set covering problems. Naval Research Logistics Quarterly **33** (1986) 241–249

22. Boros, E., Hammer, P., Ibaraki, T., Kogan, A., Mayoraz, E., Muchnik, I.: An implementation of logical analysis of data. IEEE Transactions on Knowledge and Data Engineering **12** (2000) 292–306

23. Ryoo, H., Jang, I.Y.: Milp approach to pattern generation in logical analysis of data. Machine Learning (2005) submitted.

24. Borneman, J., Chrobak, M., Vedova, G., Figueroa, A., Jiang, T.: Probe selection algorithms with applications in the analysis of microbial communities. Bioinformatics **17**(Suppl. 1) (2001) S39–S48

25. Klau, G., Rahmann, S., Schliep, A., Vingron, M., Reinert, K.: Optimal robust non-unique probe selection using integer linear programming. Bioinformatics **20**(Suppl. 1) (2004) i186–i193

26. Chaval, V.: A greedy heuristic for the set covering problem. Mathematics of Operations Research **4**(3) (1979) 233–235

27. Nemhauser, G.L., Wolsey, L.A.: Integer and Combinatorial Optimization. Wiley-Interscience Series I Discrete Mathematics and Optimization. Wiley, New York (1988)

28. Murphy, P., Aha, D.: Uci repository of machine learning databases: Readable data repository. Department of Computer Science, University of California at Irvine, CA (1994) Available from World Wide Web: http://www.ics.uci.edu/~mlearn/ MLRepository.html.
29. Heisele, B., Poggio, T., Pontil, M.: Face detection in still grey images. Technical report, MIT Artificial Intelligence Laboratory and Center for Biological and Computational Learning, Massachusetts (2000) A.I. Memo No. 1687, C.B.C.L. Paper No. 187, Data available from World Wide Web: http://cbcl.mit.edu/cbcl/software-datasets.
30. Hammer, P., Bonates, T.: Logical analysis of data: From combinatorial optimization to medical applications. RUTCOR Research Report 10-2005 (2005)

Game and Market Equilibria: Computation, Approximation, and Smoothed Analysis

Shang-Hua Teng

Department of Computer Science
Boston University, and
Akamai Technologies Inc.

I will present some recent advances in algorithmic game theory especially about Nash equilibria. As you may have already known, the notion of Nash equilibria has captured the imagination of much of the computer science theory community, both for its many applications in the growing domain of online interactions and for its deep and fundamental mathematical structures. As the complexity and scale of typical Internet applications increase, the problem of efficiently analyzing their game-theoretic properties becomes more pointed.

In particular, I will cover the recent results in settling several open questions about Nash equilibria. After a quick review the result of Chen and Deng that

> BIMATRIX, the problem of finding a Nash equilibrium in a two-person game, is a complete problem in the complexity class **PPAD** (Polynomial Parity Argument, Directed version) introduced by Papadimitriou in 1991,

I will focus on the approximation and smoothed complexity of equilibrium computation and prove the following two theorems:

- BIMATRIX does not have a fully polynomial-time approximation scheme, unless every problem in **PPAD** is solvable in polynomial time.
- The smoothed complexity of the classic Lemke-Howson algorithm, and in fact, of any algorithm for BIMATRIX is not polynomial, unless every problem in **PPAD** is solvable in randomized polynomial time.

Our results demonstrate that, even in this simplest form of non-cooperative games, equilibrium computation and approximation are polynomial-time equivalent to fixed-point computation. If time permits, I will also cover the extensions of these results to other equilibrium problems such as in trading and market economies.

Joint work with Xi Chen (Tsinghua University), Xiaotie Deng (The City University of Hong Kong). Also with Li-Sha Huang (Tsinghua University) and Paul Valiant (MIT).

M.-Y. Kao and X.-Y. Li (Eds.): AAIM 2007, LNCS 4508, p. 424, 2007.
© Springer-Verlag Berlin Heidelberg 2007

Ad Auctions – Current and Future Research

Anna R. Karlin

University of Washington

Abstract. An exploding market has emerged during the last few years on the internet, the market of sponsored search slots. Advertisers are able to buy space on the webpages produced by popular search engines and place advertisements to promote their products alongside the regular algorithmic search results. The allocation of these advertising slots and their pricing is done via auctions. Since the introduction of this concept in 1998, sponsored search has evolved into a major source of revenue for internet giants such as Google, Yahoo!, MSN and others. Its success can be attributed partly to its effectiveness as a form of highly targeted advertising, and partly to the appealing framework that allows even small-scale advertisers to use it easily and effectively while only paying when their ad is clicked upon.

Numerous interesting mathematical, algorithmic and game-theoretic questions arise when one starts to think about these auctions. How should the auctions be designed so as to maximize search engine profit? What bidding strategies should the advertisers use? What are the dynamics and convergence properties of these systems? These questions are of extreme importance to the industry as even a minor change in the framework or in the way the advertisers bid could results in millions of dollars in profit or loss for both the advertisers and the search providers.

In this talk, we survey recent research on these kinds of questions and discuss open problems in the area.

M.-Y. Kao and X.-Y. Li (Eds.): AAIM 2007, LNCS 4508, p. 425, 2007.
© Springer-Verlag Berlin Heidelberg 2007

Expressive Commerce and Its Application to Sourcing: How We Conducted $25 Billion of Generalized Combinatorial Auctions

Tuomas Sandholm

CombineNet, Inc. and Carnegie Mellon University

Abstract. Sourcing professionals buy several trillion dollars worth of goods and services yearly. We introduced a new paradigm called expressive commerce and applied it to sourcing. It combines the advantages of highly expressive human negotiation with the advantages of electronic reverse auctions. The idea is that supply and demand are expressed in drastically greater detail than in traditional electronic auctions, and are algorithmically cleared. This creates a Pareto efficiency improvement in the allocation (a win-win between the buyer and the sellers) but the market clearing problem is a highly complex combinatorial optimization problem. We developed the world's fastest tree search algorithms for solving it. We have hosted $25 billion of sourcing using the technology, and created $3.2 billion of hard-dollar savings plus numerous harder-to-quantify benefits. The suppliers also benefited by being able to express production efficiencies and creativity, and through exposure problem removal. Supply networks were redesigned, with quantitative understanding of the tradeoffs, and implemented in weeks instead of months.
URL for the paper:
http://www.cs.cmu.edu/ sandholm/Expressive%20commerce.aimag.pdf

BIO

Tuomas Sandholm is Professor in the Computer Science Department at Carnegie Mellon University. He received the Ph.D. and M.S. degrees in computer science from the University of Massachusetts at Amherst in 1996 and 1994. He earned an M.S. (B.S. included) with distinction in Industrial Engineering and Management Science from the Helsinki University of Technology, Finland, in 1991. He has published over 250 technical papers on electronic commerce; game theory; artificial intelligence; multiagent systems; auctions and exchanges; automated negotiation and contracting; coalition formation; voting; safe exchange; normative models of bounded rationality; resource-bounded reasoning; machine learning; networks; and combinatorial optimization. He has 17 years of experience building optimization-based electronic marketplaces, and several of his systems have been commercially fielded. He is also Founder, Chairman, and Chief Scientist of CombineNet, Inc., which has fielded over 400 large-scale generalized combinatorial auctions. He received the National Science Foundation Career Award in 1997, the inaugural ACM Autonomous Agents Research Award in 2001, the Alfred P. Sloan Foundation Fellowship in 2003, and the IJCAI Computers and Thought Award in 2003.

M.-Y. Kao and X.-Y. Li (Eds.): AAIM 2007, LNCS 4508, p. 426, 2007.
© Springer-Verlag Berlin Heidelberg 2007

Author Index

Lecture Notes in Computer Science

For information about Vols. 1–4418

please contact your bookseller or Springer